TRANSFORMATIVE MISSION-ORIENTED INNOVATION POLICIES

EU-SPRI FORUM ON SCIENCE, TECHNOLOGY AND INNOVATION POLICY

The aim of this series is to present some of the best and most original research emanating from the Eu-SPRI Forum on Science, Technology and Innovation Policy. The typical questions addressed by the books in the series will include, but not be limited to:

- What is the role of science, technology and innovation policy in the 21st century?
- How can policies cope with 'grand social challenges' in the areas of health, energy, security or the environment?
- Are there better ways to link governments' science and innovation policies to other public policies?
- What are the innovation policy rationales and instruments for successfully fostering competitiveness and economic growth?
- Which public policies help to proactively shape responsible and legitimate technological innovation?
- How can public sector research be made more creative and effective?
- How can more intelligent interactions be achieved between investments in research and higher education policies for universities?
- How can the forces of globalisation and localisation be balanced?

For a full list of Edward Elgar published titles, including the titles in this series, visit our website at www.e-elgar.com.

Published by
Edward Elgar Publishing Limited
The Lypiatts
15 Lansdown Road
Cheltenham
Glos GL50 2JA
UK

Edward Elgar Publishing, Inc.
William Pratt House
9 Dewey Court
Northampton
Massachusetts 01060
USA

Authorised representative in the EU for GPSR queries only: Easy Access System Europe – Mustamäe tee 50, 10621 Tallinn, Estonia, gpsr.requests@easproject.com

A catalogue record for this book is available from the British Library

Library of Congress Control Number: 2025931086

This book is available electronically in the **Elgar**online
Economics subject collection
https://doi.org/10.4337/9781803929521

MIX
Paper | Supporting responsible forestry
FSC® C013604

ISBN 978 1 80392 951 4 (cased)
ISBN 978 1 80392 952 1 (eBook)

Printed and bound by CPI Group (UK) Ltd, Croydon, CR0 4YY

Transformative Mission-Oriented Innovation Policies

Revisiting the Role of Science, Technology and Innovation in Society

Edited by

Jakob Edler

Managing Director, Fraunhofer Institute for Systems and Innovation Research ISI, Germany and Professor, Innovation Policy and Strategy, University of Manchester, UK

Mireille Matt

Senior Research Scientist, French National Institute for Research on Agriculture, Food and Environment (INRAE), France

Wolfgang Polt

Formerly Director of POLICIES, Joanneum Research, Austria

Matthias Weber

Head of Center for Innovation Systems and Policy, AIT (Austrian Institute of Technology), Austria and Professor, Innovation Policy, Université Gustave Eiffel, France

EU-SPRI FORUM ON SCIENCE, TECHNOLOGY AND INNOVATION POLICY

Edward Elgar
PUBLISHING

Cheltenham, UK • Northampton, MA, USA

Contents

Contributors

Tateo Arimoto is Principal Fellow at the Center for Research and Development Strategy (CRDS) of the Japan Science and Technology Agency (JST-CRDS) and Visiting Professor at the National Graduate Institute for Policy Studies (GRIPS). His research focuses on the science-policy-society interface, science advice and science diplomacy. He is also a Board Member of the International Network for Governmental Science Advice (INGSA) and a Fellow of the International Science Council (ISC). He has served as Director-General of the Research Institute of Science and Technology for Society (RISTEX) at JST and as Sub-Program Director of SIP Automated Driving for Universal Services (SIP-adus) at Japan's Cabinet Office.

Kathy Berger is a PhD student at Aarhus University in Denmark, who draws on her experience at AIT Austrian Institute of Technology in Vienna and in research funding and collaboration. Currently, she centres her research on transformative capacity in urban transitions. With a specific focus on transformative processes, Kathy explores how collaborative learning contributes to the reintegration of nature into urban spaces.

Wouter Boon is Professor of Innovation and Transition Studies at the Copernicus Institute of Sustainable Development, Utrecht University. His research focusses on the dynamics and governance of emerging technologies and their contribution to societal transitions, such as more sustainable healthcare and mobility. He looks at the demand side of innovation systems, specifically at responsible scaling, the role of users in innovation and demand-side innovation policy.

Christoph Brodnik is Research Knowledge Manager at the Monash Sustainable Development Institute, based in Melbourne, Australia. Christoph has an environmental science and natural resource management background and holds a PhD in Environmental Sociology. His most recent work focussed on mission-oriented research and innovation policy and programmes with a focus on portfolios and systems change. Christoph also worked closely with policymakers from the European Commission and European Innovation Agencies, as well as practitioners in different sectors conducting policy analysis and strategy development for transformation-oriented initiatives.

Susanne Bührer is Coordinator of the Business Unit Societal Change and Innovation at the Fraunhofer Institute for Systems and Innovation Research. Her research areas include the development and implementation of sophisticated evaluation and impact analyses, the conceptualisation of impact-oriented monitoring systems, the topic of women in research and innovation systems (with special consideration of Gendered Innovations) and questions of responsible research and innovation.

Stephanie Daimer is Coordinator of the Business Unit Policy for Innovation and Transformation in the Competence Center Policy and Society at the Fraunhofer Institute for Systems and Innovation Research ISI. Her research interests include the analysis and evaluation of national and European research and innovation (R&I) policies, and more recently, policies for transformative change and the role of public administration as an agent of transformation.

Michael Dinges is Head of Competence Unit Transformation Governance at AIT Austrian Institute of Technology in Vienna. His research focusses on evaluating transformation and mission-oriented research and innovation policies, the design of policy instruments and the governance of research and innovation. Most recently, Michael has led evaluation studies on the evaluation of the European Framework Programmes related to the Green Transition.

Jakob Edler has been Managing Director of the Fraunhofer Institute for Systems and Innovation Research ISI since October 2018. The German-Brit is also Professor of Innovation Policy and Strategy at the Manchester Institute of Innovation Research (MIoIR), which he led from 2011 to September 2018. His research focusses on analysing and conceptualizing governance, policy and innovation management. Edler is a member of the Fraunhofer Society Presidium, the German Science Council, the German Academy of Science and Engineering (Acatech) and the Royal Society for the Encouragement of Arts, Manufactures and Commerce (UK).

Soohong Eum is a PhD candidate at the Centre for Science and Technology Studies (CWTS) at Leiden University. His research interests encompass the impact of government priority-setting, governance structure and policy implementation on the research landscape and its contents. The primary focus of his current PhD research project is the case of South Korea's green mission, combining quantitative analysis of policy instruments and their outcomes with a qualitative examination of the epistemic system and strategic responses by researchers.

Kieron Flanagan is Professor of Science and Technology Policy at the Manchester Institute of Innovation Research, University of Manchester. His work focusses on national policies for research, technology and innovation, and on local and regional innovation policies, with a particular interest in issues of policy design, policy implementation and policy learning. He has also worked on the broader influences of the State with regard to technological change and innovation, and in particular the shaping and articulation of demand through public procurement and other interventions.

Lea Fünfschilling is Associate Professor at the Department of Sociology and CIRCLE (Center for Innovation) at Lund University. She holds a PhD in Sociology from the University of Basel in Switzerland. Her research interests are interdisciplinary, located at the intersection of sociology and science and technology studies. In particular, she has contributed to the development of a distinctly institutional perspective on sustainability transitions. She is the coordinator of the Swedish Transformative Innovation Policy Platform (STIPP; 2017–2023) and was awarded an ERC Starting Grant for the project "ENDINGS – Towards a theory of endings in innovation studies" (2023–2027).

Miriam Hufnagl is Policy and Impact Officer at Deutsches Elektronen-Synchrotron DESY in Hamburg and Research Fellow at Fraunhofer Institute for Systems and Innovation Research ISI, where she worked as a political scientist from 2009–2021, with a secondment to the German Federal Ministry of Research and Education (BMBF) in 2015–2016. Her research focusses on strategy, analyses and impact assessment in research, technology and innovation policy (mission-orientated innovation policy, large research infrastructures) as well as knowledge and technology transfer.

Matthijs Janssen is Assistant Professor at the Innovation Studies section of the Copernicus Institute of Sustainable Development, Utrecht University. He specialises in innovation policy for societal challenges, thereby straddling domains like innovation sciences, transitions, governance and evaluation. Matthijs is Co-founder and Coordinator of the Mission-oriented Innovation Policy Observatory, an initiative that unites scholars and policy practitioners in enhancing the understanding, monitoring and effective use of challenge-based innovation policies for solving complex societal problems. As Principal Scientist at Dialogic, a research-based consultancy, Matthijs has (co-)authored almost 100 policy studies.

Rainer Kattel is Professor and Deputy Director at the Institute for Innovation and Public Purpose, University College London. He has published extensively on innovation policy, its governance and specific management issues. His recent books include *The Elgar handbook of alternative theories of economic development* (ed. E. Reinert & J. Ghosh: Elgar, 2016) and *How to make an entrepreneurial state. Why innovation needs bureaucracy* (with W. Drechsler & E. Karo: Yale, 2022), the latter being the 2023 winner of the George R. Terry Book Award from the Academy of Management.

Paula Kivimaa is Research Professor in Climate and Society at the Finnish Environment Institute and Associate at SPRU, University of Sussex. She is an internationally known scholar in sustainability transitions, co-editor of the book *Innovating climate governance* (with B. Turnheim and F. Berkhout, 2018, Cambridge University Press) and a member of scientific committees advising the European Commission and the Finnish Government.

Philippe Larrue is working at the OECD Directorate for Science, Technology and Innovation (DSTI) where he leads several activities related to research and innovation governance and policy for sustainability transitions. He also regularly leads or contributes to OECD Innovation Policy Reviews. He chairs the Stakeholder Advisory Board of the European Forum for Studies of Policies for Research and Innovation (Eu-SPRI). Prior to joining OECD, he was Director of the French office of Technopolis Group and occupied various researcher positions.

Ralf Lindner is Head of the Department of Policy and Society at the Fraunhofer Institute for Systems and Innovation Research ISI. He is particularly interested in policy design and governance approaches for transformative change, responsible research and innovation, and processes of policy learning. In recent years, the main focus of his research has been on the analysis and conceptualisation of mission-oriented innovation policies (e.g., in the context of the German High-tech Strategy) and the development of practitioner-oriented approaches supporting their design, implementation and evaluation.

Mireille Matt is Senior Research Scientist at the French National Institute for Research on Agriculture, Food and Environment (INRAE) and a member of the Interdisciplinary Laboratory for Science, Innovation and Society (LISIS – East Paris area). Her research focusses on innovation studies, innovation policy studies and formative evaluation. She has contributed to the development and implementation of a real-time impact assessment method (ASIRPA real-time) to help project and programme managers navigate toward desired futures in transformative mission-oriented innovation policy contexts (such as chemical pesticide-free agriculture and climate-neutral cities).

Mariana Mazzucato (PhD) is Professor in the Economics of Innovation and Public Value at University College London (UCL), where she is Founding Director of the UCL Institute for

Innovation & Public Purpose. She is the winner of international prizes, including the Grande Ufficiale Ordine al Merito della Repubblica Italiana in 2021, Italy's highest civilian honour, the 2020 John von Neumann Award, the 2019 All European Academies Madame de Staël Prize for Cultural Values and the 2018 Leontief Prize for Advancing the Frontiers of Economic Thought. Most recently, Pope Francis appointed her to the Pontifical Academy for Life for bringing "more humanity" to the world. As well as *The entrepreneurial state: Debunking public vs. private sector myths* (2013, Anthem Press), she is the author of *The value of everything: Making and taking in the global economy* (2018, Penguin Books), *Mission economy: A moonshot guide to changing capitalism* (2021, Penguin Books) and most recently, *The big con: How the consulting industry weakens our businesses, infantilizes our governments and warps our economies* (2023, Penguin Books). She advises policymakers around the world on innovation-led inclusive and sustainable growth. Her roles have included, for example, Chair of the World Health Organization's Council on the Economics of Health for All, Co-Chair of the Global Commission on the Economics of Water and member of the South African President's Economic Advisory Council.

Jordi Molas-Gallart is Research Professor at the Spanish National Research Council (CSIC), Director of INGENIO (CSIC-UPV), a joint research centre of CSIC and the Polytechnic University of Valencia, and Visiting Fellow at SPRU (University of Sussex) where he obtained his PhD and worked for some 15 years as Researcher and Senior Lecturer. He was Editor of *Research Evaluation*, a journal published by Oxford University Press, between 2013 and 2020, and President of the European Network of Indicator Designers (ENID) between 2015 and 2019. He is a member of the Editorial Board of the journals *Research Evaluation*, *Research Policy* and *Evidence & Policy*. He has contributed to many European Commission expert groups and chaired the Science Europe working group on Research Policy and Programme Evaluation.

Kazuhito Oyamada is Fellow at the Center for Research and Development Strategy (CRDS) of the Japan Science and Technology Agency (JST). His research focusses on transformative and mission-oriented innovation policy (MOIP), and responsible research and innovation. As a policy analyst and practitioner, he has contributed to the design and implementation of several science, technology and innovation (STI) policy programmes in Japan, including the Cross-Ministerial Strategic Innovation Promotion Programme (SIP) and the Moonshot Research and Development Programme. He has also contributed to several OECD projects such as MOIP, transdisciplinary research, high-risk/high-reward research, and governance of emerging technologies.

Caetano C. R. Penna is Assistant Professor in the Economics of Technology and Innovation at Delft University of Technology. Since March 2024, he has served as Research Director at the Brazilian STI think tank Center for Strategic Studies and Management (CGEE). His research encompasses industry, technology and innovation economics and policy, with a focus on transformative and mission-oriented policies, sustainability and innovation financing and the (geo)political economy of technological change. As an independent expert of the Horizon Europe's Mutual Learning Exercise (MLE) on EU Missions, he provides expertise in funding and instrument synergies.

Wolfgang Polt was Director of POLICIES – the Institute for Economic, Social and Innovation Research of JOANNEUM Research in Vienna, Graz, and Klagenfurt (Austria) until 2024.

His research focusses on Research, Innovation and Industrial Policy and spans from analysis to strategic policy advice and evaluation of these policies. Most recently, he has contributed specifically to the conceptualization, implementation and analysis of mission-oriented policies for the OECD, the EU (e.g., in the context of the TRAMI project) and on a national level in countries like Austria and Sweden and is a university lecturer.

Dimitrios Pontikakis is an economist at the Innovation Policies and Economic Impact Unit of the European Commission's Joint Research Centre (JRC). He is currently leading a JRC Working Group on System Dynamics for System Innovation and was previously responsible for the JRC Working Group on Understanding and Managing Industrial Transitions. Dimitrios has a keen interest in transformative innovation policy, reflected in publications such as *OECD system innovation: Synthesis report*, the JRC's POINT methodology for transformative policy reviews, and most recently, the *Partnerships for regional innovation playbook.*

Douglas K. R. Robinson is a research scientist at the French National Centre for Scientific Research (CNRS) and Honorary Research Associate of the Institute for Innovation and Public Purpose (IIPP) at University College London, UK. Douglas is currently on secondment from the CNRS, taking on a role as Policy Advisor to the OECD Science and Technology Policy Division, where he is spearheading work on Technology Futures and Strategic Intelligence for STI policy, mission and challenge-oriented technology assessment, anticipatory technology governance of technologies for transitions and is part of the OECD Global Forum on Technology.

Karoline S. Rogge is Professor of Sustainability Innovation and Policy at the Science Policy Research Unit at the University of Sussex, and Deputy Head of the Competence Centre Policy and Society at the Fraunhofer Institute for Systems and Innovation Research ISI. In her research, she investigates the acceleration of sustainability transitions, with a focus on the link between policy mixes and innovation. Karoline currently leads the EMPOCI project on accelerating sustainable energy-mobility transitions, with a focus on the United Kingdom, Germany, China and the United States, complementing her recent research into the role of social innovation for accelerating energy transitions. Karoline serves on several scientific advisory boards and acts as Associate Editor for the journal *Research Policy.*

Florian Roth is Senior Researcher at the Zurich University of Applied Sciences (ZHAW) and an Affiliate at the Fraunhofer Institute for Systems and Innovation Research (ISI). He holds a PhD in Political Science from the University of Konstanz. Mr Roth's work focusses on the links between resilience, innovation and sustainability. His current research focusses on strategies to strengthen organisational and systemic capacities to manage disruptions and transformations.

Johan Schot is Professor of Global History and Sustainability Transitions at the Utrecht University Centre for Global Challenges. He is Founder and Academic Director of the Deep Transitions research project and the Transformative Innovation Policy Consortium (TIPC), as well as former Director of the Science Policy Research Unit (SPRU) at the University of Sussex Business School. Professor Schot's interests are oriented around impact-driven interdisciplinary research that strives to accelerate long-term system change and a just transition towards sustainability. Schot is a network-builder who brings together interdisciplinary research teams, policymakers, investors, governments, NGOs, the media and the corporate world.

Sarah Seus is a social scientist and researcher in the Competence Center Policy and Society of the Fraunhofer Institute for Systems and Innovation Research (ISI). Her research areas include the evaluation of science and innovation policies, specifically research on methodologies for impact evaluation (including societal impacts) of research funding; the analysis of STI policies and governance at the national and EU levels, the analysis of transition processes within the science system (e.g., transdisciplinary research) and questions of Responsible Research and Innovation (RRI).

Qi Song is a current PhD candidate at the Science Policy Research Unit (SPRU) at the University of Sussex Business School. His doctoral project, funded by the EMPOCI project, focusses on examining the role of transformative policy mixes in accelerating directed socio-technical changes, specifically in the context of multi-system e-mobility transitions in China.

Jonas Torrens is Assistant Professor at the Copernicus Institute of Sustainable Development at Utrecht University. His research focusses on societal, policy and urban experimentation with sustainability. He has worked on establishing the experimental approach of the Transformative Innovation Policy Consortium, and he is also a co-founder of the Mission-Oriented Innovation Policy Observatory at Copernicus. Currently, he is engaged in research at the intersection between futuring, transdisciplinary research and experimentation, with the aim to enhance our collective ability to navigate the polycrisis.

Elvira Uyarra is Professor of Innovation Studies at Alliance Manchester Business School (University of Manchester) where she is also Director of the Manchester Institute of Innovation Research. She is Fellow of the Regional Studies Association (RSA) and editor of the journal *Regional Studies*. Her work sits at the intersection between innovation studies, policy studies and regional studies. She researches the role of innovation and industrial policy on economic development and regional diversification, including smart specialisation and mission-oriented policies.

Renée van Dis is a post-doctoral researcher at the Laboratory for Interdisciplinary Science, Innovation and Society (LISIS) in France. She completed her PhD in sociology at the French National Research Institute for Agriculture, Food and Environment (INRAE). Her thesis explores how impact assessment in real-time (ASIRPA Real-Time approach) helps researchers to responsibilise as they are guided in envisioning their contribution to a societal mission. Originally from the Netherlands, she holds an MSc in Organic Agriculture from Wageningen University, the Netherlands, and a master's in History of Science, Technology and Society (2017) from EHESS in Paris, France.

Diana Velasco is currently the Experimentation Programme Director at the Deep Transitions Lab (DTL) in the Centre for Global Challenges at the University of Utrecht. Concurrently, she holds the position of Research Fellow at Ingenio (CSIC-UPV). With a doctoral background in Science and Technology Studies, Diana has qualifications as both a systems and computing engineer and an anthropologist. Her research delves primarily into innovation studies, transformative and mission-oriented innovation policy, transformative investment, deep transitions and the pivotal role of higher education in fostering just transitions. She has been involved in the conception, execution and evaluation of experimental engagements developed through the Transformative Innovation Policy Consortium (TIPC) and the DTL.

Prof Dr Rainer Walz is Deputy Director of the Institute and Head of the Competence Center for Sustainability and Infrastructure Systems at the Fraunhofer Institute for Systems and Innovation Research (ISI) in Karlsruhe. He studied Economics and Political Science at the University of Freiburg and at Brock University, Canada. He earned his PhD and "habilitation" in Economics. In 1991, Prof Dr Walz joined the Fraunhofer ISI. His prior employment includes the University of Wisconsin and the Enquête Commission "Protecting the Earth's Atmosphere" at the German Bundestag. He is also Adjunct Professor of Economics at the Karlsruhe Institute of Technology (KIT).

Iris Wanzenböck is Assistant Professor at the Copernicus Institute of Sustainable Development at Utrecht University. Her research is dedicated to exploring new approaches to research and innovation policies across various governance levels, spanning from the EU to national, regional and local contexts. She places emphasis on understanding the governance processes and conditions that enable mission-oriented and transformative innovation policies to effectively tackle societal challenges and support sustainable transformations. Her work integrates perspectives from geography, policy sciences and network theory to provide comprehensive insights into these complex issues.

Matthias Weber is Head of Center for Innovation Systems and Policy at AIT Austrian Institute of Technology in Vienna and Professor of Innovation Policy at Université Gustave Eiffel, Interdisciplinary Laboratory for Science, Innovation and Society (LISIS – east Paris area). His research focusses on the transformation of socio-technical systems and the role of strategic intelligence for governing research and innovation in society. He also advises governmental bodies in the EU as well as international organisations on the rationales, design and implementation of transformative and mission-oriented policies, often drawing on strategic foresight and formative evaluation methods to help navigate uncertain futures.

Joeri Wesseling works as Assistant Professor at the Copernicus Institute of Sustainable Development, Utrecht University. His work centres on (mission-oriented) innovation systems perspectives to sustainability transitions and the policy implications thereof. Sectors of study include transport and heavy industries, including fuels.

Florian Wittmann has been Senior Researcher at the Fraunhofer Institute for Systems and Innovation Research (ISI) in Karlsruhe (Germany) since 2019. His research focusses on transformative policy-making, particularly the design and governance of mission-oriented policies, as well as attempts to analyse, evaluate and monitor the implementation of such policies.

Merve Yorulmaz is a researcher at the Fraunhofer Institute for Systems and Innovation Research. Her research focusses on program evaluations, the topic of resilience, gender diversity and responsible approaches to research and innovation. She was a member of the research team providing scientific support to the German High-Tech Strategy 2025. In parallel to her work at Fraunhofer ISI, she is a PhD candidate at BETA/The University of Strasbourg, exploring the antecedents and interplay of resilience and creativity during times of crisis from a public sector lens.

Acknowledgements

As with any edited book, this volume is the result of a team effort. We would therefore like to thank all the authors of this volume. The energy and thoughtfulness they have shown in their contributions are remarkable. Moreover, the authors have also engaged in intensive debate about their contributions and the volume itself, which has contributed to making it a holistic exercise rather than an elaboration of entirely unrelated chapters.

We are also grateful for critical discussions with colleagues at the Eu-SPRI to whom we presented our hypothesis on various occasions. Above all, however, we need to thank Johanna Schuler. This book could not have materialised without her tireless work and support, as it was she who coordinated the process and pushed the editors, as well as the contributors, towards the deadlines with tireless energy and unmatched scrutiny. Finally, we thank Marianne Werder, Katja Darmstädter, Wiebke Baumann, Caden Wells, and Kim Laura Skade for their excellent technical and organisational support throughout.

Introduction: the motivation for this volume and an overview of its contributions

Jakob Edler, Mireille Matt, Wolfgang Polt and Matthias Weber

I.1 INTRODUCTION

Since the wake of industrialisation in the late eighteenth century, when the first spinning machines were driven by water frames on an industrial scale, societies and economies have been "in constant transformation" in one way or another, albeit at a slower pace than today. Driven by technological and organisational progress, disrupted and accelerated by wars and societal demands and upheaval, the economic and societal systems of the countries that we have come to label "developed" have, since industrialisation, never stood still. This economic and societal change has always been influenced by state action, both with a proactive, supporting, and enabling intent and in a reactive mode, aiming to repair collateral damage. In doing so, the state has provided key requirements for economic and societal change and tried, for better or worse, to deal with some of its major consequences.

So, why have we seen a new wave of transformative policies across the world and beyond in recent years? Why do we now need to think anew about policies for transformation? The major reason for this is, we believe, a well-justified sense of urgency. The transformations that we are dealing with today are those that aim to save the planet from climate catastrophe, to organise all aspects of our economies within planetary boundaries, to make better use of emerging and potentially disruptive technological opportunities in key systems of provision, and to do so in ways that are societally acceptable. The state can no longer only repair or try to provide the preconditions and frameworks for systems to transform while simply following a market logic. A major premise of this volume is that the state needs to revisit its responsibility. This responsibility is a big one, and it is a huge challenge. The state will not do it alone – state actors are not per se wiser than consumers, producers, engineers, and managers. However, state actors have a constitutional responsibility for and authority over the entirety of society, and are equipped with specific powers to gather and accumulate data. They therefore can and must orchestrate processes to find and promote a sufficient level of collective agreement about the direction of the societal journey and then support private initiatives to work in that direction.

Why then should we focus on science, technology, and innovation policies? Why limit the discussion, and thus the perspective of the role of the state? This is a choice that we have made in full awareness of the limitations. We acknowledge that science, technology, and innovation (STI) is only one ingredient – albeit a necessary one – in bringing about the broader socio-technical systems change that is required. However, without accelerating the generation and, in particular, the diffusion of technological and social innovations, transformations will not succeed. At the same time, we observe that numerous STI policy initiatives have been undertaken that were intended to be transformational, but these have fallen short of having a transformational effect (at least so far). In the long run, this is dangerous, as the momentum for broad, transformational dynamic shifts towards widely shared goals might be lost if we fail to mobilise the power of STI in a coherent, orchestrated manner.

Against the background of these major transformations, with this volume, we hope to contribute to a necessary and timely debate on how to address such transformations through STI policies. We do so from a specific angle by exploring whether "transformative missions" can be an appropriate approach in this context. Transformative missions are a specific type of mission that are different from other, narrower types in that their goals are broader, of a more long-term nature, and involve a great number of societal actors. They are also different from other types of "transformative policies", as they aim at providing more "directionality", imply a greater role for governments in providing this directionality, and help in the search for new governance structures and administrative capacities to implement this new policy approach.

The past few years have seen significant progress both in conceptualising the mission-oriented approach in general and in the "transformative missions" specifically. There was a period of time between the bulk of the theoretical and conceptual work being done and the first attempts at actually implementing these approaches. In the first contribution in this volume, we aim to bring together both strands. First, we look at the latest developments in conceptual thinking, where the research community has made important steps to conceptualise the new generation of innovation policies. These contributions usually provide recommendations to policymakers (ways of doing things differently, facing new challenges, and thinking differently). However, there is still a visible gap between those policy recommendations and the process of designing, formulating, and implementing policies in real settings.

Hence, the second contribution in this volume looks at bridging the gap between theoretical and conceptual discussions and the empirical process of actually implementing policies. These implementations have become widespread recently, as the geographical scope of the contributions indicates. The ambitious promise that transformative mission-oriented STI policies will generate the societal transformations necessary to overcome the current challenges has contributed to their worldwide adoption. Therefore, a main motivation of this volume is to provide a rich set of example cases of such initiatives being implemented around the world: in Korea, Japan, Canada, the United States, Australia, Austria, Belgium, Denmark, Finland, France, Germany, the Netherlands, Norway, Sweden, Spain, and the European Union. These examples take place at different levels: local, regional, national, and supranational (particularly European). They present the challenges faced in a variety of areas, such as: reducing GHG emissions ("net-zero" mission) to mitigate climate change; reducing the use of chemical pesticides in the agricultural sector down to zero; providing sustainable and tasty food in schools; developing preventive healthcare and medicine; building sustainable mobility systems; achieving the digital transformation of society; and creating more sustainable socio-technical systems in general.

This volume takes stock of how these policy approaches are succeeding and the challenges that they face. In doing so, it points to the commonalities and the inherent differences between countries and areas of implementation. It shows that transformative mission-oriented STI policies are formulated, designed, implemented, and monitored/evaluated in specific ways, depending on contextual, geographical, institutional, cultural, and historical factors. These approaches are mostly in an early, experimental stage of adoption, which implies that no stabilised or dominant governance structures or commonly accepted practice models have emerged so far. They have to cope with – and often rearrange – existing policy trajectories formed by organisational structures (ministries and funding agencies), administrative and legal rules, accounting systems, and governance models that are not yet adapted to the transformative mission-oriented STI policy approach. In response, some countries have created dedicated structures, while others have reshaped existing ones or fostered cross-ministerial and cross-sectoral coordination. The institutionalisation of the missions takes different forms in different countries: they are either centralised, i.e., owned by sectoral authorities, or by STI public agencies – or by both. The policy initiatives examined in this volume all present bold ambitions; however, they highlight the difficulties in designing, implementing, and monitoring policies with all the expected features of transformative STI mission-oriented policies. These examples underline the difficulty in being transformative: it takes time (the timing of transformation does not coincide with the timing of policymaking) and requires deep learning and capacity building over a longer period of time.

In terms of setting agendas and articulating missions, the volume provides examples of collectively developed agendas that have been established in an inclusive way by involving a wide range of actors (including citizens). As expected, missions are formulated with clear objectives and measurable, time-bound targets, but to varying degrees. The volume also discusses the geography of mission orientations: the shared visions around the mission might concern global challenges or more local ones, and agendas might be set on various scales (from supranational to local).

Transformative mission-oriented STI policies are not implemented on a blank slate, but in a landscape of existing policy rationales and mixes, which may facilitate or hamper these new policy approaches. The examples in this volume underline difficulties related to the existence of institutional rigidities and incompatibilities, misalignments of nested policy initiatives and objectives at different levels (from the strategic arena to the performance arena) and scales (from supranational to local), lack of coordination with other types of policies and funding (social, sectoral, etc.), and failures relating to interactions among a wide range of mission-relevant actors and stakeholders. Instruments used on a national and supranational scale focus on the following: supporting large infrastructures; funding to promote new STI modes, such as interdisciplinarity, transdisciplinarity, and science-technology-society networking; expanding experimentation to include new types of actors; and regulations that have been piloted experimentally to an equal extent. These policies often set up a mix of demand-side and supply-side instruments. Implementation of transformative mission-oriented STI policies in local contexts might increase legitimation, enable expected local transformations, and enhance opportunities in less-favoured regions. Instruments such as urban labs or living labs, collaborative platforms to connect multi-local experiments and certification schemes (e.g., participatory guarantee systems), are participatory and inclusive and come along with high transformative expectations.

Transformative mission-oriented STI policies also need to develop new cultures of evaluation and adopt formative evaluation approaches. The latter should enable learning about policy experiments and help with monitoring and guiding innovation experiments towards expected transformations through actors learning to navigate uncertainty. These new types of evaluation approaches are usually inclusive, participatory, and iterative. They should develop adequate frameworks, methodologies, and tools, not only to assess the results of the policies, but also to capture the strengths and weaknesses of their mission-oriented features. The idea is to integrate formative evaluation and foresight activities at each step of the policy cycle. More generally, the aim is to develop and implement a variety of forms of strategic intelligence (real-time impact assessment, foresight, understanding the landscape of the policy) to define transformative mission-oriented STI policies (policy formulation and implementation) and to steer research activities (at the project and programme level) to contribute to mission goals. One of the main challenges in building and producing a range of distributed intelligence lies in the development of monitoring, anticipatory, and strategic capacities at various policy levels (a nested approach that enables coordination between the project, programme, and mission levels) and organisations.

Designing, legitimising, and implementing transformative mission-oriented STI policies requires public administrations (ministries, local institutions, and funding agencies) to have adequate capabilities in budgeting, governing, monitoring, anticipating, and developing a new mindset focussed on systems thinking. It also requires analytical, operational, and political capacities and the creation of learning spaces to enable deep learning.

I.2 VOLUME CONTENT

The rich set of examples of experimentation presented in this volume constitutes a basis for learning about all aspects and phases throughout the policy process, and governance and capacity building. The volume serves to give an overview of the "state of the art" in terms of concepts and concrete examples of transformative mission-oriented STI policies. As the ambition of STI policies to support transformations will not vanish any time soon, we hope that the conceptual and practical lessons to be learnt from the contributions in this volume will remain relevant for years to come.

The volume is introduced by *Jakob Edler, Mireille Matt, Wolfgang Polt, and Matthias Weber*, who set the scene conceptually. In their chapter, the authors discuss a new framing of STI policy that underpins the transformational turn. In doing so, they define transformative mission-oriented STI policy as an ambitious set of state rationales and activities that push socio-technical development towards goals that are defined in societal discourses. Those goals can, and most certainly will, be contested in the context of democratic processes and they can only be achieved through mobilising and orienting market forces rather than limiting them. The authors remind us of the changing rationales of STI policy post-World War II and the various modes of STI policies that seek to influence directionality. They differentiate the role of the state in the context of the complex nature and drivers of systems change, and they present a range of critical considerations of these renewed ambitions of the state. These conceptual discussions are followed by a set of concrete reflections and suggestions in terms of policy governance, capabilities of the state, and the need for strategic intelligence and appropriate policy instrumentation. It becomes clear how challenging it is to translate good intentions to appropriate state action. They finish their conceptual contribution with a set of reflections on what really constitutes the new quality of transformative mission-oriented STI policies, i.e., orchestrating and supporting societies and economies towards defining and achieving ambitious transformative goals.

In their contribution on strategic intelligence, *Douglas Robinson, Mireille Matt, and Renée van Dis* start with the assertion that when taken seriously, missions face enormous strategic intelligence challenges in terms of their implementation, particularly when it comes to defining or agreeing on clear "goals" and thus working to steer research in a specific direction without limiting the research creativity required to achieve the goals. At the same time, missions cut across all three layers involved with steering research (as research is most often organised through funding agencies and their management): orientation, programming, and performance. These two premises combined mean that specific forms of strategic intelligence need to be deployed and used together for all three layers of steering. For long-term missions, it is crucial to have a joint understanding of how research activities will or could contribute to achieving a mission. For this, established approaches such as ASPIRPA can be mobilised. Their main function is to empower actors involved in the programme to collectively develop (potential) pathways and, in doing so, to better understand the contribution of funded research towards the mission goals, the requirements for success, and how those could be managed (e.g., by monitoring, calibrating, and tailoring research with regard to a mission). A second major function of strategic intelligence is to provide an understanding of how the system that is to be transformed might develop, what kinds of drivers might shape this process (including in interaction with other systems), and what alternative long-term scenarios might be a result. Only by looking at those various scenarios can one define what mission success would actually mean, as the mission activities are certainly not the only or the main drivers of the overall scenario development; therefore, it is important to consider what the mission activities might change in conjunction with the assumptions made in the different scenarios. Finally, as missions are situated in larger system transformation contexts, coordination mechanisms are important to ensure that the various mission activities properly feed into larger transformation dynamics to which the mission itself contributes. This broad approach of strategic intelligence highlights the specific needs of mission-oriented programmes, particularly long-term programmes.

Wouter Boon analyses a mission-oriented approach that seeks to work towards a more preventive healthcare system. This is a highly complex endeavour, framed as a challenge characterised by multiple instances of "nestedness" – mission-programme-project, multi-policy level, private-public-community – where new technologies, behaviours, practices, and regulations need to converge and form a new system.

In the specific Dutch example studied, the mission was formulated against the backdrop of many years of slow development towards more systematic prevention in healthcare. The new, mainly digital technologies were often perceived as being incompatible with the large number of regulations in this sector, and businesses often lacked a clear business case. There were also many challenging incentivising mechanisms to be put in place for individuals to adopt new technologies, as rewards were delayed and lifestyle changes were required.

The main governance challenge was the issue of coordinating multiple actor groups, both in terms of initiating and implementing supporting policies, and in terms of market actors and citizens. The governance responsibilities had been distributed across a range of actors and were concentrated in a new body. National policy initiated the mission policy, but the formulation and implementation of the mission were delegated to this new body (Health Holland), a form of public-private partnership emanating from and building upon a pre-existing institution at the public-private sector level. This body sustained and remodelled the inclusion of bottom-up initiatives, the mobilisation of further actor groups, and the coordination of initiatives to create new markets.

The mission has not been a fast success, but the first movements towards more preventive healthcare based on digital technologies were made possible through a culture of cooperation, a clear vision at the national level, and the delegation of responsibilities for coordination and implementation to a body that attained credibility by operating using established coordination mechanisms.

The South Korean example of mission-oriented policy by *Soohong Eum* demonstrates the assets and pitfalls of the governance of large-scale missions in a presidential system. Two South Korean presidents set up and implemented Green Missions in the context of international agreements and economic crises in the country. The first mission (2008) was framed mainly as an economic strategy, taking advantage of international pressures and trends towards carbon-neutral technologies. It was designed as a Keynesian infrastructure-centric economic policy and implemented through a range of top-down policy measures based on the broad presidential executive powers. The second mission (2020) was formulated and implemented more credibly as a "green" mission, supplementing economic ambitions with green targets. In both cases, the presidents were able to command mission programmes and to set up new bodies with the authority to coordinate the relevant activities of different ministries and to support implementation. In the second case, however, the process was far more inclusive, with a range of participatory processes across different stakeholder groups, guided by a joint vision. Furthermore, the implementing body applied a broad range of strategic intelligence approaches and tools to better tailor interventions and more thoroughly align investments with the green agenda. This created substantially more consensus and legitimacy compared to the first attempt. However, while the former more or less completely failed to deliver on the green agenda, the success of the latter was somewhat limited, too. The engagement proved to be too top-down, aiming to educate stakeholders rather than listen to them. In addition, there was limited time for the various policies to come into effect, as a new president in 2022 turned the attention away from the mission while keeping a range of institutional innovations for the mission implementation in place. Overall, the presidential system allows for swift, strong action, top-down steering, and the setting up of new structures to implement missions. Top-down policies, however, still need strong – and serious – stakeholder participation and broad strategic intelligence – and the time and sustainability to develop.

In their contribution on mission-oriented STI policies in Europe, *Rainer Kattel and Mariana Mazzucato* focus on the fact that mission-oriented policies are based on a strong normative ambition as well as a strong academic and epistemic underpinning – both transnational in nature. While there is already a large variety of policy innovations and adaptations as a consequence, the authors concede that there are still only limited attempts at real strategic and procedural change. Moreover, national idiosyncrasies and historic contexts still play a major role in how governance challenges are understood and addressed. In their overview of European mission-oriented policy approaches, the authors find that policymaking has taken a sharp turn towards missions. But while there is some common understanding as to why mission-oriented policies differ from traditional STI policies and of the fact that they face different kinds of governance challenges, there are clear indications that in the majority of cases, the specific challenges are tackled largely through modifying existing practices and structures, rather than by establishing truly novel institutional arrangements and policy designs. A large variety of national financing tools are mobilised in idiosyncratic ways, and common evaluation practices and guidelines are yet to be established – one cited exception being the UK's *Green Book*. This international wave of modest experimentation with adjusted processes and structures and the lack of convergence towards a dominant governance model can be interpreted as a sign of paradigmatic change within highly institutionalised national contexts. This in itself could be seen as a weak signal of major changes simply taking their time.

What is concerning, however, is that when it comes to the real implementation of those changes, Kattel and Mazzucato find a mismatch between two factors: firstly, governance challenges and ambitions in transformative mission policies and secondly, the policymaking capabilities available to address these challenges and ambitions. In other words, the creativity in adopting existing structures and policy processes is not matched by a sufficient level and quality of capacity in the public sector. For this reason, setting up and implementing transformative mission-oriented policies will, for the time being, remain a process of "muddling through"; it will continue to evolve relying on strong policy entrepreneurs and political leadership rather than on public administrations that are appropriately equipped to understand and act on the specific governance challenges of such policies. The analysis also shows the limitations of the epistemic community when it comes to supporting the transition from normative conviction to practical implementation.[1]

The contribution by *Karoline Rogge and Qi Song* deals with the effective employment of policy mixes for navigating and stimulating transformative change. The authors also focus on the capacities required by the state to govern, support, and accelerate the development, adoption, and diffusion of radically innovative solutions in and across multiple – often interacting – systems. They propose a conceptual framework that is organised along the lines of three dimensions. It distinguishes (i) three types of organisational capacities (i.e., analytical, operational, and political capacity), (ii) seven types of intervention points in policy mixes to promote sustainability transitions and, as a specific focus of their contribution, (iii) five typical challenges associated with the acceleration phase of transformative change processes. This framing allows them to systematically identify and map the transformative organisational capacity required in public policy and administration to address policy tasks derived from combining the other two dimensions of acceleration challenges and intervention points. With their focus on acceleration challenges, they stress the importance of moving beyond the experimental phase of mission-oriented policy approaches, which is crucial for meeting the ambitions of transformative missions. Their approach proposes a more comprehensive frame of reference for transformative mission-oriented policies than prevailing state-led and stakeholder-led approaches, and it offers a differentiating methodology to identify the types of critical policy capacities needed for specific missions.

In their chapter, *Rainer Walz, Susanne Bührer, and Sarah Seus* discuss a German funding programme for sustainability research and the impact it has had on transforming research practices and, in consequence, supporting system transformation towards sustainability in predefined areas. The effects of the programme on the transformation of systems are indirect, as there is no established link to the sectoral policies responsible for transforming systems. However, the R&D programme has been mobilised to further develop and accelerate specific technologies, as well as to broaden the knowledge base that supports system transformations.

In both regards, the R&D programme, which serves to support and enable transformation through research, faces the challenge of linking to sectoral policies. This is particularly important for those supported R&D activities that push for certain technologies to be developed and accelerated into the market, as corresponding demand-side measures are essential, but are developed in other ministries, often with poor or non-existing coordination. The chapter shows that change can be achieved in R&D practice in desired directions, and that funding mechanisms can make research more inter- and transdisciplinary. However, this

[1] The Transformative Innovation Policy Consortium (TIPC) is a notable positive example; however, this very example highlights the enormous, long-term effort needed to support even those policymakers that are eager and ready to work towards transformative policies.

change is limited by the "stickiness" of established institutional and disciplinary contexts of research. Furthermore, it is argued that R&D programmes to support transformation should not be linked entirely to specific missions, as this would hamper the development of long-term knowledge that might be of particular importance for system transformation. In countries in which STI policy is separated from sectoral policies, or in which sectoral policies do not have the necessary means to support sector-specific R&D, the mobilisation of R&D for missions is a strategic challenge. The conceptual considerations in the chapter by Walz et al. point to a mixed model of transformational R&D, whereby some R&D should be linked explicitly to specific missions, but in coordination with sectoral policies, or some R&D budgets should be allocated to the sectoral policies, while reserving some R&D budget for open and non-directional purposes.

Florian Wittmann, Miriam Hufnagl, Florian Roth, Merve Yorulmaz, and Ralf Lindner propose a perspective for understanding the challenges around designing and implementing mission-oriented STI policies. Their framework can be used to analyse those challenges both in transformative and accelerator missions, and both for top-down and bottom-up transformational processes. The three "translations" they propose are mission definition (from challenge to mission), mission design (from goals to instruments), and mission implementation (mission instruments into concrete actions and effects).

The main insight enabled through this conceptual framework is that we can distinguish different levels of challenges for the different translational processes as well as different kinds of actors and actor negotiations. While acknowledging the obvious shift from strategic to operational challenges when moving from the early formulation stage to the later implementation stage, the authors highlight that this distinction between strategic and operational challenges is not straightforward. Mission design, the second translation process, links the two challenges. It is a phase in which resources and actors are mobilised to turn the mission into action. However, this process still involves high requirements for coordination and delineating the boundaries of the mission in question and is thus an arena of controversial interpretations about the very nature of the mission and who is responsible for what. Furthermore, even in the implementation phase, there are coordination challenges to overcome; these may bring back major conflicts of interest between different actors responsible for different aspects of implementation. Interpretation of what needs to be done by whom may differ, and the mode of coordination in implementation has implications for the division of powers between different administrative entities. In addition, across all translation processes, established programmes and routines need to be taken into account, mobilised, and altered. The combination of reconfiguring relations between administrations and rethinking established programmes and routines thus challenges policy administration in terms of its very identity and trajectory.

Caetano Penna, Johan Schot, Diana Velasco, and Jordi Molas-Gallart present a framework for the formative evaluation of open-ended transformative missions. Formative evaluation comprises the support for the policy process through means of strategic intelligence – particularly to support joint learning along the policy cycle by drawing recursively and in real-time on the lessons learnt during implementation. The authors are interested in policy approaches that support ongoing transformations that do not have clearly defined targets nor predefined pathways, and that do not rely on promoting a limited set of technologies. Rather, those missions draw on multiple framings, and suggest solutions and alternative routes to realise them. In other words, concrete measurable targets cannot be defined before the actual policy implementation process, but are determined and, if needed, adapted in parallel with advancing along an emergent pathway. For those kinds of open processes, the authors suggest a formative evaluation approach that is integrated with policy design and implementation, is participatory and inclusive, uses a mix of methods (to be able to adjust over time), and finally builds on a nested approach to assess the interplay of various policy and transformational levels.

The authors see one major challenge that those approaches face, i.e., the need to define a theory of change that, given the openness of the transformational process, is an ongoing reflexive exercise rather than a robust *ex ante* intervention logic and model. Their heuristic used to operationalise the theory of change builds on the multi-level perspective of sustainability transitions, whereby policy levers at the three levels – niche, regime, and landscape – are translated into transformative outcomes that need to be supported by reflexivity and learning. This, as shown in the Vinnova case example, is provided through explicit and controlled policy experiments and through subsequent interactions between the evaluating team and the agency responsible for the mission. The remaining challenge of any formative evaluation is, therefore, the generalisation of lessons learnt in a specific, localised experiment for wider use in other contexts. However, the overarching lesson that holds true for almost any of the contributions in this volume is the need for developing new kinds of capacities – particularly capacities to learn, unlearn, network, interact, and communicate with communities well beyond the traditional target groups of policy agents.

The mission that *Paula Kivimaa and Jonas Torrens* focus on in their chapter is the attempt to move towards Mobility as a Service (MaaS) in Finland. The authors analyse how the system of mobility transformed to realise the planning and purchasing of seamless intermodal mobility chains. Their chapter demonstrates how directionality emerged and was shaped through the interaction and cooperation between different ministries, and how this affected cooperation among transport service providers to make seamless intermodal mobility services possible. This business model for MaaS has been taken up in other countries in the meantime, and thus become globally relevant.

The emergence of directionality is described as a gradual, multi-actor process, combining the emergent process envisaged by what they label as transformative innovation policy (TIP) with attempts to streamline actors commonly associated with mission-oriented innovation policies. In other words, it is an attempt to propose a "hybrid" model of transformative missions that provides some guidance and coherence while relying strongly on bottom-up search and experimentation processes. The emergence is understood as a rather "tentative" process for which clear targets cannot be defined *ex ante*. Instead, the authors propose six elements as being pertinent for the emergence of directionality, and which need to be used in a flexible and context-dependent way: visioning, experimentation, networking, intermediation, institutional change, contestation, and conflicts.

In terms of governance, the chapter underlines the importance of combining bottom-up and top-down elements, and how fluid that distinction is in practice. It depends on a collaborative and experimental culture, and the ability to govern conflict and contestation constructively. One interesting observation is that policy attention was proactively focused on this development model early on, with high-level agreement assured before much of the actual experimentation took place. Overall, this contribution effectively illustrates how an integrative policy approach could look – an approach that combines top-down regulatory measures and bottom-up experimental approaches to realise a mission (see Edler et al. in this volume).

In their contribution on spatial and scalar implications of missions for policy, *Elvira Uyarra, Kieron Flanagan, and Iris Wanzenböck* enrich current definitions and design principles of missions by considering their spatial dimension. Based on the claim that there is a need to move towards more desirable innovations, rather than simply a greater number of innovations, and that the state needs to take a more active role to achieve this, the authors ask how missions should be defined, implemented, and evaluated. In particular, they argue that missions are not implemented in a vacuum, but that contextual conditions matter – not least geographic or territorial specificities – and that local administrations are therefore a key element in mission-oriented policies.

The authors observe that recent work on transformative missions has often been divorced from parallel efforts to support regional innovation and highlight the increasing importance of "missions" and challenges in such endeavours. They point to the literature on smart specialisation and regional innovation systems, where transformative missions have acquired increasing traction. They criticise the idea that the missions concept may become a discretionary label and default answer to almost any kind of societal issue, ignoring the accumulated experience from pre-existing and analogous goal-oriented policies and the complex social and political issues that they raise.

First, their contribution concerns the spatial and scalar differentiation of mission-oriented policy frameworks. They propose a governance framework that distinguishes problem/solution representations (local versus global) in order to define different policy opportunity spaces for transformative missions. Micro missions, for instance, are about local, place-based solutions for local problems.

They also discuss the fact that mission-oriented policy always has spatial effects that need to be considered, even if they are not intended. Transformative missions consider transformation in the sense of tackling a pressing societal challenge, but they might also, directly or indirectly, transform places and communities. Top-down accelerator missions tend to be place-blind in theory, but in terms of their impacts, no policies are spatially neutral in practice. Moreover, the authors claim that any mission-oriented policy will inevitably influence the agglomeration of economic activity, potentially aggravating economic and societal disparities across territories. Thus, place-blind transformative policies may exclude a large proportion of communities and places if not designed in a place-sensitive way. As a pitfall, excessive "localness" may lead to fragmentation and poor take-up of solutions that could otherwise potentially be transformative.

In their chapter, *Matthijs Janssen, Iris Wanzenböck, Lea Fünfschilling, and Dimitrios Pontikakis* provide a comparative analysis of governance arrangements from seven European countries for the implementation of transformative innovation policy. They are interested in the research question of how transformative innovation policy governance arrangements unfold in different governance modes and what challenges typically arise for each arrangement. The authors distinguish three types of governance modes and how they permit public organisations to deploy structural and procedural principles in line with transformative ambitions: administration-based (i.e., guided by hierarchy), network-based (i.e., in cooperation with societal actors), or society-based (i.e., led by societal actors). There are also some hybrid examples beyond these ideal types. They further distinguish four categories of governance tasks: creating legitimacy and leadership; coordinating across multiple levels, actors, and instruments; reflexivity, learning, and experimenting; and resolving conflicts. On the basis of these two conceptual elements, they analyse seven different programmes within a matrix of "modes" and "tasks".

An important governance question that they discuss is whether governance modes can evolve gradually, and whether it is even recommended for governmental organisations to move away from administration-based governance. The authors argue that it is not necessarily problematic if national governments stick to an administration-based governance approach and focus on improving the fit between policy instruments, as long as the resulting policy mix is suitable for supporting a range of bottom-up initiatives. Moreover, the authors conclude that we should not only look at the instrumental side of policy mixes but also at multi-level arrangements with different strategic and operational roles assumed at different policy levels, and "governance mixes" combining top-down and bottom-up governance approaches. The authors thus also confirm that a "hybrid" practice is possible that combines elements of the ideal "top-down" administrative practice with more "bottom-up" network- and society-led modes.

In their contribution on the historical development and lessons that can be taken from the case of automated driving, *Kazuhito Oyamada and Tateo Arimoto* provide a detailed case study on mission-oriented STI policy in Japan. The authors give a thorough account of how mission-oriented policies have evolved in Japan over the past decades, particularly with the institutional reforms since the 2000s. It is important to strengthen the role of the Cabinet Office in cross-sectoral coordination, and that of the Council for Science and Technology Policy in particular. In instrumental terms, coordination has been achieved through large R&D programmes since the 2010s, two of which the authors analyse in more detail (automated driving systems and the Strategic Innovation Promotion Programme) as examples of the whole-of-government approach pursued in Japan.

From a governance perspective, the case of Japan is instructive in relation to the three modes of governance proposed by Janssen et al. (in this volume). Japan fits well with the administrative mode, and it seems that the country has found an effective approach for implementing missions that is compatible with the prevailing mode of political governance. Despite these promising findings, challenges remain both with regard to inter-ministerial coordination (e.g., regarding the mobilisation of demand-side policies for fostering missions) and policy implementation (e.g., regarding portfolio-based programme management, differentiated stakeholder involvement, or continuous learning and evaluation).

In their chapter on missions as boundary objects, *Matthijs Janssen, Joeri Wesseling, Jonas Torrens, Matthias Weber, and Caetano Penna* argue that missions offer interpretive flexibility but still maintain shared meaning among the actors and stakeholders of a mission. This allows different communities to engage with a mission in various types of arenas: strategic arenas where meta-narratives are developed; the programming arena where mission-specific visions are defined; implementation arenas with their focus on designing policy intervention mixes; and performance arenas where concrete solutions are found for problems. Contestation and negotiation in these arenas help to reduce the openness and uncertainty of specific missions, as well as broader umbrella missions. In this sense, the boundary-spanning role of missions can facilitate or hinder the ability to address wicked problems in a systematic and discursive manner. This is exemplified by the case of the Dutch Mission-oriented Top Sector and Innovation Policy (MTIP) umbrella framework and two of its specific missions – Circular Economy and Carbon-free Built Environment.

Achieving convergence (of different meanings associated with a mission) in and between these nested arenas is an important challenge of mission governance, alongside two other key mechanisms that the authors introduce: passage (i.e., transferring meanings from one context to another) and learning (i.e., assigning new, arena-specific meanings to a mission on the basis of learning).

Another important issue involves the relationship between the wider umbrella frameworks for a set of missions and the specific missions, i.e., whether or not the umbrella framework imposes cross-cutting rationales, monitoring frameworks, instruments, etc., on the individual missions.

The authors use two case examples to show different ways of configuring missions in the Dutch context. The insights from their theoretical framework and the empirical cases illustrate the authors' points regarding some critical challenges of mission-oriented policies. In particular, they highlight the need to balance the specificity of each individual mission with the benefits of having a common framework, depending, for instance, on the scope of a specific mission or on the level of uncertainty associated with a mission. In terms of governance, the authors advocate for a tentative governance approach and the pursuit of adaptive mission strategies. In line with this argument, the authors question the ability to define clear, ambitious, and time-bound targets in the case of wicked challenges, which require open mission-oriented approaches embedded in flexible umbrella frameworks.

The chapter by *Philippe Larrue* uses the case of net-zero (i.e., the political ambition to reduce greenhouse gas emissions and to reach net-zero emission targets) to demonstrate that the same political ambition can be addressed by pursuing quite different mission-oriented approaches and strategies. Recognising that current policy approaches are not suitable to achieve net-zero ambitions, many governments have resorted to experimenting with novel or adapted policy rationales, programmatic goals, and instrument designs. Larrue provides an overview of these mission-oriented approaches based on three main building blocks: strategic orientation, policy coordination, and policy implementation. For each of these building blocks, Larrue presents design principles that are used as an analytical grid to assess specific mission-oriented policy initiatives.

Based on an analysis of 20 concrete cases of mission-oriented policy initiatives from various OECD member countries, Larrue proposes five types of mission-oriented innovation policies and positions them in a two-dimensional framework, distinguishing the scope of a mission (narrow versus systemic) and the level of integration of a mission (low versus high). In addition to overarching governance frameworks (e.g., achieving net-zero goals) and large umbrella missions (e.g., the German High-Tech Strategy), he identifies three types of specific mission policies, each with its specific strengths and weaknesses: traditional mission-oriented funding programmes, targeted accelerator missions, and highly ambitious transformative missions.

Apart from some systemic features, Larrue argues that the practices of current net-zero mission policies are not yet up to the systemic ambitions pursued, but at best represent building blocks for the wider policy strategy required to account for the full range of determinants that influence the success of a mission ("theory of change"). In fact, most mission-oriented innovation policies are first and foremost focused on R&D, and there are very few examples of missions that join up R&I with sectoral policies.

Larrue concludes that the ability of STI authorities, in their current role as "mission champions", to attract stronger commitment from sectoral authorities that "own" the societal challenges (and hence the programmes and regulations related to these challenges) will be key to the success of missions. Such an integrated "whole-of-government" approach is needed because most current missions are not just about accelerating technological change, but also about transforming socio-technical systems. In addition, socio-technical system transformations are long-term change processes and often wicked in nature; they, therefore, call for regular reflexivity, learning, and adaptation of mission-related challenges, targets, solutions, and pathways.

The chapter by *Kathy Berger, Christoph Brodnik, and Michael Dinges* focusses on European partnerships and how they contribute to forging networks of actors from science, society, and policy to enable new forms of transformative and mission-oriented innovation. The authors analyse the transformative potential of these partnerships by examining their respective theories of change, showing what role they expect their activities to take and what contribution they expect them to have in conjunction with a wide range of other actions and intervening factors. In doing this, the authors focus on identifying transformative outcomes as indicators of longer-term transformative impacts.

Overall, European partnerships can play key roles in strategically coordinating and aligning R&I policy agendas across Europe, linking actors from different research and industrial communities, creating spaces for experimentation and policy learning, and strengthening European leadership in mission-oriented areas even beyond R&I.

Based on a qualitative text analysis of the partnership proposals in climate, energy, and mobility, the authors show that the vast majority of European partnerships' activities and outputs are geared towards "building and nurturing niches", while other transformative outcomes that are of major relevance for the wider uptake of niche solutions play only a marginal role (e.g., replication and institutionalisation). This, together with some other results, underlines that the European partnerships tend to be characterised as accelerators rather than transformative missions, even if significant differences can be found between the partnerships.

With a view to leveraging the full transformative potential of European partnerships, the authors make a number of suggestions as to how current governance rules and practices could be adjusted. In addition to better linking European partnerships to other Horizon Europe instruments (i.e., through work programmes, the European Research Council, the European Innovation Council, the Marie Sklodowska-Curie Actions, etc.), their alignment with the EU's industrial, structural, and sectoral policies could be improved. As regards the latter, there is much potential seen in better connecting to demand-side policies such as regulation, public procurement, and standardisation initiatives. Finally, European partnerships could also be encouraged to enhance their transformative effects by strengthening synergies with actions at the national and local levels.

In governance terms, this indicates that multi-level and multi-domain coordination is crucial for exploiting the transformative potential of otherwise rather narrowly defined missions, such as those developed in the context of European partnerships.

Finally, *Jakob Edler and Matthias Weber* draw broad, overall conclusions with regard to the governance of ambitious, transformative mission-oriented STI policies as formulated in the initial framework (see Edler et al., this volume). They do so by first exploiting the contributions of this volume and, in an inductive manner, deriving ten key lessons for governance in diverse contexts. They then deduce and apply a conceptual framework, defining four major governance functions. This framework allows them to condense and make sense of those ten inductive lessons and to simplify take-away messages for the future development of appropriate governance practices. On the basis of the 10 lessons, they finally develop a set of governance principles and analytical questions to support the development of a differentiated approach to mission governance. They finish their chapter with a short suggestion for a future research agenda.

PART I

TRANSFORMATIVE MISSION-ORIENTED STI POLICY: THEORY AND CONCEPT

1. Transformative mission-oriented STI policy: theoretical and conceptual rationales, intervention logics and challenges of an emerging type of STI policies

Jakob Edler, Mireille Matt, Wolfgang Polt and Matthias Weber

1.1 AN EMERGING NEW WAY OF FRAMING STI POLICY

In recent years, we have witnessed a shift not only in the framing and claims around innovation policy, but also in how science, technology, and innovation (STI) policy is framed in a broader sense. In part, STI support by the state is (again) being formulated and implemented with the claim of accomplishing specific goals and targets ("missions"), be they societal, technological, or economic developments. In this framing, missions define major problems or goals that are of high priority for society and which tend to be formulated in terms of well-defined targets and timeframes (e.g., "man on the moon by the end of the 1960s" or restricting temperature rise to two degrees Celsius by the end of the twenty-first century).

Missions come in many different forms, though it is possible to distinguish different types of missions, from large-scale technology missions to "new" missions that are derived from long-term societal challenges (Wittmann et al., 2020; Kuittinen et al., 2018, Polt et al., 2021). As knowledge and innovation are generally key to achieving missions, "mission-oriented STI policies" seek to orient the production of scientific knowledge, technology, and innovation towards those societally defined missions in a targeted and goal-oriented manner (Larrue, 2021).

The recent shift marks a profound extension of established rationales for STI policy. The term "missions" has been used before, but with different meanings than in today's debates. While in the 1950s and 1960s, the notion of missions was mainly associated with large-scale scientific or technological projects (the most emblematic examples of which were the Manhattan Project and the Apollo Programme), as early as the 1970s, Richard Nelson highlighted that the most challenging missions are not technological, but societal in nature.[1] Complementary to these technological and societal framings of missions, environmental challenges were also proposed as areas where targeted, transformative policies would be needed (Soete & Arundel, 1993). These arguments pointed to a need to extend the rationales underpinning STI policy interventions from matters of economic competitiveness or geopolitical positioning to societal and environmental concerns, and more lately to the Sustainable Development Goals of the United Nations.

[1] See the seminal work of Nelson (1977) and his distinction between technological missions and social missions, which are much more difficult to achieve.

The first fundamental difference in the new approach to missions is that STI policy is now called upon to *directly and intentionally* provide support in solving societal problems in areas that are traditionally situated in other policy domains, such as energy, mobility, health, etc. In other words, mission-oriented innovation policy transgresses the boundaries of established policy fields. Secondly, in European discourse, STI is increasingly perceived as contributing to societal developments that may entail major risks in the longer term – such as increased inequality due to uneven benefits from innovation – and has thus become normatively questioned. Thirdly, this development has challenged the legitimacy of STI funding and policy being focussed solely on technological innovations that have a predominantly economic purpose, and has triggered the search for approaches that allocate more responsibility for societally beneficial STI activities back to the state and to actors that perform R&I. The current approach being developed in response to these shifts involves characterising *new types of "transformative" missions* as politically defined areas where solutions to societal problems can be developed, as well as designing and implementing STI policies to support them.

In this chapter, we will focus on this new and transformative type of mission-oriented STI policy. We will build on the typologies of different types of missions introduced previously (Kuittinen et al., 2018; Wittmann et al., 2020; Polt et al., 2021), which distinguish transformative missions with a wide societal reach from technology-accelerating missions or scientific missions of a more targeted nature. In line with Wanzenböck et al. (2020), we thus focus on the "wicked" systemic challenges ahead and recognise that different pathways may exist to resolve them. What we will not address in this chapter is the wide spectrum of transformative innovation policies that are not guided by clearly defined societal goals but that aim to handle more open-ended transformations (Weber et al., 2021, Bergek et al., 2023). For example, these are triggered by highly uncertain technological developments (such as in the field of AI), or are set in motion to respond to unexpected crises (such as the COVID-19 pandemic).

With our take on *transformative mission-oriented STI policy*, our transformative innovation policy framework differs from others with a similar ambition that have been under discussion but which are not necessarily called "missions" and suggest different approaches to achieving their societal goals. Schot and Steinmueller (2018) proposed the notion of "transformative innovation policy" as third-generation innovation policy, which relies strongly on processes of bottom-up learning for inducing system transformations, with government taking a guiding, moderating, and enabling role. While this is certainly one possible approach to addressing missions, it tends to neglect important elements of the spectrum of policy inroads available to foster innovation and transformations in a mission-oriented way (e.g., top-down regulatory reforms, public procurement, or the establishment of dedicated public institutions). We regard these as necessary ingredients in the policy mix needed for transformative missions to be accomplished.

Mazzucato (2018) has popularised the notion of missions at the EU level based on an entrepreneurial understanding of the role of the state, advocating for clearly defined goals, targets, and guidelines to steer the realisation of missions. Her approach, however, is also a partial and rather technocratic response to societal challenges. This method may be suitable in cases where missions have clear goals and targets, where pathways can be defined, and where market creation is the central key to transformation. However, it may not be the most appropriate way forward in areas of "wicked" transformative change, where other generalisation mechanisms need to play a prominent role.

Others completely contest the ability of governments to engage successfully in this type of interventionist policy (however skilfully designed) and ascribe the main solution capabilities to markets (e.g., McCloskey & Mingardi, 2020). While we think that this type of critique overrates the problem-solving capacity of markets (after all, it is very often markets' failures that lead governments to seek more directional policies), it is certainly true that current governments have not (yet) built up the necessary capacities, governance structures, and processes to appropriately enact these new policy approaches. As it stands, we seem to be in a transition phase in which "directional" policy (albeit in new forms), after having been recognised as a necessary tool to address certain problems, needs to be re-learnt by public policy. This process obviously requires time and resources, and for the time being, the extent to which this capacity- and institution-building will succeed remains an open question.

Our approach differs from these three positions insofar as we see the bottom-up organisation of societal actors and the skilful use of market forces (Boon et al., 2022) as necessary complements to stronger directional and targeted government action. Hence, transformative mission-oriented STI policy falls into a "triangle" of these forces.

Against this backdrop, our ideal-typical conceptualisation of transformative mission-oriented STI Policies focusses on the following:

1. Public policy initiatives that *target system transformation* in relation to *wicked societal challenges* through scientific-technological and social innovations. They are *ambitious, exploratory, and groundbreaking* in nature and often require *inter- and trans-disciplinary* resources to be mobilised.
2. The initiatives display *socio-technical directionality, intentionality,* and *adaptability (agility),* which is what differentiates them from other types of innovation policy initiatives. They have *clearly defined goals* (and a well-defined timeframe) with preferably qualified and/or quantified *targets,* and their *progress is monitored* using predefined milestones; however, there is an openness towards using alternative approaches to achieve these initiatives.
3. At the same time, the initiatives are *flexible in terms of allowing for adjustment of goals, targets, and milestones* in response to improved understanding of both the challenge ahead and the set of solutions explored.
4. While the *state, as primus inter pares,* has the final responsibility for defining the missions for transformative mission policies, the identification and specification of the underlying problems to be solved are based on *socio-political discursive processes.* More than traditional innovation policy approaches, transformative mission-oriented approaches remain open to contestation and adjustment over time due to their immediate societal relevance, as well as to potential changes in circumstances during the implementation of the policy.

Transformative mission policies are enacted by a *wide range of actors* from various domains in society. In order to reconfigure existing policy trajectories, these actors are involved in *collective processes of experimentation, generalisation, and learning,* supported and guided by coherent *mixes of supply-side and demand-side policies* and *bottom-up and top-down* elements that affect innovation and transformation.

Implementing this emerging new policy approach has substantial consequences for the role and governance of STI policy, for the relationships between policy levels, for the design of instrument mixes, and for the relative role of STI policy vis-à-vis other policies. STI policy becomes increasingly legitimised through its contribution to solving such problems. The COVID-19 pandemic has highlighted the central role that STI policies play in providing solutions, but has also underlined the need to renew the policy frameworks and governmental capabilities to cope with complex and uncertain societal issues in an agile manner (Weber et al., 2021; Dreher & Weber, 2023, in this volume). Reorienting STI towards societal problems implies engagement with a much broader actor landscape (Boon & Edler, 2018) and changes in incentive structures and framework conditions, economic conditions and dynamics, actors' abilities, practices, routines, and expectations, and the ways that actors interplay with each other and with technologies. In short, to "accomplish" societal missions, STI policy needs to contribute to systems transformation, although it is never sufficient to try to achieve this alone.

As the claims of transformative STI policies are so far-reaching and ambitious but are hardly questioned (Larrue, 2021), we want to take a step back and critically revisit the emerging rationales and their implications. In this contribution, we will outline the conceptual and theoretical underpinnings of transformative STI policy. We will unpack the various rationales of transformative mission-oriented STI policies and reflect on the changing roles of government policy, taking into account the nature of system transformations. We will also utilise different theoretical schools of thought to identify and critically discuss how they contradict and complement our own take on transformative mission-oriented STI policy.

1.2 FROM TRADITIONAL STI POLICY TO TRANSFORMATIVE MISSION-ORIENTED (STI) POLICIES

1.2.1 The Changing Rationales in Perspective

In this section, we will revisit and systematise the existing and emerging rationales for STI policy from a historical perspective, but with a particular emphasis on the recent revival of mission-oriented approaches. These have complemented earlier rationales, such as those that built on a generic innovation systems framework or on enhancing competitiveness by promoting key industrial technologies, which all continue to be relevant to current STI policy debates.

From a historical perspective, it can be seen that earlier, i.e., narrower and more technology-centred, versions of mission-oriented policies were quite a prominent part of STI policies from the 1940s onwards. By and large, those "traditional" mission-oriented policies (MOP), i.e., real "moonshots", were strongly driven by technological developments and were heavily centralised, particularly in state agencies and ministries. They were implemented mostly in areas in which (i) there were no markets (except for the state), which meant that government controlled the demand side (space, defence); (ii) the private sector lacked the scale of resources (energy); and (iii) there was a "sense of (public) urgency" (JIIP et al., 2018).

Those "traditional" MOPs were largely (though not completely) abandoned in the 1980s, following some apparent government failures and a more general policy shift, which was also seen in other policy domains (e.g., in industrial policy), towards approaches based on self-organisation of markets and innovation systems.

There were several examples of failures of these types of "missions", ranging from supersonic air transport (SST) and "baroque" weapons systems to national computer industries being built up through "national champions" being fostered (see Cohen & Noll, 1991). These failures were not mere accidents (such as failures that also occur in private markets) but showed the limits of a specific type of "top-down directionality" that does not take into account the following: (i) societal acceptance, (ii) market conditions (especially demand side), and (iii) alternative options for technological development (i.e., betting on a single solution) or what is nowadays called "technological openness".

In terms of today's situation, the lessons from these policy design failures need to be reinterpreted. In the last two decades, the private sector's capacity for innovation has hugely increased in a number of critical areas (such as artificial intelligence, space flight, and communication systems), as indicated by a related shift in R&D expenditure and a superior command of technologies. At the same time, state capacities were dismantled from the 1990s onwards, as evidenced by the diminished importance of government research institutions, for example. Thus, policies today must take these shifts in innovation capacity into account and seek to incentivise strong private sector involvement in the pursuit of societal missions. In addition, new forms of steering capacity need to be developed that are better suited to address complex transformative problems.

The reduction of government capacities in most countries has been accompanied by a shift in government policies away from trying to influence the *direction* of technological change and instead focussing on influencing the *rate* of innovation. This approach, while being successful in improving innovation systems in general, has increasingly been called into question for its insufficient contribution to societal demands in addressing "grand societal challenges". Against this background, rationales for new types of mission-oriented policies were developed, very explicitly already in the early 1990s (Soete & Arundel, 1993), but coming back strongly in the mid-2010s (for many, see Kuittinen et al., 2018, Mazzucato, 2018).

These new claims regarding the contribution of STI to addressing societal challenges were based on the argument that the prevailing policy approach is unable to help resolve key long-term challenges that our societies are confronted with, and that policy rationales need to evolve accordingly. Key problems to be addressed relate to path dependencies that have emerged in historically bound contexts, and that have created interdependencies between technologies, infrastructures, behaviour, and the mindsets and organisational and institutional frameworks in which they are embedded. This situation called for an extension of the fundamental lines of reasoning to justify policy interventions in the processes of research and innovation. Until about 10 years ago, market and (structural) system failures were the recognised intervention rationales. Those rationales, however, have been increasingly complemented by "transformative" system failures as equally valid rationales for policy intervention. Weber and Rohracher (2012) distinguish four subtypes of transformative system failures: directionality failures, demand articulation failures, policy coordination failures, and reflexivity failures. Schot and Steinmueller (2018) interpret this development as part of a more fundamental shift towards a new third generation of innovation policy, which they see as being embedded in a global long-term process of "deep transition" towards a huge new surge of development (Schot & Kanger, 2018).

In the meantime, several attempts have been made to move from these new basic rationales and concepts to more specific questions of how they could be turned into governance principles and policy instruments. Steward (2012) suggested that the agendas need to shift from traditional macroeconomic and micro-organisational perspectives to the meso level of socio-technical systems of production and consumption, where networks and actor configurations are the key entry points for policy interventions. These interventions combine both technology-driven supply-side measures and challenge-led demand-oriented measures. This meets the general tendency in research and innovation policy to "resurrect the demand side" of innovation (Edler & Georghiou, 2007).

The need to move from general principles to implementing specific governance and instrumental measures is also reflected in the deficits that can be seen in the EC's Horizon 2020 framework programme. This programme has turned its focus towards addressing societal challenges to create tangible changes at the instrument and project levels. This motivated the call for missions as a novel instrument in the framework programme Horizon Europe (European Commission, 2017). Mazzucato (2018) proposed a very specific approach to missions in Horizon Europe in order to address these deficits – an approach that relied on defining clear goals, targets, and pathways and enabling the state to take an entrepreneurial role. However, it did not capture the need to cater to a range of transformative mission-oriented approaches to account for the diversity in missions. Recent work by Larrue (2021) has demonstrated this diversity by screening mission-oriented innovation policies at the national level. Meanwhile, concepts are being developed that seek to operationalise transformative innovation policy, strongly advocating for a broader, socio-technical understanding of transformation compared to innovation policy approaches (Diercks et al., 2019). Others stress the need for long-term thinking combined with agility on the part of the major actors involved, particularly the state, to adjust to transformation dynamics in a flexible manner (Weber et al., 2021).

1.2.2 The Nature of System Changes and the Role of Governance and the State

Despite the call for creating general guiding principles for more transformative and mission-oriented STI policy, the aforementioned shift to the meso level implies that systems, societal challenges, and transformation processes differ. Socio-technical systems have not only emerged in specific historical contexts, but they are also often bound by location and show marked thematic and sectoral differences in their patterns, their underlying mechanisms of change, and the relative role of policy (Coenen et al., 2012).

A crucial challenge, therefore, is to overcome these different kinds of path dependencies by mobilising STI policy as well as sectoral and wider framework-setting policies. These path dependencies can take various forms: physical infrastructures represent rigid system backbones, as do organisational and institutional structures. At the micro-level, mindsets and routinised behaviour can equally be a barrier to transformation, and vested interests tend to protect the status quo with more rigour – and blackmailing power – than potential winners fighting for the transition. Some of these insights are not new (Dosi, 1982; Geels, 2004), but with the urgency of transformations, they have moved to the centre stage of policy debates.

These systems, understood as configurations of social and technological elements, show complex behaviour. They perform specific societal functions and are characterised by continuity and delimitation from their environment (Borrás & Edler, 2014). Rather than offering a single magic wand to bring about change, transforming these systems requires multiple coherent entry points. This is why multi-actor processes that aim to design and implement reasonably coherent strategies have become so popular in recent years (Avelino & Wittmayer, 2016). Similarly, the complexity involved means that processes of continuous learning are required in order to adapt to uncertain and emerging patterns of change, which can hardly be planned in advance (Pahl-Wostl et al., 2010; Purkus & Lüdtke, 2020). Learning in specific niches, however, will not automatically lead to system transformations unless complemented by punctuated decisions at key moments during a change process in order to establish and consolidate new structures and institutions (Berkhout et al., 2004). These brief considerations indicate how daunting the challenge for policy is in general, and for STI policy in particular, when it comes to making agile contributions to shaping system changes.

An additional aspect is the observation that policy cultures differ as much as socio-technical systems. These cultures have historically experienced equal growth and shown clear differences in preferences for certain governance styles at both national levels (Andersen, 2019; Larrue, 2021) and sectoral levels (Cairney, 2021). While it is important to take these cultural differences into account, the need for transformative policies may nevertheless require moving beyond these established governance styles in any political system. The aforementioned features of system transformations indicate that there are a multitude of possible entry points for policy interventions and their policy mixes (Kern et al., 2019). There are also different promising governance practices for bringing these interventions about, including "bottom-up" experimentation and learning (Schot et al., 2019), proactive market-forming interventions (Mazzucato, 2018) and "top-down" regulatory and other demand-side measures (Edler & Georghiou, 2007).

So, our take on transformative and mission-oriented STI policies aims at systematising the understanding of these various policy challenges, governance styles, and policy options, while recognising that the entry points and policy approaches to be pursued in any particular case will depend to a significant extent on both the system specificities and the political cultures in which they are embedded. What is similar, however, is the need for achieving policy coherence to trigger transformative change.

This also suggests *a pragmatic, rather than an ideological, take on the roles of the state in system transformations*. Mission orientation may suggest a specific framing for STI policy ambitions, and it can draw on the multitude of roles that the state may adopt in view of some basic types of socio-technical system transformations, such as those suggested by Borrás and Edler (2020), whereby the basic fit of different roles depends on the context for and type of systems transition. Furthermore, these roles may also change over the course of a system transformation process as part of a higher-order learning process in the governance of change.

1.3 CRITICAL CONSIDERATIONS OF TRANSFORMATIVE MISSION-ORIENTED STI POLICIES

The move toward transformative mission-oriented STI policy marks a substantial shift in the definition of the role of the state and therefore is not uncontested and faces criticism from various angles. This criticism is as varied as the different basic understandings of transformative mission policies.

Firstly, there is opposition from neoliberal thinkers who object to Mazzucato's (2013) notion of the "entrepreneurial state" (Karlson et al., 2021; McCloskey et al., 2020). They raise fundamental arguments against what they perceive as an overly interventionist policy approach that would constrain both the freedom of science and the process of exploration and discovery in free markets. Secondly, there are more moderate critiques from the field of evolutionary economics. These critiques point to the potential limitations of state intervention in evolutionary processes and question the ability of governments to provide guidance in efficient and effective ways. Thirdly, this latter criticism is underpinned by political scientists who argue that the potential of the state to intervene and provide direction varies greatly in relation to different systems' transitions. Some transitions develop through market-driven dynamics that are often difficult to channel through state action (Borrás & Edler, 2020). Furthermore, some political scientists see the state as being intellectually and operationally overstretched, claiming that the capabilities needed to operationalise, implement, and govern transformative STI policy today are not sufficient in most public administrations (Lindner et al., 2016; Kattel & Mazzucato, 2018).

The most fundamental criticism targets the very understanding of the role of the state in market economies. Neoliberal economists strongly object to the rediscovery of the state giving direction to technological and economic development. Why, they ask, should the state and its governance capabilities be better equipped than the bottom-up dynamics of the market to find solutions for urgent societal problems (McCloskey & Mingardi, 2020; McCloskey et al., 2020)? According to this line of thought, an interventionist state, seeking to give direction to innovation activities, would constrain both the freedom of science and the process of exploration and discovery in free markets, as it would limit the creativity and governing power of the marketplace. At the same time, proponents of such an interventionist state would strongly exaggerate its coordinating and forward-looking capabilities. While aiming at innovation-based solutions for societal problems, the likelihood of efficient and effective solutions to challenges would be diminished, openness towards technologies and innovation would be constrained, and with it the ability of societies to truly tackle societal problems.

This is not the place to engage in a fundamental debate between neoliberalists and state interventionists. Notwithstanding the well-known fundamental philosophical and ontological differences between these two camps, it appears that the neoliberal criticism has a fundamentally different starting point and continues to ignore apparent failures of markets and innovation systems. It does not start with the urgency of societal problems that have developed in market and state-led economies alike. It does not discuss the societally detrimental effects of market economies going unchecked by societal forces moderated through the state. It also does not acknowledge the power of societal discourse in democratic processes for defining desirable transformations that the state can and should support – rather than steer. It does not acknowledge that on this legitimate basis, there is a mix of instruments to support directionality, such as regulation, incentives, deliberation and co-generation of solutions, and a range of instruments to help accelerate the diffusion of desirable innovations. While all of these instruments intervene in the market, they do not abandon market forces but rather nudge them in directions defined in a democratic process. The fundamental neoliberal critique therefore fails to develop an alternative framework, given the urgency of societal, climate, and environmental needs and the necessary coordination on the pathway towards desirable transformations.

However, a number of elements of this criticism have also been raised by scholars who, in principle, acknowledge the need for directionality and transformations supported by the state. Firstly, there is still a rather basic critique that transformative mission-oriented STI policy continues to be *technologically deterministic*. The more these policies determine pathways for concrete technological solutions and prioritise certain innovations over others, the more they will constrain the creativity of the market and the development of alternative innovative solutions in the future. Some critics of current mission and transformative policies thus call for technological openness and policy interventions that, while giving direction through targets, deliberation, and (potentially) price mechanisms (such as a CO_2 tax), keep markets broadly intact (Schäuble, 2021). Secondly, debates are increasingly taking place over the capacities of the state and non-state actors to fully understand transformations and wicked problems and the systemic implications of transformations. These implications amount to transformation failures that need to be overcome, especially as the existing role of the state may be part of the problem. Defining the interventions that may most effectively tackle transformation failures is consequently an enormous challenge in terms of capacity and governance. Some critics, therefore, fear the state will become overstretched intellectually and operationally, and claim that the *capabilities needed to operationalise, implement, and govern transformative STI policy today are not sufficient* in most public administrations (Lindner et al., 2016). There is some rather recent literature on transformative capacities that tries to overcome this critique and suggest pragmatic ways to build up necessary capacities, including, but not limited to, state actors (Hölscher, 2020; Borrás, 2021; Borrás et al., 2023).

Thirdly, this latter criticism is underpinned by scholars, mainly political scientists, who argue that all too often, transformative mission-oriented STI policies are *not contextualised enough*. The potential of the state to intervene and provide direction varies widely for different systems' transitions, depending on the subject area (mobility, health, energy, climate, etc.) and given institutional policy settings. Some transitions develop in market-driven dynamics that are often hard to channel through state action, while others are very much triggered by state action from the start (Borrás & Edler, 2020). In both situations, the leverage of the state to make a difference through transformative mission-oriented STI policy differs considerably. Still, neither policymakers nor proponents of certain transformations are sufficiently differentiating the relative role of the state and the relative role of related STI policies.

The above-mentioned critical approaches to transformative mission-oriented STI policies cannot be dismissed easily. Indeed, a purely technocratic approach that does not mobilise a vast array of societal actors and change their behaviour is bound to fail. Likewise, in the past, attempts to implement policies without taking market mechanisms and market dynamics into account have (at least partly) failed. A number of projects in the 1970s and 1980s (Concorde and the fast breeder reactor being often cited as examples) are testimony to such government failures. And it is certainly true that most governments – at least for the time being – lack (or have lost) the knowledge and capacity for this emerging type of policy approach.

However, none of these issues cannot be overcome: citizen engagement is increasingly being sought as part of transformative policies, a large number of actors are often involved both in designing as well as implementing policies (including industry – often in the form of institutionalised PPPs or civil society – in urban missions, for example), and current and future market dynamics are being both addressed and taken into account when designing policies. Hence, the main difficulty continues to be in designing governance structures, policy mixes, and instruments for meeting the transformative challenges and creating sufficient levels of public acceptance while at the same time asserting the pitfalls of potential government failure.

1.4 FROM RATIONALES TO POLICY INTERVENTION

As described above, transformative mission-oriented STI policy calls for the capacity to be built up that is currently lacking or only partially available in government. In the following section, we will describe some of the issues that these government capacities have to tackle. In doing so, we will take a critical look at the concrete implementation of policies *through governance practices, capabilities, and strategic intelligence* as well as new forms of *instrumentation* that have been introduced in recent years.

1.4.1 Policy Governance, Capabilities and Strategic Intelligence

Above all, deliberate transformative mission policies strive to *mobilise a broad range of governmental and non-governmental actors* that, one way or the other, need to co-define and co-own the transformation and support it. Next to widening the range of policy options considered necessary for triggering transformative and goal-oriented change as part of a coherent policy mix, the biggest challenge in governing transformative mission-oriented policies may be *coordinating within government and with other broader stakeholder groups*, building up related capabilities, and supporting strategic intelligence. Even for matters of instrumentation, it is less the invention of new instruments that is being called for, but rather the mobilisation and orchestration of existing instruments for transformative purposes.

In other words, beyond engaging in broader societal discourse whereby stakeholders are enrolled in transformative agendas, there are *important demands on state actors to coordinate within government* in much more extensive and systematic ways compared to traditional STI policies. In terms of mission-oriented policy, the state is the ensemble of public actors – at various levels – who contribute to a given mission. This means that those involved need the capabilities to map critical governance structures and relevant state actions, and to mobilise different specialised policies and instruments and find appropriate ways to organise interplay between them.

Policy coordination is indeed needed to define a mission, but also to ensure coherence between different policy areas and policy levels and their concrete actions (Ulmanen et al., 2022). *Horizontal, vertical, multi-level, and temporal policy coordination are regarded as important requirements for all successful mission-oriented policies.* Horizontal policy coordination refers to the necessity to combine STI policy with sectoral policies (transport, health, agriculture, energy policies) and cross-cutting policies (tax policy, trade policy, competition policy). Defining and achieving missions also requires vertical policy coordination (between ministries and implementing agencies), as well as coordination between multiple levels of government (European, national, regional, and local) (Wittmann et al., 2020; Larrue, 2021; Polt et al., 2021). Another challenge around coordination is dual timing: the timing of the policy versus the timing of the transformation. The sequence of the intervention and the modification and adaptation of the policy mix should result in continuous improvements and the harmonisation necessary to foster the envisioned transformations (Sartorius & Zundel, 2005). Policy mixes should be iteratively adapted along the transformation pathway.

Different *mechanisms for coordination are conceivable, from delegation to centralised mission agencies and mission boards or cross-ministerial mission committees.* But no matter which governance model is chosen, it remains a considerable challenge to combine the power of specialisation within individual units with the need for coordination and holistic thinking in mission policies – or a whole-of-government approach (Peters, 2018).

The multi-faceted nature of policy coordination, in turn, calls for *institutional and regulatory experimenting* to overcome path dependencies and established traditions in policy fields that have developed over time (Weber et al., 2021). More fundamentally, mission orientation unleashes a debate about the roles of a widening range of actors in R&I as well as in R&I

policy. As a consequence, established governance frameworks are being contested along with new modalities.

For many reasons, the design and implementation of transformative mission-oriented policies pose challenges that necessitate a broad range of governance capabilities and support from distributed strategic intelligence (Kuhlmann and Rip, 2001). Firstly, on the *strategic level*, policymakers and stakeholders alike need to be able to analyse and understand the very system that is influenced through the mission-oriented policy and to orchestrate societal discourse to define the mission. Secondly, on the *conceptual level*, as each mission will target different systems, the capability is needed to define appropriate governance approaches and policy mixes for specific missions. Thirdly, capabilities are needed on the *operational level* to implement a policy or instrument mix and monitor and assess progress in terms of systems change (see also Kastrinos & Weber, 2020; Weber et al., 2021). As mission-oriented policies reach out to various societal arenas and economic sectors, strategic intelligence needs to reach out beyond the innovation domain to address developments, outcomes, and impacts in the respective target areas. In sum, mission-oriented policies call for systematic reflexive and interconnected governance (Voß & Kemp, 2006) to be widely embedded in innovation systems and beyond (Lindner et al., 2016), along with appropriate strategic capabilities.

In what follows, we will elaborate on the capability- and intelligence-related requirements at these three levels. While several of the requirements apply to more than one level, others may be specific to one level only. Firstly, mission policies are highly politicised and remain in constant flux; they thus require governance capabilities and processes to generate legitimacy and engage in collective sense-making through the entire policy cycle, and at strategic, conceptual, and operational levels. A first policy challenge mainly at the strategic level in any form of liberal democracy is the inherent politicisation of priority-setting. To choose a mission means to prioritise a societal challenge over others. Compared to market failure and system failure approaches, the demand for legitimacy thus changes, increases, and is more contested in decisions over priorities with a transformative ambition. Societal acceptance of the choice of missions and the transformation process that is envisaged are critical (Borrás & Edler, 2014).

Further, on the conceptual level, the definition of the problem and possible solution pathways is not a technocratic act defining one clearly demarcated and defined mission. Rather, as missions deal with *"wicked" problems, they are shaped and developed in discursive "framing" processes* (Wanzenböck et al., 2020) that close down periodically at critical points when missions are politically defined and redefined. But missions – and their implementation – will always be open to interpretation and challenged by a more diverse set of actors compared to traditional STI policies. They will tend to evolve over time, and the conceptual ideas as to how those missions should be supported will remain contested (Kuhlmann & Rip, 2014). This brings in political power dynamics that are often overlooked (Meadowcroft, 2009), as well as a requirement to moderate plurality rather than seek comprehensive consensus (Grin et al., 2010). In order to avoid stalemate situations and open up novel perspectives for incumbent actors, it is also important to consider, and possibly compensate for, those who may potentially lose out as a result of a transformation. Against this background, state actors need critical networking and orchestration capabilities to create open and appropriate actor networks, to moderate conflicting interests and complex discourses, and to translate those discourses into policies. Thus, strategic approaches for collective sense-making, i.e., various forms of participatory co-design and deliberative processes, remain a constant and important feature at all levels of mission policies.

A related challenge is the increased complexity of appropriate policy design and implementation (Haddad et al., 2022) *calling for specific conceptual and analytical capabilities*. To work towards a mission requires inducing and orienting systems change, rather than merely generating and diffusing distinct innovations to create economic welfare. To contribute to a mission, mission policies need to advance the underlying socio-technical systems in directions that promise to remedy the defined problem to some extent. It thus means a change of routines and practices, institutions (including values and regulations), infrastructures, necessary capabilities, etc. The selection process of the market may provide important innovations, but on its own, it will not be able to generate this kind of system change. This, in turn, poses challenges at two levels. Firstly, there needs to be the conceptual capability to define the basic logic of state intervention in existing systems. What is the concrete role of state and non-state actors, and to what extent can they intervene in socio-technical systems? As stated above, this is extremely complicated, as the societal challenges underlying missions differ considerably in "nature, scope, and scale" (Wanzenböck et al., 2020), as do the possibilities of state intervention (Borrás & Edler, 2020). Secondly, defining innovation policies and further policies to support changes to system conditions is a multi-dimensional and complex problem. It asks for the analytical capacity to identify and understand hampering and supporting system conditions and non-linear system dynamics. This also includes the capacity to monitor those systems changes, including innovation supply conditions, broader market and system societal attitudes, and the willingness to support the transformation.

This is where *systematic support through evaluation* is required. In order to transform systems under considerable uncertainty and in actors' landscapes that are more heterogeneous than in traditional STI policies, strong formative evaluation is necessary. The formative evaluation needs to build in reflexivity and enable joint learning from the outset and throughout the entire policy process (Molas Gallart et al., 2021; Belcher & Hughes, 2020b). Contrary to Molas-Gallart et al. (2021), we argue that the multi-level perspective alone, with its focus on bottom-up learning processes, is not enough to capture the richness of mechanisms at play in socio-technical transformation strategies and policies. The capabilities discussed so far would be part of it. However, formative evaluation would have to go a step further; it would have to include a wider variety of mechanisms for scaling and embedding new solutions. What is key here is the systematic, method-based effort to collectively understand the possible effects of policy interventions in conjunction with other transformational mechanisms; this is also increasingly being piloted in transformational RTD programmes (Deutsch et al., 2021; Matt et al., 2023).

For mission programmes, *monitoring approaches* should be *anticipatory* and based on endogenous futures thinking that will not only allow expected transformations and impacts to be defined but also the impact-generating mechanisms (e.g., the evolving network of actors, critical tipping points, impeding factors, and the role of intermediaries) and the milestones along the pathway towards impacts. Real-time impact monitoring (Matt et al., 2023) should be *participatory* to define the possible futures but also to create the distributed intelligence necessary to provide rapid feedback about emergent issues, patterns of change, critical incidences, and possible bifurcations. A joint exercise such as this builds credibility and legitimacy for the intervention – and helps to make its limitations clearer. As significant changes in socio-technical systems will be needed to achieve missions, and given the very long time frames that missions have, monitoring exercises should be iterative and adaptive and should tolerate ambiguity and some confusion. At each iteration, goals, targets, milestones, and expected transformations will be questioned. Rather than relying solely on quantitative measures, they will be based on qualitative data and moderated discourse.

As any policy intervention also needs also to be justified and accounted for, *summative ex-post evaluations of mission-oriented policies remain important.* Such summative elements need to be part of an overall formative evaluation strategy and be mobilised at critical decision points in the course of implementing a longer-term mission. The process of measuring progress will rely on the milestones defined by the anticipatory exercises. It will have to capture changes in socio-technical systems and focus on behavioural and institutional change as much as on concrete changes in inputs and outputs (Belcher et al., 2020a), reconfigurations of networks of actors, and visions of intended transformations. Importantly, this approach should also critically reflect on the changes within the government in terms of coordination and complementary actions across administrative units, as this is required for effective implementation. The behavioural effects of transformative policies also need to be taken into account.

Furthermore, the mission and its policies will have to be revisited repeatedly to *enable joint learning and to readjust policy interventions* as needed, and reassess the envisaged goals and milestones through a discursive process. By its very nature, the *measurement* of progress will not rely on quantifiable variables but *will mobilise qualitative analysis* in order to increase the plausibility of transformational mechanisms. This will enable the final impact – the state of the mission – to be traced back to the interventions. Those evaluating mission policies will need to learn from a few examples of innovation policy mix analyses (Cunningham et al., 2016) and, more importantly, from transition studies, in order to analyse the interplay of policies in much more sophisticated ways (Rogge & Schleich, 2018; Rosenow et al., 2017) than are currently used.

1.4.2 Policy Instrumentation: The Importance of Combined Policies

As with any policy approach, the choice and mix of instruments must fit the task. Thus, the instrumentation deployed for transformative mission-oriented policies must take into account some general requirements derived from societal challenges. These include the involvement of many different actors, who can contribute their various perspectives about the problems to be solved and the type of solution needed. Moreover, the instrument mix for transformative mission-oriented policies needs to be able to handle uncertainty, non-linearity, emergence, bifurcations, path dependencies, and generalisation.

In addition to these general principles, there are also instrumental requirements that need to be addressed. Beyond the well-documented challenges of policy mixes for innovation (Flanagan et al., 2011; Cunningham et al., 2016), three characteristics of transformative mission-oriented STI policies stand out.

Firstly, instruments are needed to orchestrate the definition of a transformation and its institutionalisation in public discourse and coordinated action. *Building common visions* of the desired system transformations is an important precondition for tackling directionality failures (Weber & Rohracher, 2012). However, creating these shared perspectives and expectations requires collective coordination as well as dedicated efforts to identify and support the specific pathways and agendas to follow. Thus, *reflexive and participatory processes* are needed to build a shared vision, enable open coordination, and define joint agendas for action.

Secondly, the policy measures should *mix supply-push and demand-pull instruments* (Larrue, 2021), address all stages of the innovation cycle as required (from research to market creation, including demonstration) and cut across various policy fields, sectors, and disciplines. In particular, the systematic inclusion of demand-based instruments provides support for the contributions that innovation can make towards the transformation. Instruments such as living labs, public-private cooperation, experimental approaches for integrating consumers and producers in open innovation initiatives, and public procurement for innovation and instruments that support creative niches (Kivimaa & Kern, 2016) could help to overcome demand articulation failures (Weber & Rohracher, 2012). Rogge & Reichardt (2016) state that policy mixes for sustainability transitions should also involve systemic instruments (Smits & Kuhlmann, 2004; Wieczorek & Hekkert, 2012). In addition, Kivimaa & Kern (2016) propose that policy mixes for sustainability transitions should include instruments to help destabilise currently existing parts of the system. The destructive part of the policy mix is often based on non-STI policies. These include a whole range of instruments such as taxes, import restrictions, regulations, legislative structural reforms, phase-out measures (reduced support for dominant, but declining technologies), social policies, or policies that help to avoid negative consequences of destabilisation and support new key actors or networks. In this chapter, we contend that steering and progressively stabilising new development paths towards missions will require a coherent portfolio of STI instruments and non-STI instruments, emergence and destabilisation-related instruments, and top-down and bottom-up initiatives that are deployed and coordinated at different levels of interventions and over time. The key factor is not so much the type of instruments that are included in the policy mix, but rather the features of the policy mix, which should be compatible with the governance approach being taken for the mission. To cope with uncertainty, directionality, and complexity, the instrument mix should be flexible, revisable, dynamic, and open (Kuhlmann & Rip, 2018) and induce collective experimentation, learning, reflexivity, and reversibility.

Thirdly, transformation-oriented instruments should support and enable system actors to contribute to a transformation or adjust to its consequences. In particular, transformations of socio-technical systems depend on the *diffusion, scaling up, embedding, or generalisation of experiments* and innovation activities to maximise desirable societal impacts and achieve the defined mission (Sengers et al., 2021; Wigboldus et al., 2016). A variety of concepts exist to capture the journey and process that will transform, mainstream, institutionalise, and embed innovation experiments and outcomes. Sengers et al. (2021) consider four mechanisms of embedding that take into account temporal, spatial, and structural dimensions: replication and proliferation; expansion and consolidation; challenging and reframing; and circulation and anchoring. In general, there is a weak understanding of the generalisation mechanisms that could be supported by transformative mission-oriented policies. However, these concepts and mechanisms highlight specific aspects that should be considered: the need to combine market and non-market dynamics; the co-evolution of the social, technological, and institutional dimensions of innovations; the spatial, temporal, and structural features of the process; and the need to create new types of stakeholders and networks of actors. Generalisation will require "top-down" interventions by transformative mission-oriented policies that could take the form of new rules and regulations, the creation of structures and infrastructures, support for new types of networks, platforms, and alliances, as well as "bottom-up" interventions aiming to support experimentation and learning with niche applications.

As with more traditional policy mixes in innovation policy (Cunningham et al., 2016), the complex mix of instruments used to influence transitions, however, is most often not the result of a careful ex-ante design but rather an ex-post approach to making sense of a messy process (Kivimaa & Kern, 2016; Rogge & Reichardt, 2016). Regarding the coordination and management of mixes, Larrue (2021) provides some interesting empirical lessons about the implementation of transformative mission-oriented innovation policies in various countries. Transformative mission-oriented STI policies mainly mobilise existing direct interventions, and very few have established new modes of intervention. However, there is now a broader range of policy instruments within the mixes that aim to solve a given challenge. The elaboration by Larrue (2021) shows that the width and diversity of the instrument mixes and the level of integration of instruments are related to the type of transformative mission-oriented innovation programmes or frameworks. It further highlights that, in general, missions addressing wicked problems require scientific, technological, behavioural (individual and collective), and institutional changes that call for a more distributed organisation. Furthermore, reflecting on existing transformative mission approaches, one can conclude that the following represent effective modes of promoting various actors' commitments to mission-oriented STI policies: strategic agendas co-produced with public and private partners, endorsement and support by public policies, multi-year agreements that formalise the actors' respective contributions to the mission, and the engagement of users in projects and of partners in governance. Finally, greater coordination and sharing of information between policy bodies will improve instrument mixes' consistency, effectiveness, and efficiency by increasing positive interactions and reducing negative, unintended interactions.

The main challenges in policy implementation (Larrue, 2021; Haddad et al., 2022) concern directionality (maintaining the mission focus during implementation) and funding issues, but also the instrument mix and its evolution (connecting supply-side and demand-pull instruments, supporting market take-up, and including non-STI policy instruments such as regulations, price mechanisms, and public procurement).

1.5 CONCLUSIONS

The contribution of this conceptual chapter is to improve our understanding of the need for and challenges of state action in driving transformation that is valuable for society. We have focussed on transformative mission-oriented STI policies, i.e., policies that seek transformation and that do so by explicitly defining and redefining certain missions or goals together with other systems actors and mobilising instruments of science, technology, and innovation policy to work towards those goals. The chapter has focussed on the role of the state and policy as we witness *governments' increased ambitions* to mobilise STI for transformation towards clearly defined goals that are (as of yet) not matched by a *sufficient understanding of the specific limits, challenges, and responsibilities of the state.* In discussing this, however, we have argued that neither a technocratic state-centred approach (Mazzucato, 2018), nor a mainly bottom-up approach that gives the state the responsibility for supporting societal dynamics (Schot & Steinmueller, 2018b), nor a market-driven approach with the state having basic regulatory functions, are appropriate on their own to mobilise STI for transformation.

We have not outlined a grand design for transformative mission-oriented STI policies; we do not believe in a simple functional framework that links certain missions to certain policies and governance approaches. Rather, we have argued for a *pragmatic approach of trial and error, of engaging, designing, implementing, and redesigning policies for transformations.* In defining the role of the state, it is important to consider the specificities of political cultures and socio-technical systems and to understand the need for multiple entry points. What we want to propose is a menu of options and a systematic frame for making good use of them, but not a single "recipe".

Firstly, our discussion showed that *the intervention rationales for transformative mission-oriented STI policies are far less clear-cut than the STI policy rationales that have been established for decades.* Further reflection and, crucially, empirical work will be needed to understand how the roles of the state, with all the coordination and strategic intelligence requirements, can best be performed to mobilise, rather than impede, societal dynamics in desired directions. What we do believe, however, is that the above-mentioned critical approaches to transformative mission-oriented STI policy must be taken into account to avoid government failure. Transformative mission-oriented STI policy has to operate in the triangle of government directionality, market dynamics, and societal actors and behaviour. It will only succeed if it manages to align these three forces.

Secondly, in order to support this policy learning in general, *courage is needed for experimentation in terms of governance structures, incentive systems, and instrumentation.* New ways may be tried out to organise the design and set-up of interventions between established policymaking bodies; innovation policy may be defined and utilised as supporting policies on which other policy domains can draw; new agencies may bundle the expertise and budgets of existing domain ministries for a specific time span; and to initiate certain dynamics, more novel market-creating policies may be linked explicitly to the generation of technological and non-technological innovation through such coordination mechanisms. Many more entry points for experimentation and new thinking are needed, and possible.

Thirdly, there is *great potential for the state to mobilise and learn from evaluation.* Our discussion has shown that supporting transformative mission-oriented STI policies through evaluation not only requires much more ambitious formative approaches that still have to be developed but that as opposed to traditional innovation policies, they cannot rely solely on ex ante "intervention logics" or "theories of change". Instead, the mobilisation of the system and its change, and of the corresponding transformative innovation pathways, will need to be monitored in real time. The contribution of interventions to those changes and the expected transformations should be assessed in relation to the desired directionality and their scope of impact. A continuous interplay of formative and summative elements would add reflexivity to the system and allow for adjustment of policy interventions.

At the same time, however, the critique of interventionist, transformative mission-oriented policies as discussed in this chapter, needs to be taken seriously. There is a very tangible danger of overburdening the state. Transformation will not happen as a top-down execution of a grand policy plan to "accomplish a mission"; it will only happen if market forces are also mobilised to support change, if there are incentives in the marketplace that pull in the same direction, and if citizens have a sense of ownership of the transformation. The *main challenge for the state will be to find the right balance* between taking a moderating, mobilising, incentivising, interacting role, and letting markets find and diffuse creative solutions, and allowing citizens to engage in social innovation and change of practice without being managed from the top down.

Finally, the current hype for transformative mission-oriented policies should be an encouragement to support the development of novel policy approaches through STI policy research. The most pressing research challenges are related to the appropriate policy mixes, i.e., the set of instruments to be used, the governance structures, and processes, the role of governmental and non-governmental actors, and the knowledge and intelligence needed to inform policy-making before, during, and after policy implementation. Having a good conceptual understanding of the specificities of this (still emerging) type of policy is necessary to take a sensible approach to solving its challenges. This conceptualisation aims to support this understanding and lay the foundations for evidence- and concept-based policy-making.

REFERENCES

Andersen, M. S. (2019). The politics of carbon taxation: How varieties of policy style matter. *Environmental Politics, 28*(6), 1084–1104.

Avelino, F., & Wittmayer, J. M. (2016). Shifting power relations in sustainability transitions: A multi-actor perspective. *Journal of Environmental Policy & Planning, 18*(5), 628–649.

Belcher, B. M., Davel, R., & Claus, R. (2020a). A refined method for theory-based evaluation of the societal impacts of research. *MethodsX, 7*, 192–203.

Belcher, B. M., & Hughes, K. (2020b). Understanding and evaluating the impact of integrated problem-oriented research programmes: Concepts and considerations. *Research Evaluation, 30*(2), 1–15.

Bergek, A., Hellsmark, H., & Karltorp, K. (2023). Directionality challenges for transformative innovation policy: Lessons from implementing climate goals in the process industry. *Industry and Innovation, 30*(8), 1110–1139.

Berkhout, F., Smith, A., & Stirling, A. (2004). Socio-technological regimes and transition contexts. *System Innovation and the Transition to Sustainability: Theory, Evidence and Policy, 44*(106), 48–75.

Boon, W., & Edler, J. (2018). Demand, challenges, and innovation. Making sense of new trends in innovation policy. *Science and Public Policy, 45*(4), 435–447.

Boon, W. P. C., Edler, J., & Robinson, D. K. R. (2022). Conceptualizing market formation for transformative policy. *Environmental Innovation and Societal Transitions, 42*, 152–169.

Borrás, S. (2021, June). *Public leadership in the governance of transition* [Presentation at the Eu-SPRI Conference]. Oslo.

Borrás, S., & Edler, J., 2014. Introduction: on governance, systems and change, in: Borrás, S., Edler, J. (Eds), *The Governance of Socio-Technical Systems. Explaining change.* Edward Elgar Pub., Cheltenham (U.K.), Northampton, MA, USA, pp. 1–22.

Borrás, S., & Edler, J., 2014. The governance of change in socio-technical and innovation systems: three pillars for a conceptual framework, in: Borrás, S., Edler, J. (Eds), The Governance of Socio-Technical Systems. Explaining change. Edward Elgar Pub., Cheltenham (U.K.), Northampton, MA, USA, pp. 23–48.

Borrás, S., & Edler, J. (2020). The roles of the state in the governance of socio-technical systems' transformation. *Research Policy, 49*(5), 103971.

Borrás, S., Haakonsson, S., Poulsen, R. T., Pallesen, T., Hendriksen, C., Somavilla, L., Kugelberg, S., Larsen, H., & Gerli, F. (2023). *The transformative capacity of public sector organizations in sustainability transitions: A conceptualization.* Innovation Studies 2023/02, CIRCLE, University of Lund.

Bugge, M. M., Coenen, L., & Branstad, A. (2018). Governing socio-technical change: Orchestrating demand for assisted living in ageing societies. *Science and Public Policy, 45*(4), 468–479.

Cairney, P. (2021). The concept of a sectoral policy style. In M. Howlett & J. Tosun (Eds.), *The Routledge handbook of policy styles* (pp. 77–88). Routledge.

Coenen, L., Benneworth, P., & Truffer, B. (2012). Toward a spatial perspective on sustainability transitions. *Research Policy, 41*(6), 968–979.

Cohen, L., & Noll, R. (1991). *The technology pork barrel*. Brookings Institution.
Cunningham, P., Edler, J., Flanagan, K., & Larédo, P. (2016). The innovation policy mix. In J. Edler, P. Cunningham, A. Gök, & P. Shapira (Eds.), *Handbook of innovation policy impact* (pp. 505–542). Edward Elgar Publishing.
Deutsch, L., Belcher, B., Claus, R., & Hoffmann, S. (2021). Leading inter-and transdisciplinary research: Lessons from applying theories of change to a strategic research program. *Environmental Science & Policy, 120*, 29–41.
Diercks, G., Larsen, H., & Steward, F. (2019). Transformative innovation policy: Addressing variety in an emerging policy paradigm. *Research Policy, 48*(4), 880–894.
Dosi, G. (1982). Technological paradigms and technological trajectories: A suggested interpretation of the determinants and directions of technical change. *Research Policy, 11*(3), 147–162.
Dreher, C., & Weber, M. (2023). Agile ful-politik: Disruptiven und transformativen Wandel politisch gestalten. In A. Botthof, J. Edler, K. Hahn, H. Hirsch-Kreinsen, M. Weber, & J. Wessels (Eds.), *Transformative und agile innovationssysteme. Neue praktiken und innovationspolitische herausforderungen* (pp. 293–318). Campus.
Edler, J. (2019, October). *From enabling to delivering: Understanding the implications of mission oriented innovation policy*. Technical University Munich.
Edler, J., Cunningham, P., & Gök, A. (2016). *Handbook of innovation policy impact*. Edward Elgar Publishing.
Edler, J., & Georghiou, L. (2007). Public procurement and innovation – Resurrecting the demand side. *Research Policy, 36*(7), 949–963.
European Commission: Directorate-General for Research and Innovation. (2017). LAB – FAB – APP: investing in the European future we want: report of the independent High Level Group on maximising the impact of EU research & innovation programmes. Publications Office. https://data.europa.eu/doi/10.2777/477357.
European Commission: Directorate-General for Research and Innovation, Türk, A., Arrilucea, E., Skov Kristensen, F., Unger, M., Domini, A., Polt, W., Zee, F., Goetheer, A., Fisher, R., Lehenkari, J., Pelkonen, A., Chicot, J., & Kuittinen, H. (2018). *Mission-oriented research and innovation: Inventory and characterisation of initiatives*. Final report, European Commission, Brussels.
Flanagan, K., Uyarra, E., & Laranja, M. (2011). Reconceptualising the "policy mix" for innovation. *Research Policy, 40*(5), 702–713.
Geels, F. W. (2004). From sectoral systems of innovation to socio-technical systems: Insights about dynamics and change from sociology and institutional theory. *Research Policy, 33*(6–7), 897–920.
Grin, J., Rotmans, J., & Schot, J. (2010). *Transitions to sustainable development: New directions in the study of long term transformative change*. Routledge.
Haddad, C. R., Naki′c, V., Bergek, A., & Hellsmark, H. (2022). Transformative innovation policy: A systematic review. *Environmental Innovation and Societal Transitions, 43*, 14–40.
Hölscher, K. (2020). Capacities for transformative climate governance: A conceptual framework. In K. Hölscher & N. Frantzeskaki (Eds.), *Transformative climate governance. A capacities perspective to systematize, evaluate and guide climate action* (pp. 49–96). Palgrave Macmillan.
Karlson, N., Sandström, C., & Wennberg, K. (2021). Bureaucrats or markets in innovation policy? A critique of the entrepreneurial state. *The Review of Austrian Economics, 34*(1), 81–95.
Kastrinos, N., & Weber, K. M. (2020). Sustainable development goals in the research and innovation policy of the European Union. *Technological Forecasting and Social Change, 157*, 120056.
Kattel, R., & Mazzucato, M. (2018). Mission-oriented innovation policy and dynamic capabilities in the public sector. *Industrial and Corporate Change, 27*(5), 787–801.
Kern, F., Rogge, K. S., & Howlett, M. (2019). Policy mixes for sustainability transitions: New approaches and insights through bridging innovation and policy studies. *Research Policy, 48*(10), 103832.
Kivimaa, P., & Kern, F. (2016). Creative destruction or mere niche support? Innovation policy mixes for sustainability transitions. *Research Policy, 45*(1), 205–217.
Kuhlmann, S., Boekholt, P., Georghiou, L., Guy, K., Héraud, J.-A., Laredo, P., Lemola, T., Loveridge, D., Luukkonen, T., Polt, W., Rip, A., Sanz-Menendez, L., & Smits, R. (1999). *Improving distributed intelligence in complex innovation systems*. Final report of the Advanced Science & Technology Policy Planning Network (ASTPP).

Kuhlmann, S., & Rip, A. (2014). *The challenge of addressing Grand Challenges*. Report to the European Research and Innovation Area Board.
Kuhlmann, S., & Rip, A. (2018). Next-generation innovation policy and grand challenges. *Science and public policy*, *45*(4), 448–454.
Kuhlmann, S., Stegmaier, P., & Konrad, K. (2019). The tentative governance of emerging science and technology – A conceptual introduction. *Research Policy*, *48*(5), 1091–1097.
Kuittinen, H., Polt, W., & Weber, K. M. (2018). Mission Europe? A revival of mission-oriented policy in the European Union. In Council for Research and Technology Development (Ed.), *Re: thinking Europe. Positions on shaping an idea* (pp. 197–212). Holzhausen.
Larrue, P. (2021). The design and implementation of mission-oriented innovation policies: A new systemic policy approach to address societal challenges. *OECD Science, Technology and Industry Policy Papers*.
Lindner, R., Daimer, S., Beckert, B., Heyen, N., Koehler, J., Teufel, B., et al. (2016). Addressing directionality: Orientation failure and the systems of innovation heuristic. Towards reflexive governance. *Fraunhofer ISI Discussion Papers – Innovation Systems and Policy Analysis*.
Matt, M., Robinson, D. K. R., Joly, P.-B., van Dis, R., & Colinet, L. (2023). ASIRPA Real-Time in the making or how to empower researchers to steer research towards desired societal goals. *Research Evaluation*, *32*(2), 412–425.
Mazzucato, M. (2013). *The entrepreneurial state. Debunking public vs. private sector myths* (Rev. ed). Anthem Press.
Mazzucato, M. (2018). *Mission-oriented research & innovation in the European Union*. European Commission.
McCloskey, D. N., & Mingardi, A. (2020). *The myth of the entrepreneurial state*. American Institute for Economic Research.
Meadowcroft, J. (2009). What about the politics? Sustainable development, transition management, and long term energy transitions. *Policy Sciences*, *42*(4), 323–340.
Molas Gallart, J., Boni Aristizábal, A., Schot, J., & Giachi, S. (2021). A formative approach to the evaluation of Transformative Innovation Policy. *Research Evaluation*, *30*(4), 431–442.
Nelson, R. R. (1977). *The moon and the ghetto: An essay on public policy analysis*. Norton & Co Publishers.
Pahl-Wostl, C., Holtz, G., Kastens, B., & Knieper, C. (2010). Analyzing complex water governance regimes: The management and transition framework. *Environmental Science & Policy*, *13*(7), 571–581.
Peters, B. G. (2018). The challenge of policy coordination. *Policy Design and Practice*, *1*(1), 1–11.
Polt, W., Ploder, M., Breitfuss, M., Daimer, S., Jackwerth, T., & Zielinski, A. (2021, July). *Politikstile und Politikinstrumente in der F&I-Politik*. Studien zum deutschen Innovationssystem. http://hdl.handle.net/10419/231475
Purkus, A., & Lüdtke, J. (2020). A systemic evaluation framework for a multi-actor, forest-based bioeconomy governance process: The German Charter for Wood 2.0 as a case study. *Forest Policy and Economics*, *113*, 102113.
Rogge, K. S., & Reichardt, K. (2016). Policy mixes for sustainability transitions: An extended concept and framework for analysis. *Research Policy*, *45*(8), 1620–1635.
Rogge, K. S., & Schleich, J. (2018). Do policy mix characteristics matter for low-carbon innovation? A survey-based exploration of renewable power generation technologies in Germany. *Research Policy*, *47*(9), 1639–1654.
Rosenow, J., Kern, F., & Rogge, K. (2017). The need for comprehensive and well targeted instrument mixes to stimulate energy transitions: The case of energy efficiency policy. *Energy Research & Social Science*, *33*, 95–104.
Sartorius, C., & Zundel, S. (Eds.) (2005). *Time strategies, innovation and environmental policy*. Edward Elgar Publishing.
Schäuble, W. (2021). *Grenzerfahrungen. Wie wir an Krisen wachsen*. Siedler.
Schot, J., & Kanger, L. (2018). Deep transitions: Emergence, acceleration, stabilization and directionality. *Research Policy*, *47*(6), 1045–1059.
Schot, J., Kivimaa, P., & Torrens, J. (2019). *Transforming experimentation: Experimental policy engagements and their transformative outcomes*. TIPC Research Report, Brighton.

Schot, J., & Steinmueller, W. E. (2018b). Three frames for innovation policy: R&D, systems of innovation and transformative change. *Research Policy*, *47*(9), 1554–1567.

Sengers F., Turnheim B., & Berkhout F. (2021). Beyond experiments: Embedding outcomes in climate governance. *Environment and Planning C: Politics and Space*, *39*(6), 1148–1171.

Smits, R., & Kuhlmann, S. (2004). The rise of systemic instruments in innovation policy. *International Journal of Foresight and Innovation Policy, 1*(1–2), 4–32.

Soete, L., & Arundel, A. (1993). *An integrated approach to European innovation and technology diffusion policy (a Maastricht memorandum)*. EUR (Luxembourg).

Steward, F. (2012). Transformative innovation policy to meet the challenge of climate change: Sociotechnical networks aligned with consumption and end-use as new transition arenas for a low-carbon society or green economy. *Technology Analysis & Strategic Management*, *24*(4), 331–343.

Torrens, J., & Schot, J. (2017, February). *The roles of experimentation Transformative Innovation Policy*. Transformative Innovation Policy Consortium, Research Brief.

Ulmanen, J., Bergek, A., & Hellsmark, H. (2022). Lost in translation: Challenges in creating new transformative innovation policy practices. *PLOS Sustain Transform 1*(10), e0000031.

Voß, J. P., & Kemp, R. (2006). Sustainability and reflexive governance: Introduction. In J. P. Voß, D. Bauknecht, and R. Kemp (Eds.), *Reflexive governance for sustainable development* (pp. 3–28). Edward Elgar Publishing.

Wanzenböck, I., Wesseling, J. H., Frenken, K., Hekkert, M. P., & Weber, K. M. (2020). A framework for mission-oriented innovation policy: Alternative pathways through the problem–solution space. *Science and Public Policy*, *47*(4), 474–489.

Weber, K. M., & Rohracher, H. (2012). Legitimizing research, technology and innovation policies for transformative change: Combining insights from innovation systems and multi-level perspective in a comprehensive "failures" framework. *Research Policy, 41*(6), 1037–1047.

Weber, M., Biegelbauer, P., Brodnik, C., Dachs, B., Dreher, C., Kovaéc, M., et al. (2021, August). *Agilität in der F&I-Politik: Konzept, definition, operationalisierung*. Studien zum deutschen Innovationssystem. http://hdl.handle.net/10419/231476

Wieczorek, A. J., & Hekkert, M. P. (2012). Systemic instruments for systemic innovation problems: A framework for policy makers and innovation scholars. *Science and Public Policy*, *39*(1), 74–87.

Wigboldus, S., Klerkx, L., Leeuwis, C., Schut, M., Muilerman, S., & Jochemsen, H. (2016). Systemic perspectives on scaling agricultural innovations. A review. *Agronomy for Sustainable Development*, *36*(3), 1–20.

Wittmann, F., Hufnagl, M., Lindner, R., Roth, F., & Edler, J. (2020). Developing a typology for mission-oriented innovation policies. *Fraunhofer ISI Discussion Papers – Innovation Systems and Policy Analysis*.

2. From formulation to implementation: conceptualizing mission-oriented policies as multi-stage translation processes

Florian Wittmann, Miriam Hufnagl, Florian Roth, Merve Yorulmaz, Ralf Lindner and Stephanie Daimer

2.1 INTRODUCTION

In recent years, mission-oriented approaches have attracted considerable attention within innovation policy debates and beyond. Mission-oriented innovation policies (MOIP) aim to effectively address pressing societal challenges and propel forward the transformation of socio-technical systems. While Borrás and Edler (2020) acknowledge that "demand on the state to develop transformative policies is normatively reasonable" (p. 1), Janssen (2021, p. 5) observes a shift "from policy for knowledge to policy for solutions", which raises expectations. The Organisation for Economic Co-operation and Development (OECD) recently admonished the "over-optimism" about what MOIP can achieve (Larrue, 2021, p. 12), contending that so far "their effectiveness [...] is still to be proven". Therefore, even though the popularity of MOIP among policymakers and scholars has grown constantly, it seems that the elaborate scholarly debate is not always well-attuned to address the challenges of actual policymaking. In other words, theory is running ahead of practice.

Recent research has defined several key elements of successful MOIP: a legitimate and urgent societal challenge to address, ambitious mission goals, effective measures, and coordination structures (Mazzucato, 2018, 2019). However, a conceptual framework that covers all these elements is still lacking. Consequently, discussions have often centred on select aspects of MOIP without integrating other perspectives and accounting for the dynamics of MOIP implementation (Janssen et al., 2021). This chapter aims to address these shortcomings by proposing a process-oriented analytical perspective on MOIP, which considers the different actors and levels of contestation involved.

We argue that missions are shaped by three interrelated yet distinct translation processes related to (1) mission definition, (2) mission design, and (3) mission implementation (cf. Wittmann et al., 2021a). By disentangling mission dynamics and exploring the relationships between different translation processes, we present a conceptual framework that serves multiple purposes. Firstly, it provides policymakers with a better understanding of the dynamics and pitfalls of missions, indicating critical points in the realisation process that require special attention. Secondly, the conceptualisation supports researchers who are working with MOIP by integrating insights from policy studies, public administration research, and other strands of research to increase analytical clarity and rigor. Finally, conceptualising missions as multiple translation processes can contribute to an improved understanding of their potential effects and therefore may contribute to better assessing the impacts of missions – an aspect that still needs to be addressed (Amanatidou et al., 2014; Janssen et al., 2021; Wittmann et al., 2021d).

The chapter is structured as follows: the next section highlights the need for a better understanding of the dynamics of MOIP, focussing on the different actors and contestations involved. Seeking to enhance analytical rigor, we subsequently argue for a process-oriented perspective on MOIP that manifests itself in multiple translation processes. This provides more granularity and a more comprehensive picture of the specific choices and challenges involved at each procedural stage. Moving towards an integrated perspective, the following section focusses on the interactions between different translation processes. The final section summarises the key insights and outlines the implications for research and policy practice.

2.2 A PROCESS-ORIENTED PERSPECTIVE ON MOIP

The concept of MOIP bears the promise of shifting science, technology, and innovation (STI) policies towards a broad and transformative approach that spans sector-specific policies and addresses grand societal challenges (Kuhlmann & Rip, 2014). As Cantner and Vannuccini (2018, p. 845) argue, "for any policy intervention in this context, it is not so much the intensity of innovation activities that matters, but rather their direction". This directionality is the defining characteristic of mission orientation. To this end, MOIP seeks to mobilise actors and resources in a cross-sectoral and interdisciplinary way (Lindner et al., 2021). Over the last few years, many countries and international organisations have decided to utilise STI policies to realise missions related to issues such as climate change mitigation, digital transformation, and public health.

Despite its growing popularity, previous research has pointed to several issues related to MOIP that pose significant challenges for STI practitioners and scholars alike. The main issues relate to the high complexity and ambitious requirements of missions, the heterogeneity of mission types and goals (Polt et al., 2019; Wanzenböck et al., 2020; Wittmann et al., 2021a), unclear boundaries (Hekkert et al., 2020; Wesseling & Meijerhof, 2020; Wittmann et al., 2020b) and insufficient governance arrangements and policy instruments (Janssen, 2020; Wittmann et al., 2021c). While the importance of different dynamics and negotiations for missions within "innovation bureaucracies" (Kattel et al., 2019), as in public actors striving for positive change, has been acknowledged (Kattel & Mazzucato, 2018; Larrue, 2021; Lindner et al., 2021), a more comprehensive understanding of the dynamics that characterise the setup of missions and how these elements relate to each other is still missing.

In this regard, we argue in favour of a process-oriented perspective that can help to unravel the different challenges that are commonly associated with missions.

Firstly, missions are often perceived from a perspective of novelty, highlighting the differences from earlier generations of innovation policy. However, this perspective overlooks the starting point of missions. The novelty of the approach does not imply that missions are emerging in a vacuum, establishing a new field of activities that is unconnected to existing policies. Instead, missions usually develop in the context of established policies (Larrue, 2021), debates and practices that shape the context for missions (Edler & Salas Gironés, 2020). In many instances, innovation policies combine characteristics from different generations of innovation policies (Arnold et al., 2019, p. 53) and display policy layering. This implies that in practice, despite the shift towards new mission orientation, missions typically rely on established instruments and include features of existing policies and strategies. Therefore, understanding how missions interact with existing policies and instruments is key for the analysis of missions.

Secondly, MOIPs are commonly accompanied by a language of cooperation and actor mobilisation. Even though these are important elements, such language may result in overlooking the fact that "missions emerge as a negotiated outcome between different interests, concerns and imperatives – they are neither apolitical in their formulation, nor neutral in their conduct" (Janssen et al., 2021, p. 440). Introducing directionality imposes a politicisation of MOIP compared to traditional innovation policies (Boon & Edler, 2018; Breitinger et al., 2021; Hekkert et al., 2020), resulting in cognitive, material, and normative grounds for contestation, as actors may have different understandings, perceptions, and interests. The way the emerging issues are addressed shapes the outcomes and dynamics of missions in terms of goals, actor mobilisation, etc.

Thirdly, missions, despite the formulation of goals to be achieved, should not be considered static policies. Instead, missions are likely to develop over time, thereby being affected by the specific context and the existing understandings of the underlying problems and possible solutions (Wanzenböck et al., 2020). In this vein, "it is essential to understand missions not as rigid policies with linear impact dynamics, but as dynamic and experimental approaches that can evolve over time as new conditions and insights emerge" (Lindner et al., 2021, p. 31).

Fourthly, research has pointed to an enormous empirical diversity of missions. In this regard, a process-oriented perspective can contribute to a better understanding of the sources of variety of missions (cf. Wittmann et al., 2021a). Recent attempts to categorise missions advocated for distinguishing between different ideal types of accelerator and transformer missions (Kuittinen et al., 2018; Polt et al., 2019; cf. Wittmann et al., 2021a for a stronger emphasis on governance); the former generally representing cases that "can be understood as missions that seek to find an answer to a challenge with a relatively confined scope (e.g., Moonshot, research in one particular field), but do not aim for a comprehensive system change" (Wittmann et al., 2020a, p. 13), whereas the latter transformer missions more generally "aim to achieve a comprehensive change affecting a socio-technical system as a whole and are therefore not limited to scientific progress and regulatory changes" (Wittmann et al., 2020a, p. 14). Even within single national frameworks, such as the German High-tech Strategy 2025, very different missions and ways of linking them to societal challenges exist, pointing to different levels of ambition, scope, understandings of drivers for change, etc.

Finally, the concept of mission orientation in STI policy not only imposes a new type of objective (focussed on solving societal challenges, potentially transformative) but also imposes new requirements for policymaking which challenge established routines (for a detailed overview, see Lindner et al., 2021). An output-oriented perspective falls short of understanding the underlying dynamics shaping missions, such as an increased need for coordination across different levels, public actors, and domains, which is necessary for the legitimisation and mobilisation of private and public actors, as well as openness to experimentation, self-reflection, and policy-learning.

So far, the complex connections and negotiation processes between different elements of MOIP have received rather limited attention. Acknowledging different perspectives, belief systems, and normative assumptions of actors as part of complex contestation and translation processes might contribute to understanding missions in their complexity and analytically distinguishing between different processes.

2.3 CONCEPTUALISING MOIP AS MULTI-STAGE TRANSLATION PROCESSES

In order to incorporate the different processes of negotiation and contestation at different levels, we propose a conceptual framework of missions as multiple connected translation processes. The concept of translation processes has been coined by Callon (1986) and Latour (1986, 1987), who understand translations "as a complex process of negotiation during which meanings, claims and interests change and gain ground" (Waeraas & Nielsen, 2016, p. 237). This idea has been transferred to the process of policymaking and programme realisation, taking into consideration the "ideation, negotiation and decision-making that is driven and governed by multiple actors" (Kroll, 2019, p. 637). In the case of MOIP, various national ministries as well as supra-national or sub-regional authorities (regional, local), specialised agencies, representatives of affected industry branches, civil society organisations, and other actors may be involved in creating and bringing MOIP into practice.

Kroll (2019) links the idea of translation processes with the concept of multiple, loosely coupled streams of implementation and policy debate, each driven by its own path-dependent logic. This perspective emphasises the process-character of policies that lead new impulses "into an existing path-dependent system of narratives and support policy practice" (Kroll, 2019, p. 637) between the level of strategic priority and agenda setting (sphere of high politics/ political discourse involving parliament, secretaries of state, and high-level ministry officers), the level of topic orientation ("mission owners" i.e., political administration, ministerial authority, work level ministry officers), and the level of actual implementation (executive level of agencies/implementation of programmes).

The translation process approach yields three important insights for understanding MOIP. Firstly, it points to the importance of embedding the analysis in the existing institutional and political context. Secondly, the idea of path-dependency points to the fact that translation processes are connected and that, in many cases, once taken, decisions will be costly to reverse (cf. also Diercks, 2019). Thirdly, it reveals the necessity of taking into account policy capacities (Borrás, 2019; Wu et al., 2017) as moderating factors shaping these translation processes. Whereas it is clear that MOIP imposes comprehensive capacity requirements on involved actors (Lindner et al., 2021, pp. 12–15), the extent to which these capacities are available shapes the character of missions and influences the translation processes.

Applying the concept of translation processes to MOIP, we claim that MOIP entails three main translation processes along the way from deriving a mission out of a societal challenge to finally implementing policies that deal with this challenge:

1. *Mission formulation*: Translation of a societal challenge into a specific mission with dedicated priorities/goals.
2. *Mission design:* Translation of mission goals into a specific set of instruments, activities, and coordination structures.
3. *Mission implementation*: Translation of mission activities and inputs into concrete actions and effects.

The case of the German High-Tech mission strategy to combat cancer serves as a brief illustrative example of how we perceive these processes of multiple translations: departing from the underlying societal challenge (aging society, increasing cancer cases), it prioritises different aspects to tackle this problem (reducing avoidable cancer cases, better therapies). These goals are subsequently translated into a mission design comprising a set of instruments (funding for cancer-related research, promoting patient involvement through changes in funding schemes, fostering awareness campaigns advocating for prevention) and a dedicated governance structure (National Decade against Cancer, involving partners and supporters). Ultimately, these instruments need to be directed to working towards the desired changes (increased participation in early detection measures, more efficient therapies).

Each translation process unfolds in a different institutional arena and is characterised by specific dynamics of negotiations among various actors. This tripartite perspective builds upon Larrue (2021), who distinguishes between strategic orientation, policy coordination, and policy implementation. The proposed framework may add valuable insights for MOIP, which typically cuts across different topics and institutional responsibilities and involves a wide range of actors. Conceptualising missions as multiple translation processes can help to unravel these complex dynamics. As Kroll (2019) concludes, this is crucial since the interplay of these different processes shapes the realisation and effects of a policy in the end. Therefore, the concept of translation processes can help us better understand both dimensions of negotiation: within different topic-specific arenas and between different levels of responsibility and power.

In the following section, we elaborate on the three main translation processes in greater detail, discussing key actors, central issues of contestation, influencing factors, and required capacities that are summarised in Table 2.1.

2.3.1 Mission Formulation: Translating Challenges into Goals

While missions and societal challenges are often understood as directly connected, the relationship is actually more complex in practice, as missions represent only a facet of the societal challenge and interpret it in a specific and often selective manner. No mission is equivalent to the underlying societal challenges, but prioritises and weighs certain aspects it seeks to achieve against different priorities and scope. Accordingly, the first translation process entails selecting and emphasising core aspects of societal challenges while also defining boundaries and excluding other aspects related to a societal challenge. Larrue (2021, p. 9) describes this as a narrowing-down process that typically starts from what he refers to as "challenge areas" rather than from concrete missions.

Whereas the importance of reasonable mission goals has been increasingly emphasised by research (Janssen et al., 2021; Wittmann et al., 2021a), "[t]he relation between the nature of challenges and the design of MOIPs is still largely unexplored" (Larrue, 2021, p. 34). The process of mission formulation ultimately seeks to answer and cover several core aspects: Which societal challenge should be overcome by the mission? What is the underlying understanding of the problem and its possible solution? What kind of change or transformation is intended, and by whom?

The process of negotiating mission goals is closely linked to the strategic level of policymaking, deciding on the priorities of a mission and pointing to the importance of actor groups and processes as key factors driving mission formulation. However, even if this step is usually associated with some kind of public actor agency, missions might also be influenced by other key stakeholders to ideally foster commitment and joint ownership, and create an impetus to achieve the aspired change (transformation) right from the start. These stakeholders may be involved to varying degrees in the formulation process in different functions (input provider,

Table 2.1 *Key characteristics of translation processes*

-	*Mission formulation*	*Mission design*	*Mission implementation*
Main function	Narrowing down societal challenge to specific mission goal(s)	Choosing an adequate instrument mix/coordination structures fit for purpose to meet the goal(s)	Implementing instruments and activities within the mission
Level	Strategic level	Policy coordination	Policy implementation
Key actors	High-level politics, representatives of responsible ministries, key stakeholders, society	Mission owners, i.e., involved ministerial authority, work-level ministry officers	Executing administration, funding agencies, etc.
Issues of negotiation and contestation	Directionality, scope, level of ambition Priority and agenda setting Stakeholder involvement and representation	Actor and resource mobilisation/ involvement Mix/types of instruments Redesign vs. development of new instruments Coordination structures	Administration and implementation of instruments Role of monitoring, flexibility and learning Role of civil servants
Influencing factors	Ideational frames Political and institutional context Existing public discourses and awareness (urgency, legitimacy)	Ideational frames Belief systems Existing policies and structures Autonomy and power of actors Participating actors	Resources for learning/ evaluation Belief systems Administrative cultures
Key capacities	Communicative and knowledge absorptive capacities	Knowledge absorptive and organisational capacities	Organisational and communicative capacities

Source: Authors' own elaboration.

co-creator). Additionally, this step triggers questions about the choice of stakeholders who contribute to mission formulation and the absorptive and communicative capacity of public actors (Borrás, 2019) to deal with inputs.

Policy literature in general and research on agenda setting in particular have pointed to the diverse challenges arising in this context that can shape outcomes (Kuhlmann & Rip, 2018; Schot & Steinmueller, 2018). A broader involvement of actors may increase legitimacy and ensure a broader mobilisation, therefore also increasing the possible success of MOIP in general through joint ownership and leadership. Involving a higher number of actors may also lead to a shift in priorities, watering down the ambitions or delaying the formulation process. Empirical practice has demonstrated different approaches even within the same national strategy (Wittmann et al., 2020b). For example, the missions of the German High-Tech strategy range from a top-down mode of mission formulation by the involved ministries to a joint declaration of goals by key stakeholders in the National Decade against Cancer (as the main pillar of the mission on combating cancer). A more expert-driven approach has been pursued in the Horizon Europe programme of the European Union. In this case, the process of narrowing down mission areas and proposing mission goals has been delegated to mission boards, partly at least in order to insulate the formulation process from political influence (Mazzucato, 2019, pp. 7–8).

A second issue of negotiation is the question of the scope, ambition, and directionality of missions. The perception of urgency and legitimacy in society are critical aspects for missions that formulate ambitious and comprehensive goals (Janssen et al., 2020). However, MOIPs with transformative goals pose a particular challenge to risk-averse authorities who try for "policy innovations" (Howlett, 2014). As argued by Braams et al. (2021), as a result of an extensive analysis of transition and public administration literature, tasks related to socio-technical transitions (like most MOIPs with a transformative ambition) go beyond established models of public administration. Such far-reaching transformative missions incur considerable costs and redistributive consequences for some stakeholders, resulting in potential opposition to these changes (Wittmann et al., 2021a).

The mission "Achieving substantial greenhouse gas neutrality in industry" in the German High-Tech Strategy 2025 exemplifies this case. Whereas the mission goals themselves are ambitious, forcing substantial changes to be made in the industrial sector, the scope of the mission is limited to only a few industries, related mainly to the production of steel, aluminium, and cement. In this context, the mission strives mainly for the development and implementation of technical innovations. At the same time, other possible goals, e.g., related to the demand side of industrial production, are beyond the scope of the mission (Wittmann et al., 2020b).

The negotiation of mission goals takes place at the political-strategic level. This means it does not happen in a vacuum, but is embedded in a broader social, economic, and institutional context. Edler and Salas Gironés (2020) have demonstrated the importance of ideational factors in shaping missions across different countries, focussing on the issue of mobility transition in Germany, the Netherlands, and the United Kingdom. They show that the same overarching challenge is approached in different ways depending on the national context, reflecting country-specific priorities (Larrue, 2021).[1] As argued by Wanzenböck et al. (2020), perceptions of the underlying problem and possible solutions may or may not converge. This implies that in different contexts the process of mission formulation may have varying starting points. At the same time, this points to the fact that negotiation processes may feed back into the perception of the overall societal challenge and its understanding. As a consequence, the mission formulation process may contribute to converging perceptions among stakeholders, shifting priorities from the start of the process.

[1] A similar pattern can be found looking beyond the case of MOIP. Most countries have followed a mitigation strategy in order to address the challenge of climate change, in effect focussing on reducing greenhouse gas emissions. In contrast, only a few countries have adopted a strong climate change adaptation approach. An exception in this regard is the Netherlands, which have a long tradition of adaptation strategies that cover several policy fields, including innovation policies (Knowledge for Climate, 2014).

2.3.2 Mission Design: Translating Goals into Policy Instruments and Activities

The second translation occurs when mission goals are operationalised to come up with a specific set of policy instruments. This translation process also entails the development of structures for communication and coordination between the entities in charge of these instruments. In contrast to the process of mission formulation at the strategic level, mission design takes place at a lower level, with involved ministries and other stakeholders representing different subsystems that contribute resources to the mission (mission owners). To put missions into action, an understanding of the instruments and coordination structures that are necessary to achieve the postulated goals is needed. This translation step is centred on the identification, selection, and mobilisation of resources and inputs as well as on the alignment of these activities towards the earlier defined mission goal or goals. Also needed to put missions into action and manage the process are absorptive and organisational capacities (cf. Borrás, 2019). At the same time, it creates the need for delineating the boundaries of a mission at an instrument level, defining activities that contribute to the mission goals and are part of the governance arrangement of a mission and those that characterise the wider socio-technical system. Negotiation processes are therefore likely to centre around several aspects.

First, since MOIPs are characterised by a broader cross-sectoral approach, they require a high level of commitment from different actors and their readiness to contribute to the postulated goals. A good example in this respect is the way the Dutch Topsector approach was further developed and crafted into the "Mission-oriented Topsector and Innovation Policy" (MTIP) strategy to create joint ownership and leadership. This included joint agreements on mission content and funding after all relevant ministries, knowledge partners (e.g., academies, universities), and industry representatives had held structured negotiations on common goals and committed resources (Janssen, 2020). In contrast, the German High-Tech Strategy in most missions relies only on the mobilisation of resources from Federal ministries (Wittmann et al., 2021c).

Second, complex policy interventions generally require a bundle of measures as missions are more than a mere elaboration of instruments but are an integrated mix of instruments. Therefore, missions are supposed to apply a variety of policy instruments that "span different stages of the innovation cycle from research to demonstration and market deployment, mix supply-push and demand-pull instruments and cut across various policy fields, sectors and disciplines" (Larrue, 2021, p. 11). Key questions in this context are what kind of instruments (Hufnagl, 2010) are combined for what purpose, i.e., how the instruments are supposed to contribute towards goal achievement.

Finally, missions are accompanied by the need to establish coordination structures that bring together actors from different sub-systems and provide the foundation for the governance of the mission. Such organisational decisions, similar to autonomy over spending, touch upon power interests of involved actors and their ability to exercise control over the implementation process (cf. Moe, 1989).

The translation process of mission design may be affected by different aspects: for instance, actors may interpret differently the way to achieve the postulated goals. Edler and Salas Gironés (2020) point to the importance of different policy traditions, beliefs about the role of the state, and ownership of a problem. This aspect is closely linked to the understanding of the possible solutions (Wanzenböck et al., 2020) and the possibility that public actors may take different roles in transformation processes (Borrás & Edler, 2020). In particular, it touches upon the question of the extent a mission may go beyond the confines of STI policy and employ additional means such as regulation in order to facilitate the required changes (see, e.g., Expertenkommission Forschung und Innovation, 2021 for a discussion), or as Lascoumes and Le Gales (2007, p. 3) argue, whether instruments are not purely technocratic but are associated with different societal values and interpretations.

Moreover, missions are rarely built from scratch but are embedded into an established field of policy measures and activities (Larrue, 2021, p. 9), so that missions are likely to be a combination of past and present policy instruments or are built around existing policies. Empirical works highlight that missions often rely on previously existing policies that are only selectively accompanied by new instruments (Janssen, 2020; Wittmann et al., 2021c). This ties in with the observation of Rose (1990, p. 263) that "policy-makers are heirs before they are choosers". Generally, the input of MOIPs in many instances will be characterised by different modes of institutional change, such as the layering or refurbishing of different existing policies and activities just like in previous policy generations (cf. e.g., Hacker, 2004; Thelen & Streeck, 2005; Kern & Howlett, 2009; Mahoney & Thelen, 2010).

Therefore, mission design is not only about designing new interventions, but also about purposefully combining and adjusting existing interventions. Once again, the Dutch MTIP strategy serves as an illustrative case.

Rather than adding more instruments to the policy mix, the emphasis in implementing a national MOIP has been put on setting up the coordination mechanisms that allow organizations to make better use of available instruments. In this case, "better" would refer to innovation capacities being mobilised to contribute to solving societal challenges (as prioritised in missions) rather than for yielding innovative output per se (Janssen, 2020, p. 22).

2.3.3 Mission Implementation: Translating Instruments into Practice

The policy instruments and activities that were developed during the design phase need to be brought into practice in order to move missions toward the anticipated goals. The final translation process focusses on the implementation of this policy portfolio. As noted by Pressman and Wildavsky (1984), policy goals or objectives do not necessarily have to translate into the expected results but are shaped by the implementation process. This means that the translation process is not limited to questions of efficiency and effectiveness of instruments, but also includes links between implementation and initial goals, and the coordination between different policy instruments and activities.

In contrast to the process of mission design that takes place mainly at the level of mission owners, such as involved ministries or stakeholders, the implementation refers to a broader set of actors including funding agencies and "street-level bureaucrats" (Lipsky, 1980) who are responsible for carrying out these activities. In fact, these administrative actors may perceive goals differently, have their own agendas, may alternate or even sabotage policies during the implementation phase (Brehm, 1997; Kam, 2000) or might be rooted in administrative cultures that conflict with the new requirements (Braams et al., 2021).

While the existence of sufficient capacity at the individual, organisational, and systemic levels is a prerequisite for sound implementation and simultaneous coordination of different activities (cf. Considine et al., 2014; Wu et al., 2017), MOIP impose additional new requirements for the actors involved (Lindner et al., 2021), which are likely to become points of contestation reaching beyond capacity constraints. Missions usually entail a mix of established and new policies, so that the alignment of existing instruments with the newly formulated mission goals is a key aspect of mission implementation. Typically, this requires changes at different levels, such as the adaptation of programme goals or new criteria for eligibility, selection, and assessment. Yet these changes may not fully materialise in the expected way, as actors tend to rely on established routines and ideas that may not be compatible with the new postulated mission goals. Also, when new instruments are introduced in the wake of mission design, the implementation plays an important role; new instruments typically leave significant room for interpretation by actors, which can lead to "bureaucratic drift" (Shepsle, 1992). The implementation of systemic instruments, which have become increasingly popular (Smits & Kuhlmann, 2004), especially requires new capabilities of implementing agencies and individual bureaucrats to reach across ministerial pillars and take a broader perspective of the impacts of their policies. In this regard, Daimer et al. (2012, p. 184) acknowledge that these instruments "need to develop, e.g., new methods to measure new impact types or to mobilise and rethink beliefs, tacit assessments, emotions or behavioural patterns".

Moreover, MOIPs require an understanding of implementation from the perspective of an instrument mix that constantly needs to be adapted and optimised. Due to their complexity, transformative systemic innovation policies have to put a particularly strong emphasis on experimentation and policy learning (Grillitsch et al., 2019; Weber et al., 2021). However, such paradigm shifts may conflict with the risk-averse cultures of administrative actors, who often tend to focus on minimising mistakes and failures (cf. Lindner et al., 2021, p. 31). Different understandings on when and how to revise or even terminate unsuccessful instruments and how these insights may feed back into the process of mission design and mission formulation may be a key negotiation issue among the involved actors. Closely related to this is also the increasing demand for formative evaluation approaches to support transformative policies (Magro & Wilson, 2019; Molas-Gallart et al., 2021; Wittmann et al., 2021d). Since such approaches entail the promise to actively support the formulation, design, and implementation of missions, they are not only capacity-intensive but also may face discussions about the necessary degree of openness and accountability.

2.4 TOWARDS A PERSPECTIVE OF INTERCONNECTED TRANSLATION PROCESSES

In the previous sections, we presented the three main translation processes constitutive for MOIP: mission formulation, design, and implementation. Each of these processes involves specific actors, areas of contestation, capacities, and context conditions. While highlighting the analytical value of distinguishing these different processes, we believe that only an integrated view of the entire process provides a holistic picture of the dynamics of MOIPs. Thereby, we acknowledge that missions hardly follow a simple linear logic from design to effect but rather characterised by multiple interactions and feedback loops between different translation processes, resulting in a downward and upward logic (see Figure 2.1). Mission design and mission formulation especially, but also mission design and implementation in many instances, might be understood as partly co-evolving processes, reflecting the dynamic character of missions (cf. Janssen et al., 2021). In this section, we describe possible interactions, spillovers, and feedback dynamics that may arise from these translation processes.

Mission formulation is closely linked with the processes of mission design and implementation, or as Winter (Winter, 2012, p. 258) argues, "the roots of implementation problems can often be found in the prior policy formulation process". The development of a powerful narrative and well-designed mission formulation process is a key prerequisite for actor and resource mobilisation and can therefore shape mission design and implementation (Lindner et al., 2021). A compelling mission goal may strengthen the legitimacy and therefore enforce a higher commitment of involved actors to contribute to both the goals and a broader instrument mix. In contrast, missions with poorly defined and ambiguous goals may not only limit the ability to mobilise actors for a shared goal but also create room for negotiation and contestation, providing actors with agency to interpret the goals according to their own preferences and resulting in a cascade of problems from formulation to implementation. An example in this regard may be the German High-Tech Strategy where weaknesses in the formulation process were considered a major obstacle for actor mobilisation and policy design (Roth et al., 2021).

In turn, mission formulation processes can hardly be thought of as independent from the feasible options for mission design, implementation, and assessment. In other words, if policymakers do not have suitable instruments at their disposal for implementing an ambitious mission in a comprehensive fashion, this certainly feeds back into mission formulation by predefining a level of ambition that can plausibly be reached. In a similar way, long-standing existing policy approaches may shape the understanding of goals and solutions, resulting in missions being centred on these policies.

Mission implementation may also necessitate an adjustment of goals and mission design, making them more or less ambitious or even altering their character altogether. We contend that the success or failure of specific instruments may feed back into the (re)formulation of mission objectives and instrument mixes. For example, if a policy instrument fails to trigger behavioural change on the societal level needed for system transformation (e.g., in the context of decarbonisation), mission goals may be retrospectively cut back in order to fit a more narrowly technology-focussed, accelerator-type mission. In a similar vein, a narrowing down of the portfolio approach might occur once promising solutions emerge during implementation.

Finally, the effects of MOIPs may feed back into the process in different ways. Positive results may reinforce the positive narrative of a mission and its legitimacy and credibility of intention, as well as strengthen actor commitment and mobilisation. Positive results may also contribute to an increasing convergence on the perspective of solutions and problems, shifting missions towards a higher level of alignment (cf. Wanzenböck et al., 2020). Needless to say, that besides a virtuous circle such as this, missions may also be also affected by a vicious circle of non-materialising results, which may undermine a mission's legitimacy and decrease actor mobilisation while opening up the discussion about goals or the legitimacy of a mission in general.

2.5 CONCLUSION

This chapter proposes to conceptualise the realisation of MOIP as multiple translation processes, starting from a grand societal challenge and implementing policy instruments that work towards desired mission effects. This conceptualisation analytically disentangles different steps and dynamics at different levels that characterise MOIP and thereby allow for a

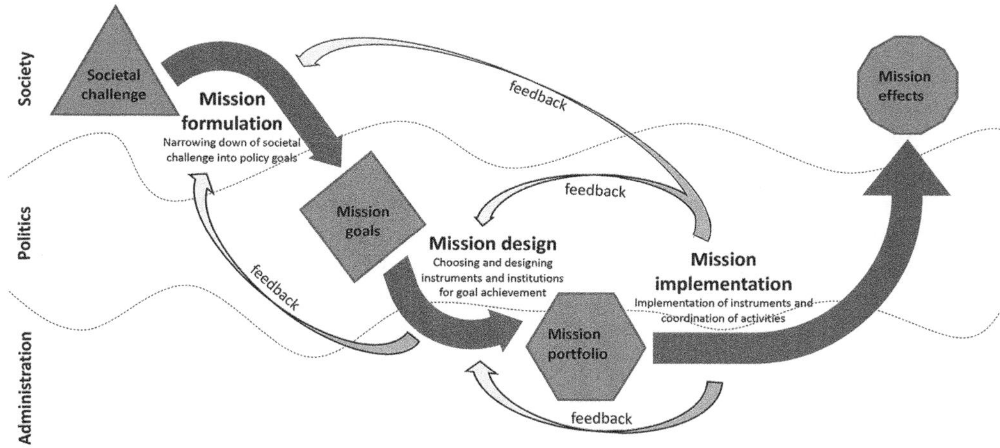

Source: Authors' own elaboration.

Figure 2.1 Translation processes and interdependencies

reduction of the complexity of MOIP from a process-oriented perspective. The direct negotiation processes of actors at different levels are associated with the three main translation processes of mission formulation, design and implementation. However, we also perceive these processes as interconnected throughout the entire mission process (feedback): the results of one translation process have profound implications for the subsequent development of another. Problems related to the choice of a mission, such as a poor mission definition, may also "cascade" down to the anticipated impact. While this is not to deny that missions may evolve over time, e.g., because of learning processes or external factors, it points to the importance of path dependencies that influence the development of missions. This conceptualisation can be considered particularly useful for different groups, including innovation researchers, ministerial authorities/policymakers, and actors involved in the impact assessment of MOIP.

Firstly, for researchers studying the realisation and impacts of MOIP, this chapter provides an analytical clarification for systematically disentangling the dynamics in MOIP at different levels and identifying the specific dynamics related to them. In doing so, it raises awareness of the importance of path dependencies in this process, highlighting the need for a comprehensive understanding of the factors shaping MOIP. The process-oriented perspective more closely connects the study of MOIP to the strand of literature on policy studies that can provide useful insights into the different dynamics and processes that characterise complex policies such as MOIP. This includes research dealing with different elements of the policy cycle (cf. Jann & Wegrich, 2017), the debate on the composition of and necessity for systemic instruments (Daimer et al., 2012; Smits & Kuhlmann, 2004; Wieczorek & Hekkert, 2012), policy coordination (Braun, 2008; Lindner, 2012), and reflexive governance (Borrás, 2011; Lindner et al., 2016).

Secondly, the framework can support attempts at assessing the impact of MOIP by researchers and possibly the executive administration. The understanding of missions as multiple connected translation processes can be a useful perspective for addressing the challenges of conceptualising and measuring the impacts of MOIP (Amanatidou et al., 2014; Hekkert et al., 2020, p. 77; Weber & Polt, 2014, p. 9), as proposed in a recent framework (Wittmann et al., 2021b). In this context, a process-oriented perspective allows us to better understand and systematise the different interdependencies between the mission stages and the materialisation of impacts. We then built on the idea of Kroll (2019), who traces deviations from the anticipated outcomes in EU Regional Policy back to multiple sources that need to be kept analytically apart. He argues that attempts to measure impacts against initial ambitions might fall short of anticipating the process of translation and interpretation, calling for a process-oriented perspective. Otherwise, he concludes, "it remains impossible to tell whether an innovation strategy was already flawed in ambition, wrong decisions were taken on thematic areas of action, or if, indeed, one is witnessing a technical failure in the choice of measures" (Kroll, 2019, p. 636).

This chapter points to the pitfalls of an approach that is too narrow and ignores the dynamics that may affect the materialisation of mission effects and ultimately mission impact at different stages. A mission undermining the key pillars of the MOIP approach, such as the mobilisation of private actors and resources, will face considerable challenges from the very beginning. An understanding of how the impacts of a mission can materialise should therefore not depart from mission goals but should incorporate the process and context of mission formulation into its perspective in order to adequately capture the context in which a mission emerged. By doing so, it can help to contribute to answering the stronger need for ex-ante evaluation and a different, more formative focus of evaluation (Amanatidou et al., 2014; Teirlink et al., 2011; Weber & Polt, 2014).

Finally, for policymakers, our framework clarifies the different procedural steps and points to the specific challenges and pitfalls connected with each of them. Thereby, it can support the process by making the possible consequences of decisions more explicit. Against the background of Larrue's comparative analysis that "[f]ew of MOIP initiatives have set objectives that have the expected mission characteristics: clear, bold and inspirational, with wide societal relevance, ambitious but realistic, targeted, measurable, time-bound and solution neutral" (Larrue, 2021, p. 9), the path dependency of such decisions outlined in this chapter echoes the call (Janssen et al., 2020; Wittmann et al., 2021a) for a careful planning of the mission process from the very beginning.

REFERENCES

Amanatidou, E., Cunningham, P., Gök, A., & Garefi, I. (2014). Using evaluation research as a means for policy Analysis in a "new" mission-oriented policy context. *Minerva, 52*(4), 419–438.

Arnold, E., Aström, T., Andréasson, H., Nielsen, K., Wain, M., Tofteng, M., & Røtnes, R. (2019). *Raising the ambition level in Norwegian innovation policy: Final report.* Technopolis Group.

Boon, W., & Edler, J. (2018). Demand, challenges, and innovation. Making sense of new trends in innovation policy. *Science and Public Policy, 45*(4), 435–447.

Borrás, S. (2011). Policy learning and organizational capacities in innovation policies. *Science and Public Policy, 38*(9), 725–734.

Borrás, S. (2019). Domestic capacity to deliver innovative Solutions for grand Social challenges. In D. Stone & K. Moloney (Eds.), *The Oxford handbook of global policy and transnational administration* (pp. 182–199). Oxford University Press.

Borrás, S., & Edler, J. (2020). The roles of the state in the governance of socio-technical systems' transformation. *Research Policy, 49*(5), 103971.

Braams, R. B., Wesseling, J. H., Meijer, A. J., & Hekkert, M. P. [Marko P.] (2021). Legitimizing transformative government. *Environmental Innovation and Societal Transitions, 39*, 191–205.
Braun, D. (2008). Organising the political coordination of knowledge and innovation policies. *Science and Public Policy, 35*(4), 227–239.
Brehm, J. (1997). *Working, shirking, and sabotage: Bureaucratic response to a democratic public. Michigan studies in political analysis.* University of Michigan Press.
Breitinger, J. C., Edler, J., Jackwerth-Rice, T., Lindner, R., & Schraad-Tischler, D. (2021). *Good-practice-beispiele für missionsorientierte innovationsstrategien und ihre umsetzung: Ergebnispapier 1: Innovation for transformation – Wie die Verbindung von Innovationsförderung und gesellschaftlicher Problemlösung gelingen kann.*
Callon, M. (1986). Some elements of a sociology of translation: Domestication of the scallops and the fishermen of St. Brieuc Bay. In J. Law (Ed.), *Power, action, and belief: A new Sociology of knowledge?* (pp. 196–229). Routledge & Kegan Paul.
Cantner, U., & Vannuccini, S. (2018). Elements of a Schumpeterian catalytic research and innovation policy. *Industrial and Corporate Change, 27*(5), 833–850.
Considine, M., Alexander, D., & Lewis, J. M. (2014). Policy design as craft: Teasing out policy design expertise using a semi-experimental approach. *Policy Sciences, 47*(3), 209–225.
Daimer, S., Hufnagl, M., & Warnke, P. (2012). Challenge-oriented policy-making and innovation systems theory: Reconsidering systemic instruments. In Fraunhofer Institut für System- und Innovationsforschung (Ed.), *Innovation system revisited - Experiences from 40 years of Fraunhofer ISI research* (pp. 217–234). Fraunhofer Verlag.
Diercks, G. (2019). Lost in translation: How legacy limits the OECD in promoting new policy mixes for sustainability transitions. *Research Policy, 48*(10), 103667.
Edler, J., & Salas Gironés, E. (2020, June 5). *How do framing and ideas influence the design of missions? A comparative analysis between Germany, the Netherlands, & the United Kingdom: EU-SPRI virtual session on Shaping System Transitions - Insights from practice.*
Expertenkommission Forschung und Innovation (2021). *Gutachten zu Forschung, Innovation und technologischer Leistungsfähigkeit Deutschlands 2021.*
Grillitsch, M., Hansen, T., Coenen, L., Miörner, J., & Moodysson, J. (2019). Innovation policy for system-wide transformation: The case of strategic innovation programmes (SIPs) in Sweden. *Research Policy, 48*(4), 1048–1061.
Hacker, J. S. (2004). Privatizing risk without privatizing the welfare State: The hidden politics of social policy retrenchment in the United States. *American Political Science Review, 98*(2), 243–260.
Hekkert, M. P., Janssen, M. J., Wesseling, J. H., & Negro, S. O. (2020). Mission-oriented innovation systems. *Environmental Innovation and Societal Transitions, 34*, 76–79.
Howlett, M. (2014). Why are policy innovations rare and so often negative? Blame avoidance and problem denial in climate change policy-making. *Global Environmental Change, 29*, 395–403.
Hufnagl, M. (2010). *Dimensionen von Policy-instrumenten – eine Systematik am Beispiel Innovationspolitik.* Fraunhofer Verlag.
Jann, W., & Wegrich, K. (2017). Theories of the policy cycle. In F. Fischer & G. J. Miller (Eds.), *Handbook of public policy analysis* (pp. 69–88). Routledge.
Janssen, M. (2020). *Post-commencement analysis of the Dutch "Mission-oriented Topsector and Innovation Policy" strategy.* Copernicus Institute of Sustainable Development, Utrecht University. Mission-Oriented Innovation Policy Observatory (MIPO).
Janssen, M. (2021). Legitimation and effects of mission-oriented innovation policies: A spillover perspective. *Hacienda Pública Española/Review of Public Economics*, 2171.
Janssen, M. J., Torrens, J., Wesseling, J. H., & Wanzenböck, I. (2021). The promises and premises of mission-oriented innovation policy – A reflection and ways forward. *Science and Public Policy, 48*(3), 438–444.
Janssen, M. J., Torrens, J., Wesseling, J., Wanzenböck, I., & Patterson, J. (2020). *Position paper: "Mission-oriented innovation policy observatory".* Copernicus Institute of Sustainable Development, Utrecht University.
Kam, C. (2000). Not just parliamentary "Cowboys and Indians": Ministerial responsibility and bureaucratic drift. *Governance, 13*(3), 365–392.

Kattel, R., Drechsler, W., & Karo, E. (2019). *Innovation bureaucracies: How agile stability creates the entrepreneurial state* (Working Paper Series IIPP WP 2019–12). UCL Institute for Innovation and Public Purpose.

Kattel, R., & Mazzucato, M. (2018). Mission-oriented innovation policy and dynamic capabilities in the public sector. *Industrial and Corporate Change*, *27*(5), 787–801.

Kern, F., & Howlett, M. (2009). Implementing transition management as policy reforms: A case study of the Dutch energy sector. *Policy Sciences*, *42*(4), 391–408.

Knowledge for Climate (2014). *Innovation in climate adaptation: Climate adaptation in the Netherlands.*

Kroll, H. (2019). How to evaluate innovation strategies with a transformative ambition? A proposal for a structured, process-based approach. *Science and Public Policy*, *46*(5), 635–647.

Kuhlmann, S., & Rip, A. (2014). *The challenge of addressing Grand Challenges: A think piece on how innovation can be driven towards the "Grand Challenges" as defined under the prospective European Union Framework Programme Horizon 2020.* European Research and Innovation Area Board (ERIAB).

Kuhlmann, S., & Rip, A. (2018). Next-generation innovation policy and grand challenges. *Science and Public Policy*, *45*(4), 448–454.

Kuittinen, H., Skov Kristensen, F., Pelkonen, A., Lehenkari, J., Goetheer, A., van der Zee, F., Arrilucea, E., Unger, M., Türk, A., Polt, W., Fisher, R., Domini, A., Chicot, J., Terziev, N., Vincze, M., Taranic, I., Lykogianni, E., & Misojcic, M. (2018). *Mission-oriented research and innovation: Assessing the impact of a mission-oriented research and innovation approach: Final report.* European Commission, Directorate-General for Research and Innovation.

Larrue, P. (2021). *The design and implementation of mission-oriented innovation policies: A new systemic policy approach to address societal challenges* (OECD Science, Technology and Industry Policy Papers No. 100). OECD.

Lascoumes, P., & Le Gales, P. (2007). Introduction: Understanding public policy through its instruments? From the nature of instruments to the sociology of public policy instrumentation. *Governance*, *20*(1), 1–21.

Latour, B. (1986). The powers of association. In J. Law (Ed.), *Power, action, and belief: A new sociology of knowledge?* (pp. 261–277). Routledge & Kegan Paul.

Latour, B. (1987). *Science in action: How to follow scientists and engineers through society.* Harvard University Press.

Lindner, R. (2012). Cross-sectoral coordination of STI-policies: Governance principles to bridge policy-fragmentation. In Fraunhofer Institut für System- und Innovationsforschung (Ed.), *Innovations systems revisited – Experiences from 40 years of Fraunhofer ISI research* (pp. 275–289). Fraunhofer Verlag.

Lindner, R., Daimer, S., Beckert, B., Heyen, N., Köhler, J. H., Teufel, B., Warnke, P., & Wydra, S. (2016). Addressing directionality: Orientation failure and the systems of innovation heuristic: Towards reflexive governance. *Discussion Papers Innovation System and Policy Analysis, 52.*

Lindner, R., Edler, J., Hufnagl, M., Kimpeler, S., Kroll, H., Roth, F., Wittmann, F., & Yorulmaz, M. (2021). *Mission-oriented innovation policy: From ambition to successful implementation* (Policy Brief 02–2021). Fraunhofer Institut für System- und Innovationsforschung ISI.

Lipsky, M. (1980). *Michael Lipsky. Street-level bureaucracy: Dilemmas of the individual in public services* (Vol. 10). Russell Sage Foundation.

Magro, E., & Wilson, J. R. (2019). Policy-mix evaluation: Governance challenges from new place-based innovation policies. *Research Policy*, *48*(10), 103612.

Mahoney, J., & Thelen, K. A. (Eds.). (2010). *Explaining institutional change: Ambiguity, agency, and power.* Cambridge University Press.

Mazzucato, M. (2018). Mission-oriented innovation policies: Challenges and opportunities. *Industrial and Corporate Change*, *27*(5), 803–815.

Mazzucato, M. (2019). *Governing missions in the European Union.* Report for the European Commission.

Moe, T. M. (1989). The politics of bureaucratic structure. In J. E. Chubb & P. E. Peterson (Eds.), *Can the government govern?* (pp. 267–3329). Brookings Institution.

Molas-Gallart, J., Boni, A., Giachi, S., & Schot, J. (2021). A formative approach to the evaluation of Transformative Innovation Policies. *Research Evaluation*, *30*(4), 431–442, https://doi.org/10.1093/reseval/rvab016.

Polt, W., Weber, M., Biegelbauer, P., & Unger, M. (2019). *Matching type of mission and governance in mission-oriented R&I policy: Conceptual improvement and guidance for policy: Eu-SPRI Conference. Rome, 06.06.2020.* 2019 EU-SPRI Conference – Science Technology and Innovation Policies for Sustainable Development Goals. Actors, Instruments and Evaluation.

Pressman, J. L., & Wildavsky, A. B. (1984). *Implementation: How great expectations in Washington are dashed in Oakland* (3rd ed.). Oakland Project series. University of California Press.

Rose, R. (1990). Inheritance before choice in public policy. *Journal of Theoretical Politics*, *2*(3), 263–291.

Roth, F., Lindner, R., Hufnagl, M., Wittmann, F., & Yorulmaz, M. (2021). *The future of mission-oriented policies: Final report of the Scientific Support Action to the German High-Tech Strategy 2025* (Vol. 1). Fraunhofer Institut für System- und Innovationsforschung ISI.

Schot, J., & Steinmueller, W. E. (2018). Three frames for innovation policy: R&D, systems of innovation and transformative change. *Research Policy*, *47*(9), 1554–1567.

Shepsle, K. A. (1992). Organizational design and political control of administrative agencies: A comment: Bureaucratic drift, coalition drift, and time consistency. *Journal of Law, Economics, and Organization*, *8*(1), 111–118.

Smits, R., & Kuhlmann, S. (2004). The rise of systemic instruments in innovation policy. *International Journal of Foresight and Innovation Policy*, *1*(1/2), 4–32.

Teirlink, P., Verbeek, A., Delanghe, H., Heijs, J., Sachwald, F., Bayhan, D., Bukulmez, E., Ozdemir, A. H., Gok, A., Edler, J., Baanante, I., Moya, E., Gauci-Borda, I., Elias, B., Dinges, M., & Niederl, A. (2011). *Optimising the research and innovation policy mix: The practice and challenges of imapct assessment in Europe. Findings from FP7 OMC-net project 234501*, European Commission, Brussels.

Thelen, K. A., & Streeck, W. (2005). *Beyond continuity: Institutional change in advanced political economies.* Oxford University Press.

Waeraas, A., & Nielsen, J. A. (2016). Translation theory "translated": Three perspectives on translation in organizational research. *International Journal of Management Reviews*, *18*(3), 236–270.

Wanzenböck, I., Wesseling, J. H., Frenken, K., Hekkert, M. P [Marko P.], & Weber, K. M. (2020). A framework for mission-oriented innovation policy: Alternative pathways through the problem–solution space. *Science and Public Policy*, *47*(4), 474–489.

Weber, M., Biegelbauer, P., Brodnik, C., Dachs, B., Dreher, C., Kovaéc, M., Pulenkova, E., Schartinger, D., & Schwäb, C. (2021). *Agilität in der F&I-Politik: Konzept, definition, operationalisierung* (Studien zum deutschen Innovationssystem 8–2021).

Weber, M., & Polt, M. (2014). Assessing mission-orientated R&D programs: Combining foresight and evaluation. *Fteval - Journal for Research and Technology Policy Evaluation, 39*, 5–10.

Wesseling, J., & Meijerhof, N. (2020). *Development and application of a Mission-oriented Innovation Systems (MIS) approach* (Working Paper).

Wieczorek, A. J., & Hekkert, M. P. (2012). Systemic instruments for systemic innovation problems: A framework for policy makers and innovation scholars. *Science and Public Policy*, *39*(1), 74–87.

Winter, S. C. (2012). Implementation. Introduction. In B. G. Peters & J. Pierre (Eds.), *The SAGE handbook of public administration* (pp. 255–264). SAGE Publications.

Wittmann, F., Hufnagl, M., Lindner, R., Roth, F., & Edler, J. (2020a). *Developing a typology for mission-oriented innovation policies* (Discussion Papers Innovation System and Policy Analysis No. 64). Fraunhofer Institute of Systems and Innovation Research ISI.

Wittmann, F., Hufnagl, M., Lindner, R., Roth, F., & Edler, J. (2021a). Governing varieties of mission-oriented innovation policies: A new typology. *Science and Public Policy*, *48*(5), 727–738.

Wittmann, F., Hufnagl, M., Roth, F., Lindner, R., & Kroll, H. (2021b). *A framework for formative evaluation and impact assessment of mission-oriented innovation policies: Final report of the Scientific Support Action to the German High-Tech Strategy 2025* (Vol. 2). Fraunhofer Institute of Systems and Innovation Research ISI.

Wittmann, F., Hufnagl, M., Roth, F., Yorulmaz, M., & Lindner, R. (2021c). *Second mission analysis report of the Scientific Support Action to the German Hightech Strategy 2025: Zooming in: Translating missions into policy instruments*. Fraunhofer Institut für System- und Innovationsforschung ISI.
Wittmann, F., Roth, F., & Hufnagl, M. (2020b). *First mission analysis report of the Scientific Support Action to the German Hightech Strategy 2025: Setting the stage: Positioning the missions in the socio-technical system*. Fraunhofer Institute of Systems and Innovation Research ISI.
Wittmann, F., Yorulmaz, M., & Hufnagl, M. (2021d). *Impact assessment of mission-oriented policies: Challenges and overview of selected existing approaches*. Project deliverable. Fraunhofer Institut für System- und Innovationsforschung ISI.
Wu, X., Ramesh, M., & Howlett, M. (2017). Policy capacity: A conceptual framework for understanding policy competences and capabilities. *Policy and Society*, *34*(3–4), 165–171.

3. Missions as a boundary object in umbrella frameworks: the case of the Dutch mission-oriented top sector and innovation policy

Matthijs Janssen, Joeri Wesseling, Jonas Torrens, Matthias Weber and Caetano C. R. Penna

3.1 INTRODUCTION

Mission-oriented innovation policies (MOIP) have a prominent place in ongoing debates about innovation policies for addressing societal challenges (Mazzucato, 2018; Larrue, 2021). While these debates generate different visions of what such policies might look like (Mazzucato et al., 2020; Alves et al., 2021), it is typically understood that they need to serve a clear, ambitious, and time-bound societal goal. It appears that there is merit in scrutinising how mission goals are framed, how they relate to higher-order societal goals, and who is engaging with them (Janssen et al., 2021; Parks, 2022), instead of focussing exclusively on how policy instruments are to be selected and designed.

Relying on missions to coordinate innovation activities foregrounds initiatives in the areas of steering directions and promoting alignment – in policies, investments, and practices on the ground. Steering is sometimes presented as state-led and top-down, which may be unsuitable for wicked societal challenges, as it could overwhelm policy capacity and the political willingness to impose clear directions (Wanzenböck et al., 2020). Wicked challenges are societal problems with an uncertain, contested, and complex nature. In the context of these problems, diverse, bottom-up experimentation efforts are required in order to identify promising solutions (Alford & Head, 2017; Hekkert et al., 2020). In such cases, the challenge is to mobilise and unite actors' efforts around a mission without imposing top-down directionality too strongly. Instead, directional objectives will need to be constructed iteratively in the course of time as a result of learning processes about both problem definition and possible solutions.

To address this challenge, we propose the consideration of missions as boundary objects (BO) (Janssen et al., 2023), i.e., objects with enough plasticity – or interpretive flexibility – to be adaptable to specific contexts, yet sufficiently robust to maintain a common identity across these contexts (Star & Griesemer, 1989; Star, 2010). This perspective highlights how openness in framing a mission allows the engagement of different stakeholder types or communities on their own terms and in line with their own interpretations and interests, while fostering dialogue to develop more cohesive relationships in the scope of a given societal challenge.[1] The BO perspective has significant but underexplored implications for how missions may be used

[1] Communities are characterised by their similar interests and ways of thinking. They may include different stakeholder groups, such as civil servants, scientists, entrepreneurs, managers, and interest groups, as well as different epistemic communities across and within those stakeholder groups (Haas, 1992).

as devices for spurring and combining activities that address societal problems. To further explore the analytical and practical applicability of this view, there is a need for more empirical work on how missions operate as boundary objects.

This chapter will explore policy strategies for addressing multiple missions in different problem areas, also known as "umbrella missions" (JIIP, 2018) or "umbrella programmes" (Larrue, 2021). Frameworks of this kind – umbrella frameworks as we will call them – comprise overarching goals and an overarching "common" governance structure, which together serve to support a number of missions that target predominantly different societal and (in some cases) technological challenges. Individual missions often rely on additional governance processes and policy measures, especially if they are the responsibility of actors that are not in charge of the umbrella framework. For instance, the German HighTech Strategy 2025, launched by the Federal Ministry of Education and Research (BMBF), draws not only on the BMBF's targeted R&D support but also on other ministries' policies and programmes in support of the 12 national missions (Wittmann et al., 2021). The mission pillar in the EU Horizon Europe framework is another example, as it lays out the structures for adapting research and innovation policies to address five very different societal challenges (European Commission, 2021).

Studying these kinds of frameworks as boundary objects is particularly relevant. In itself, an umbrella framework already represents a certain view on why to engage in mission-oriented innovation policy, but as it covers goals in different societal domains, it must afford divergent views on how MOIPs can drive change – and through which actions and solution directions this would need to happen. Nevertheless, previous research has neglected the interactions between umbrella frameworks and the various missions they support and has mostly failed to address the potential for accommodating contestation and negotiation. Taking a BO perspective, we explore this interface by asking the following questions:

1. *What subjects provoke disagreements among the stakeholders engaged in defining the umbrella framework's ambitions and principles, individual missions, and how these intersect?*
2. *How do the common policy elements in an umbrella framework relate to differences in how strategies and goals of individual missions are understood by different stakeholders?*

This chapter is an empirical exploration of how missions, understood as boundary objects, facilitate or hinder the ability to address wicked problems in a systematic manner. The analysis centres on the case of the Dutch Mission-oriented Top Sector and Innovation Policy (MTIP) umbrella framework and two of its missions – CE and BE. Rather than treating the case as an instance of best practice, we examine the coordination issues related to (mis)alignment to elicit possible policy responses. This study also provides a detailed description of how the umbrella framework is organised since few concrete examples of such strategies have been documented.

Our results reveal divergent interpretations of issues concerning the motivation of the MOIP, the type of novelty sought, and the actors to be included in development and diffusion processes. Those responsible for umbrella frameworks covering multiple missions should manage the degree of consensus on such topics, as this influences the suitability of particular policy setups and the desirable degree of openness and inclusion. Missions can be helpful as a common reference point. They also facilitate social learning, even when relying on heterogeneous policy instruments (e.g., short-term versus long-term measures or variety- versus synergy-oriented measures) that are not fully streamlined.

The remainder of this chapter is structured as follows. Section 3.2 elaborates on the theoretical concepts on which this study relies, notably the BO perspective and the notion of umbrella frameworks. Section 3.3 describes which methods were used for studying this in the empirical setting of the Dutch approach to MOIP. In relation to the MTIP and two of its missions, Section 3.4 reports on the interpretations that are emerging in the various arenas in which contestation and negotiation take place. Section 3.5 discusses these findings and explores the most notably contested subjects and the influence of certain features of the umbrella framework. Policy implications are outlined in section 3.6 and conclusions are drawn in section 3.7.

3.2 THEORY

Missions are meant to achieve some degree of coordination not only in research and innovation activities but also in mobilising the necessary activities in distinct sectors. How this coordination is achieved is a matter of debate. Key propositions regarding missions, as in Mazzucato's (2018) approach, place much emphasis on setting bold and ambitious goals that need to be achieved within a fixed timeframe. This overemphasis on setting goals and achieving them in a timely fashion is problematic because it presumes a great deal of knowledge of future challenges and the ability to reach consensus, while not being explicit about how such consensus is achieved.

Instead, once it is acknowledged that the formulation and pursuit of missions are embedded in (and evolve from) systems with heterogeneous actors who each hold certain beliefs and preferences, missions can be understood as the negotiated outcomes of contestation processes (Janssen et al., 2021; Parks, 2022). Accordingly, Janssen et al. (2023) have proposed using a boundary object perspective to analyse the processes of goal setting and mission coordination undertaken by different stakeholders and communities involved in mission-oriented policy arenas. This perspective builds on four interrelated policy arenas involving changing groups of communities that aim to fulfil a certain policy task. Based on the policy level distinctions introduced by Barre et al. (2013), the policy arenas range from strategy design, programme definition, and policy implementation to activity performance.

In the strategic arena, communities advance possible policy rationales for pursuing a mission. These rationales can fit different "metanarratives" for adhering to MOIP, much like Schot and Steinmueller's (2018) three frames of innovation policy. According to the challenge-led R&D policy metanarrative (first frame), MOIP translates to "mission-oriented research policy"; the systemic industrial policy metanarrative (second frame) is associated with "mission-oriented industrial policy"; and "transformative missions" fit the transformative innovation policy metanarrative (third frame). Meanwhile, the formulation of specific missions and strategy goals can be understood as the primary outcome of the programming arena. In this arena, debates centre primarily on how to formulate goals that can mobilise different actors and how to scope the actions that correspond to a mission. The question of which policies and instruments to rely on and how to target them is decided in the implementation arena. Finally, in the performance arena, actors experiment with developing and diffusing actual solutions for the prioritised problems. The degree of convergence that is achieved can differ for each of the arenas, just like the specific interpretation(s) on which the communities converge.

Treating missions as boundary objects makes it possible to unpack the "coordination task" of missions in these different arenas. It gives rise to an understanding that on the one hand, the coordination process requires a balance of constructive ambiguity and moments in which closure is achieved, and on the other hand, that boundary objects become less plastic when mission agreements are more formalised (i.e., clear directionality). The BO framework identifies three key mechanisms by which a boundary object may induce broadly supported coordination across and within mission arenas. These are visualised in Figure 3.1. *Convergence /divergence* refers to the breadth of different meanings that communities in one arena give to the mission aspect that is at stake (e.g., the strategy or the vision). The more communities converge on a similar meaning, the more broadly supported the mission-oriented approach will be. *Passage* refers to the transfer of meanings from one context to another. One arena might, for example, converge on a mission meaning that is then passed on to another arena via arena-spanning boundary object carriers like strategy documents or monitoring frameworks. Finally, *learning* can take place within arenas, such as new lessons that create new, arena-specific mission meanings. Learning can also take place across arenas in the form of deliberate reflexivity on whether mission meanings are coherent across arenas and by sharing lessons across arenas to enable or maintain (re)alignment of mission meanings (Janssen et al., 2023).

A central aspect of this perspective is the nestedness of different arenas and the potential alignment or misalignment that arises at these arenas' interfaces. If negotiations and interpretations in the vision arena converge on a different meaning from that of the implementation arena (or there is no convergence in the latter arena at all), a situation might emerge in which the policy instruments that are deployed do not support the kind of innovation processes or innovation types that the vision demands. This issue of misalignment is particularly relevant when considering the phenomenon of umbrella frameworks as a means of structuring specific missions. Despite various references in earlier literature (JIIP, 2018; Larrue, 2021), there is no established definition of what an umbrella framework for MOIP entails. We have conceptualised it by linking it to the four arenas of Figure 3.1. Figure 3.2 shows three archetypal ways in which umbrella frameworks can relate to the shaping of individual missions.

In one possible model, the umbrella framework is akin to a meta-strategy that is formed in yet another different arena from the other four arenas (panel a in Figure 3.2). In this model, the umbrella framework primarily presents a common headline for the MOIP but without imposing commonalities in terms of the policy objectives and policy approaches that are to be pursued. The umbrella framework is then merely a canopy that specifies a basic governance structure (and perhaps a set of broadly scoped mission topics or areas) for allowing stakeholders in the other arenas to define the concrete goals, strategies, and policy instrumentation of actual missions. How these mission aspects are defined depends on what interpretations are passed on via "vertical passage" between the arenas.

Another way that an umbrella framework could relate to individual missions is by effectively forming the outcome of contestation and negotiation in the strategic arena (panel b). In that case, the umbrella framework itself reflects the rationale for having different policies for different missions, thus posing a metanarrative that dictates the reasons for engaging in MOIP. In this model, individual missions can still vary in terms of what policies they implement, but the logic for having those policies (i.e., the motivation for steering innovation towards addressing a societal goal) is similar for all the missions under the umbrella.

Finally, there is a model in which the umbrella framework also contains more operational elements for supporting various missions (panel c). This could be a common rationale or

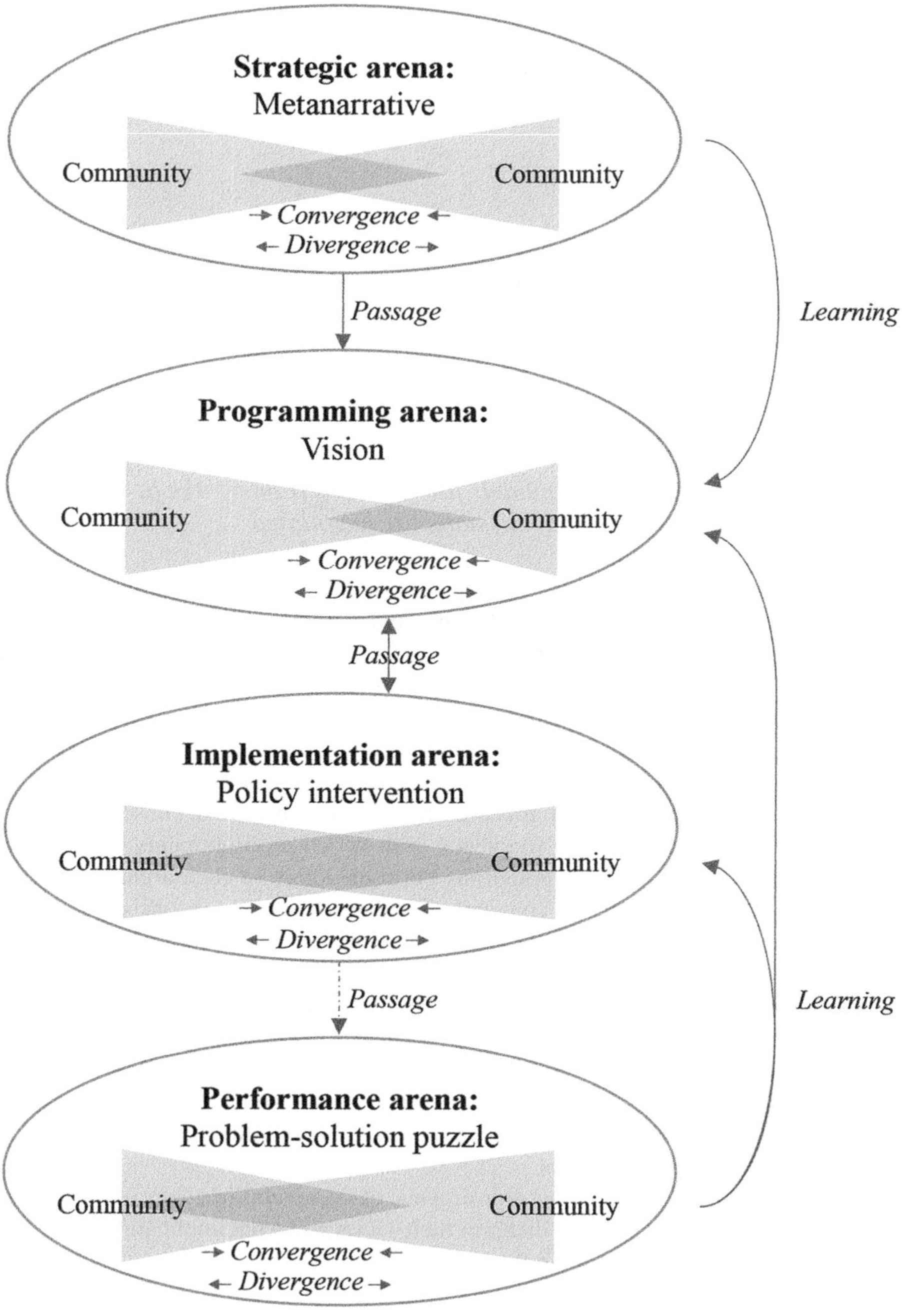

Source: Figure adopted from Janssen et al., 2023.

Figure 3.1 *Mechanisms (in italics) involved in how missions can drive coordination across communities and arenas*

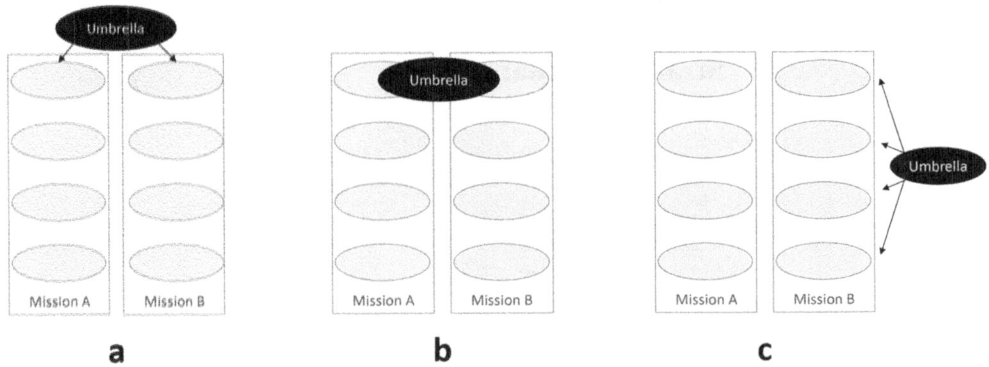

Source: authors' own elaboration.

Figure 3.2 *Panels a–c: Three models for umbrella frameworks in relation to the four arenas in which missions are negotiated*

common instruments and monitoring frameworks, running alongside equivalents (of such mission aspects) articulated in the arenas of individual missions. Therefore, umbrella frameworks of this kind consist of a policy strategy and policy package that influence the formation of interpretations in all arenas and in all missions via horizontal "passage".

Applying the BO perspective when analysing actual cases raises doubts regarding the idea that the goals of missions and the criteria and scopes of supporting policy instruments will automatically be consistent with the policy intervention rationales of the umbrella framework. In fact, the BO perspective highlights the way cooperation without consensus is possible within these different arenas and the frequency with which misalignments between arenas arise, given the complexity of the negotiations between actors in each arena. Convergence could occur when communities are confronted with each other's inconsistent views; for instance, when they are asked to develop a joint vision on how to address a societal challenge (via which strategy, vision, etc.). However, convergence can even arise in one arena without the interactive processes of negotiation and contestation taking place, for instance, when non-interacting communities in one arena are influenced by the passage of meaning from other arenas. While getting communities to talk to each other might increase the likelihood that they can work out differences in their respective views, convergence is by no means guaranteed. Also, when communities interact directly, it is possible that their contrasting views will only diverge further.

To enhance our understanding of how convergence of interpretations does or does not arise, this study examines why stakeholders differ in their views and what implications these differences have for the common elements in the umbrella framework.

3.3 METHODS

To study how different communities give meaning to mission strategies and mission goals, we revisit a case studied as part of earlier research (notably Janssen, 2020) on the Dutch MTIP and two of its underlying 25 missions: one on CE and one on BE. The MTIP is an umbrella framework that not only provides a basic governance structure for the definition and pursuit of a number of concrete missions, it also involves numerous policy instruments and monitoring efforts. It exhibits characteristics of the third model shown in Figure 3.2, which allows us to study the relationship between interpretations in the MTIP and developing interpretations in the arenas of the CE and BE missions.

The underlying data for studying this consists of document research and 20 purposively sampled interviews conducted between June and October 2020. The interviewees represent the different communities involved in the "delegated" mission governance. The interview questions addressed the organisation and performance of several aspects of governance, including exploring problem/solution directions, deciding what to support, and reflexivity on this issue. The resulting data was used in an explorative (open) analysis that served to understand interpretation/negotiation and (mis)alignment. For this chapter, the existing material has been reinterpreted along the lines of diagnostic questions for studying missions as boundary objects (see Janssen et al., 2023, Table 2). Moreover, we have relied on secondary sources that were published since the interviews took place to provide more recent information and to close any gaps from the interviews in relation to the diagnostic questions.

To address the main research questions, the analysis reported in the next section focusses on which communities are engaged in a certain arena, which interpretation of a mission strategy or goal they are advancing, what objects institutionalise and carry these interpretations, how interpretations pass across arenas, how interpretations converge or diverge in each arena and (as far as this can be observed) how the ensuing contestation of different views plays out, and finally, what learning processes occur in the context of mission interpretation. In this analysis, convergence is understood as a trend towards communities having similar views on a mission aspect (strategy, vision, policy instruments, or the potential of certain solutions to address a societal problem). Divergence happens when some communities adhere to a view that moves away from how others interpret that mission aspect. Both convergence and divergence are outcomes rather than actions, unlike learning. Learning is a deliberate action aimed at determining which interpretations are prevalent in an arena in order to adjust the scoping of mission aspects. For example, new insights into potentially feasible solutions and what hinders them might constitute a reason to adjust a mission's vision or policy interventions.

3.4 BOUNDARY OBJECT OBSERVATIONS FOR MTIP AND THE TWO MISSIONS

3.4.1 The Mission-oriented Top Sector and Innovation Policy

3.4.1.1 Background

In the Netherlands, the development of the MOIP umbrella framework was based on the 2012 top sectors policy launched by the Ministry of Economic Affairs and Climate Policy (MinEZK) and the Ministry of Education, Science and Culture. At the heart of this policy strategy are nine "top sectors"; techno-economic domains characterised by a strong research and innovation profile (Janssen, 2015). Each domain has a "top team" that includes representatives from the science and business communities and relevant specialist departments. These top teams are tasked with increasing innovation opportunities for their domain and can rely on the Top Consortia for Knowledge and Innovation (TKIs), which act as secretariats. The TKIs deal with matters such as network activities and roadmap development to support innovation ecosystems, identify innovation ambitions and bottlenecks, and formulate Knowledge and Innovation Agendas (KIAs). The top teams use their KIAs to influence research funding, regulations, export promotion, and curriculum development. To align research planning in the Dutch innovation system with the KIAs, Knowledge and Innovation Contracts (KICs) are signed once every four years by the top sectors, relevant ministries, and knowledge partners such as professional research organisations. Additionally, several policy schemes explicitly serve the top sectors and the top sector policy. Therefore, the umbrella framework influences the direction of a broad range of policy and regulatory instruments.

Originally, the top sector approach aimed to increase private contributions to public-private research and, in time, to strengthen the competitiveness of the Netherlands. However, over the years, there has been a gradual shift towards an explicit normative framing, focussing collective innovation efforts on societal challenges. This was partially caused by following the normative changes in European R&D policy from economic competitiveness towards "grand societal challenges" and "societal missions". Moreover, some sector-specific ministries wanted to influence such innovation directions in the top sectors relevant to them. In mid-2017, the top sectors were asked to draft the 2018–2021 KIAs with a focus on societal challenges and key technologies. By mid-2018, the MinEZK announced that these topics would guide a new generation of top sectors policy. As of 2019, this movement has continued the trend towards Mission-Driven Top Sectors and Innovation Policy, adopting the "mission" language more explicitly.

How quickly the top sector approach is morphing into the MTIP differs across top sectors and therefore across missions. At present, there are four overarching mission topics, each made up of five to eight concrete missions and combined with a policy track to promote key technologies. Each mission topic is managed by a number of ministries committed to achieving the mission goals, with the vision for the missions themselves often originating from public consultations.

Both in terms of governance and policy instruments, the MTIP is still firmly anchored in the top sectors policy. The shift to the MTIP implies that changes are being made to these existing structures, such as the creation of "mission teams". In missions that rely heavily on a single top sector, these teams are positioned alongside the top teams. For example, the CE mission (see section on The MTIP mission on circular economy (CE) below) can call on several top sectors for innovations, whereas the mission team has a more central position.

3.4.1.2 The MTIP's mission strategy as a boundary object

Figure 3.3 provides a non-exhaustive overview of which mission object "carriers" (e.g., reports, strategies, objectives, calls, policies, etc.) allow different stakeholder groups to act upon the MTIP.

In the strategic arena, the inspiration for modelling the new national innovation strategy (partially) on missions has come mainly from the example set by the EU's Horizon Europe programme. Although high-level policy officials allegedly consulted Mariana Mazzucato four to five years earlier, the 2019 Letter to Parliament outlining the MTIP[2] refers to her *Research*

---

[2] https://www.rijksoverheid.nl/documenten/kamerstukken/2019/04/26/kamerbrief-over-missiegedreven-topsectoren-en-innovatiebeleid

	Stakeholder groups			
Arena type	STI academics	R&I policy makers	Sectoral policy makers	Societal stakeholders / sectoral academics
Strategic arena	Reports by EC; parliamentary motions; Letters to Parliament			
Programmatic arena		Mission goals; Knowledge and Innovation Agendas; Multi-annual Mission-oriented Innovation Programs (MMIPs)		
Implementation arena		New 'Mission teams' at the core of the Topsector governance structure; NWO calls; PPP R&D allowance		
Performance arena		Monitoring and examples of innovation projects fitting R&D&I policies and agendas		

Source: Authors, adapted from Janssen et al. 2023.

Figure 3.3 *Examples of BO carriers for bridging communities at different policy level arenas related to the MTIP*

and innovation in the European Union report (Mazzucato, 2018) rather than to other parts of her work describing the rationale and imperatives for mission-oriented innovation policy. The letter, which was both inspired by and inspired a series of reports (e.g., the Rathenau Institute for science studies and technology assessment[3]) mostly speaks about parliamentary motions aimed at intensifying the link between innovation policy and (grand) societal challenges. In particular, the MinEZK's commitment to national climate goals appears to have influenced the focus on wicked societal problems. Remarkably, the documents outlining the strategic view behind the MTIP do not draw explicitly from the rich Dutch tradition of transition management and transition policy (Kern & Smith, 2008; Loorbach & Rotmans, 2010). Instead, the documents underscore how articulating societal demands allows for synergetic collaborations, which contributes to solving the societal problem while enhancing economic growth. Notably, from the beginning, "key enabling technologies" have been a core element of the MTIP. This shows that the dominant metanarrative here is best described using the mission-oriented industrial policy perspective (see section 3.2).

This scope is also reflected in the question of which communities have most clearly influenced the MTIP strategic arena. Interviewees refer mainly to economists and STI-minded policy bodies, institutes, or think-tanks like the Dutch Advisory Council for Science, Technology and Innovation, the European Commission (EC), and the Organisation for Economic Co-operation and Development (OECD). While the latter two bodies mainly frame missions in relation to guiding R&D policy, the national think-tanks mostly highlighted the (presumably not fully exploited) potential to address societal challenges.[4] Thus, mission interpretations clearly passed from the EU strategic arena to the Dutch. The EU and Dutch strategic

[3] The Rathenau Institute combined insights from several of its reports in its own letter to Parliament of 11 June 2019, titled "Mission-oriented innovation policy requires collaboration between local and national levels".

[4] See, for instance, the AWTI report (7 February 2020) titled "Enhance the role of science, technology and innovation in societal transitions".

arenas converged on a similar interpretation. This interpretation was notably carried by the Mazzucato (2018) EU report and the 2019 Dutch Letter to Parliament and institutionalised by similar epistemic communities that mainly included the STI-oriented policymakers and researchers. Sector-specific policymakers, transition scholars, and societal stakeholders had less impact.

In the programming arena, the debates were concerned primarily with developing a vision that focussed on which changes the MTIP mission should make and how such changes could be made. As highlighted above, this arena retained many of the pre-existing participant groups involved in the top-sector approach. The ministries' mission goals, as listed in the letter to Parliament of July 2019, constitute the key BO carriers that convey the MTIP's objectives. Some goals followed from existing frameworks like the Climate Agreement goals (notably the missions in the "Energy Transition and Sustainability" topic), while others have been formulated exclusively for the MTIP. After the societal goals were created, stakeholders involved in writing KIAs were invited to describe how their plans related to the goals. Moreover, broader KIA research lines were operationalised to create multi-annual mission-oriented innovation programmes (MMIPs), which give more specific guidance to innovation efforts (compared to the KIAs from the top-sector era). Due to the tight link between the missions and the KIAs/MMIPs, the communities that deliberated on solution directions included R&I policymakers, sector-specific policymakers, and societal stakeholders. According to interviewees, for several mission topics, these parties intended to open up communications with professionals, citizens, and community organisations instead of continuing to engage primarily with the industry-science complex. In this matter, sector-specific policymakers have been especially vocal. As far as they are concerned, defining and completing the mission is more about diffusion and implementation than about innovation. Hence, a broader range of communities were involved in the programming arena. These communities held intensive negotiations around the main BO carriers for the KIA and MMIPs. In a sense, the narrower mission interpretation passed from the strategic arena and was broadened at the programming level.

The main communities involved in negotiating concrete policy interventions in the implementation arena are the original participants in the renewed governance structure designed by the MinEZK's R&I policymakers. In the newly established mission teams, different scientific, sector-specific, and business communities are represented. Moreover, opportunities for negotiation and contestation were broadened by establishing the various advisory and decision-making bodies that these teams can rely on. Decisions on who participates in these bodies are likely to affect how missions are interpreted and contested.

The calls and programmes created for MMIPs are the central BO carriers in this arena. This concerns research funding calls by the Dutch Research Council (I), which also involve transdisciplinary network events like sandpits and programmes for steering public-private R&D collaboration projects that are subsidised via an allowance scheme (see Janssen et al., 2020). The former invites varied perspectives on the underlying societal challenge; the latter aims to develop and test specific technology-based solutions. So, at the implementation level, mission interpretations are renegotiated between a diverse range of professional and epistemic communities. This suggests that convergence due to passage from strategic and implementation arenas is limited in this arena.

Finally, the performance arena is the most concrete arena in which the MTIP umbrella framework provides guidance for actions. In this arena, activity is associated mostly with innovation and demonstration projects that are supported directly through the MTIP's policy instruments. The monitoring framework created by the MinEZK and the policy execution agency RVO.nl[5] is a primary BO carrier designed to keep track of project results, such as publications and patents. Similarly, the teams and support bodies operating in the governance structures track progress by referring frequently to the project portfolio, which shows progress in relation to the realisation of the MMIPs. The relatively narrow focus on innovation projects as the main vehicles for driving change appears to be advocated by communities composed of policymakers from R&I and other relevant sectors. Within these communities, the prevalence of actors who are directly involved with proposing solutions (i.e., academics and entrepreneurs in those sectors) varies. The precise role played by sector-specific stakeholders varies greatly depending on how concrete experimental activities as found in the projects are scoped. However, for some technocratic missions, their influence on visions seems larger than their actual contribution to searching for solutions. On the whole, we observe more convergence in the scope of actual projects than one would expect based on the diverging views. This is particularly noticeable within the programming arena and is due largely to the passage of certain views that arise in the implementation arena, especially when it comes to what kinds of projects to support (i.e., innovation projects are often based on technology projects). At this point, it is too early to see how this impacts deliberate learning efforts.

3.4.2 The MTIP Mission on Circular Economy (CE)

3.4.2.1 Background

One of the missions in the overarching topic of "Energy Transition and Sustainability" (ET&S) is to achieve a fully CE by 2050 and a 50 per cent reduction in resource usage by 2030. This is the only topic that the Ministry of Infrastructure and Water Management (rather than MinEZK) is primarily responsible for. Like other Dutch examples, this mission stems from an agenda that pre-dates the MTIP, having started more or less concurrently with the government-wide CE programme "The Netherlands Circular in 2050" created in September 2016.[6] Following the ensuing 2017 raw materials agreement (signed by more than 400 partners), five transition agendas were drafted in 2018. These agendas related to biomass and food, plastics, manufacturing industries, construction, and consumer goods and formed the basis for the further elaborated Implementation Programme Circular Economy (UPCM) 2019–2023. In addition, a CE agenda was also drawn up in 2017 as part of the National Science Agenda (NWA). The KIA CE drafted in 2019 is based largely on that agenda and defines three MMIPs: design for circularity; circular raw material chains and processes; and trust, behaviour, and acceptance.

3.4.2.2 The CE mission as a boundary object

Figure 3.4 shows an overview of exemplary BO carriers and involved communities in the Dutch CE mission.

---

5 Ministry of Economic Affairs and Climate Policy (2018). *Mission-oriented innovation policy with impact*. MinEZK. https://www.rijksoverheid.nl/documenten/kamerstukken/2018/07/13/kamerbrief-naar-missiegedreven-innovatiebeleid-met-impact

6 Ministry of Infrastructure and Water Management (14 September 2016). National Programme Circular Economy.

<table>
<tr><th></th><th colspan="4">Stakeholder groups</th></tr>
<tr><th>Arena type</th><th>STI academics</th><th>R&I policy makers</th><th>Sectoral policy makers</th><th>Societal stakeholders / sectoral academics</th></tr>
<tr><td>Strategic arena</td><td></td><td colspan="2">Letters to Parliament, on MTIP and Circular Economy policy</td><td></td></tr>
<tr><td>Programmatic arena</td><td></td><td colspan="3">Transition agendas Circular Economy, Circular Economy Implementation Programme 2019-2023</td></tr>
<tr><td>Implementation arena</td><td></td><td></td><td colspan="2">Sectoral procurement and diffusion programs; Versnellingshuis accelerator</td></tr>
<tr><td>Performance arena</td><td></td><td></td><td colspan="2">Circular economy practices and experiments, national and regional</td></tr>
</table>

Source: Authors, adapted from Janssen et al. 2023.

Figure 3.4 Examples of BO carriers for bridging communities at different policy level arenas related to the CE mission

The MTIP Letters to Parliament (Section 3.4.1) have also influenced the strategic arena in defining the CE policy approach. However, this mission is not just an implementation of the MTIP framework and its techno-economic focus. Rather, as it is rooted in a series of preceding agendas and strategies, it follows a distinctive logic that is shown in the priorities laid out by the Infrastructure and Water Management Ministry in additional Letters to Parliament about the CE. An overview letter published in May 2021[7] attributed only minor importance to the MTIP and positioned it as a policy framework for addressing particular research and innovation questions (also of a non-technological nature). The nationwide effort to achieve a CE is far more comprehensive, reflecting a transformative framework rather than a mission-oriented R&D or industrial strategy framework. This is demonstrated by the fact that the Netherlands Environmental Assessment Agency (PBL) is regarded as the most central monitoring actor in the network of collaborating organisations, rather than the CE mission team. The above indicates that there was minor passage of mission meanings between the MTIP and the CE missions. Instead, the broad group of professional and epistemic communities involved in the CE mission defined different meanings for the mission concept. These meanings were formalised in the BO carrying overview letter.

In addition, the participant group in the programming arena is skewed towards the societal stakeholder side when compared to the communities involved in the MTIP. Sector-specific researchers (not STI researchers), firms, and societal actors (e.g., housing corporations) were most deeply involved in the priority domains targeted by the five transition agendas developed several years before the introduction of the MTIP. The leadership provided by these groups, with minor involvement of R&I or sector-specific policymakers, does not imply consensus or a lack of contestation in this arena. The openness of the CE mission is reflected in the different interpretations of what its objectives are, such as reducing greenhouse gases; solving housing and infrastructural challenges; spurring the transition to circular agriculture; preventing biodiversity loss; and reducing critical resource supply risks. The leading interpretation impacts

7 https://www.rijksoverheid.nl/onderwerpen/circulaire-economie/documenten/kamerstukken/2021/05/21/kabinetsreactie-op-de-pbl-integrale-circulaire-economie-rapportage

the choice of which solutions to promote. Solutions range from technological approaches such as mechanical and chemical recycling to behavioural solutions such as reducing and reusing. Hence, there is no convergence in the programming arena on which specific mission policy approach should be taken and which solutions should be supported. However, the mission BO is still compelling enough to legitimise and create coherence at the higher level of aggregation. This suggests that reflexive learning is particularly important in helping the mission meaning evolve and converge over time. To some extent, this learning process is taking place in the form of the Integral Circular Economy Reports published by the PBL every two years.[8] Since those reports describe the progress in the development and diffusion of different CE strategies, they are an important basis for reconsidering which solutions and efforts to prioritise.

In the implementation arena, concrete policy interventions include procurement and diffusion programmes created by various national and subnational policy bodies. The mission also builds on knowledge diffusion platforms and implementation-support programmes like the Versnellingshuis, which is an accelerator for circular entrepreneurship. These initiatives typically originate from citizen movements and academic associations. While the ministries of Infrastructure and Water Management and the MinEZK provide financial support, the STI academics and R&I policymakers are not heavily involved in influencing the scope and workings of such initiatives. In this arena, contestation relates to different interpretations and preferences regarding both problems or solutions and the position of different initiatives with respect to each other. For instance, the Mission Team's decision to launch a new national platform, CircularityNL, provoked fierce debate regarding its relationship to existing hubs that coordinated actors within the CE. It appears that the lack of convergence in the programming arena is continued in the implementation arena, as a result of the many different communities and interests involved. The lack of convergence is also caused by the complexity of the interdependent societal problems and solutions that the mission captures, which require more reflexive learning in order to converge on mission meanings.

Finally, actors in the performance arena focus on actual projects contributing to the mission goal. However, unlike MTIP, this extends beyond innovation and demonstration projects (with the associated MTIP monitoring framework acting as a mission carrier) to the varied CE practices and experiments emerging at local levels. Many Dutch regions encourage bottom-up activities that are undertaken by citizens and firms with the aim of reducing waste and closing resource loops. Facilitating such activities puts a spotlight on them (e.g., through CE promotion campaigns) and increases the likelihood that others will adopt similar practices. In addition, the national government runs a website with exemplary CE practices. These examples aim to demonstrate how the CE should be interpreted (including in other arenas). The performance arena is populated mostly by sector-specific policymakers (e.g., on spatial planning or waste management) and private and public societal stakeholders, so contestation and negotiation often arise in the context of which practices contribute most to the overarching goal. In relation to CE, this often involves debates on whether to also target geographically distributed supply chains. Although the precise mission solutions are still contested in the performance arena, convergence has occurred in relation to the mission and a suitable learning approach (i.e., focussing on a decentralised project organisation that collectively shares best practices).

---

[8] https://www.pbl.nl/en/publications/circular-economy-progress-report-2022

3.4.3 The MTIP Mission on a Carbon-free Built Environment (BE)

3.4.3.1 Background

Another mission that falls under ET&S is to strive for a carbon-free built environment by 2050. The mission's subgoals are: to disconnect 30,000–50,000 houses per year from the natural gas infrastructure by 2021 and 200,000 per year before 2030; to ensure that 1.5 million houses and 15 per cent of utility buildings and societal real estate are natural gas-free by 2030; and to raise sustainable energy production above 20 per cent of local energy consumption within the built environment. These goals follow from the Dutch Climate Agreement of 2017. In addition, the mission is connected to the "construction agenda" for accelerating public and private investments in innovation in the construction sector. This agenda is backed by the ministries of Interior Affairs (BZK) and Infrastructure and Water (I&W) and acts as a platform for promoting the CE transition agenda on construction. Nevertheless, there are substantial differences between the BE and CE missions' frameworks, support, and involved communities.

The main governance elements consist of MMIPs, a mission team, and a TKI. Three MMIPs, part of the KIA for the ET&S topic (called IKIA), specify how to meet the goals. These MMIPs were assigned a budget of 250 million euros as part of the Dutch Climate Agreement. The MMIPs' topics are related to the acceleration of energy renovations in the built environment; sustainable heat and cold in the built environment (including greenhouse horticulture); and electrification of the energy system in the built environment. Furthermore, the mission is supported by the ET&S MMIP on a robust and socially supported energy system. As with all other missions, there is a mission team for planning policy instruments according to the MMIPs. It is supported on an operational level by a pre-existing "Urban Energy" TKI.

3.4.3.2 The BE mission as a boundary object

In line with the previous two sections, Figure 3.5 shows exemplary boundary object carriers and the communities involved in the carbon-free built environment mission.

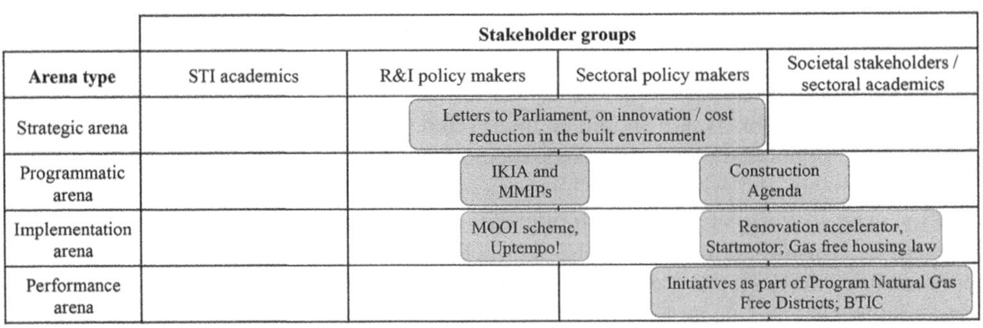

Source: Authors, adapted from Janssen et al. 2023.

Figure 3.5 Examples of BO carriers for bridging communities at different policy level arenas related to the BE mission

Boundary object carriers in the strategic arena articulate both the reasons for using innovation policies and the ways in which such policies can be used for societal purposes. The most notable examples of this are the Ministry of BZK's December 2019 letters to parliament on "Cost reduction and innovation in the construction sector"[9] and "Implementation of the Climate Agreements on the built environment".[10] While these letters outline their connection to broader policy frameworks like the Climate Agreement and the MTIP, they also reflect the Ministry's own interpretation and prioritisation. For instance, there is a strong focus on the short-term goal of disconnecting existing houses from the natural gas infrastructure while keeping housing affordable. Given the urgency, the letters assign a minor role to innovations driven by the MTIP. Other factors like standardisation, industrialisation (e.g., prefab elements), better collaboration along the value chain, and pooling demand are expected to play a bigger role. Once again, this widened focus corresponds to a transformative framework, rather than any other innovation policy framework. Even in relation to the goal of disconnecting 200,000 houses by 2030, for which the development of novel solutions could be expected, technological innovation is only one of many possible roads to success being considered. The MMIPs are often mentioned as a source of novel technologies, but according to the sector-specific and R&I-minded policymakers at the BZK, the mission is focussed primarily on reaching sufficient scale to allow innovative or non-innovative solutions to really take off. Therefore, the different communities only converge to a limited extent on a shared view of how the mission relates to achieving the formulated goals.

The programming arena's boundary object carriers each express a vision on how the carbon-free built environment goals and subgoals can be achieved. This concerns both the MMIPs and the construction agenda, each of which have a different origin and, therefore, scope. The MMIPs are part of the IKIA that was formulated in response to the Climate Agreement. Since the translation from built environment goals to a research and innovation agenda was led by R&I actors with a prominent position in the innovation system and in the top sectors, the IKIA and MMIPs are generally seen as continuations of existing programmes. That continuation implies that they are still rooted in considerations about the competitive capabilities of the Dutch economy and the associated potential for successfully exporting new technologies. The construction agenda proposed by sector-specific stakeholders, on the other hand, lines up roughly with the BZK's strong focus on spurring innovations that are ready to be scaled up (possibly due to the passage mechanism). Moreover, it highlights solutions that think of business models or institutional changes as a core strategy instead of just an add-on for a new technology. This split in visions suggests there is little convergence in the programming arena. Or, to put it differently, there is currently convergence around two rather distinct interpretations of what kinds of innovations to focus on.

The policy interventions that support a mission also indicate what meaning and urgency the mission can have. In this case, the implementation arena seems to contain two sets of instruments.

The first category consists of policies programmed by the mission team and the Urban Energy TKI. In addition to the generic policies offered as part of the MTIP umbrella

[9] https://www.rijksoverheid.nl/documenten/kamerstukken/2019/12/17/kamerbrief-kostenreductie-en-innovatie-in-de-bouw

[10] https://www.rijksoverheid.nl/documenten/kamerstukken/2019/12/17/kamerbrief-over-uitwerking-klimaatakkoord-gebouwde-omgeving

framework (i.e., the INWO grants and the PPP allowance scheme), this category also includes a new instrument that was launched through a collaboration between the ministries of BZK and EZK. The Mission-driven Research, Development and Innovation (MOOI) scheme for mission-oriented research and innovation encourages multidisciplinary groups comprised of at least three organisations to propose projects in which various technological and non-technological sub-solutions are combined, and activities relating to the commercialisation and societal acceptance of these solutions are explored. With a minimum of 2 million euros per project (max. 65 per cent subsidised), budgets are approximately 10 times larger than budgets supported by regular policies. Similarly, the "Uptempo!" programme funded by BZK and executed by two TKIs aims to drive the implementation of new and innovative solutions by conducting multidisciplinary learning-by-doing research on the upscaling of energy innovations for the built environment. Policies in this category place more emphasis on adoption and diffusion.

The BZK is the main organisation responsible for the second category of policies, which includes the law on natural gas-free housing. The initiatives governed by this law include the Startmotor programme, which aims to disconnect at least 100,000 housing corporation dwellings from the natural gas grid by 2023. The programme has a "Climate Envelope" budget of 500 million euros (2019–2023), which vastly exceeds the funding available for innovation. The aim of the programme is to use this budget to create the market opportunities needed to drive the development of novel solutions. The programme's renovation accelerator unites housing corporations with the construction sector and engineering firms with a view to spreading new knowledge that has originated in other innovation policies.

As the latter example shows, the various policies are to some extent complementary as they stretch over different innovation development stages. While the policies mobilise very diverse resources, they also raise concerns about the lack of clarity on which solutions to prepare in response to all of the upscaling funding. Because initiatives like the Startmotor programme and the new natural gas-free housing law are technology-neutral, R&I-oriented stakeholders complain that the demand side is not converging on clear directions that are required to narrow their focus to a few promising solution paths.

Finally, the mission scope in the performance arena is reflected in actual innovation projects and initiatives deployed as part of the Programme on Natural-Gas-Free Districts,[11] for example. This programme collects the findings from 50 living labs (user-centred research and innovation environments aimed at fostering interaction and learning) across the country to demonstrate the solutions that districts are working on and the problems they have encountered. Typically, these relate to organisational issues like citizen participation, financing, orchestrating change processes, interconnecting challenges, and legal affairs. The programme has its own knowledge and learning programme; however, it lacks formal channels for communicating the demand and bottleneck issues that it encounters to other arenas where research and innovation programmes are discussed. An exception to this can be found in the Building Technology and Innovation Centre (BTIC), an initiative that links the development and diffusion of innovations. The BTIC emerged out of the construction agenda (its scope is not influenced by passage based on MTIP documents) and unites knowledge partners, industry organisations from the construction sector, a housing corporation, and policy partners including three ministries and a TKI. It drives the creation of groups that contribute to the BE

11 https://aardgasvrijewijken.nl/default.aspx

mission by running comprehensive research projects, addressing topics like the energy transition, digitalisation, circularity, infrastructure replacement, and climate adaptation. It recognises that human capital could be a critical factor for mission completion and so it encourages educational institutes to adapt their curricula. The fact that the BTIC's structure and activities resemble a top sector or a mission team suggests that it is filling a gap left behind by the MTIP, a factor that occasionally causes confusion as to who is coordinating the mission. This points to a lack of convergence around the execution of the mission strategy, but does not necessarily imply that there is strong contestation regarding the types of solutions that are supported by either the mission team or the BTIC. Nonetheless, learning is likely to be hampered if coordination is distributed over different intermediating entities.

3.5 COMPARISON OF OBSERVATIONS ACROSS MTIP AND THE TWO MISSIONS

In this section, we will discuss how convergence and divergence of mission interpretations play out differently for the MTIP umbrella framework and two of the underlying missions. We will focus specifically focus on the inherent dualism of boundary objects, which allows for different interpretations while uniting different communities. Section 3.5.1 provides an overview of the policy issues for which there tend to be divergent interpretations and how this relates to misalignment risks. Section 3.5.2 addresses the relation between common policy elements in an umbrella framework and differences in how mission strategies and mission goals are understood. In particular, this section discusses the possible ways that the study discovered the value of missions to connect seemingly unorganized communities, policies, and investments.

3.5.1 Contested Policy Issues

When the different interpretations of the mission strategy and individual missions are considered carefully, four strategic policy issues stand out.

3.5.1.1 Competitiveness-oriented versus challenge-oriented innovation missions

One major source of contestation is the alleged objective of pursuing a mission-based innovation approach. The MTIP strategy aims mostly to drive innovation for competitiveness, which leads to visions that highlight capabilities that can be leveraged in order to fulfil societal demands that also represent promising market opportunities. In contrast, the BE and especially CE missions focus on achieving a certain societal goal regardless of whether solutions can be exported or need to be imported and what this means for profit potential (e.g., solar technology for local energy production). In terms of strategic arena metanarratives, the MTIP itself can thus be regarded as a blend of a frame 1 and frame 3 innovation policy initiative, while the BE and CE missions are clearer examples of frame 3 innovation policy initiatives. A possible example of the resulting misalignment in this setup is that when moving from strategic to operational (i.e., programming, implementation, and performance) arenas, policy interventions and performance-level activities may focus on practices that help to meet the societal goal but ignore competitiveness, or vice versa.

The type of innovation ambition that is being prioritised also influences policy aspects like knowledge sharing. From a challenge-oriented perspective, it is desirable for knowledge and experiences to be circulated widely to exchange lessons and accelerate upscaling. From a competitiveness perspective, knowledge sharing is less desirable when spread to international markets.[12]

3.5.1.2 Technological versus social innovation

The place of social innovation in addressing societal challenges is also a point of contestation. In some arenas and communities, there is a clear bias towards technical innovation. This bias is potentially caused by routines and policy legacies that are more tuned to such a scope (first or second frame of innovation policy thinking, rather than transformative) and inherited from previous policy frameworks and communities (i.e., top sectors). Views on what types of innovation are feasible and legitimate differ between communities, and if this remains implicit, it can perpetuate misalignment (e.g., between support policies and roadmaps/field-level experiments), which can in turn hamper progress. In those cases, it is good if differences come to the surface, and a more tentative policy approach that maintains a broad policy scope is taken, as this facilitates reflexive learning and reconsideration of communities' mission interpretations. Furthermore, technological and social innovation should not be seen as alternatives, but rather as complementary. A better understanding of synergies is needed to address wicked problems.

In MMIPs, attention is given to social and institutional elements, but mostly as add-ons to technological innovations rather than stand-alone solutions. The CE mission tends to be open to relatively diverse types of innovation, while for BE, the set of possible solutions is more limited. This is also visible in the aforementioned MOOI scheme, which still focusses on R&D projects but now with more room for parties to diffuse (apply) or adopt (purchase) the solutions. Another finding from comparing the two missions is that in some domains, there are broadly accepted ways of discussing whether solutions are ready for diffusion, which makes it easier for groups to prioritise action. The use of technology readiness levels (for technology-based solutions), for instance, can effect convergence, as it frames expectations on what type of innovation to pursue. However, in other domains that rely more on social innovation, those indicators are not as effective.

3.5.1.3 Supply- versus demand-oriented innovation policy

It is frequently suggested that policymakers should pursue missions by "streamlining" policies to simplify the policy mix. It seems that a plethora of policies are emerging in each of the missions, which aim to cover the different dimensions of the mission's societal problem and different stages of the innovation process. The notion of streamlining suggests a degree of simplicity and ease of access to these policies. However, the actual result is a very complex and nearly incomprehensible matrix of policies/support schemes.

In our cases, we have found that the MinEZK focusses on supply measures and hopes to put other ministries in charge of arranging favourable diffusion conditions. This brings two risks: other ministries might avoid engaging with innovation altogether as this is the realm of the

---

[12] While such a purely competitiveness-oriented perspective seems inappropriate for MOIPs targeted at addressing urgent societal challenges, scoping the MOIP in economic terms might also provide the favourable business perspectives that could contribute to its success in evoking unprecedented innovation and diffusion efforts.

EZK or, conversely, they may go too far and also focus on (early-stage) technological innovation, leading to inefficiencies and strained resources at the diffusion stage. In any case, the ability to mobilise demand-side, sector-specific policy developments is an essential element of mission orientation.

One of the first impressions that arises when studying BE is that a continuous stream of policies is emerging (also rooted in policy contexts like the construction agenda and the climate envelope) in tandem with MTIP. This is less evident for the CE mission. The mission team positions itself as the entity that is ultimately in charge of completing the mission, whereas the ministry goes to the other extreme and considers innovation just one of many potential avenues for meeting its goals.

3.5.1.4 Variety-oriented versus selectivity-oriented innovation policy

Different views also exist on whether MOIP should lead to the exploration of more diverse solutions or focus on creating synergies for a select number of solutions.

The MTIP showed how the mission concept in the strategic arena was interpreted as a matter of strengthening innovation opportunities by aligning capacities (and system building) around a few core topics. However, in the MTIP's programming and instrumental arenas, the communities translated the relatively open MTIP strategy in a different way in relation to their own particular contexts, implicitly favouring the search for more (instead of less) variety in solutions that might contribute to a mission goal. The pattern is reversed for the two missions. The sector-specific policymakers who dominate the strategic arena regard missions as vehicles for accelerating the search for new solution directions that need to be proven in markets. In the more operational arenas, there is a preference for driving change by building momentum around several specific solutions.

What is probably most striking here is that, although the neoliberal views of the economists and STI scholars that design the MTIP favour diversity, they insist on giving strong directions in this case. Conversely, sector-specific policymakers engaging with the CE and BE missions are not normally concerned with how solutions perform in markets (they "select" solutions based on functional criteria rather than commercial viability), but now the pressure is rising to a point that they might be afraid of making decisions. These trends suggest that these different epistemic communities may gradually converge on a similar view of whether to operate in a variety-oriented or selectivity-oriented manner, but both sides might also takepositions that are opposite to each other and to their own usual views.

3.5.2 Missions as BOs for Interlinking Policies within an Umbrella Framework

As the comparative observations have shown, different epistemic communities (consisting of varied stakeholders with shared views) can have different interpretations of an umbrella framework and the rationales and workings of individual missions within that framework. For missions to drive change, it is essential to acknowledge these communities' respective standpoints and understand the extent to which MOIP development and implementation processes can support convergence.

Looking at the particular subjects that provoke disagreements among stakeholders involved in the Dutch version of MOIP, a major object of strategic contestation concerns the tension between exploring a wide range of new solutions (enabled by missions that mobilise and connect a diverse range of stakeholders) and building momentum for the development and diffusion of a select number of more established solutions. All four contestation issues discussed in section 3.5.1 relate to this overarching coordination challenge because, in the case of the MTIP, creating variety is associated mostly with facilitating the process of supplying technological innovations in particular, which can boost economic competitiveness. The alternative would be a demand-side policy approach that also calls for social innovations to enhance adoption of solutions for meeting societal challenges, rather than achieving export success.

Different understandings about what the umbrella framework aims to engender and how mission goals can be met translate into different views on how policy should be organised. Interviewees whose reasoning is based on a diffusion perspective consistently call for a long-term approach to programming that relies on sustained policy support for a select number of solution directions that have the best potential for achieving social goals within the set timeframes. Other stakeholders regard missions as devices for crafting a much looser type of coherence between policy measures, assuming that market creation should result in some solution directions becoming dominant in the portfolios of different types of innovation policy support instruments (without a need for top-down portfolio management or formal integration of those instruments).

Regarding the research question as to which common elements to incorporate in an umbrella framework, it is important to start by asking why the design of that framework matters.

If the framework is only an extra arena for appointing mission areas that provide guidance mostly via "vertical passage", it may be possible for stakeholders in more operational areas to adhere to slightly different interpretations in the future. In that case, the programming, implementation, and performance levels could remain aligned with each other while not necessarily being in line with the umbrella framework and strategic arena. Because the processes of convergence (within arenas) and passage (across arenas) unfold differently for each mission, the umbrella framework might end up encompassing missions that are significantly different in terms of their rationales, governance, instruments, and solution approaches.

For an umbrella framework like the MTIP that directly influences the more operational arenas (through MTIP instruments, monitoring frameworks, etc.), there seems to be less room for mission interpretations to differ from each other and from the logic imposed by the umbrella framework. On the one hand, such a setup allows more scope for actively steering innovation activities via "horizontal passage" from the MTIP to the individual missions. However, on the other hand, this might result in structural power imbalances that hamper the process of working out differences through contestation and negotiation. As long as the MTIP of the MinEZK (and the views embedded in it) is regarded as the cornerstone of Dutch attempts to use innovations for tackling wicked challenges, stakeholders in various arenas may be reluctant to propose divergent views or to participate at all.

A comparison of the interpretations in the Dutch MTIP and two of its missions suggests that the choice of common framework elements should also depend on the properties of the various missions it encapsulates, in particular how wicked these missions are.

One viewpoint that has been expressed in this context is that umbrella frameworks should streamline measures in such a way that projects do not need to requalify when moving to policies that support more mature solutions. On the other hand, there are also arguments for permitting openness in the policy mix. Deviating from fully planned integrated policy approaches (adapted to major investments in specific solutions) can be a particularly effective solution for wicked challenges that are not easily translated into a select set of specific solutions. In such cases, there can be merit in sticking to a "fragmented" policy mix that offers room for supporting different solution directions (while also eliminating some possibilities). In that case, missions as boundary objects play a role in ensuring coherence between innovation activities supported by different types of policy options. By serving as common points of reference, the missions can guide the focus of (and thus enhance alignment in) innovation efforts that are not deliberately orchestrated from the outset.

3.6 POLICY IMPLICATIONS

Some of the common policy features in an umbrella framework may work better for some missions than for others. One advantage of standardisation is relative clarity. However, the downside of standardisation is that it can result in a mismatch between some unique features of particular missions. Therefore, the resulting challenge is to "customise" generically defined elements in the mission strategy and to design an umbrella "boundary" framework with recognisable yet flexible elements so that specific elements can be adapted to a given mission. For instance, in contrast to "moon" missions, which have limited uncertainty and contestation along with comparatively clear agendas, in "ghetto"-type (transformative) missions, it cannot be assumed that the mission goals themselves generate collective mobilisation (i.e., help to overcome the inertia stemming from actors not knowing which of multiple competing solution directions to support). This implies that the MOIPs for pursuing such a mission do not require strong mandates, clear goals, and careful planning, but rather governance approaches that make it possible to manage a variety of different approaches in parallel. In other words, a portfolio-based approach (and the governance thereof) to exploring alternatives is needed.

One aspect of missions that requires particular attention is the extent of the scope. The broader the scope of the mission, the harder it can be to inspire and engage actors. A mission that is highly open and thus allows anyone to interpret it in a different way can paradoxically fail to resonate with the unique interests of actors. This is likely to reduce the potential for creating momentum. However, if a mission is defined too narrowly, then there is a risk that inclusiveness and mobilisation will be too limited to ensure that the societal challenge to which the mission is tied is addressed appropriately. As such, striking the correct balance in this context is a very delicate matter. Therefore, as a general principle, we recommend aiming for sufficiently concise and targeted missions in order to generate momentum by inspiring a diverse range of actors. If actors involved in the more specifically defined problem and/or potential solutions think that they can have a larger impact when engaging with the mission, this may well encourage them to participate. If a mission strategy commenced with the launching of broad mission areas, the process of defining more specific underlying missions can depend on the interpretation and negotiation processes that follow the mission launch. If communities unite on distinct, clearly delimited interpretations, these can be incorporated when refining mission visions (cf. the key mechanism of learning).

These considerations point to the importance of adaptive strategies for transformative mission-oriented policies. This suggests that it should be expected that mission goals and plans will be further refined once the nature of the mission challenge is better understood and the possibilities offered by alternative solutions are better known, which is very much in line with concepts such as goal-oriented modulation (Kemp et al., 2007), "governance on the inside" (Smith & Stirling, 2007) and tentative governance (Kuhlmann et al., 2019). For instance, since the start of the climate debates, we have learnt much more about the mechanisms that underpin climate change, and we have also gained a better understanding of the performance and emission-reducing potential (and side effects) of various technological and non-technological options. Tuomi (2012) goes even further in arguing that due to ontological unpredictability, the possibilities of new technological options cannot be anticipated and that this challenge is crucial in the context of longer-term change processes such as those needed in the case of transformative missions. Again, the implication is that caution is required when addressing urgent needs by sticking to a static umbrella framework and associated mission-specific roadmaps without ever reconsidering mission visions and policies in light of new insights as to the feasibility of new and existing solutions. Stability is key for creating visions of the future that actors can respond to (even without central orchestration). However, it does come with the pitfall of ignoring major new possibilities or threats.

3.7 CONCLUSION

This study builds on the idea that regarding missions from a boundary object perspective helps to diagnose coordination and (mis)alignment issues. Previous literature tends to present the optimistic belief that just having a clear, ambitious, and timebound goal will almost automatically mobilise and align various stakeholders. However, missions targeted at wicked societal challenges are often open-ended and have umbrella frameworks that only specify "mission areas" rather than establishing clear directions. A BO perspective acknowledges the uncertain, contested, and subjective nature of missions, as it focusses attention on how differently communities perceive a mission. In order to be able to specify more targeted strategies and programmes at a later stage, it is important to make these differences transparent by opening up the possibility of exploring alternative pathways.

In the case of the Dutch MTIP and two of its missions, it has been shown that communities have divergent views on strategic issues such as the type of mission (policy) objective to concentrate on, the type of innovation that should be supported, the dominant force to target (solution supply or demand), and the importance of either boosting variety in novel solutions or supporting synergies and momentum by selecting some more established options.

As the MTIP and the two missions were only recently put in place, there is little knowledge available concerning learning activities and the ways in which different communities work out the observed differences in interpretations. However, the BO perspective becomes more important when considered in light of the encountered degree of openness. It recognises that plasticity allows stakeholders to contribute to the same goal without actively collaborating or even agreeing with each other (e.g., on a certain framing). Even when the issue in question is approached from different perspectives, fragmented "bottom-up" initiatives by different communities can still advance the development and diffusion of innovative solutions if they accept the mission as a valid form of guidance towards the goal of addressing the focal challenge.

The downside of working on missions as open concepts is that there are no guarantees that all efforts will be fully aligned. Openness can potentially create problems, such as the development of solutions that get stuck in the prototyping stage (i.e., no diffusion occurs) or that do not converge towards a dominant design. Conversely, a policy could also fail by accommodating, promoting, and selecting the solutions at too early a stage, i.e., by giving in to vague promises and failing to consider the high levels of uncertainty involved. The upside of openness is that it prevents situations where progress is stifled precisely because actors cannot agree on a clear direction. Missions that serve as a common point of reference may result in uncoordinated yet mutually reinforcing and legitimised investments, which reveal ultimately which solutions are most promising. Therefore, allowing for openness can be a deliberate strategy for exploring the mission preferences of different communities (spanning industries and citizens). While ensuring convergence might often be regarded as a starting point for formulating policies for wicked challenges, it can also occur as the result of running policies that allow different communities to engage with an open concept.

To sum up, MOIP includes the deployment of policy instruments and governance structures. Typically, governance structures serve to make stakeholders decide on directions, which can then be supported through the policy instruments. However, it is also through the activities supported by these instruments that societies can learn about the potential of solutions. In the absence of well-defined and uncontested views on how to tackle a mission, the ensuing openness enables different communities to contribute (based on their own interpretation of the mission) to experimentation with novel solutions. The freedom that this model provides can prevent paralysis caused by the inability to make firm decisions on what to focus on, a typical feature of umbrella frameworks that encapsulate wicked missions for which the state has only assigned problem areas rather than prioritising concrete solutions.

REFERENCES

Alford, J., & Head, B. W. (2017). Wicked and less wicked problems: A typology and a contingency framework. *Policy and Society*, *36*(3), 397–413.

Alves, A. C., Vonortas, N. S., & Zawislak, P. A. (2021). Mission-oriented policy for innovation and the fuzzy boundary of market Creation: The Brazilian shipbuilding Case. *Science and Public Policy*, *48*(1), 80–92.

Barre, R., Henriques, L., Pontikakis, D., & Weber, K. M. (2013). Measuring the integration and coordination dynamics of the European Research Area. *Science and Public Policy*, *40*(2), 187–205.

European Commission (2018). *Mission-oriented research & innovation in the European Union: A problem-solving approach to fuel innovation-led growth.* Directorate-General for Research and Innovation. European Commission, Brussels.

European Commission (2021). *European missions, Communication from the Commission to the European Parliament, the Council, the European Economic and Social Committee and the Committee of the Regions.*

Haas, P. M. (1992). Introduction: Epistemic communities and international policy coordination. *International Organization*, *46*(1), 1–35.

Hekkert, M. P., Janssen, M., Wesseling, J. H., & Negro, S. (2020). Mission-oriented innovation systems. *Environmental Innovation and Societal Transmissions*, *34*, 76–79.

Janssen, M. (2015). Cross-specialization and structural holes: The case of the Dutch Topsectors. *Papers in Evolutionary Economic Geography*, *15*, 19.

Janssen, M. (2020). *Post-commencement analysis of the Dutch "Mission-oriented Topsector and Innovation Policy" strategy.* MIPO, Utrecht University.

Janssen, M., Bogers, M., & Wanzenböck, I. (2020). Do systemic innovation intermediaries broaden horizons? A proximity perspective on R&D partnership formation. *Industry and Innovation*, *27*(6), 605–629.

Janssen, M., Torrens, J. C. L., Wesseling, J., & Wanzenböck, I. (2021). The promises and premises of mission-oriented innovation policy: A reflection and ways forward. *Science and Public Policy*, *48*(3), 438–444.

Janssen, M. J., Wesseling, J., Torrens, J., Weber, K. M., Penna, C., & Klerkx, L. (2023). Missions as boundary objects for transformative change: Understanding coordination across policy, research, and stakeholder communities. *Science and Public Policy*, *50*(3), 398–415.

JIIP (2018). *Mission-oriented research and innovation: Inventory and characterisation of initiatives.* Final report, Joint Institute for Innovation Research. European Commission.

Kemp, R., Loorbach, D., & Rotmans, J. (2007). Transition management as a model for managing processes of co-evolution towards sustainable development. *The International Journal of Sustainable Development & World Ecology, 14*(1), 78–91.

Kern, F., & Smith, A. (2008). Restructuring energy systems for sustainability? Energy transition policy in the Netherlands. *Energy Policy, 36*(11), 4093–4103.

Kuhlmann, S., Stegmaier, P., & Konrad, K. (2019). The tentative governance of emerging science and technology – A conceptual introduction. *Research Policy, 48*(5), 1091–1097.

Larrue, P. (2021). The design and implementation of mission-oriented innovation policies: A new systemic policy approach to address societal challenges. *OECD Science, Technology and Industry Policy Papers, 100.* OECD.

Loorbach, D., & Rotmans, J. (2010). The practice of transition management: Examples and lessons from four distinct cases. *Futures, 42*(3), 237–246.

Mazzucato, M. (2018). Mission-oriented innovation policies: Challenges and opportunities. *Industrial and Corporate Change, 27*(5), 803–815.

Mazzucato, M., Kattel, R., & Ryan-Collins, J. (2020). Challenge-driven innovation policy: Towards a new policy toolkit. *Journal of Industry, Competition and Trade, 20*(2), 421–437.

Parks, D. (2022). Directionality in transformative innovation policy: Who is giving directions? *Environmental Innovation and Societal Transitions, 43*, 1–13.

Schot, J., & Steinmueller, W. E. (2018). Three frames for innovation policy: R&D, systems of innovation and transformative change. *Research Policy, 47*(9), 1554–1567.

Smith, A., & Stirling, A. (2007). Moving outside or inside? Objectification and reflexivity in the governance of socio-technical systems. *Journal of Environmental Policy & Planning, 9*(3–4), 351–373.

Star, S. L. (2010). This is not a boundary object: Reflections on the origin of a concept. *Science, Technology, & Human Values, 35*(5), 601–617.

Star, S. L., & Griesemer, J. R. (1989). Institutional ecology, "translations" and boundary objects: Amateurs and professionals in Berkeley's Museum of Vertebrate Zoology, 1907–39. *Social Studies of Science, 19*(3), 387–420.

Tuomi, I. (2012). Foresight in an unpredictable world. *Technology Analysis & Strategic Management, 24*(8), 735–751.

Wanzenböck, I., Wesseling, J. H., Frenken, K., Hekkert, M. P., & Weber, K. M. (2020). A framework for mission-oriented innovation policy: Alternative pathways through the problem–solution space. *Science and Public Policy, 47*(4), 474–489.

Weber, K. M., & Rohracher, H. (2012). Legitimizing research, technology and innovation policies for transformative change: combining insights from innovation systems and multi-level perspective in a comprehensive failures framework. *Research Policy, 41*, 1037–1047.

Wittmann, F., Hufnagl, M., Lindner, R., Roth, F., & Edler, J. (2021). Governing varieties of mission-oriented innovation policies: A new typology. *Science and Public Policy, 48*(5), 727–738.

PART II

TRANSFORMATIVE MISSION-ORIENTED STI POLICIES IN DIFFERENT POLITICAL SYSTEMS AND ON DIFFERENT LEVELS

4. Mission-oriented innovation policies in Europe: from the normative to the epistemic turn?

Rainer Kattel and Mariana Mazzucato

4.1 INTRODUCTION: THE RETURN OF MISSIONS

While up until the 1980s, mission-oriented innovation policies played a critical role in the post-war recovery and growth of developed and developing countries, they became decidedly less important in the wake of Thatcherite-Reaganite economic policies and public sector reforms (Foray et al., 2012; Mazzucato, 2016). However, during the 2010s and early 2020s, missions have made a strong comeback: governments in Europe and elsewhere are increasingly turning their attention towards tackling "grand challenges" or "wicked issues" such as climate change through science, technology, and innovation (STI) policies (Borrás & Edler, 2020; Mazzucato et al., 2020). The key driver behind challenge-driven STI policy initiatives is the ambition of achieving a particular type of economic growth (e.g., smart, inclusive, and sustainable). This constitutes a direct admission that economic growth not only has a rate but also a direction, which means that it can have multiple different directions (Mazzucato, 2017). Policymakers have a choice, for instance, of how, and how quickly, they aim to decarbonise certain economic sectors. These alternative pathways include diverse trade-offs, spillovers, and dynamic efficiency changes across sectors. Thus, as Stirling (2009) argues, it is not about picking the one and only direction of change, but a wider directionality that includes and enables multiple competing and complementary pathways and experimentation along the way.

In policy practice, the European Union has arguably been the leading policy actor in bringing missions back. For over a decade, the EU has attempted to reorient its STI agenda towards grand challenges (Soete et al., 2017). From the Lund declaration of 2009 onwards, the EU has attempted to tackle what has been called the "orientation failure" inherent in its innovation policies (Daimer et al., 2012). In 2021, the EU's Horizon Europe R&D funding programme commenced. With this programme, the EU has made a commitment to providing 53.5 billion euros in funding through Horizon Europe's second pillar, "Global Challenges and European Industrial Competitiveness", and its five mission areas[1] by 2027. Another aim of the pro-

[1] One of the authors, Mariana Mazzucato, wrote two key reports for the European Commission in the run-up to the launch of the Horizon Europe programme (Mazzucato, 2018, 2019). The mission areas are as follows: Adaptation to Climate Change – support at least 150 European regions and communities in becoming climate resilient by 2030; Cancer – working with Europe's Beating Cancer Plan to improve the lives of more than 3 million people by 2030 through prevention, cure, and solutions for living longer and better; Restore our Ocean and Waters by 2030; 100 Climate-Neutral and Smart Cities by 2030; A Soil Deal for Europe – 100 living labs and lighthouse projects to lead the transition towards healthy soils by 2030. The European Commission's official missions

gramme is to encourage member states to refocus their STI policies according to the mission-oriented approach (Mazzucato, 2018; European Commission, 2023). This "normative turn" in STI policies is likely to be reinforced by COVID-19 responses that promise to build back better (Mazzucato et al., 2021) and the various implications of the war in Ukraine. However, while the EU and other leading economies have promised to increase public spending on the various challenges that they face, the debate on effectively aligning and absorbing these funds is only beginning (Mazzucato et al., 2023). Policy directionality is vital, but having the capacity to implement and achieve ambitious goals is equally important.

Thus, while the normative turn and its focus on the directionality of STI (and other) policies have been discussed at length, there seems to be much less discussion around the fact that we can arguably also see an epistemic turn, or at least the beginnings of one, in how policymakers conceptualise policy challenges and the impact of policy instruments. More precisely, the epistemic turn refers to the way in which public institutions create and use the knowledge necessary to design and implement transformative policies. For instance, this includes the way in which public institutions conceptualise the value that they create, the kinds of evaluation and appraisal methodologies that they use, and how and with whom they collaborate in gathering evidence and intelligence. In the broadest sense, the epistemic turn denotes shifts and changes in how public organisations learn and build their capacities.

Missions are used not only in the STI context, but also in the wider discourse around public sector reform, most notably in digitalisation, but also in the wide and amorphous area of public sector innovation (van Buuren et al., 2020; Kattel & Takala, 2021). The public sector's adoption of working practices from (strategic and user-centric) design and agile software development methodologies has driven the emergence of new public organisations in the form of public sector design, digital, and innovation labs (Hill, 2015; Tõnurist et al., 2017; Mergel et al., 2020). These working practices focus on agile processes such as prototyping and experimentation and rely on epistemological frameworks from action research and ethnography, which complement prevalent economic theories and public policy analysis. The resulting new agencies attempt to bring experiential, experimental, and techno-solutionist "hacker" routines to public organisations (van Buuren et al., 2019).

In some cases, we can see both the normative and epistemic turns converge within a single agency, as exemplified by the Swedish innovation agency, Vinnova. Over the past few years, Vinnova has focussed on designing missions that go beyond STI activities and on using strategic design approaches as their main tools in the process (Hill, 2022; Kattel, 2022). Likewise, it can be argued that the accelerator labs created by the United Nations Development Programme (UNDP) in 115 countries in recent years represent a similar convergence of challenge-driven or mission-oriented policies and strategic design-based tools (Altman & Nagle, 2020). The influx of various new practices, from systems thinking to strategic design, and the creation of policy and innovation labs to embed such new practices in the public sector signal attempts to shift the focus of public agencies towards citizen needs, which often span the boundaries of individual ministries. While in many instances, such practices and organisations remain at the margins of the public sector, in some cases they lead to wide-ranging changes in civil service or public service provision (Mergel et al., 2020). Above all, these practices are aimed at changing the way policymakers and public organisations learn (i.e., gather and analyse information,

homepage is here: https://research-and-innovation.ec.europa.eu/funding/funding-opportunities/funding-programmes-and-open-calls/horizon-europe/eu-missions-horizon-europe_en and appraise and evaluate policy interventions). Strategic design and similar approaches represent an attempt to quicken learning feedback cycles through small-scale experiments. This should make it possible to avoid multi-year policy design and implementation cycles where learning only comes at the end of the process.

In academic discourse in particular, we can see a convergence of the normative and epistemic turns. Thus, for instance, Janssen et al. (2021) argue that "missions are subject to problem-based governance, which encompasses the various efforts focused on directly adapting socio-economic systems dealing with the societal challenge... Missions are also in dialogue and tension with the structures and arrangements involved in innovation governance..., which impact upon the rate, direction and quality of activities in the innovation system" (Janssen et al., 2021; Björk et al., 2022). Hekkert et al. (2020) have defined a mission-oriented innovation system as "the network of agents and set of institutions that contribute to the development and diffusion of innovative solutions with the aim to define, pursue and complete a societal mission" (Hekkert et al., 2020). Similarly Tödtling et al. (2021) have applied a mission-oriented logic to regional innovation systems whereby mission orientation "constitutes the wider regional (territorial) framework, reflecting the capacity of regions to address various and partly interrelated challenges" (Tödtling et al., 2021). These arguments indicate that for ambitious policy goals such as missions, it is necessary to rethink some of the foundations that public organisations use as a basis when gathering and analysing information and appraising and evaluating policy interventions – i.e., the epistemic foundations of policymaking.

The aim of this chapter is to investigate the emerging mission-oriented innovation practice in Europe, with a particular focus on the question of whether the normative and epistemic turns converge as the abovementioned conceptual developments suggest. The chapter bases its perspective on Nelson's admonition that while we can send a man to the moon, we must continue to grapple with solving the problems on Earth (Nelson, 1977). Nelson implied that we need a different epistemic framework in order to tackle socioeconomic challenges. This chapter explores whether the emerging practice in Europe is any closer to discovering these new epistemic frameworks. We will show that while there is plenty of evidence of the normative turn, the epistemic one seems to be much rarer in actual policy practice.

We are focussing on Europe, as it is arguably the trailblazer in such policy practices, while also offering a variety of capitalist and politico-administrative systems. To answer these questions, the chapter draws on a systematic literature review, interviews with policymakers in selected countries, and discussions from a number of workshops with European policymakers. As many such policies have only been initiated during the past few years, the chapter is exploratory in nature; the aim is to provide a snapshot of emerging practices and issues.

The chapter is structured as follows: the next section briefly introduces the research design, followed by a summary of the main findings; the chapter then concludes with a discussion of the emerging challenges relating to mission-oriented innovation policies in the European context.

4.2 RESEARCH DESIGN

The chapter takes an interpretivist and hermeneutical approach to understanding knowledge in social sciences as a process of sense-making (Maykut & Morehouse, 2002; Drechsler,

2004). In this case, the sense-making includes a literature review, interviews with policymakers, and participatory observations from workshops with policymakers.

The literature review draws on both academic databases and the rapidly expanding analytical collections developed by the OECD (2021), the European Commission,[2] and academic research institutions such as Fraunhofer (Wittmann et al., 2021), the Mission-Oriented Innovation Policy Observatory (Hekkert et al., 2020)[3] and the Mission-Oriented Innovation Network at the UCL Institute for Innovation and Public Purpose (UCL IIPP).[4] It is important to note that the chapter does not cover actual policy documents as these have been discussed in great detail in Fisher et al. (2018), Bertelsmann Stiftung (2021), and Larrue (2021). Instead, the chapter utilises a literature review approach to supplement the abovementioned policy reviews.

The scope of the literature review consists of academic papers, book chapters, and publications issued between 1990 and 2020. The initial search was carried out on titles and abstracts using Boolean search strings.[5] The search strings were clustered around two sets of issues: first, articles discussing innovation and innovation policy, missions, societal challenges, sustainable development goals (SDGs), transformation, and transition; and second, articles discussing STI organisations, new ways of working (e.g., service design), and capabilities. The aim of using multiple keywords was to ensure that a broad range of literature from the past two decades and across various disciplines was included. In summary, the screening and selection processes identified a total of 660 references and excluded 389 references, yielding a corpus of 271 references for inclusion in qualitative synthesis as of August 2022.

The academic literature review was complemented by 23 interviews with stakeholders in Sweden, the United Kingdom, and Latvia,[6] by key takeaways from a series of workshops conducted by the OECD Observatory of Public Sector Innovation (OPSI) and UCL IIPP in 2020 and 2021;[7] and by workshops carried out by UCL IIPP's Mission-Oriented Innovation Network during 2022 and 2023.[8]

The choice of countries covers both traditional types of capitalism in the sense of varieties of capitalism literature (liberal and coordinated market economies; Hall & Soskice, 2001) and

---

2 For further background and case studies by the European Commission, see https://research-and-innovation.ec.europa.eu/funding/funding-opportunities/funding-programmes-and-open-calls/horizon-europe/eu-missions-horizon-europe_en

3 The observatory is located at Utrecht University. See https://www.uu.nl/en/research/copernicus-institute-of-sustainable-development/mission-oriented-innovation-policy-observatory

4 For more information, see https://www.ucl.ac.uk/bartlett/public-purpose/partnerships/mission-oriented-innovation-network-moin

5 The data sources used in this study include the following electronic databases and manual sources: the core electronic databases were Scopus, Web of Science, EBSCO, ProQuest, and Wiley; manual searching was conducted via Google Scholar and hand-scanning of related websites, e.g., OECD, European Union, Nesta UK, etc.

6 The interviews were carried out in 2021 and focussed on lead agencies (UKRI in the UK, Vinnova in Sweden, and the Investment and Development Agency of Latvia). The UK case has been discussed in detail in McLaren and Kattel (2022) and the Swedish case in Kattel (2022).

7 Over 150 civil servants, economists, business leaders, and policymakers from 36 countries took part. For further details, see the summary in Conway (2020).

8 The workshops are captured in case studies and blogs available at https://www.ucl.ac.uk/bartlett/public-purpose/partnerships/mission-oriented-innovation-network-moin

a hybrid type from Central and Eastern Europe (Karo, 2011). The three countries also follow different administrative traditions (Peters, 2021): Anglo-American, Scandinavian, and mixed traditions. In addition, in the European Union innovation scoreboard, these countries fall into three different performance groups (leaders, strong performers, and moderate performers, respectively). Thus, these three countries should provide diverse input for discussing mission-oriented innovation policies in Europe.

4.3 MAIN FINDINGS

This section is divided into three segments, based on the dominant keywords and concepts found in the literature review, the interviews, and workshops:

1. What are the main drivers of mission-oriented policies?
2. What kinds of support structures are used to facilitate missions?
3. What are the specific tools and methods used in missions?

4.3.1 What Are the Main Drivers of Mission-Oriented Policies?

In emerging policy practice, governments are utilising mission-oriented innovation to target a wide range of challenges, from basic scientific advances (e.g., combating cancer) to modifying human behaviour (e.g., food systems). This has led to various attempts to describe mission-oriented innovation types, with a particular focus on the kinds of challenges that the missions attempt to tackle (Fisher et al., 2018; Wittmann et al., 2021). According to these typologies, missions are attempting either to accelerate existing STI practices (e.g., supporting applied research in batteries for sustainable transportation) or to set "transformative" goals that aim to radically change existing systems (e.g., supporting the transition to a circular economy). Another way to differentiate between mission types is to look at how and by whom missions are designed and implemented. Thus, for instance, the OECD's recent research categorises missions by their level of intervention (i.e., centre of government, ministry/agency, programme, and ecosystem) (Larrue, 2021). Mission taxonomy can also be approached from an implementation perspective, which would include top-down coordination missions (such as Germany's High-Tech Strategy 2025), sector-specific coordination missions (such as the Topsectoren approach in the Netherlands), place-based vision- or consensus-building missions (such as Missions València 2030), and user-centric behavioural missions (such as the food and street missions developed by Vinnova in Sweden).

While these typologies are useful for understanding the rapidly evolving mission landscape, it is also clear that there is no common classification system for missions. On the contrary, missions seem to have become an umbrella term that covers multiple transformation- and challenge-oriented policy approaches. However, there are several common drivers behind missions.

First, it is increasingly clear to many decision-makers that societal challenges are "wicked". As Wanzenböck et al. (2020) have elaborated, challenges can be described as wicked when they are characterised by high degrees of contestation, complexity and uncertainty. While it is unclear whether such characteristics are useful in actual policy practice (Levin et al., 2012; Peters, 2017; Peters & Tarpey, 2019), the contestation, complexity, and uncertainty underlying such challenges as the climate emergency perhaps make "wickedness" the key driver behind the adoption of missions as a policy approach.

Second, siloed policy design and implementation processes are perceived as key obstacles in tackling the aforementioned complexity of policy challenges. By contrast, the mission-oriented approach offers a new way of coordinating policies within the STI field and in a broader sense (Kattel & Mazzucato, 2018; Wittmann et al., 2021). It is important to note that missions do not impose a top-down approach to STI or any other policy arena, but rather seek to balance overall goal-setting with bottom-up experimentation and open-ended pathways for achieving the goals (Mazzucato, 2021). This is why, in practice, mission design often focusses on stakeholder engagement and co-creation. Slotting missions into existing policymaking infrastructure, e.g., by tasking existing STI agencies with the implementation of missions, runs the risk of "mission washing", whereby existing policies receive a new label. As such, coordination remains one of the key challenges in mission practice. The EU's Horizon programme is a good example of this phenomenon, as missions do form part of one of its pillars, but they are not coordinated with other instruments such as the EU's Green Deal.

Third, there is a growing sense of the ineffectiveness of some existing policy tools, particularly in the area of STI policies (OECD, 2015). Specifically, in recent decades, STI policy has been guided by the idea of increasing the (external) competitiveness of economies, which has led to analytical frameworks that seek a better understanding of the impact of individual policy measures (Kattel et al., 2018, 2020; Tödtling et al., 2021). This approach has relied on the idea of market failure as the dominant justification for STI interventions. However, it has been shown that market failure-based interventions are prone to two main problems: First, they tend to focus on individual measures rather than policy mixes (Edler et al., 2016) and thus create a false sense of causality; and second, the market-failure approach reinforces a reactive focus on STI policies, which makes the coordination of different policies particularly challenging. In this context, a mission-oriented approach is understood as an ex ante market-shaping policy that makes it possible to achieve the common goal of designing and implementing a policy portfolio comprising various actions oriented towards a wider directionality.

4.3.2 What Kind of Support Structures are Used to Facilitate Missions?

Mission-oriented innovation policy is supported and delivered by three policy structures, which are often interlinked: institutional entrepreneurship, specific funding mechanisms, and procurement.

4.3.2.1 Institutional entrepreneurship

As Grillitsch et al. argue, "institutional entrepreneurs understood as actors who initiate changes that lead to a divergence from existing institutions are thus essential for providing directionality" (Grillitsch et al., 2018). Accordingly, one of the key challenges for mission-oriented innovation is promoting institutional entrepreneurship in order to change the status quo in terms of policy practice and actively introduce overtly normative policy goals. In practice, overcoming this challenge requires various different forms of institutional entrepreneurship. It is important to note that institutional leadership can emerge through the repurposing of existing institutions or the creation of new institutions. For instance, KfW, a German promotional bank, plays an important role in national mission practices, while in Scotland, the newly created Scottish National Investment Bank has taken up a similar role.

As mission-oriented innovation often targets wicked policy issues involving coordination challenges, the creation of new coordination structures is one of the most frequently used support mechanisms for missions. Thus, for instance, the European Union has created mission boards for each of its five missions.[9] However, mission governance can be implemented and supported by developing or repurposing existing coordination mechanisms. For example, the Netherlands introduced the Topsectoren approach in 2012 to strengthen coordination and collaboration between various STI system actors. The nine sectors that were selected stemmed from R&D- and export-intensive domains such as agrifood logistics, life sciences and health, and high-tech systems and materials.[10] While originally the primary goal of this approach was to align the activities of research institutes more closely with the knowledge demands of innovative firms, it gradually shifted towards more transformative goals, as the Mission-oriented Topsector and Innovation Policy (MTIP) now contains 25 missions across four themes. While much of the Topsector governance has evolved into MTIP governance, perhaps the most marked difference is the creation of mission teams: "They are positioned as the engines for driving changes, as formally their tasks include the developing, executing and organising – through engaging various ecosystem actors – of both the missions and the multi-annual innovation programs" (Janssen, 2020). Similarly, the UK's 2017 Industrial Strategy introduced the Industrial Strategy Challenge Fund and 23 goal-oriented challenges, each led by a challenge director from the industry (McLaren & Kattel, 2022).

Creating new leadership posts within existing organisations can also provide institutional entrepreneurship for mission directionality. For example, the Swedish innovation agency Vinnova created a new post for the director of strategic design in order to rethink its approach to innovation policy. Vinnova, established in 2001, can be seen as a relatively standard innovation policy agency that advises the government on innovation policy and designs and implements innovation support measures (Chaminade & Edquist, 2006). However, once it turned its attention to mission-oriented innovation policies in the late 2010s, Vinnova took quite a different approach from standard innovation agencies. It created a director of strategic design position and hired a designer for this post. In light of the criticism directed at typical technocratic policy design processes, the agency developed a detailed plan for more engaged innovation practices in the Swedish missions (Hill, 2022). Taking the mission themes of healthy sustainable mobility and healthy sustainable food as their starting points, Vinnova coordinated intensive co-design sessions across Sweden, where up to 400 different stakeholder organisations engaged in "actors workshops" (Hill, 2020). As with the Dutch case, the Swedish application of missions can also be understood as the layering of missions into existing, evolving policy practice and an effort to move beyond the competitiveness paradigm (Rohracher et al., 2022).

When we look at mission-oriented innovation through the broader lens of public sector reform, it becomes apparent that many governments are attempting to establish innovation and policy labs as institutional entrepreneurs. This reflects the public sector's adoption of new working practices from (strategic) design and agile software development practices from the private and third sectors. As recent studies have shown, these practices are mostly adopted

---

9 For further details, see https://research-and-innovation.ec.europa.eu/funding/funding-opportunities/funding-programmes-and-open-calls/horizon-europe/eu-missions-horizon-europe/mission-boards_en

10 The original sectors and processes are briefly described here: https://www.topsectoren.nl/innovatie

by new, often peripheral public organisations in the form of public sector design, digital, and innovation labs (Hill, 2015; Bason, 2017; Tõnurist et al., 2017; Mergel et al., 2020). The working practices focus on agile processes such as prototyping and experimentation and rely on epistemological frameworks from action research and ethnography rather than economics or public policy analysis (van Buuren et al., 2020). In addition to Sweden, Latvia has also made an explicit decision to use design methodologies to develop its mission approach. Some local governments, such as the London borough of Camden, have also adopted design as the key practice in mission development. This indicates that public organisations require different epistemic frameworks and related capacities in order to integrate experimentation into their mission practice at a deep level.

4.3.2.2 Funding mechanisms

There is emerging evidence that public funding for missions yields the largest multiplier compared to other policy interventions (Deleidi et al., 2020; Dosi et al., 2021). However, governments are funding missions through multiple avenues, namely the following:

1. Repurposing or upgrading existing funding mechanisms or institutions to make them more mission-oriented or to ensure that they serve a specific purpose in a mission-oriented policy mix.
2. Introducing new funding mechanisms.
3. Creating entirely new funding institutions, such as mission-oriented public banks.
4. And finally, particularly for green missions, "greening" central banks and financial regulations: this has become a pivotal means of supporting the sustainability transition.

One example of repurposing or upgrading is the aforementioned policy evolution that has taken place in the Netherlands, where sector-specific coordination and cooperation policies have gradually evolved into a mission-oriented policy mix that targets a variety of missions (Janssen, 2020). Similarly, in 2017, the UK government launched a new Industrial Strategy that included four mission areas (clean growth, ageing society, future of mobility, and artificial intelligence and data economy) and introduced the Industrial Strategy Challenge Fund (ISCF), which focusses on 23 challenges across the mission areas.[11] The ISCF is, in turn, part of a larger National Productivity Investment Fund that was established in 2016. It is also notable that, in this case, mission funding evolved within the context of a wider change in the STI funding and institutional landscape (McLaren & Kattel, 2022).

Similarly, the European Investment Bank (EIB) has become one of the main providers of long-term financing and risk-sharing in the EU; as such, it has been taking the lead in key policy areas, such as climate action finance (by committing itself to dedicating 50 per cent of its financing resources to climate action by 2030) (Mazzucato & Mikheeva, 2020). Furthermore, the EIB is the lead implementing and advisory partner in the InvestEU Programme, which is one of the key components of the EU financing framework for 2021–2027.

However, governments can also repurpose existing institutions, such as state-owned companies (Gasperin et al., 2021). For instance, the Danish public energy company Ørsted has adopted a wholesale renewable energy strategy. By the end of 2019, Ørsted had invested

[11] For further details on the ISCGF, see https://www.ukri.org/our-work/our-main-funds/industrial-strategy-challenge-fund/

DKK 193 billion in renewable energy and has been hailed as the "first green energy major" (Sheppard, 2020). With this shift, the company revamped the industrial offshore wind landscape in Denmark (Algers & Kattel, 2021; Voldsgaard & Rüdiger, 2021).

Governments can also create entirely new funding mechanisms. For example, in 2021 Denmark's Innovation Agency published a call for roadmaps for mission-driven green partnerships. The fund encourages all relevant stakeholders across the Danish research and innovation system to come together to contribute their expertise and propose a realistic and robust path towards the development of cutting-edge solutions within the four missions – ranging from strategic research to commercialisation, with a focus on short-, mid-, and long-term impact.[12] The call is being financed by DKK 700 million in public funds.

When it comes to establishing new mission-oriented institutions, Scotland has created an excellent example in its "mission-led development bank providing patient capital to build a stronger, fairer, more sustainable Scotland",[13] founded in 2020. With its capital comprising £2 billion in public funds, the bank focusses on the following three missions: achieving a just transition to net-zero carbon emissions by 2045, extending equality of opportunity through improving places by 2040, and harnessing innovation to enable our people to flourish by 2040 (Mazzucato & Macfarlane, 2019).

One topic that is receiving growing attention in the academic debate is the role of central banks and financial regulators in addressing climate-related financial risks (Campiglio et al., 2018). Since the 1990s, central banks have narrowed their mandates to focus on price stability and limited their interventions to adjustments of the reference interest rate. However, since the Global Financial Crisis of 2008, central banks have increasingly used a wider range of "unconventional" measures, including quantitative easing and various other short- and longer-term liquidity programmes to stimulate the economy. The extensive use of these tools – which are often associated with sector-specific conditions (e.g., certain refinancing lines offered by the European Central Bank (ECB) are only available to Eurozone banks if they commit to financing SMEs or other non-financial corporations) – has raised questions about central banks' market neutrality and independence. This, in turn, has led to suggestions that central banks could do more to direct finance towards green growth. Despite the great potential in this area, research suggests that the ECB's corporate bond purchases mirror the investment choices of financial markets, which means that, to date, they have mostly favoured large, carbon-intensive companies (Matikainen et al., 2017; Jourdan & Kalinowski, 2019). In fact, the ECB portfolio is currently contributing significantly to biodiversity loss (Kedward et al., 2021).

4.3.2.3 Procurement

For the first generation of mission-oriented policies, public procurement was a key means of supporting public (military) missions (Mowery, 2012). The current wave of mission-oriented thinking allocates an even more extensive role to public procurement. As argued by Lember et al., "public procurement is increasingly seen as a horizontal policy measure that should

[12] Further details are available here: https://innovationsfonden.dk/en/news-press-jobs/now-open-call-roadmaps-mission-driven-green-partnerships

[13] The quote is from the website of the new public bank: https://www.thebank.scot. See also Mazzucato and Macfarlane (2019).

be applied across the public sector and regardless of the characteristics or missions of public agencies" (Lember et al., 2015; Mazzucato, 2020).

The idea is that by placing sophisticated demands on markets, preferably through functional requirements and standards (Edquist & Zabala-Iturriagagoitia, 2020), public procurers can create strong incentives for private providers to come up with new solutions or to upgrade their production-related processes in order to cope with government requirements.

In theory, procurement can be used in at least two ways in the context of missions: as part of the policy mix or as an organisational practice for achieving policy goals.

In the first case, procurement is related directly to mission implementation, such as in sustainability-focussed missions that are implemented explicitly through specific, targeted green procurement practices, for example, or tenders that are aimed at inducing increased private R&D spending in carbon-neutral technologies. This use of procurement corresponds to the normative approach to policymaking: the normative policy goals are expressed through the demand created via procurement (Mazzucato, 2020).

Alternatively, procurement can serve missions as an organisational practice for relating to markets. Thus, for instance, public digital agencies' agile procurement practices help them to achieve their policy goals (Fernandez-Monge Cortazar et al., 2021; Kattel & Takala, 2021). This use of procurement corresponds better with epistemic changes in policymaking: procurement enforces different feedback loops for policymakers.

In mission practice, the normative approach is used almost exclusively, particularly in the area of green public procurement (GPP) (Cheng et al., 2018). However, the field is under-theorised and under-researched (Testa et al., 2012; Chiappinelli & Zipperer, 2017).

4.3.3 Emerging Tools and Methods

In principle, as there are multiple pathways for using the mission-oriented approach in public policy, a wide variety of policymaking tools and methods can be deployed. However, there are a number of quite specific tools and methods that are increasingly favoured by public organisations when it comes to implementing missions.

First, many mission-oriented approaches attempt to bring new ways of working into the public sector, particularly in terms of experimentation and stakeholder engagement. As mentioned above, such approaches rely on strategic design and similar, mostly qualitative, tools and methods.

Second, missions typically use a portfolio approach when it comes to investing in projects. Rather than viewing policy effectiveness through the lens of individual interventions and the associated risk mitigation activities, a mission-oriented approach seeks to tackle uncertainty by supporting multiple solutions to a policy problem.

Third, governments are seeking ways to evaluate missions on the basis of new frameworks and methodologies that focus on public value and spillover. This may include qualitative approaches such as systems thinking and quantitative approaches such as investment multipliers and agent-based modelling.

All three sets of tools and methods rely on the key differentiation between market-failure fixing (cost-benefit analysis) and market shaping as a means of justifying public sector activities (Kattel et al., 2018, 2020; Mazzucato & Ryan-Collins, 2019). Table 4.1 summarises these two approaches to policy justification and demonstrates how they lead to very

Table 4.1 *Market-failure versus market-shaping approaches to policy*

Key element	Market-fixing approach	Mission-oriented approach
Justification for the role of government	Market or coordination failures: • Public goods • Negative externalities • Imperfect competition/information	All markets and institutions are co-created by public, private, and third sectors. The role of government is to ensure that markets support public purposes.
Policy appraisal	Ex ante cost-benefit analysis (CBA) – allocative efficiency assuming static general relationships, prices, etc.	Focussed on achieving missions through systemic change – dynamic efficiency (including innovation, spillover effects, and systemic change).
Underlying assumptions	Possible to estimate reliable future value using discounting/monetisation of externalities/risk assessment; the system is characterised by equilibrium behaviour.	Future is uncertain because of potential for novelty and non-marginal change; the system is characterised by complex behaviour.
Policy evaluation	Focus on whether a specific policy solves market failure and whether government failure is avoided (Pareto efficient).	Ongoing and reflexive evaluation of whether the system is moving in the direction of the mission via achievement of intermediate milestones. Focus on the portfolio of policies and interventions and their interaction.
Approach to risk	Highly risk-averse; optimism bias assumed.	Failure is accepted and encouraged as a learning device.

Source: Adopted from Kattel et al., 2018.

different assumptions about policy processes and tools, including measurement and evaluation frameworks.

4.3.3.1 New ways of working and knowing

In theory, a market-shaping approach utilises quite a different epistemology of what and how we know about policymaking processes and their impact. Consequently, policymakers have begun to focus on learning from and incorporating new methods and analytical tools such as strategic design, complexity economics, foresight, and policy labs (Tõnurist et al., 2017; Mergel, Ganapati, & Whitford, 2020; van Buuren et al., 2020). As we have shown above, Vinnova coordinated intensive co-design sessions across Sweden, where up to 400 different stakeholder organisations engaged in "actor workshops". The underlying principle for the extensive engagement process is to generate a different kind of knowledge base for innovation policy action; this knowledge base should be closer to "users" of specific policy outcomes, such as school pupils, teachers, and parents. Latvia has taken a similar, design-driven mission creation approach, focussing on user workshops and policy tools that are open to experimentation (such as using regulatory sandboxes for mission implementation). However, most governments are still developing and implementing missions in predominantly traditional ways.

Public sector capacity is typically defined as the set of skills, capabilities, and resources necessary to perform policy functions, which range from the provision of public services to policy design and implementation (Wu et al., 2018). Contemporary missions assume that the public sector possesses capacities for planning for the long term, for large-scale coordination, and agile experimentation (Kattel, 2022; McLaren & Kattel, 2022). Such complex sets of capacities indicate the changing role of the state in socioeconomic transitions (Borrás & Edler, 2020) and the importance of "reflexive governance" (Rip, 2006). As summarised by Fisher et al. (2018), "mission-oriented R&I initiatives must be reflexive and flexible enough so that they can be reassessed and adapted to new developments and challenges, or possibly ceased". Green transitions and missions in particular have focussed on "transformative capacities" (Borrás et al., 2023) and, more specifically, on what transformation means for street-level bureaucrats (Braams et al., 2023).

However, it must be noted that mission-oriented policies are implemented in the context of specific countries, which means that capacities reflect existing strengths and weaknesses. For instance, in countries with strong innovation systems such as Sweden, policymakers are faced with coordinating a number of policy actors from central and local governments and engaging with current industry leaders. By contrast, countries with less developed innovation systems, such as Latvia, may be able to move more quickly as stakeholder groups are smaller. On the other hand, these countries have fewer actors that can assist with implementation in both the public and private sectors.

While in academic discussions, much attention is given to the capacity for missions, in policy practice, it is rare that public organisations in charge of missions invest in learning and training at an organisational and individual level.

4.3.3.2 New approaches to appraisal and evaluation

As Janssen et al. (2021) argue, "it seems disproportionate to base evaluations entirely on the extent [to which] mission goals are achieved", and accordingly, there is a "need to develop reflexive and formative evaluation approaches for assessing and adapting the ways missions relate to systemic transformation" (Janssen et al., 2021; Sharpe et al., 2021).

For example, in 2020, the UK government added a chapter on how policymakers could evaluate transformational change to its guidelines for policy evaluation and appraisal (known as the *Green book*). The new guidelines for transformational change evaluation draw on the idea of system mapping: "Where significant transformational change is an objective, it is important to map the key systems effects and research the likelihood, magnitude, and location of tipping and leverage points." Similarly, Sharpe et al. call for a "risk opportunity analysis" – "a more general form of cost-benefit analysis appropriate for situations of non-marginal change, heterogeneous actors, and fundamental uncertainty" (Sharpe et al., 2021).

Another important method of appraising and evaluating dynamic efficiencies, and thus also missions, is the application of public value in this context. For instance, building on the market failure concept, Bozeman has developed the idea of public value mapping, which focusses explicitly on public value failures that occur "when core public values are not reflected in social relations, either in the market or public policy" (Bozeman, 2002). While these are essentially negative assessment criteria, public value mapping can also be used to gauge the market-shaping impact of public organisations. A well-known example that could be mentioned here is the UK public broadcaster, the BBC, which uses public value tests to understand and justify its impact on the economy and society (Mazzucato et al., 2020).

However, despite strong theoretical arguments for developing new tools, mission practices still predominantly tend to use existing monitoring and evaluation tools (Rohracher et al., 2022).

4.4 DISCUSSION: "BIG SCIENCE FOR BIG PROBLEMS" VS "SMALL PIECES LOOSELY JOINED"?

The previous generation of mission-oriented policies was often characterised by a "big science for big problems" mindset, which was epitomised in the Moonshot mission (Ergas, 1987). The current iteration of missions is based on a much more nuanced theoretical scaffolding that emphasises the social and intractable nature of challenges that cannot be tackled without wide-ranging experimentation (Mazzucato, 2021). In practice, this seems to mimic a "small pieces loosely joined" landscape, to use a phrase from David Weinberger's eponymous book, *Small Pieces Looseley Joined: A Unified Theory of the Web.* In some cases, such as Vinnova, this seems to be a conscious choice: missions are not practised as large, state-driven projects but rather as society-wide efforts (to transform mobility, for instance) comprising multiple smaller tests and experiments that are aimed at learning what works in different contexts (for instance, in transforming mobility) and that will be scaled up in due course. In this case, the lead agency's role is that of mission orchestration (Danish Design Center, 2021). In practice, this type of role seems to be quite challenging, as typical innovation agencies do not possess strong capacities when it comes to stakeholder networking and ecosystem management.

Academic discourse and policy practice certainly converge around the normative turn, particularly in STI policies. One would struggle to find a European country that is not attempting to implement a version of missions – at least on paper and/or in policy rhetoric – although actual policy practice tends to be more uneven. One could also argue that the COVID-19 pandemic and the war in Ukraine have reinforced the need for normative frames, as evidenced by the massive recovery and green transition investments being made on both sides of the Atlantic. However, the first signs of "mission creep" are starting to emerge, as some public organisations and policymakers are using missions as a label for their existing policies and ideas without actual engagement with the underlying concepts.

On the other hand, the epistemic turn predicted by academic discourse can rarely be found in policy practice. Some interesting attempts have been made (e.g., Vinnova's focus on strategic design), but for the most part, missions are being implemented through business-as-usual practices. As discussed above, missions are mostly being implemented by existing organisations and often by augmenting and upgrading existing interventions, which means a certain degree of incrementalism should be expected. Large-scale experimentation within missions seems to present a particular challenge for public organisations.

In light of this, one might ask whether mission practice has moved beyond the challenge posed by Richard Nelson in the late 1970s: Why can we go to the moon, but not solve the problem of ghettos (Nelson, 1977)? As discussed above, Nelson's contention is that in order to apply the mission approach to socioeconomic challenges, academic and policy practices need to be based on a new epistemological footing. While one can read Nelson's call for action as a way of bringing social science's epistemological foundations closer to those of natural science, it is also possible to make the exact opposite argument: social sciences need epistemological foundations that are aligned with the need for experimentation posed by policy practice. In any case, the mission practice seems to lag severely behind in this regard.

One potential reason for this lag could be the lack of discussion around the impact of various public administration reforms (e.g., the new public management reforms of the 1990s and 2000s) on the capacities of STI agencies (Kattel et al., 2023). The new public management (NPM) reforms that the majority of Western governments implemented in the 1990s and early 2000s – and in some cases are still implementing – focussed on increasing public sector effectiveness by introducing various business practices in public organisations (Hood, 1991; Drechsler, 2005; Lapuente & Van de Walle, 2020). In many instances, such reforms led to an excessive emphasis on short-term savings, narrow performance targets, and overly decentralised public actors (Pollitt & Bouckaert, 2011). As a correction to NPM reforms, public organisations have sought to re-emphasise the idea of the public value that they are providing to citizens, including via STI policies (Bozeman, 2002; Bozeman & Fukumoto, 2019; Mazzucato & Ryan-Collins, 2019), and how the value is co-created with citizens and other stakeholders (Osborne et al., 2016). Particularly responsible research and innovation approaches to STI have sought to incorporate norms and values into innovation discourse, thus bringing societally desirable innovation outcomes, or directionality, into the limelight (Stilgoe et al., 2013; Rip, 2016). However, much less attention has been paid to how NPM and similar reforms have changed the epistemological foundations of STI and other public sector agencies. And while the mission-oriented approach can be seen as a way of building stronger "reflexive" capacities within public organisations (Rip, 2006), this is rarely the case in policy practice. Most missions either rely on incremental change in existing governance structures or are layered on top of existing policies and institutions.

4.5 CONCLUSION: WHERE DO MISSIONS GO FROM HERE?

Mission-oriented policies have taken Europe by storm. However, far from being a uniform policy practice, missions are being implemented in widely different ways, depending on contextual factors in most cases. The economist Hyman Minsky once quipped that there are as many types of capitalism as there are varieties of Heinz pickles – namely, 57. As it turns out, one could say the same thing about mission-oriented innovation policy: there are as many varieties as there are countries designing and implementing it. The widespread application of this approach is vindicating the decision to place missions at the heart of the normative turn in STI and public policymaking at large. This chapter is concerned with the question of whether this normative focus on directionality is accompanied by the epistemic turn that is required – as has been argued by academics since the late 1970s – to enable policy learning from a wide range of experimentation under the umbrella of the mission. While there are some interesting practical examples of this shift in actual missions, most are implemented through existing institutions and policy schemes, with the result that missions are understood through existing epistemological and analytical tools. As such, many missions are constrained to business-as-usual practice and incremental shifts. Indeed, there is a risk that the concept of missions as a shift towards understanding policymaking as a bottom-up experimentation aimed at solving the deeper challenges societies face will be overrun by political events such as inflation, the cost-of-living crisis, and the return of more traditional, mercantilist industrial policies.

Policymaking organisations need to take mission implementation much more seriously, both in terms of investing in capability development and, in particular, in learning how to work in different ways. The shift cannot be expected to be easy, yet unless the capability and analytical gaps are consciously targeted, experimentation within missions will continue to be impeded, and the success and impact of many missions may be much lower than expected as a consequence.

ACKNOWLEDGEMENT

The research for this chapter has been partially supported by the Horizon 2020 project Growinpro, the Organisation for Economic Co-operation and Development (OECD) Observatory of Public Sector Innovation, and Vinnova. We would like to thank Alberto Inda Razo, Diana Morales Aristizabal, and Miran Lorca De Urarte for their help with the research for this chapter, and participants at the Eu-SPRI 2021 conference for their feedback on an earlier version of it.

REFERENCES

Algers, J., & Kattel, R. (2021). *Equinor and Ørsted: How industrial policy shaped the Scandinavian energy giants.* https://www.ucl.ac.uk/bartlett/public-purpose/publications/2021/mar/equinor-and-orsted-how-industrial-policy-shaped-scandinavian-energy-giants

Altman, E. J., & Nagle, F. (2020). Accelerating innovation. *MIT Sloan Management Review, 61*(4), 24–30.

Bason, C. (2017). *Leading public design.* Policy Press.

Bertelsmann Stiftung (2021). *Good practices in mission-oriented innovation strategies and their implementation.* https://www.bertelsmann-stiftung.de/en/publications/publication/did/good-practices-in-mission-oriented-innovation-strategies-and-their-implementation

Björk, A., Gronchi, I., Leppänen, J., Sarkia, K., Taddesse, R., & Vourdaki, A. (2022). *Missions for governance. Unleashing missions beyond policy.* Demos Helsinki. https://demoshelsinki.fi/wp-content/uploads/2022/10/Missions-for-Governance-Demos-Helsinki.pdf

Borrás, S., & Edler, J. (2020). The roles of the state in the governance of socio-technical systems' transformation. *Research Policy, 49*(5). https://doi.org/10.1016/j.respol.2020.103971

Borrás, S., Haakonsson, S., Taudal Poulsen, R., Pallesen, T., Hendriksen, C., Somavilla, L., Kugelberg, S., Larsen, H. & Gerli, F. (2023). *The transformative capacity of public sector organizations in sustainability transitions: A conceptualization. Papers in innovation studies* [Preprint]. https://ideas.repec.org//p/hhs/lucirc/2023_002.html

Bozeman, B. (2002). Public-value failure: When efficient markets may not do. *Public Administration Review, 62*(2), 145–161. https://doi.org/10.1111/0033-3352.00165

Bozeman, B., & Fukumoto, E. (2019). Public values theory: What is missing? *The American Review of Public Administration, 49*(6), 635–648. https://doi.org/10.1177/0275074018814244

Braams, R. B., Wesseling, J. H., Meijer, A. J., & Hekkert, M. P.. (2023). Civil servant tactics for realizing transition tasks understanding the microdynamics of transformative government. *Public Administration.* https://doi.org/10.1111/padm.12933

van Buuren, A., Lewis, J. M., Peters, B. G., & Voorberg, W.. (2019). *Improving public policy and administration: Exploring the potential of design.* https://doi.org/10.1332/030557319X15579230420063

van Buuren, A., Lewis, J. M., Peters, B. G., & Voorberg, W. (2020). Improving public policy and administration: Exploring the potential of design. *Policy & Politics, 48*(1). https://www.ingentaconnect.com/content/tpp/pap/2020/00000048/00000001/art00001

Campiglio, E., Dafermos, Y., Monnin, P., Ryan-Collins, J., Schotten, G., & Tanaka, M. (2018). Climate change challenges for central banks and financial regulators. *Nature Climate Change*, *8*(6), 462–468. https://doi.org/10.1038/s41558-018-0175-0

Chaminade, C., & Edquist, C. (2006). Rationales for public policy intervention from a systems of innovation approach: The case of Vinnova. *CIRCLE working papers* [Preprint], (2006/04). https://citeseerx.ist.psu.edu/viewdoc/download?doi=10.1.1.905.6090&rep=rep1&type=pdf

Cheng, W., Appolloni, A., D'Amato, A., & Zhu, Q. (2018). Green Public Procurement, missing concepts and future trends – A critical review. *Journal of Cleaner Production*, *176*, 770–784. https://doi.org/10.1016/j.jclepro.2017.12.027

Chiappinelli, O., & Zipperer, V. (2017). Using public procurement as a decarbonisation policy: A look at Germany. *DIW Economic Bulletin*, *7*(49), 523–532. https://hdl.handle.net/10419/172946

Conway, R. (2020, December 14). Reflections from the 2020 Mission Oriented Innovation Bootcamp. *UCL Institute for Public Purpose and Innovation Blog*. https://rowan-conway.medium.com/reflections-from-the-2020-mission-oriented-innovation-bootcamp-83d513f9fed0

Daimer, S., Hufnagl, M., & Warnke, P. (2012). Challenge-oriented policy-making and innovation systems theory. In *Innovation system revisited: Experiences from 40 years of Fraunhofer ISI research* (pp. 217–234). Fraunhofer Institut für System- und Innovationsforschung -ISI-, Karlsruhe.

Danish Design Center (2021). *A design-driven approach to launching and driving missions*. https://ddc.dk/tools/missions-playbook-a-design-driven-approach-to-launching-and-driving-missions/

Deleidi, M., Mazzucato, M., & Semieniuk, G. (2020). Neither crowding in nor out: Public direct investment mobilising private investment into renewable electricity projects. *Energy Policy*, *140*, 111195. https://doi.org/10.1016/j.enpol.2019.111195

Dosi, G., Lamperti, F., Mazzucato, M., Napoletano, M., & Roventini, A. (2021). *Mission-oriented policies and the "Entrepreneurial State" at work: An agent-based exploration*. https://www.ucl.ac.uk/bartlett/public-purpose/publications/2021/jun/mission-oriented-policies-and-entrepreneurial-state-work-agent-based

Drechsler, W. (2004). Natural vs. social sciences: On understanding in economics. In *Globalization, economic development and inequality: An alternative perspective* (p. 71). Edward Elgar Publishing.

Drechsler, W. (2005). The rise and demise of the new public management. *Post-Autistic Economics Review*, *33*(14), 17–28. https://www.paecon.net/PAEReview/issue33/Drechsler33.htm

Edler, J., Cunningham, P., Gök, A., & Shapira, P. (2016). *Handbook of innovation policy impact*. Edward Elgar Publishing.

Edquist, C., & Zabala-Iturriagagoitia, J. M. (2020) Functional procurement for innovation, welfare and the environment: A mission-oriented approach. *Papers in Innovation Studies*, *2020*(1). Lund University, CIRCLE – Center for Innovation, Research and Competences in the Learning Economy. https://ideas.repec.org/p/hhs/lucirc/2020_001.html

Ergas, H. (1987). Does technology policy matter? In B. Guile & H. Brooks (Eds.), *Technology and global industry*. National Academic Press.

European Commission (2023). *EU Missions two years on: Assessment of progress and way forward*. European Commission. https://research-and-innovation.ec.europa.eu/system/files/2023–07/com_2023_457_1_en_act_part1.pdf

Fisher, R., Chicot, J., Domini, A., Polt, W., Türk, A., Unger, M., Kuittinen, H., Arrilucea, E., van der Zee, F., Goetheer, A., Lehenkari, J., Pelkonen, A., & Kristensen, F. S. (2018). *Mission-Oriented Research and Innovation Inventory and characterisation of initiatives (Final Report)*. https://doi.org/10.2777/697082

Foray, D., Mowery, D., & Nelson, R. (2012). Public R&D and social challenges: What lessons from mission R&D programs? *Research Policy*, *41*(10), 1697–1702.

Gasperin, S., Dosi, G., Mazzucato, M., & Roventini, A. (2021). *Strategic missions and policy opportunities for State-owned enterprises*. https://www.ucl.ac.uk/bartlett/public-purpose/publications/2021/mar/strategic-missions-and-policy-opportunities-state-owned-enterprises

Grillitsch, M., Hansen, T., Coenen, L., Miörner, J., & Moodysson, J. (2018). Innovation policy for system-wide transformation: The case of strategic innovation programmes (SIPs) in Sweden. *Research Policy*, *48*. https://doi.org/10.1016/j.respol.2018.10.004

Hall, P., & Soskice, D. (2001). *Varieties of capitalism*. Oxford University Press.

Hekkert, M. P., Janssen, M. J., Wesseling, J. H., & Negro, S. O. (2020). Mission-oriented innovation systems. *Environmental Innovation and Societal Transitions*, *34*, 76–79. https://doi.org/10.1016/j.eist.2019.11.011

Hill, D. (2015). *Dark matter and Trojan Horses: A strategic design vocabulary*. Strelka Press.

Hill, D. (2020, May 1). *Realizing mission-oriented innovations in a fast-moving world*. International Science Council.https://council.science/current/blog/realizing-mission-oriented-innovations-in-a-fast-moving-world/

Hill, D. (2022). *Designing mission. Mission-oriented innovation in Sweden – A practice guide by Vinnova. Stockholm*. https://www.vinnova.se/en/publikationer/mission-oriented-innovation---a-handbook-from-vinnova/

Hood, C. (1991). A public management for all seasons? *Public Administration*, *69*(1), 3–19. https://doi.org/10.1111/j.1467-9299.1991.tb00779.x

Janssen, M. (2020). *Post-commencement analysis of the Dutch "Mission-oriented Topsector and Innovation Policy" strategy*. Utrecht University. https://www.uu.nl/sites/default/files/Post-commencement%20analysis%20of%20the%20Dutch%20Mission-oriented%20Topsector%20and%20Innovation%20Policy.pdf

Janssen, M. J., Torrens, J., Wesseling, J. H., & Wanzenböck, I. (2021). Promises and premises of mission-oriented innovation policy – A reflection and ways forward. *Science and Public Policy*, *48*(3), 438–444.

Jourdan, S., & Kalinowski, W. (2019). *Aligning monetary policy with EU's climate targets*. Veblen Institute for Economic Reforms and Positive Money Europe.

Karo, E. (2011). The evolution of innovation policy governance systems and policy capacities in the Baltic States. *Journal of Baltic Studies*, *42*(4), 511–536.

Kattel, R. (2022). *Dynamic capabilities of the public sector: Towards a new synthesis* [Working Paper]. UCL Institute for Innovation and Public Purpose. https://www.ucl.ac.uk/bartlett/public-purpose/publications/2022/mar/dynamic capabilities public sector towards new synthesis

Kattel, R., Haverkamp, K., Kalema, N., & Kofler, J. (2023). *New public management and innovation policy: A systematic literature review*. UCL Institute for Innovation and Public Purpose [Preprint]. https://www.ucl.ac.uk/bartlett/public-purpose/publications/2023/mar/new-public-management-and-innovation-policy-systematic-literature-review

Kattel, R., & Mazzucato, M. (2018). Mission-oriented innovation policy and dynamic capabilities in the public sector. *Industrial and Corporate Change*, *27*(5), 787–801. https://doi.org/10.1093/icc/dty032

Kattel, R., Mazzucato, M., Ryan-Collins, J., & Sharpe, S. (2018). *The economics of change: Policy and appraisal for missions, market shaping and public purpose*. UCL Institute for Innovation and Public Purpose. https://www.ucl.ac.uk/bartlett/public-purpose/publications/2018/jul/economics-change-policy-and-appraisal-missions-market-shaping-and-public

Kattel, R., & Takala, V. (2021). *Dynamic capabilities in the public sector: The case of the UK's Government Digital Service*. UCL Institute for Innovation and Public Purpose [Preprint]. https://www.ucl.ac.uk/bartlett/public-purpose/publications/2021/jan/dynamic-capabilities-public-sector-case-uks-government-digital-service

Kedward, K., Ryan-Collins, J., & Buller, A. (2021). *Quantitative easing and nature loss: Exploring nature-related financial risks and impacts in the European Central Bank's corporate bond portfolio*. https://www.ucl.ac.uk/bartlett/public-purpose/publications/2021/jul/quantitative-easing-and-nature-loss

Lapuente, V., & Van de Walle, S. (2020). The effects of new public management on the quality of public services. *Governance*. https://doi.org/10.1111/gove.12502

Larrue, P. (2021). *The design and implementation of mission-oriented innovation policies: A new systemic policy approach to address societal challenges*. https://doi.org/10.1787/3f6c76a4-en

Lember, V., Kattel, R., & Kalvet, T. (2015). Quo vadis public procurement of innovation? *Innovation: The European Journal of Social Science Research*, *28*(3), 403–421.

Levin, K., Cashore, B., Bernstein, S., & Auld, G. (2012). Overcoming the tragedy of super wicked problems: Constraining our future selves to ameliorate global climate change. *Policy Sciences, 45*(2), 123–152. https://doi.org/10.1007/s11077–012–9151–0

Matikainen, S., Campiglio, E., & Zenghelis, D. (2017). *The climate impact of quantitative easing*. https://doi.org/10.13140/RG.2.2.24108.05763

Maykut, P., & Morehouse, R. (2002). *Beginning qualitative research: A philosophical and practical guide*. Routledge.

Mazzucato, M. (2016). From market fixing to market-creating: A new framework for innovation policy. *Industry and Innovation*, *23*(2), 140–156. https://doi.org/10.1080/13662716.2016.1146124

Mazzucato, M. (2017). *Mission-oriented innovation policy: Challenges and opportunities, UCL Institute for Innovation and Public Purpose*. https://www.ucl.ac.uk/bartlett/public-purpose/publications/2017/sep/mission-oriented-innovation-policy-challenges-and-opportunities

Mazzucato, M. (2018). *Mission-oriented research & innovation in the European Union: A problem-solving approach to fuel innovation-led growth*. Publications Office of the European Union. http://op.europa.eu/en/publication-detail/-/publication/5b2811d1–16be-11e8–9253–01aa75ed71a1/language-en

Mazzucato, M. (2019). *Governing missions in the European Union*. European Commission. https://research-and-innovation.ec.europa.eu/knowledge-publications-tools-and-data/publications/all-publications/governing-missions-european-union_en

Mazzucato, M. (2020). *Mission-oriented public procurement: International examples*. https://www.ucl.ac.uk/bartlett/public-purpose/publications/2020/dec/mission-oriented-public-procurement-international-examples

Mazzucato, M. (2021). *Mission economy*. Allen Lane. https://www.penguin.co.uk/books/315191/mission-economy/9780241419731

Mazzucato, M., Carreras, M., & Mikheeva, O. (2023). *Steering economic recovery in Europe. Lessons for governing the Recovery and Resilience Facility*. https://policycommons.net/artifacts/3447040/steering-economic-recovery-in-europe/4247208/

Mazzucato, M., Conway, R., Mazzoli, E. M., Knoll, E., & Albala, S. (2020). *Creating and measuring dynamic public value at the BBC*. UCL Institute for Innovation and Public Purpose [Preprint]. https://www.ucl.ac.uk/bartlett/public-purpose/publications/2020/dec/creating-and-measuring-dynamic-public-value-bbc

Mazzucato, M., Kattel, R., Begovic, M., & Quaggiotto, G. (2021). *COVID-19 and the need for dynamic state capabilities: an international comparison*. United Nations Development Programme and UCL Institute for Innovation and Public Purpose. Available at:https://www.ucl.ac.uk/bartlett/public-purpose/sites/public-purpose/files/54241_-_undp_wp-covid19_state_ressilience-v51.pdf

Mazzucato, M., Kattel, R., & Ryan-Collins, J. (2020). Challenge-driven innovation policy: Towards a new policy toolkit. *Journal of Industry, Competition and Trade*, *20*(2), 421–437. https://doi.org/10.1007/s10842-019-00329-w

Mazzucato, M., Kattel, R., Ryan-Collins, J., & Sharpe, S. (2020). *Alternative policy evaluation frameworks and tools*. https://www.ucl.ac.uk/bartlett/public-purpose/publications/2020/nov/alternative-policy-evaluation-frameworks-and-tools

Mazzucato, M., & Macfarlane, L. (2019). *A mission-oriented framework for the Scottish National Investment Bank*. https://www.ucl.ac.uk/bartlett/public-purpose/publications/2019/mar/mission-oriented-framework-scottish-national-investment-bank

Mazzucato, M., & Mikheeva, O. (2020). *The EIB and the new EU missions framework*. https://www.ucl.ac.uk/bartlett/public-purpose/publications/2020/nov/eib-and-new-eu-missions-framework

Mazzucato, M., & Ryan-Collins, J. (2019). *Putting value creation back into "public value": From market-fixing to market-shaping*. https://www.ucl.ac.uk/bartlett/public-purpose/publications/2019/jun/putting-value-creation-back-public-value-market-fixing-market-shaping

McLaren, J., & Kattel, R. (2022). *Policy capacities for transformative innovation policy: A case study of UK Research and Innovation* [Working Paper]. UCL Institute for Innovation and Public Purpose. https://www.ucl.ac.uk/bartlett/public-purpose/publications/2022/mar/policy-capacities-transformative-innovation-policy-case-study-uk-research

Mergel, I., Ganapati, S., & Whitford, A. B. (2020). Agile: A new way of governing. *Public Administration Review*, *81*(1), 161–165. https://doi.org/10.1111/puar.13202

Monge, F., Barns, S., Kattel, R and Bria, F. (2022). A new data deal – The case of Barcelona. UCL Institute for Innovation and Public Purpose. https://www.ucl.ac.uk/bartlett/public-purpose/publications/2022/feb/new-data-deal-case-barcelona

Mowery, D. C. (2012). Defense-related R&D as a model for "Grand Challenges" technology policies. *Research Policy*, *41*(10), 1703–1715. https://doi.org/10.1016/j.respol.2012.03.027

Nelson, R. R. (1977). *The moon and the ghetto: An essay on public policy analysis* (1st ed.). W. W. Norton & Company.
OECD (2015). *Systems innovation: Synthesis report.* OECD. http://www.pte.pl/pliki/2/1/OECD%20System.pdf
Osborne, S. P., Radnor, Z., & Strokosch, K. (2016). Co-production and the co-creation of value in public services: A suitable case for treatment? *Public Management Review, 18*(5), 639–653. https://doi.org/10.1080/14719037.2015.1111927
Peters, B. G. (2017). Full article: What is so wicked about wicked problems? A conceptual analysis and a research program. *Policy and Society, 36*(3), 385–396. https://doi.org/10.1080/14494035.2017.1361633
Peters, B. G. (2021). *Administrative traditions: Understanding the roots of contemporary administrative behavior.* Oxford University Press. https://doi.org/10.1093/oso/9780198297253.001.0001
Peters, B. G., & Tarpey, M. (2019). Are wicked problems really so wicked? Perceptions of policy problems. *Policy and Society, 38*(2), 218–236. https://doi.org/10.1080/14494035.2019.1626595
Pollitt, C., & Bouckaert, G. (2011). *Public management reform: A comparative analysis – new public management, governance, and the neo-Weberian state.* Oxford University Press.
Rip, A. (2006). A co-evolutionary approach to reflexive governance – and its ironies. In J. P. Voß, D. Bauknecht, & R. Kemp (eds), *Reflexive governance for sustainable development.* Edward Elgar Publishing, pp. 82–100.
Rip, A. (2016). The clothes of the emperor. An essay on RRI in and around Brussels. *Journal of Responsible Innovation, 3*(3), 290–304. https://doi.org/10.1080/23299460.2016.1255701
Rohracher, H., Coenen, L., & Kordas, O. (2022). Mission incomplete: Layered practices of monitoring and evaluation in Swedish transformative innovation policy. *Science and Public Policy*, scac071. https://doi.org/10.1093/scipol/scac071
Sharpe, S., Mercure, J. F., Vinuales, J., Ives, M., Pollitt, H., Knobloch, F., Nijsse, F. J. M. M., & Grubb, M. (2021). *Deciding how to decide: Risk-opportunity analysis as a generalisation of cost-benefit analysis.* https://www.ucl.ac.uk/bartlett/public-purpose/publications/2021/jan/deciding-how-decide-risk-opportunity-analysis-generalisation-cost-benefit
Sheppard, D. (2020, February 4). Can Orsted be the first green energy supermajor? *Financial Times.* https://www.ft.com/content/74b377c8-4435-11ea-abea-0c7a29cd66fe
Soete, L., Foray, D., Licht, G., Llerena, P., Martinez-Ros, E., Mazzucato, M., Renda, A., Serger, S. S., & Weresa, M. A. (2017). *Towards a mission-oriented research and innovation policy in the European Union; An ESIR memorandum.* https://doi.org/10.2777/715942
Stilgoe, J., Owen, R., & Macnaghten, P. (2013). Developing a framework for responsible innovation. *Research Policy, 42*(9), 1568–1580. https://doi.org/10.1016/j.respol.2013.05.008
Stirling, A. (2009). *Direction, distribution and diversity! Pluralising progress in innovation, sustainability and development.* STEPS Centre. https://opendocs.ids.ac.uk/opendocs/handle/20.500.12413/2458
Testa, F., Iraldo, F., Frey, M., & Daddi, T. (2012). What factors influence the uptake of GPP (green public procurement) practices? New evidence from an Italian survey. *Ecological Economics, 82*, 88–96. https://doi.org/10.1016/j.ecolecon.2012.07.011
Tödtling, F., Trippl, M., & Desch, V. (2021). New directions for RIS studies and policies in the face of grand societal challenges. *Vienna: Geography of Innovation and Sustainability Transitions – GEIST Working Paper series.* https://epub.wu.ac.at/8125/
Tõnurist, P., Kattel, R., & Lember, V. (2017). Innovation labs in the public sector: What they are and what they do? *Public Management Review, 19*(10), 1455–1479. https://doi.org/10.1080/14719037.2017.1287939
Voldsgaard, A., & Rüdiger, M. (2021). *Innovative enterprise, industrial ecosystems and sustainable transition: The case of transforming DONG energy to Ørsted.* https://doi.org/10.1007/978-3-030-72579-2_160
Wanzenböck, I., Wesseling, J. H., Frenken, K., Hekkert, M. P., & Weber, K. M. (2020). A Framework for Mission-oriented Innovation Policy: Alternative Pathways through the Problem-solution Space. *Science & Public Policy*, 47: 474–89.

Wittmann, F., Hufnagl, M., Lindner, R., Roth, F., & Edler, J. (2021). Governing varieties of mission-oriented innovation policies: A new typology. *Science and Public Policy*, scab044. https://doi.org/10.1093/scipol/scab044

Wu, X., Howlett, M., & Ramesh, M. (Eds.) (2018). *Policy capacity and governance: Assessing governmental competences and capabilities in theory and practice*. Palgrave Macmillan.

5. The formative evaluation of open-ended transformative missions: the Vinnova experience in the food system

Caetano C. R. Penna, Johan Schot, Diana Velasco and Jordi Molas-Gallart

5.1 INTRODUCTION

In the past decade, the focus of innovation policy has moved from increasing the rate of technological change to shifting its direction and, ultimately, transforming socio-technical systems – such as those that provide energy, mobility, and food – in a sustainable direction. Scientific progress and innovation are no longer regarded as ends in themselves, but as means towards promoting social and environmental welfare. In this context, policymakers began to embrace the idea of innovation *missions* to orient policies, inspired by the major projects of the twentieth century that sought to accomplish formidable technological feats – namely, the Manhattan Project (development of the atom bomb) and the Apollo Programme (putting a human on the moon and bringing them back safely). However, contemporary missions to transform socio-technical systems are very different from the technological missions of the past (Foray et al., 2012; Soete & Arundel, 1993). This is due to many factors, namely because previous missions were insulated from society, with the result that the economic diffusion of their results was not a core objective; while for contemporary missions, the diffusion of results and solutions to sustainability challenges is at their core. For example, it is not enough that hydrogen fuel cell cars are technically feasible; they need to substitute for conventional internal combustion engine cars in mass consumer markets and therefore be economically viable and socially acceptable.

Recent contributions have proposed new ways to conceptualise different types of mission-oriented policies (MOP) by identifying successive generations of such policies, which each require distinctive institutional capacities and capabilities (Mazzucato & Kattel, 2020); discussing how missions are affected due to convergent or divergent views (framings) on problems and solutions (Wanzenböck et al., 2020); or analysing the implications of a mission's focus (on scientific-technological progress or behavioural change) and its requirements for governance and coordination (Wittmann et al., 2020). While the resulting typologies differ, they all include a specific category of contemporary missions. Due to divergent framings, a combination of behavioural and scientific-technological ambitions, and uncertainty about possible solutions, these missions require new types of institutional capacities and capabilities and entail highly complex policy design, implementation, and governance processes. They can be regarded as *open-ended transformative mission-oriented policies*, for which conceiving end states and their measurements are elusive tasks. Instead, the emphasis in this kind of MOP is on using emergent processes, experimentation, and learning as a way to develop multiple solutions and open up fruitful directions for the transformative change of socio-technical systems.

This chapter focusses on the challenge of designing, implementing, and monitoring open-ended transformative mission-oriented policies. It presents a case study of the Swedish innovation agency Vinnova and its approach to promoting new transformation directions for the food socio-technical system in Sweden through the establishment of missions. In the next section, we briefly characterise this type of mission to propose that its design and implementation benefit from a flexible formative monitoring, evaluation, and learning (MEL) approach. In Section 5.3, we present the MEL approach developed by the *Transformative Innovation Policy Consortium* (TIPC), of which Vinnova is a part. This approach, based on socio-technical transitions theory and anchored on the notion of a transformative theory of change, was applied (with Vinnova) as a tool to aid the agency's mission approach. Section 5.4 develops the case study, looking at how Vinnova used TIPC's MEL approach to complement their own way of designing missions. The results reveal aspects critical to designing, implementing, and monitoring an open-ended MOP and show that this specific type of transformative mission is more useful as a flexible policy framework to open up directionalities and promote experimentation and learning. Therefore, they are better understood not as end targets that symbolise a transformed socio-technical system but as means (possible pathways) to transformation. Section 5.5 elaborates on this conclusion, drawing implications for policymaking and discussing areas for future research.

5.2 A CHARACTERISATION OF OPEN-ENDED TRANSFORMATIVE MISSION-ORIENTED POLICIES

In the 2010s, an academic debate on the appropriateness of mission-oriented innovation policies to address socio-environmental challenges commenced. Mazzucato (2016, p. 141) put forth the initial tenets for a renewed mission-oriented approach to address societal challenges:

> Missions imply setting directions of change... Missions should be broad enough to catalyse many different sectors (the man on [the] moon mission required a dozen sectors to engage) but concrete enough to translate into specific problems to solve, so that progress toward the mission can be evaluated on a continual basis.

She recognises, however, the difficulty in translating societal challenges into concrete missions, which should therefore be done by the State "together with other stakeholders" (2016, p. 153), to draw on their distinctive capacities and capabilities. Thus, Mazzucato establishes four principles for diagnosing challenges. Missions should: (a) "be well defined"; (b) "not comprise a single R&D or innovation project, but a portfolio of such projects"; (c) "result in investment across different sectors and involve different types of actors"; and (d) promote "joined up policy-making" (2018a, p. 805). Such initial thinking led to the milestone publication of the *Mazzucato report*, which laid the groundwork for the current European Commission's mission-oriented policy impetus. Its definition of what constitutes appropriate missions to guide innovation policy is based on five criteria (Mazzucato, 2018b, pp. 14–15):

1. Being bold and inspirational, with wide societal relevance

2. Setting a clear direction, which means setting targeted, measurable, and time-bound goals
3. Being ambitious but with realistic research and innovation actions
4. Fostering cross-disciplinary, cross-sectoral, and cross-actor innovation
5. Requiring multiple, bottom-up solutions

Therefore, despite acknowledging the difficulty of rendering socio-environmental challenges into concrete missions and calling for stakeholder participation in the process, the *Mazzucato report*, inspired by the "old" mission-oriented approach, still called for the establishment of unequivocal "targeted, measurable and time-bound" objectives. Such an approach can be fruitful for certain challenges that have a widely accepted and (to a large extent) indisputable framing – meaning a clear identification of the causes, consequences, and solutions to the challenge. Examples of such a challenge are removing plastics from the ocean and waterways (Miedzinski et al., 2019) or increasing electricity generation and consumption from renewable sources[1] such as wind turbines (Wanzenböck et al., 2020). However, this approach may be of problematic application in the case of open-ended missions for which multiple framings are possible and solutions are not clearly identified, rendering multiple directions of change (emergent solutions) as not only possible but necessary for unpredictable, hard-to-control, endless, and evolving socio-technical systems.

What are the distinctive features of these types of missions? Our characterisation of open-ended transformative mission-oriented policies draws on the seminal work of Soete and Arundel (1993) as a starting point. In the third chapter of the *Maastricht memorandum*, the authors examine "the use of innovation, technology and other complementary policies to support the goal of environmentally sustainable development", an issue they select for its long-term orientation, complexity ("the development and diffusion of innovations to help meet this goal, cannot be reached without a wide range of policies, the involvement of many different economic actors and changes to existing economic, social and cultural institutions"), and the necessity of involving public and private agents (pp. 47–48). They propose a characterisation of "old" mission-oriented policies (see the left-hand column in Table 5.1): lack of support for diffusion of results, strictly and clearly defined technical goals, top-down technocratic governance, centralised state-led control, and limited participation of a small number of private agents who can develop radical innovations. In contrast, Soete and Arundel (1993) argue, while environmental missions emerge as a new focus for STI policy, this would only "superficially" mean "a return to the emphasis in the 1950s and 1960s on public goals that were met through mission-oriented projects", because "there is a fundamental difference between older mission-oriented projects [...] and new projects to support environmentally sustainable development" (1993, p. 50).

In their view, the complexity of and participatory requirements for meeting environmental missions dictate a *systemic* approach to innovation development and policy, moving away from the linear model that informed the original mission-oriented programmes focussed on well-defined governmental targets. Such a systemic approach would be more suited to the characteristics of socio-environmental challenges (see the right-hand column in Table 5.1): the need to actively encourage the diffusion of results, which in turn, implies that economic

[1] Although which sources should be considered renewable (or clean) remains disputed.

Table 5.1 Old versus new missions

Old: Defence, nuclear and aerospace	New: Socio-environmental challenges
Diffusion of the results outside of the core of participants is of minor importance or actively discouraged.	Diffusion of the results is a central goal and is actively encouraged.
The mission is defined in terms of the number of technical achievements, with little regard to their economic feasibility.	The mission is defined in terms of economically feasible technical solutions to specific societal problems.
The goals and the direction of technological development are defined in advance by a small group of experts.	The direction of technical change is influenced by a wide range of actors, including government, private firms, and consumer groups.
Centralised control within a government administration.	Decentralised control with many agents involved.
Participation is limited to a small group of firms due to the emphasis on radical technologies.	Emphasis on the development of both radical and incremental innovations in order to permit many firms to participate.

Source: Version by Mazzucato & Penna, 2016 of Table 5 in Soete & Arundel, 1993, p. 51.

aspects are of paramount importance for mission accomplishment; participatory governance; decentralisation of control; and wide participation of many agents who develop not only radical but also incremental solutions to the challenge.

From Soete and Arundel's initial characterisation, it follows that to solve socio-environmental challenges, it would not be enough for a mission-oriented policy to lead to the demonstration of the technical feasibility of a radical solution – for instance, the production of energy with solar panels. The solution will only be effective if the policy is accompanied by incentives to diffuse the technology so that it substitutes for the old technologies, for example, solar energy production replacing fossil energy production. An additional hurdle on top of the technical feasibility and economic viability is that new technological solutions also need to be *socially acceptable* (Schot, 1992). For example, solar energy farms need to overcome the "NIMBY" ("not in my backyard") resistance. We also note that other types of innovations, not only radical and incremental technologies but also behavioural and social innovations, may represent practical solutions to new missions. The starting and end points in "new" missions are different from the "old" missions. The mission is set by the government's interest in responding to external and internal pressures that do not necessarily respond to societal needs. In this case, diffusion and social acceptability are not targets, although those can be developed in the process.

This characterisation of socio-environmental challenges assumes that solutions for a given socio-environmental problem have already been understood or conceived and, furthermore, that stakeholders hold "convergent views" (Wanzenböck et al., 2020) on the appropriate solution(s). However, this assumption is problematic for certain challenges such as climate change or obesity and for challenges that are in the infancy of their "issue life-cycle" (Geels & Penna, 2015). This criticism was raised by Steward (2012), who proposed a "transformative innovation policy model" as an alternative to the mission-oriented policy model. Recognising that the "arrival of the idea of [socio-technical] transition onto the policy stage was [...] accompanied by a renewed interest in a more mission oriented state led innovation policy" (2012, p. 333), he contends that the MOP model is inappropriate for "broad societal problems with no straightforward technological solution", so that "solutions would require a diversity of technology and social change and [cross] the production and consumption divide" (2012, p. 336). He thus calls for a new *transformative* approach to innovation policy.

Schot and Steinmueller (2018) took up the proposals by Steward (2012) and provided a detailed conceptualisation of a transformative frame for innovation policy. In contrast to the two previous frames (innovation for growth and diffusion-oriented/systemic innovation), Frame 3 directly addresses societal goals, giving innovation policy a primary focus on the transformation of socio-technical systems, as conceptualised in the literature of sustainability transitions. Therefore, transformative innovation policy seeks to address the "transformational failures" as identified by Weber and Rohracher (2012). By tackling societal challenges first and foremost, Frame 3 assumes that explicit attention dedicated to social and environmental welfare in the innovation development process will both increase productivity and decrease inequality.

Diercks, Larsen and Steward (2019) made a further step in the definition of a transformative approach to innovation policy, arguing that the new mission-oriented policy model represented only a "narrow" type of transformative innovation policy (see the bottom-right quadrant in Figure 5.1),[2] due to its underlying understanding of the innovation process as narrowly focussed on "triple-helix" agents and supply-side instruments. In contrast, a "broad" transformative innovation policy would have a wider understanding of the innovation process (see the top-right quadrant in Figure 5.1), whereby a diverse group of public and civil society bodies would play an active role and demand-side tools would be deployed on top of supply-side ones. The authors define the broad type of transformative innovation policy in terms of three aspects: "(1) targeting policy domains beyond economic and industrial policy; (2) including policy objectives dealing with a broad range of societal challenges; and (3) a policy logic that challenges a strong pro-[technological-]innovation bias." (Diercks et al., 2019, p. 890). Crucially for our discussion, the authors emphasise that in transforming socio-technical systems, the ex ante definition of solutions to be pursued is challenging, as they "require more than achieving specific (mission-oriented) scientific, technological breakthroughs" (2019, p. 891). In light of this, not all societal challenges would be amenable to a mission-oriented policy approach with clearly "targeted, measurable and time-bound" goals.

The issue of defining missions in terms of problems and solutions for a given challenge is explored by Wanzenböck et al. (2020), who propose a two-dimensional typology to contextualise different types of missions. This contextualisation is based on whether different views (framings) of problems and solutions are convergent (uncontested, well-defined, and informed) or divergent (contested, complex, and uncertain). Divergent views on solutions and problems would result in *disorientation* over mission-definition (the authors suggest it to be the case of "sustainable agriculture"); divergent views on solutions and convergent views on problems would result in *problems in search of a solution* (e.g., "obesity"); convergent views on solutions and problems would result in *alignment* (e.g., "smoking ban, CCTV, and wind

[2] It may be argued that the old MOP model sits somewhere in the bottom-left quadrant, while the diffusion-oriented/systemic policy model sits in the top-left quadrant.

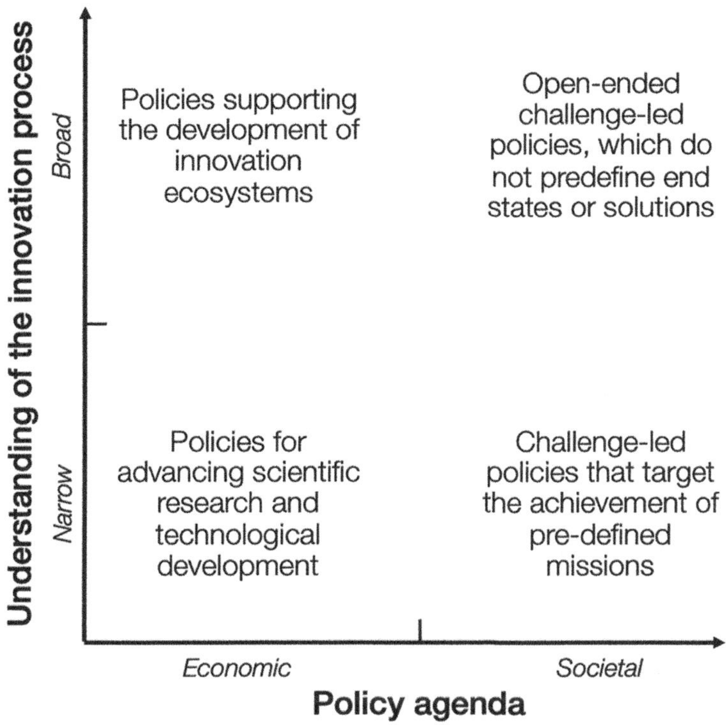

Source: Based on Diercks et al., 2019, p. 883.

Figure 5.1 Diercks et al.'s (2019) typology of innovation policies

energy"); and, finally, convergent views on solutions and divergent views on problems would result in *solutions in search of a problem* (e.g., "self-driving cars").

Both typologies (Diercks et al., 2019; Wanzenböck et al., 2020) underscore our argument that, in many instances, it is difficult to clearly diagnose a societal challenge and translate it into unequivocal policy missions, which is a key element in Mazzucato's (2018a, 2018b) widespread approach. If contested or complex societal challenges are to be addressed through a mission-oriented approach, there is a need to develop a new approach to its design and implementation – an approach that is open-ended in terms of its solutions and transformative in its framing. Instead of summative or stage-gate evaluation, the lack of clear end-states, and therefore of ex ante and definite impact indicators, means that transformative open-ended mission-oriented policies require formative evaluation methods that benefit from experimentation and support learning (which we will discuss in the next section).

We do not claim that "missions" should be abandoned as a model for science, technology, and innovation policies. On the contrary, we agree they establish a useful "boundary object" to which different stakeholders can relate and thus be attracted to the policy process, each contributing with their perspective, resources, and capabilities in different arenas of the process (Janssen et al., in this volume). Furthermore, the notion of "societal" or "transformative" missions emphasises the importance of shifting innovation policy towards Frame 3 thinking, whereby societal challenges are confronted upfront. To transform a socio-technical system or multiple systems, a *transformative* mission-oriented policy combines elements from systemic, behavioural, and technological missions: state-led stewardship and bottom-up experimentation; radical innovation and incremental innovation; attention to technical viability, economic feasibility, and social acceptability; and embedding of the process and outcomes within society.

Box 5.1 sums up our characterisation of open-ended transformative innovation policy. These characteristics and requirements mean that designing and implementing transformative socio-technical missions can benefit from socio-technical transitions theory, which offers an understanding of how socio-technical systems are transformed and move from one configuration to another.

BOX 5.1 COMPARISON BETWEEN THE NEW MISSION-ORIENTED POLICY APPROACH (MOP) AND A TRANSFORMATIVE APPROACH TO OPEN-ENDED MISSIONS

The new mission-oriented policy (MOP) approach has an economic growth bias as its origin and focusses mostly on technological innovation. A *transformative* MOP has a dual societal and environmental focus, is agnostic about growth (which could be a consequence of addressing societal challenges), and should encompass broader types of innovation (beyond technologies).

The new MOP approach still targets products and process innovations (micro level) interested in the notion of technical change and industrial change at a mezzo-cross-sectoral level. In a *transformative* MOP, the levels of transformation move from the micro and sector-specific level and from a focus on radical and incremental product and process innovation to the socio-technical system change level. A *transformative* MOP thus adopts a multidimensional approach to transformation, understood as a "socio-technical system transition". From a transitions theory perspective, while MOP tends to emphasise the creation of new innovative alternatives at the socio-technical "regime" level, a *transformative* MOP promotes the creation of the new (niches) as well as the destabilisation of the old (regimes).

The new MOP model is conceived as a state-led, top-down initiative, with few concrete examples of initiatives that include relevant participatory processes (despite early and contemporary calls for inclusion, bottom-up approaches, etc.). For a MOP to be *transformative*, participatory processes must come to the fore, promoting *open-ended*, emergent transformations.

The MOP model implies that the change process starts with the "translation" of the societal challenge into missions. However, there are challenges that do not have clear framings or straightforward and unique solutions, so this translation process is all but straightforward. This is especially true when the transformation entails changes in multiple socio-technical systems. For such *transformative* processes surmounted by uncertainty and contestation, the transformative open-ended mission-oriented policy process modulates and coordinates ongoing change initiatives by supporting experimentation across socio-technical systems and embedding learning mechanisms in the process.

Therefore, such policies stretch ongoing transformation initiatives to overcome transformational failures.

The new MOP model proposes predefined and specific end-states ("mission targets"), establishing a single direction for change. A *transformative open-ended* MOP works with an open-ended search process that leads to a diversity of end-states and directionalities. Transformative open-ended MOPs acknowledge that uncertainty is part of the transformation process, which will (and may) not be reduced through rigid policy recipes.

Note: * The concepts of socio-technical regimes and niches will be discussed in Section 5.3.
Source: Authors' elaboration.

5.3 A FORMATIVE EVALUATION APPROACH TO OPEN-ENDED TRANSFORMATIVE MISSIONS

5.3.1 Formative Evaluation of Transformative Innovation Policy

Summative evaluation approaches focus on the results of an intervention, often to provide external accountability. They seek to evaluate "the appropriateness, relevance, efficiency and impact of an intervention in order to provide accountability to the government, taxpayers and society more broadly" (Arnold et al., 2018, p. 39). Yet, the techniques used to deliver on these goals (for instance, experimental and quasi-experimental techniques based on bibliometric or economic data) tell us little about the mechanisms that turn an intervention into a success or failure and are of limited use for learning and adaptation. These methods are of problematic application to Frame 3 innovation policies given the complex nature of the processes they try to trigger and the lengthy time horizon of transformation processes, which tend to appear long after the initial policy intervention (Molas-Gallart et al., 2021).

In contrast to the decisive character of summative approaches, *formative* evaluations focus on monitoring developments, instituting learning, and improving practices of policy implementation and their outcome (Edler et al., 2012). Formative approaches are more appropriate for evaluating transformative innovation policy, due to the emphasis on learning, the generation of strategic capacity and capabilities for policy design and implementation, and the need to account for contextual factors, wider impacts, and non-technological change (Amanatidou et al., 2014). Based on this understanding, Molas-Gallart et al. (2021) propose a formative approach to evaluating transformative policy experiments. They draw on socio-technical transitions literature to identify principles of transformative change[3] that may guide the evaluation of transformative innovation policy. This in turn results in six guidelines:

[3] (a) consideration of directionality, (b) accounting for the inclusion of many stakeholders, (c) explicit focus on societal challenges, (d) aiming for systemic change, (e) importance of deep learning and reflexivity, and (f) embracing conflict.

1. Adopt a formative approach to evaluation
2. Integrate evaluation with policy design and implementation
3. Ensure the evaluation process is inclusive and participatory
4. Use a mix of methods and techniques
5. Use a nested approach to assess multi-level transformative innovation policies
6. Use a flexible Theory of Change.

These principles can be used to evaluate any level of transformative innovation policy (e.g., the macro policy level, the programmatic level, or the project experiment level). The use of a flexible Theory of Change (ToC) is particularly relevant in this context. ToCs represent and operationalise the change that is expected to be triggered by a specific intervention; it is "an account of what is expected to happen" (Molas-Gallart et al., 2021, p. 435). Therefore, a ToC provides a structured way of considering how change occurs as a result of an intervention, specifying underlying assumptions about a change process. In practice, a ToC shows the expectation of how *inputs* (capacity/resources and capabilities) will be mobilised through explicit *activities*, resulting in immediate *outputs*, intermediary *outcomes*, and, eventually, socio-economic *impacts*.

The approach developed by Molas-Gallart et al. (2021) is based on the transition theory. As the authors explain:

> Our approach is similar to theory-oriented approaches in that it revolves around a description of an expected process of change that "consider programmes in their context, which includes actors' environments… and public service culture and behaviour" (Stame, 2004, p. 63). But unlike most theory-oriented approaches in which the evaluator builds the theory behind the programme interpreting the understanding of what may happen offered by actors involved in the intervention (Stame, 2004), we actively use transitions theory to coproduce with the policy actors a ToC that focuses on transformative changes. (p. 435)

Open-ended transformative mission-oriented policies inherit key characteristics of transformative innovation policy: directionality towards addressing societal challenges, inclusiveness and participation, complex and ill-defined problems and solutions, the importance of learning, etc. Therefore, the formative evaluation approach of Molas-Gallart et al. (2021) is well-suited for appraising transformative innovation policies. As such, this approach was adopted to support Vinnova's mission-oriented strategy for transforming the food socio-technical system (see Section 5.4).

5.3.2 The Transformation of Socio-technical Systems: Stylised Facts as Transformative Outcomes

In the context of an open-ended transformative mission-oriented policy, the formative evaluation approach focusses on outcomes instead of impact, since the latter cannot be attributed directly to the intervention in complex (unpredictable, hard-to-control, endlessly evolving) systems (Molas-Gallart et al., 2021). Instead of measurable indicators for clearly defined missions (and of impact indicators), the formative evaluation approach introduces a new layer in the Theory of Change – what Ghosh et al. (2021) call "transformative outcomes", which represent stylised facts about socio-technical transition processes.

Drawing from the Multilevel Perspective (MLP) on Socio-Technical Transitions (Geels, 2002; Geels & Schot, 2007) (Figure 5.2) and the Strategic Niche Management approach (Kemp et al., 1998), Ghosh et al. (2021) identify three core transformative processes in socio-technical transitions: (1) building or constructing niches; (2) accelerating and embedding niches in the regime; and (3) opening up the existing regime by unlocking path dependencies. In the MLP, niches are protected institutional spaces where multi-dimensional practices are in flow and not yet stabilised. Regimes are stable socio-technical configurations of rules and tend to follow path-dependent trajectories. The landscape comprises macro processes – such as long-term and slow-moving trends like climate change or rapid external shocks like the COVID-19 pandemic – on which regime and niche actors have little agency to change (at least in the short term), but which directly influence the context for niches and regime.

In each of those three transition processes, four systemic "transformative outcomes" are identified: they are general intermediary outcomes that enable emergent transformations. Thus, Ghosh et al. (2021) propose the existence of a set of 12 transformative outcomes:

A. *Niche building*

1. Shielding: The establishment of environments that protect against external interference (e.g., from incumbent actors), where innovations can thrive.
2. Learning: A prerequisite for learning is actionable knowledge, which is a type of practical learning to solve transformational problems (Luederitz et al., 2017). Whether learning will happen in the face of actionable knowledge, however, depends on the existence of appropriate capacities.
3. Networking: Networks are critical for transformative change (Farla et al., 2012), as they may lead to the pooling of resources and enable the diffusion of novel ideas, technologies, and practices.
4. Navigating expectations: Since innovation is uncertain, managing expectations (e.g., through shared visions and narratives) contributes to guiding and accelerating innovation towards a desired direction.

B. *Niche expansion and embedding*

1. Upscaling: Defined as increasing user adoption, upscaling leads to the establishment of a critical mass for transformative change. It includes a broad set of user preferences and technologies, policy measures, industry strategies, and cultural meanings and symbols.
2. Replicating: Replicate niche spaces in new socio-spatial and cultural contexts that are adapted to local conditions.
3. Circulating: Circulation of resources (ideas, rules, products, tools, people) beyond original niches. The circulation of such resources triggers learning processes that allow for the embedding of niches in local contexts.
4. Institutionalising: Understood as the stabilisation of shared formal and informal rules. This means the niche upscales to become the new regime (Fuenfschilling & Truffer, 2014).

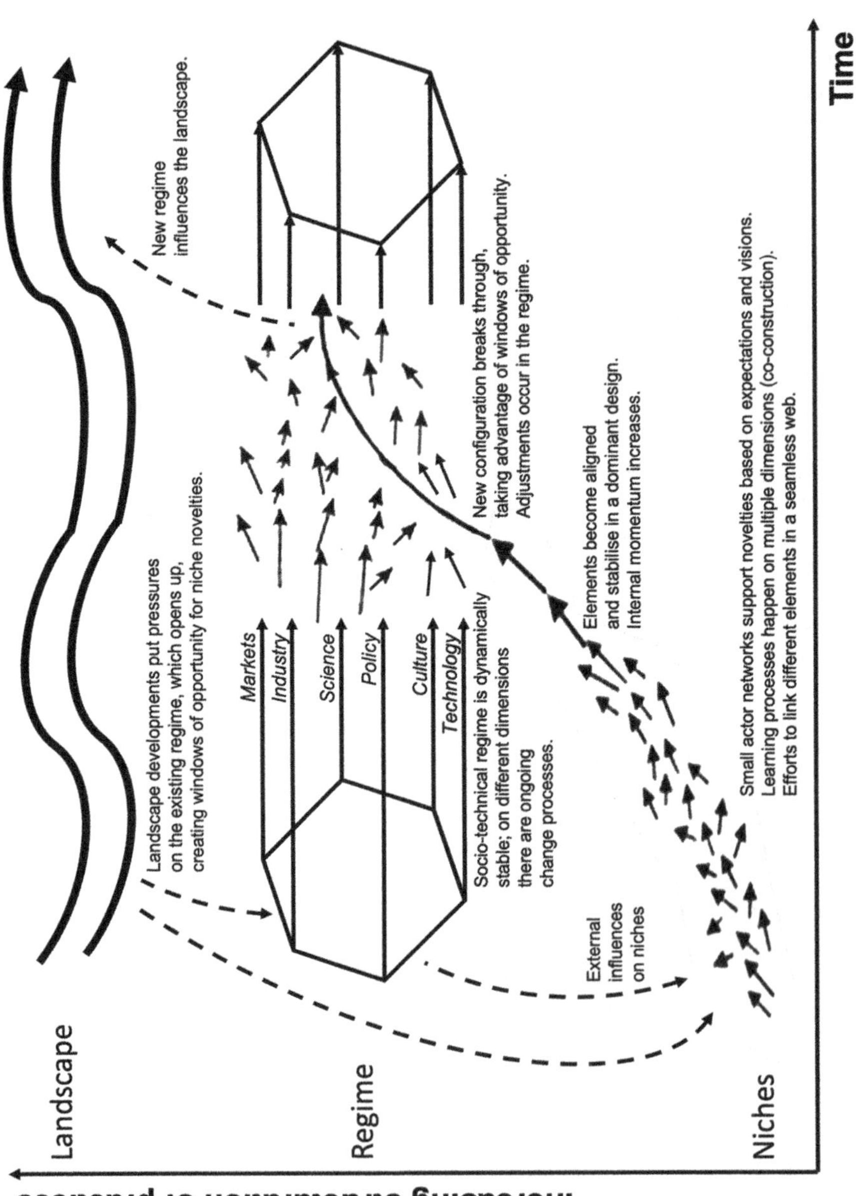

Source: Based on Geels, 2002; Geels & Schot, 2007.

Figure 5.2 The multilevel perspective (MLP) on socio-technical transitions

C. *Destabilisation and opening up of regimes*

1. De-aligning and destabilising regimes: The decline of established socio-technical regimes is the flip side of innovation development and diffusion. Destabilisation is the unlocking of institutional configurations and the break with path dependencies that create unsustainable trajectories of socio-technical systems (Turnheim & Geels, 2013). Destabilisation is often at the core of calls for "comprehensive" policy and instrument mixes to foster socio-technical transitions.
2. Unlearning and deep learning in regimes: These refer to the unlearning and "deep learning" of regime actors, necessary to break path dependencies and institute alternative rules for solving systemic problems.
3. Strengthening niche-regime interactions: Regime interactions with niches tend to contribute to the opening up of regimes and are often associated with the dedication of more regime resources to empower niches and make them more competitive.
4. Changing perceptions of landscape pressures: Landscape dynamics influence "how, in which direction and with what speed systemic transformations unfold" (Ghosh et al., 2021, p. 9), potentially destabilising regimes, but in a process that is subject to the ways regime actors perceive and react to these landscape pressures.

Transformative outcomes are also understood as "leverage points" in a process of transforming socio-technical systems. If a transformative outcome is achieved and maintained, it tends "to elicit a larger process that, in the long run, transforms a system" (Alvial-Palavicino et al., 2021, p. 27).

The incorporation of transformative outcomes into a theory of change results in what we may call a "transformative theory of change". As we will show in our case study, part of the formative evaluation of an open-ended transformative mission-oriented policy is the co-construction of a transformative ToC with those responsible for designing and implementing the policy (e.g., policymakers and civil servants), further identifying how the expected outcomes of the intervention can be mapped against the 12 theoretical transformative outcomes (Molas-Gallart et al., 2021). Of course, some outcomes may be absent from a specific policy. The specific application of this approach depends on whether the policy is at an early stage of designing the intervention process or whether it is an ongoing intervention and the approach is applied during implementation (Alvial-Palavicino et al., 2021). In the next section, we explain how the formative evaluation approach was applied in the Vinnova case.

5.4 CASE STUDY: VINNOVA'S TRANSFORMATIVE MISSION-ORIENTED APPROACH TO TRANSFORMING THE FOOD SYSTEM

5.4.1 The Vinnova Approach to Missions and the Food Mission

Vinnova, founded in 2001, is Sweden's innovation agency designed to build innovation capacity and competitiveness along with the different innovation system actors. It emerged when Frame 2 of STI policy was becoming mainstream. More recently, the agency started to orient its initiatives to meet the agenda set by the Sustainable Development Goals (SDGs), creating opportunities, fostering collaborations, and testing new approaches to meet critical societal challenges.

In this context, Vinnova started to develop its own mission-oriented approach, based on strategic design principles and tested from 2019 onwards. The approach focusses on a systemic redesign of the cultures of decision-making, which can be both at an individual and an institutional level. It recognises uncertainty by prototyping in the space of "unknown unknowns" to mitigate risk while delivering ambitious change. It distances itself from design thinking by focussing on the context without a predefined end product in mind and delivering a more open-ended and strategic design (Hill, 2012). In this vein, strategic design can support policymaking beyond the optimisation of known solutions and provide a systemic approach to exploring new alternatives to problems defined in a participatory way, where different tools and methods are used to engage with a diverse group of actors (Van Buuren et al., 2020). The agency decided to pursue two missions, initially defined as:

1. Ensure that every street in Sweden is healthy, sustainable, and vibrant by 2030 ("urban mission").
2. Ensure that every student in Sweden eats sustainable and tasty school food by 2025 ("food mission").

The missions were initially closely defined following the time-bound criteria proposed by the *Mazzucato report* (Mazzucato, 2018b). Our case study focusses on the food mission. We will present the process followed to discuss how the mission became open-ended and experimental, later aided by the formative evaluation approach to transformative mission-oriented innovation policies.

Starting in 2019, Vinnova stepped in as a policy intermediary between different food-related government agencies, working closely with the Swedish Food Agency. As a product of long trust-building and collaborative work, the two agencies shaped the food mission as a joint initiative. A dedicated team was formed within Vinnova, which engaged with a counterpart team from the Swedish Food Agency to define the food mission and the experimentation phase. Vinnova led the process, proposing and enabling the methodology by funding the whole-system mapping. The purpose was: (a) to have a comprehensive and co-creative way of looking at Sweden's food (socio-technical) system, and (b) to propose an experimental policy design bringing together different national agencies (such as those responsible for Food, Agriculture, Procurement, Health, Youth and Civil Society Affairs, and Environmental Protection). In terms of governance, the Swedish Food Agency coordinated the entire process (Sjöblom, 2021).

Vinnova's mission-oriented methodology consists of four stages (see Figure 5.3). The initial stage is to build a collective understanding of the system and find entry points or angles for intervention. The methods are workshops with different stakeholders, individual dialogues, and desk analysis of relevant documents. During this phase of the food mission, the Vinnova team identified four sub-areas of (or entry angles to) the food system: (a) new farming; (b) new food; (c) new retail; and (d) new school food (as an access point to the delivery of public meals). School food was chosen for the development of the first food system mission, based on the assumption that it could leverage the transformation of the entire socio-technical

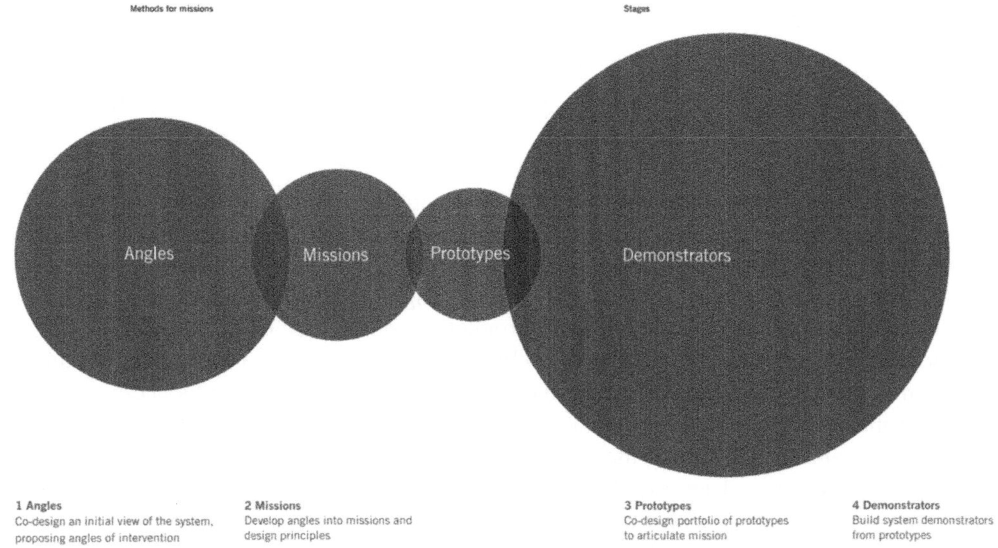

Source: Hill, 2022, pp. 140–41.

Figure 5.3 Vinnova's mission definition and implementation methodology

system. The rationale is related to the EU Commission's strategy "from farm to fork",[4] in which Sweden participates, which highlights the importance of using procurement to redirect public meals so that they are in line with sustainability goals. School food would also have a critical impact on children's eating preferences, habits, daily nutritional needs, and ultimately, health. Adding to the fact that school food accounts for almost half of all the meals served in Sweden every day, these aspects came together to form a mission in this area (Sjöblom, 2021).

The second phase of Vinnova's mission approach consists of defining the mission itself. Based on the results of the angles, the food mission process was based on the working hypothesis that the school meals system could act as a lever to transform the entire system. During this stage, a participatory process (congruent with an inclusive and bottom-up approach from the mission definition) was developed through design workshops with actors directly involved with school meals, such as chefs, teachers, parents, other public agency representatives (such as the national food, public health, and procurement agencies), and food providers. The initial mission statement was to *"ensure that every student in Sweden eats healthy, sustainable and tasty school food by 2025"*. But throughout the workshops, the timebound mission statement was modified in favour of *"Every child in Sweden should eat sustainable and good school food"*, showing an evolution to an open-ended mission that fosters transformations at different levels. Uncertainty was recognised, not only in the complexity of the challenge but also in the need for commitment from other agencies, which crosses budget allocations. In this sense, the mission has helped to create a shared vision, becoming a "boundary object" (Janssen

[4] More information about the strategy can be found here: https://ec.europa.eu/food/horizontal-topics/farm-fork-strategy_en

et al. in this volume) to help with gathering efforts from different government agencies and stakeholders.

In addition to the design workshops, the group mapped a school meals system to understand the challenges, actors, and change pathways required to transform the system. Through the system mapping, the team first identified leverage points and actors in order to define how to start the prototype phase.

The second part of the whole process (phases 3 and 4) started in 2021. Vinnova and the Swedish Food Agency opened a country-wide call for municipalities to propose school meal system-related prototypes addressing at least two of the eight identified leverage points:[5] (a) the involvement of students and school staff in the design of the school meal system; (b) procurement and purchase of sustainable food, products, and services; (c) a connection between the school meal and the pedagogical task; (d) building competencies on environmental, nutritious, and appetising food; (e) design of the meal environment; (f) an evidence-based goal for school meals for the whole of Sweden; (g) strategies, implementation, and follow-up of the objectives; and (h) the school meal system as a workshop for new sustainable solutions. Four municipalities spread across different geographical areas of the country were selected to become project partners. These municipalities committed to developing their prototypes for at least one year, with the possibility of extending the process further. The prototypes were related to curriculum changes in the schools around the relationship with food, the procurement process, local production of sustainable food, and changes in the school kitchen and restaurant. The municipalities received support for mapping their local school food system, funding for testing their prototypes, engagement from the national government to support policy changes, and cross-learning sessions with the other experimental environments developed in the other municipalities. This approach to prototypes for system transformation is seen by Vinnova as a low-cost, low-risk way of learning how a mission can be solved. In this regard, the mission approach supports experimentation and encourages learning to overcome transformational failures.

The final stage comprises assembling the prototypes into system demonstrators. The precise way this phase will be deployed was under discussion at the moment of writing, but we anticipate that it could serve as a holistic showcase of system transformation that could trigger – in terms of transformative outcomes – circulation and replication of change at the national level.

5.4.2 Introducing Formative Evaluation in Vinnova's Mission-Oriented Approach

Vinnova is part of the Transformative Innovation Policy Consortium (TIPC),[6] a group of researchers, policymakers, and funding agencies working together to advance the new transformative framing for science, technology, and innovation (STI) policy. From 2020, Vinnova and TIPC researchers started an experimental policy engagement (EPE) to explore the rationale, practice, and feasibility of a formative evaluation approach to enhance Vinnova's mission-oriented methodology. By working together on strengthening policy design and implementation practices, a core team of Vinnova officers and TIPC researchers sought to

---

5 More information about the process is on the Swedish Food Agency website: https://www.livsmedelsverket.se

6 More information is available at: https://www.tipconsortium.net/

assess whether Vinnova's approach to missions could foster systems transformation and derive learning from a formative assessment. Consistent with the TIP formative evaluation principles (see Section 5.3.1), the team included the evaluation practice in the mission design and implementation, making the evaluation permanent and present during the whole cycle. The goal was to foster reflexivity and strengthen the transformative potential of Vinnova's mission-oriented approach to the food system in the process. The formative evaluation approach was concretely implemented through a series of interactions between the TIPC researchers and the Vinnova mission team. The interactions intended to explicitly define the mission theory of change, applying it to the transformative outcomes and developing an evaluation plan based on specific monitoring and learning strategies (MEL plan).

TIPC's formative evaluation methodology used the transformative outcomes as a guide for evaluation, instead of evaluating emerging outcomes as the mission advanced without any specific steering direction, which is more attuned with the mission-oriented methodology and strategic design approach utilised by Vinnova. The assumption behind the TIP evaluation approach is that transformative outcomes are conducive to transformative change and thus should guide the activities enacted by the mission while the intervention is still ongoing. In this sense, the mission's ToC was to maximise the transformative potential presented in the ToCs. The ToC provided a guiding framework to intentionally intervene in the transformation processes (Van Es et al., 2015). The ToC construction went backwards from the transformative outcomes to the expected immediate results (outputs) led by activities contributing to the expected systemic change.

The ToC definition truly fostered learning and reflexivity within both teams. Tension emerged between Vinnova's approach to drawing transformation pathways as they surfaced from the previous phases and TIP's evaluation approach that proposes a flexible ToC to be included in the intervention, which implies the definition of transformative outcomes. The discussions led to the partial development of a ToC focused on the school food mission but without a full specification of outputs and activities. This process was addressed for 15 months to clarify the purposes of the formative evaluation and a ToC was developed: describe the desired change in medium- and long-term perspectives; identify transformative outcomes; specify underlying assumptions; define a MEL plan that includes areas of inquiry and indicators; and visualise a flexible ToC that is focussed around the TOs.

Vinnova's team defined three medium-term outcomes for their mission, aiming to bring about significant changes in individuals and organisations at the national level. With a strong focus on fostering openness and promoting continuous learning among government agents and incumbents, their ToC centred around unlocking transformative processes at the regime level (see Section 5.3.2 on transformation of socio-technical systems). These objectives served as crucial milestones in realising the desired transformation of the Swedish food system as a whole, with a specific focus on the school meals system (see Table 5.2).

Through differ ent discussions, the Vinnova team made explicit their assumptions and why they focussed on working towards incumbents' transformation. However, it was clear that without building, expanding, and embedding niches, the mission would not deliver the proposed intended impact. The call for the municipalities' prototypes was a first step towards that purpose, but a further deployment of resources and instruments would need to be advanced in order to ensure that "*every child in Sweden should eat sustainable and good school food*". A second phase of the prototypes focused on food niche identification and nurturing was

Table 5.2 Vinnova's food mission outcomes, underlying assumptions and associated transformative outcomes

Outcomes	**Assumptions**	**Transformative outcomes**
New understanding of the food system in terms of agents and configurations	The food system of Sweden is characterised by a low capacity for innovation and renewal. This is partly an issue of rigid and conservative ways of working within the sector	Unlearning and deep learning in regimes Strengthening regime-niche interactions
Policy and business actors within the food system change their perception and behaviour towards the relationship between the three dimensions of sustainability	As long as economic growth and productivity are seen as the overarching goals by the dominant policy and industry actors of the sector, it will be challenging to enable rapid transformative change towards a fully sustainable food system	Changing perceptions of landscape pressures Unlearning and deep learning in regimes
Government agencies are working together to produce alternatives for new sustainable food systems	To enhance the possibility of changing the food system, different government agencies must work together with a joint vision on where to go	De-aligning and destabilising regimes

Source: TIPC/Vinnova EPE Teams.

published in 2022.[7] The call focusses on mobilising networks of niche actors to transform the Swedish food sector, connect them, circulate good practices, learn from them, and also have a sense of the actors' expectations.

The process improved and strengthened Vinnova's strategic design approach to open-ended missions by explicitly considering the concepts and transformative outcomes of socio-technical transition. By means of this formative evaluation, Vinnova was able to identify the macro process of destabilising and unlocking regimes as a key for the success of its transformative mission. Visually, Vinnova introduced a socio-technical transition layer, or directionality, to its mission-oriented model (Figure 5.4). The goal of food system transformation is portrayed as a systemic experimentation process, enabling spaces not only to create innovations (prototypes) but also to destabilise and change incumbent actors' behaviours, perceptions, and ways of working. By enabling prototypes and further demonstrators, niches are identified, nurtured, and scaled up. Interestingly, these regime destabilisation and niche-nurturing ambitions were not explicit in the initial ToC defined by the TIPC-Vinnova team.

5.5 CONCLUSION

This chapter discussed the challenge of carrying out open-ended transformative mission-oriented policies. Open-ended missions differ from traditionally defined innovation missions in

[7] More information at: https://www.vinnova.se/en/m/sustainable-food-systems/a-new-recipe-for-the-food-system/

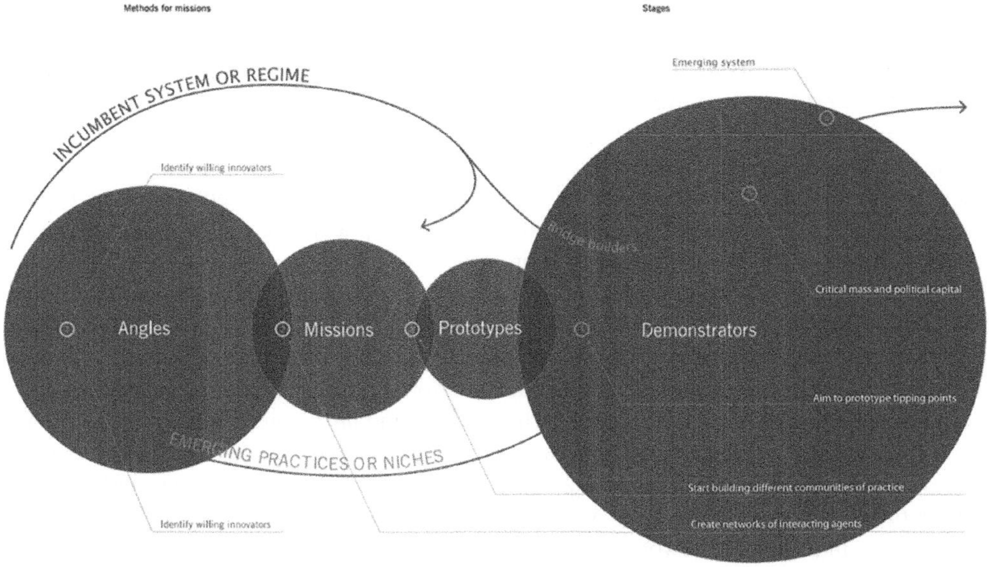

Source: Hill, 2022, 142–43.

Figure 5.4 *Vinnova's mission definition and implementation methodology with a new socio-technical transition directionality*

that they cannot anticipate end states as measurable targets, because solutions are not unequivocal or consensual. Therefore, it is not feasible to establish a realistic timeframe for the accomplishment of these solutions. Given the characteristics of open-ended transformative MOPs, we argued that a formative evaluation approach is an adequate and useful tool for its design, implementation, and continuous monitoring, so as to promote participation, learning, and reflexivity.

We explored the application of a specific formative evaluation methodology for transformative innovation policy (Molas-Gallart et al., 2021) using a case study of Vinnova in its endeavour to develop and implement its own mission approach to transforming the food socio-technical system from within. Vinnova's eventual open-ended mission definition evolved from a time-bound mission definition of transforming the sub-area of school foods in a sustainable way by a certain endpoint. Indeed, Vinnova's approach does not align with the traditional top-down definition of missions, as they define the mission from the bottom up, through a nuanced understanding of the system (system mapping), identifying entry points (angles), and levers for change together with key stakeholders of the system. Vinnova's mission strategy could therefore be understood as an ecosystem-based mission or open-ended transformative mission approach. In this approach, transformation pathways remain open through the initial phases, developing from angles into missions and into prototypes (experimental solutions), which are later aggregated into larger demonstrator programmes.

Vinnova engaged in a policy experiment process with the Transformative Innovation Policy Consortium (TIPC) to implement the formative evaluation approach, with the goal to catalyse the transformative potential of its mission methodology. Initially, Vinnova was reluctant to define a direction of change based on the socio-technical transition theory, so that even a flexible ToC seemed at odds with its emergent strategic design approach. While this caused tensions in the process, making the implementation of a transformative ToC lengthy, those tensions were tackled through a series of interactions between the TIPC and Vinnova teams, resulting in a recognition of the value of embedding certain theoretical processes and constructs in its mission approach (from the socio-technical transitions theory, such as regime destabilisation and niche nurturing and building). This deep learning at Vinnova is an important outcome of the formative evaluation approach.

However, there is one remaining issue to be tackled during the demonstrator phase: how to assemble different prototypes into full demonstrators and how to circulate and replicate those prototypes in different regions will require specific ToCs. For instance, they will need to be tailored to regional characteristics and account for scalability to enable systemic change, since currently, the ToCs are specific for niche creation and nurturing, while systemic change occurs at the regime level when niches grow and become institutionalised. Furthermore, more work is needed at this stage to translate these lessons into actual metrics and indicators to monitor the demonstration projects against the mission statement. Currently, most efforts in developing indicator frameworks for sustainability attempt to capture impacts, which are achieved only in the long term and for which it is difficult to attribute causality as stemming from immediate action. Instead, a framework for monitoring and evaluating formative systemic transformations should lead to process and (mid-term) outcome indicators. The nested approach to assess multi-level transformative innovation with a concise set of indicators that respond to the flexible ToCs would be a good way to move this endeavour forward. We propose that this be done using participative methodologies to identify signs of change and translate those into indicators that respond to the ToCs.

Some issues were not examined in our case study. For instance, how Vinnova's food mission may serve to assemble (or is related to) a policy mix around the mission statement for the food sector, including traditional "top down" elements such as sector-specific policies and regulations, was not examined. We therefore did not discuss how the formative evaluation framework could contribute to unlocking learning at a higher level of policymaking, beyond Vinnova's arena of operation.

Finally, another aspect not discussed in this chapter relates to the types of institutional capacities and capabilities an organisation like Vinnova needs to develop (or rely on externally) in order to design, implement, and monitor an open-ended transformative mission. One hypothesis is that the traditional set of institutional capacities (state, administrative, and policy capacities) is still relevant but must be complemented by a new "transformative" capability set to unlock internal organisational processes (e.g., within Vinnova) and the mission deployment with external actors (including policymakers and other agencies leading mission formation), which would include:

1. Capabilities to deepen learning and unlearning: Shifting niche and regime actors' mental frameworks, values, norms, and ways of doing to a new set of practices and routines;
2. Networking capabilities: Such as the ability to generate interdisciplinary, intersectoral, and multi-stakeholder collaboration and cooperation vertically and horizontally;
3. Relational governance capabilities: To promote decentralisation in decision-making with convergence mechanisms toward shared goals;

4. Social appropriation capabilities: The ability to communicate and disseminate information. Once the challenges are appropriated by local agents and translated into specific needs, there are higher chances to gain long-term commitment and acceptance.

This list is not exhaustive, being based on initial insights from the Vinnova case study. Therefore, the examination and identification of new institutional capabilities for transformative change (and the process of transforming public agencies themselves) and analysing how they contribute to open-ended transformative missions (as well as to more traditional top-down challenge-led missions) should be a relevant area for future research in this era of challenge-led innovation policies.

REFERENCES

Alvial-Palavicino, C., Matti, C., & Witte, J. (Eds.). (2021). *MOTION handbook: Developing a transformative theory of change*. Transformative Innovation Policy Consortium.

Amanatidou, E., Cunningham, P., Gök, A., & Garefi, I. (2014). Using evaluation research as a means for policy analysis in a "new" mission-oriented policy context. *Minerva*, *52*(4), 419–438.

Arnold, E., Åström, T., Glass, C., & de Scalzi, M. (2018). *How should we evaluate complex programmes for innovation and socio-technical transitions?* Brighton: Technopolis Group.

Diercks, G., Larsen, H., & Steward, F. (2019). Transformative innovation policy: Addressing variety in an emerging policy paradigm. *Research Policy*, *48*(4), 880–894.

Edler, J., Berger, M., Dinges, M., & Gök, A. (2012). The practice of evaluation in innovation policy in Europe. *Research Evaluation, 21*(3), 167–182.

Farla, J., Markard, J., Raven, R., & Coenen, L. (2012). Sustainability transitions in the making: A closer look at actors, strategies and resources. *Technological Forecasting and Social Change*, *79*(6), 991–998.

Foray, D., Mowery, D. C., & Nelson, R. R. (2012). Public R&D and social challenges: What lessons from mission R&D programs? *Research Policy*, *41*(10), 1697–1902.

Fuenfschilling, L., & Truffer, B. (2014). The structuration of socio-technical regimes – Conceptual foundations from institutional theory. *Research Policy*, *43*(4), 772–791.

Geels, F. W. (2002). Technological transitions as evolutionary reconfiguration processes: A multi-level perspective and a case-study. *Research Policy*, *31*(8–9), 1257–1274.

Geels, F. W., & Penna, C. C. R. (2015). Societal problems and industry reorientation: Elaborating the Dialectic Issue LifeCycle (DILC) model and a case study of car safety in the USA (1900–1995). *Research Policy*, *44*(1), 67–82.

Geels, F. W., & Schot, J. (2007). Typology of socio-technical transition pathways. *Research Policy*, *36*(3), 399–417.

Ghosh, B., Kivimaa, P., Ramirez, M., Schot, J., & Torrens, J. (2021). Transformative outcomes: Assessing and reorienting experimentation with transformative innovation policy. *Science and Public Policy*, *48*(5), 739–756.

Hill, D. (2012). *Dark matter and trojan horses: A strategic design vocabulary*. Strelka.

Hill, D. (2022) *Designing missions: Mission-oriented innovation in Sweden – A practice guide by Vinnova*. Stockholm: Vinnova.

Janssen, M., Wesseling, J., Torrens, J., Weber, K. M., & Penna, C. C. R. (2025). Missions as a boundary object in umbrella frameworks: The case of the Dutch Mission-oriented Top Sector and Innovation Policy. In J. Edler, M. Matt, W. Polt & M. Weber (Eds.). *Transformative mission-oriented innovation policies: Revisiting the role of science, technology and innovation in society*. Edward Elgar Publishing.

Kemp, R., Schot, J., & Hoogma, R. (1998). Regime shifts to sustainability through processes of niche formation: The approach of strategic niche management. *Technology Analysis & Strategic Management*, *10*(2), 175–198.

Luederitz, C., Schäpke, N., Wiek, A., Lang, D. J., Bergmann, M., Bos, J. J., et al. (2017). Learning through evaluation – A tentative evaluative scheme for sustainability transition experiments. *Journal of Cleaner Production*, *169*, 61–76.

Mazzucato, M. (2016). From market fixing to market-creating: A new framework for innovation policy. *Industry and Innovation*, *23*(2), 140–156.

Mazzucato, M. (2018a). Mission-oriented innovation policies: Challenges and opportunities. *Industrial and Corporate Change*, *27*(5), 803–815.

Mazzucato, M. (2018b). *Mission-oriented research & innovation in the European Union: A problem-solving approach to fuel innovation-led growth.*

Mazzucato, M., & Kattel, R. (2020). Grand challenges, industrial policy, and public value. In A. Oqubay, C. Cramer, H.-J. Chang, & R. Kozul-Wright (Eds.), *The Oxford handbook of industrial policy* (pp. 311–337). Oxford University Press.

Mazzucato, M., & Penna, C. C. R. (2016). Beyond market failures: The market creating and shaping roles of state investment banks. *Journal of Economic Policy Reform*, *19*(4), 305–326.

Miedzinski, M., Mazzucato, M., & Ekins, P. (2019). A framework for mission-oriented innovation policy roadmapping for the SDGs: The case of plastic-free oceans. UCL Institute for Innovation and Public Purpose, Working Paper Series (IIPP WP 2019-03).

Molas-Gallart, J., Boni, A., Giachi, S., & Schot, J. (2021). A formative approach to the evaluation of Transformative Innovation Policies. *Research Evaluation*, *30*(4), 431–442.

Schot, J. (1992). Constructive technology assessment and technology dynamics: The case of clean technologies. *Science, Technology, & Human Values*, *17*(1), 36–56.

Schot, J., & Steinmueller, W. E. (2018). Three frames for innovation policy: R&D, systems of innovation and transformative change. *Research Policy*, *47*(9), 1554–1567.

Shapira, H., Ketchie, A., & Nehe, M. (2017). The integration of design thinking and strategic sustainable development. *Journal of Cleaner Production*, *140*, 277–287.

Sjöblom, J. (2021). *Steg för steg mot ett hållbart skolmåltids-system. Ett missionorienterat pilotprojekt. Delrapport Med Beskrivning Och Lärdomar Från Projektet "Ett Nytt Recept För Skolmåltider" 2019–2021.* Stockholm: Vinnova.

Soete, L., & Arundel, A. (1993). *An integrated approach to European innovation and technology diffusion policy: A Maastricht memorandum.*

Stame, N. (2004). Theory-based evaluation and types of complexity. *Evaluation*, *10*(1), 58–76.

Steward, F. (2012). Transformative innovation policy to meet the challenge of climate change: Socio-technical networks aligned with consumption and end-use as new transition arenas for a low-carbon society or green economy. *Technology Analysis & Strategic Management*, *24*(4), 331–343.

Turnheim, B., & Geels, F. W. (2013). The destabilisation of existing regimes: Confronting a multi-dimensional framework with a case study of the British coal industry (1913–1967). *Research Policy*, *42*(10), 1749–1767.

Van Buuren, A., Lewis, J. M., Guy Peters, B., & Voorberg, W. (2020). Improving public policy and administration: Exploring the potential of design. *Policy & Politics*, *48*(1), 3–19.

Van Es, M., Guijt, I., & Vogel, I. (2015). *Theory of change thinking in practice: A stepwise approach.* Hivos.

Wanzenböck, I., Wesseling, J. H., Frenken, K., Hekkert, M. P., & Weber, K. M. (2020). A framework for mission-oriented innovation policy: Alternative pathways through the problem-solution space. *Science and Public Policy*, *47*(4), 474–489.

Weber, K. M., & Rohracher, H. (2012). Legitimizing research, technology and innovation policies for transformative change: Combining insights from innovation systems and multi-level perspective in a comprehensive "failures" framework. *Research Policy*, *41*(6), 1037–1047.

Wittmann, F., Hufnagl, M., Lindner, R., Roth, F., & Edler, J. (2020). Developing a typology for mission-oriented innovation policies. *Fraunhofer ISI Discussion Papers: Innovation Systems and Policy Analysis, 64.*

6. Challenges and opportunities of mission-oriented innovation policies for realising transformative agendas: general principles and application to the case of the fight against climate change

Philippe Larrue

6.1 INTRODUCTION: WHY ARE MISSION-ORIENTED POLICIES NEEDED?

6.1.1 Mounting Societal Challenges Call for New STI Governance and Policy Frameworks

In recent years, there has been a growing consensus among policymakers and analysts that existing STI governance and policy frameworks are unable to address the systemic challenges posed by climate change (IPCC, 2022; OECD, 2022a). It has become increasingly clear that "change-as-usual" is no longer an option. However, current efforts to translate net-zero commitments into tangible actions (Jeudy-Hugo et al., 2021) have fallen significantly short of what is required to achieve the relevant targets (World Resources Institute, 2020). If countries merely implement their 2030 pledges and maintain the same pace, global warming is projected to increase between 2.4 and 2.6°C by 2100 (UNEP, 2021). Moreover, public expenditure on research, development, and deployment of low-carbon technologies as a percentage of GDP has stagnated since around 2010, even though achieving the net-zero goals would necessitate a tripling of investments in clean energy and infrastructure by 2030 (IEA, 2022). This trend coincides with a slowdown in patenting activities related to low-carbon technologies (Cervantes et al., 2023).

To limit global warming to 1.5°C, various dimensions of change are required, involving different mitigation approaches and a mix of social and technological innovations at different stages of development and diffusion (IPCC, 2018). Achieving the 2030 emission reduction targets necessitates both the widespread adoption of existing technologies and the advancement and scaling up of technologies that are still in the laboratory or demonstration phase. Furthermore, these scientific and technological advancements must be complemented by behavioural, regulatory, political, and social changes. Coherent and coordinated systemic changes across various domains and levels of society are essential for successful sustainability transitions.

6.1.2 Traditional Policy Landscapes Are Not Fit to Achieve Net Zero

In recent decades, policy and governance frameworks have become increasingly fragmented due to the specialisation of individual policy instruments. Policy initiatives targeting market failures have focused on specific areas, such as private R&D, feasibility studies, and knowledge enhancement in certain disciplines or sectors (Mazzucato, 2016). Furthermore, current STI policy mixes rely heavily on indirect incentives, such as R&D tax credits, that struggle to strategically align research activities with priorities, despite some experimentation with "green R&D tax credits" (OECD, 2022b).

Addressing climate change requires comprehensive government support that spans disciplinary, sector-specific, and policy silos (OECD, 2019). Interventions that are aimed at technological innovation must be part of broader socio-economic and political changes if a genuine socio-technical transformation is to be achieved (Geels & Schot, 2007; Schot & Steinmueller, 2018). Without a coordinated framework, the existing, dispersed policy mixes cannot drive the systemic changes needed for a successful transition to a net-zero future (OECD, 2021).

As such, tackling challenges on an unprecedented scale calls for effective strategic orientation and holistic coordination of STI interventions. Recent experiences like the COVID-19 pandemic have highlighted the importance of collective action frameworks with clear objectives (OECD, 2020).

6.1.3 A Wealth of Mission-Oriented Policy Experimentations to Learn From

In recent years, mission-oriented innovation policies (MOIPs) have gained significant attention. Although initially derived from space and military programmes like the Manhattan or Apollo projects, the concept of mission orientation has evolved to encompass all systemic interventions aimed at addressing societal challenges. The OECD defines MOIPs as a coordinated package of policy and regulatory measures that have been specifically tailored for the purpose of mobilising science, technology, and innovation in order to address well-defined objectives that relate to a societal challenge, within a defined timeframe. These measures can span different stages of the innovation cycle (from research to demonstration and market deployment), mix supply-push and demand-pull instruments, and extend across various policy fields, sectors, and disciplines (Larrue, 2021). Numerous countries have adopted MOIPs through pilot programmes and experiments, as can be seen from the approximately 40 initiatives that the OECD has identified as being closely aligned with the definition of MOIPs.

It is too early to gauge the impact of most MOIP initiatives, and one has to be prudent in drawing conclusions at such an early stage in MOIP policy history. Most of these initiatives were launched between 2018 and 2020 and are only just entering their implementation phase. Additionally, there are fewer earlier MOIPs that predate the conceptualisation of the policy approach and those that do exist differ from the ideal MOIP described in the definition. However, the available evidence offers valuable insights on multiple issues, ranging from how to codevelop a strategic agenda to the different ways of integrating policy instruments across administrative silos.

6.1.4 Objectives and Content of the Chapter

This chapter aims to present the conceptual framework and results of the work that has been conducted at the OECD since 2019 under the guidance of its Committee for Scientific and Technological Policy (CSTP) to analyse mission-oriented policies and their additionality when it comes to addressing climate change. Following this introduction, a second section will outline the definition and ideal type of MOIPs, as well as variations on this ideal type. The third section will present the theory of change (ToC) used to analyse net-zero missions and the key results stemming from the analysis of net-zero missions using this framework. The concluding section lists the remaining knowledge gaps and resulting priorities for future work on missions.

The work of the OECD CSTP on missions complements the rapidly growing body of literature on missions in a variety of ways:

1. Policymakers' continued interest in this policy approach has allowed the OECD to follow a progressive and cumulative approach. Since 2019, the OECD has used, deepened, and extended a single underlying framework, which started with a definition of MOIPs and building block and design principles for them. Later, the OECD built on this to develop a MOIP Theory of change (ToC) that links the design principles to their expected effects. More recently, this work has been prolonged to develop monitoring and evaluation tools.
2. As is the case with other forms of transformative policies, the evolution of MOIPs relies on reflexive mechanisms whereby reflective learning occurs between policy practitioners on the one hand and researchers and other experts on the other. These mechanisms are essential to understanding and improving the policies. The Weberian "ideal type" approach that underpins the MOIP analytical framework is proving to be particularly effective in this context, as it requires constant interaction between theory and practice in the context of MOIPs.
3. This approach also makes it possible to combine positive and normative perspectives on missions. MOIPs (like innovation systems) are both a conceptual construct and a form of policy implemented in a growing number of countries. The MOIP ideal type and ToC are normative, as they synthesize the state of knowledge on what is expected in terms of how MOIPs should be designed (in the case of the ideal type) and what effects they should produce (in the case of the ToC). They are also concretely anchored in policy practices, meaning that any policymaker can use them to interrogate their own way of "doing missions". Therefore, the aim of the MOIP ideal type and the ToC is to provide governments both with a map for positioning their efforts and their results in an encompassing framework and with a compass for guiding their actions so as to best leverage the potential of this approach.

6.2 WHAT ARE MISSION-ORIENTED POLICIES? COMMON PRINCIPLES AND DIVERSITY IN THIS POLICY APPROACH

This section presents the analytical framework that has formed the basis of the CSTP's work on missions over the last four years. This framework is undergoing incremental enhancements and improvements, most notably with a view to developing action-oriented tools for helping countries to implement missions.

6.2.1 The Three Building Blocks of Mission Orientation

In principle, there are three main dimensions that distinguish MOIPs from more traditional STI policy approaches (see Figure 6.1): a strategic orientation towards collectively agreed objectives; the coordination of policies to align the strategies and plans of the various actors that contribute to developing the components needed to achieve the collective objectives (e.g., knowledge, technologies, funding, skills, regulations, markets); and the implementation of policies via a comprehensive mix of policy interventions and diverse initiatives so as to support a wide range of activities.

Various mechanisms and policies are used to ensure that these dimensions are implemented in all national innovation systems. The main novelty of the mission-oriented policy approach lies in the proactive and intentional integration of these three dimensions into a dedicated collective institutional framework in order to tackle a selected challenge. Concretely, a mission-oriented policy is a "proactive platform for collective actions" that articulate, for each selected challenge, a collectively developed agenda, a dedicated structure of governance to take and monitor the effects of common or mutually consistent decisions and, finally, a tailor-made and integrated policy mix.

6.2.2 The Mission-Oriented Policy "Ideal Type"

The three dimensions can be further disaggregated into "design principles" that are expected from policy interventions matching the MOIP definition (Figure 6.2). As only few initiatives, if any, can "tick all the boxes", the MOIP design principles should be considered as applying only to the *ideal type* of MOIPs. In practice, the design principles are found in different combinations and applied in various ways; this reflects the significant variations in MOIP design depending on country- and topic-specific factors and various other contingencies. The design principles are normative in that they help set a direction for policy change. At the same time,

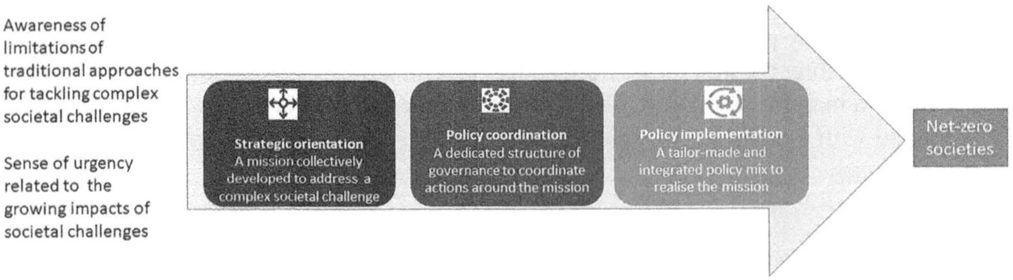

Source: Author's own elaboration.

Figure 6.1 *MOIPs: as an integrated framework for steering, coordinating, and implementing collective action towards net zero*

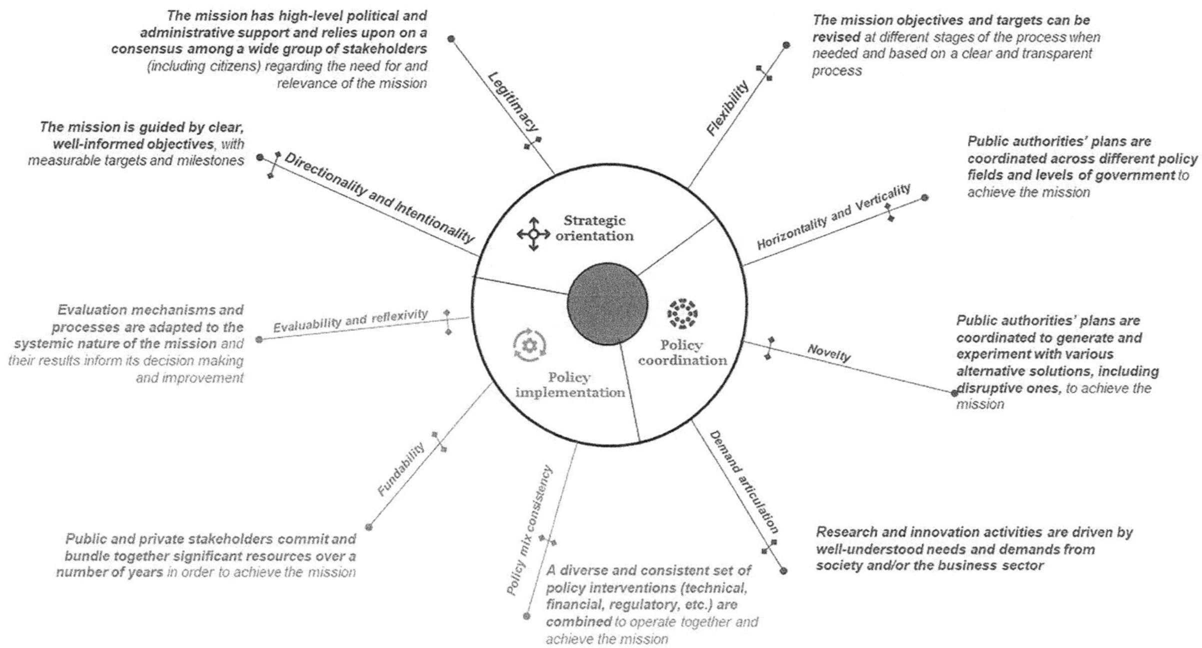

Source: Author's own elaboration.

Figure 6.2 The design principles of mission-oriented policies

however, they are flexible enough to allow for different types of initiatives that are suited for different (mission and governance) contexts.

Although initially developed as a means of mapping practices in different MOIP initiatives (Larrue, 2021), the design principles model was later used to create a mission self-assessment tool. This tool was tested and improved in the context of the Austrian mobilisation around the five EU missions, in collaboration with Austrian public authorities, mission stakeholders, and experts.[1] Based on this framework, the self-assessment tool was used to promote collective thinking on the extent to which and under what conditions the policy and governance framework set up by Austrian authorities for each of the five EU missions has been effective in the following areas:

1. Developing a collectively relevant and legitimate strategic agenda with clear objectives and targets (strategic orientation).
2. Aligning the plans of the public and private actors towards the common objectives (policy coordination).
3. Implementing and monitoring all necessary policy and regulatory measures on a systemic basis to achieve the common objectives (policy implementation).

6.2.3 Variation Around and Towards the Mission-Oriented Policy "Ideal Type"

As typical in this type of Weberian approach, the main added value lies not only in the conceptual design of the "ideal type", but also in how different MOIP initiatives differ from this ideal type in practice. Understanding the reasons for these variations provides critical input on the barriers that these initiatives face, the different strategies that underpin their adoption and the influence of the underlying topic-related and/or national contexts. Analysing the evolution of specific initiatives along each of the ideal type design principles is even more instructive. This involves posing questions such as: How do these initiatives increase their legitimacy by securing high-level back-up? How do they improve the horizontal and vertical coordination? How do they develop new monitoring mechanisms that are adapted to their systemic nature? These changes result from policy learning, the gradual building of trust between the partners, the acquisition of new skills in mission organisations and, more generally, from a slow and painstaking process of enlisting of new actors (and therefore additional expertise, resources, and intervention instruments) in these new initiatives.

These variations give rise to different types of MOIP initiatives. One of the most well-known typologies of missions distinguishes between acceleration and transformative missions and sub-groups of these types (Kuittinen et al., 2018; Wittmann et al., 2021). Based on

[1] This exploratory work was conducted over 2022–2023 by the OECD Mission Action Lab, with financial and technical support from the Austrian Federal Ministry of Education, Science and Research (BMBWF) and in collaboration with mission experts from Joanneum Research's Institute for Economic, Social and Innovation Research (POLICIES), notably Michael Ploder and Wolfgang Polt, and AIT's Center for Innovation Systems & Policy (Matthias Weber). This project included a survey of each EU mission community in Austria, focus groups and a final roundtable. The OECD Mission Action Lab (MAL) is a joint initiative created in 2021 by the OECD Directorate for Science, Innovation and Technology, the OECD Observatory of Public Sector Innovation, and the OECD Development Co-Operation Directorate.

in-depth case study research on over 20 policy initiatives, the OECD has observed four main MOIP types (Larrue, 2021). Each type involves specific trade-offs between mission characteristics that produce their distinctive strengths and weaknesses (Table 6.1).

The different types of MOIPs vary notably in their scope and level of integration (Figure 6.3). They can be distinguished from the most traditional type of policy framework, which is composed of individual instruments that each have a single goal (e.g., increasing R&D investment) and are aimed at alleviating specific market failures (e.g., underinvestment in R&D due to the low appropriability and high uncertainty of research results). These policies, which are the result of decades of influence by the New Public Management doctrine, are characterised by a narrow scope and a low level of integration in the instrument portfolio. In large "umbrella missions", multiple public interventions that address various needs for achieving a common goal are collated within a single institutional vehicle (which may or may not be officially referred to as a mission). However, these interventions are loosely integrated, and for the most part, they involve an exchange of information between the various responsible authorities and collective ex-post monitoring rather than ex ante collective decision-making and real joint action. On the opposite corner of the diagram, some initiatives use a closely integrated portfolio of instruments to pursue focussed and well-delineated goals; in most cases, these goals are related to the development of green technological innovations. Finally, a few missions combine a broad systemic scope and a significant level of integration, allowing for joint steering of activities along a collectively defined pathway in order to address a certain challenge. With their significant levels of integration, the latter two types of mission are closer to the ideal type.

To date, few mission initiatives (if any) have succeeded in achieving strong integration on a wide scope, as this requires many actors and policy instruments. The following two main pathways towards transformative missions have been identified:

1. The evolution of large umbrella missions towards increased integration: this requires time, as it involves building trust between the parties concerned and reforming many incentives and formal and informal processes.
2. Broadening the scope of acceleration missions to establish larger, more systemic national missions while maintaining their close coordination: this pathway has been used in Norway, for example.

6.3 WHAT ARE MISSION-ORIENTED POLICIES EXPECTED TO DELIVER?

This section presents the framework that was used to guide the fieldwork conducted on a comprehensive pool of missions aimed at reducing GHG emissions.

6.3.1 The Theory of Change for Analysing Net-zero Missions: What Are the Outputs and Outcomes of Mission-Oriented Innovation Policies for Net Zero?

Our knowledge is still limited when it comes to the extent to which, how, and under what conditions mission-oriented policies do produce their expected impact. At present, the latest knowledge on missions focusses on different designs of MOIP types. Some of their processes

Table 6.1 *Basic characteristics of the four main types of MOIPs*

Type	Leadership	Missions	Examples	Typical strengths	Typical weaknesses
1. Overarching mission-oriented strategic frame-works	• Centre of government • High-level cross-governmental body	• Multiple missions or mission areas • Pursuing ambitious challenges • Long-term horizon	• Horizon Europe's missions (EU) • Mission-driven Top Sectors policy (Netherlands) • High-Tech Strategy 2025 (Germany) • France 2030 Acceleration Strategies (France) • Hydrogen Earthshot (United States) • Moonshot R&D Program (Japan)	• Visibility and inclusiveness • Broad scopeSystemic intervention	• Loose coordination, umbrella type • Higher transaction costs
2. Challenge-based programmes and schemes	• Agency	• Focussed • Seeking acceleration of innovation • Mid- to long-term horizon	• Pilot-E (Norway) • Industrial Strategy Challenge Fund (United Kingdom) • Science Foundation Ireland's Innovative Prize (IE)	• Coordination of public support along the innovation chain • Hands-on management of projects/project portfolios • Articulation of supply and demand for knowledge	• Niche solutions, scale-up required • Focussed on technological/ engineering solutions

Type	Leadership	Missions	Examples	Typical strengths	Typical weaknesses
3. Topic-specific mission-oriented programmes	• Ministry • Agency • Large research institutes	• Build upon existing programmes to ensure more effective targeting and coordination • Mix of societal and competitive challenges	• Cross-ministerial Strategic Innovation Promotion Program (Japan) • CSIRO's missions (Australia) • Flagship Programs (Finland) • NRC Challenge program (Canada)The Alchemist (Korea) • Building of Tomorrow/ Cities of the Future (Austria)	• Continuity with former topic-specific research and innovation programmes	• Path dependencies limit the adoption of mission-oriented policy practices (result-driven, cross-sector coordination) • Tendency to remain focussed on research
4. Ecosystem-based mission programmes	• Ministry • Agency	• Innovation agenda developed by the innovation actors themselves, with neutral support from public authorities	• InnoMissions (Denmark) • Growth Engines (Finland) • Strategic Innovation Areas (Wallonia, Belgium) • Strategic Innovation Programmes (Sweden)	• Strong legitimacy, ownership, and commitment of all actors towards the codeveloped strategic agenda • Role of public authorities as orchestrators/ facilitator of ecosystem formation	• Less challenge-/ mission-based. Often very open at the start • Risk of capture of missions by and in favor of leading/ historical players

Source: Author's own elaboration.

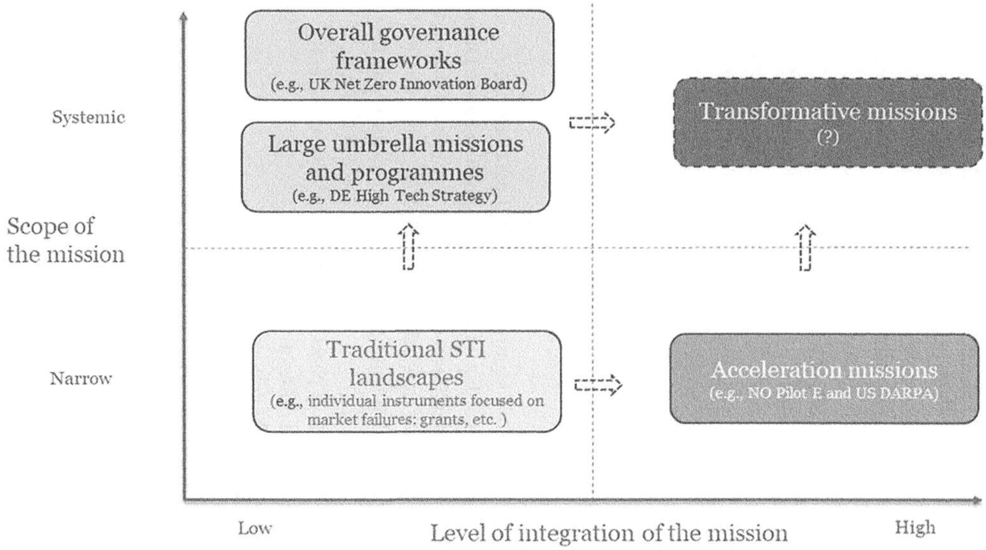

Source: Author's own elaboration.

Figure 6.3 *Scope and level of integration of policy frameworks*

have also been analysed through case studies, focussing on the definition of their objectives and strategic agenda rather than on their implementation, as most of these policies are still at an early stage. Although a "theory of change" for mission-oriented policy is certainly needed, it has yet to be formulated for the most part (Janssen et al., 2021; Hekkert et al., 2020).

This chapter presents a ToC for net-zero missions, making explicit the expected causal relationships between the problems they tackle and their desired goals, inputs, outputs, outcomes, and impact (Figure 6.4). It is composed of several stages that connect the nature of the missions to the impact they are expected to produce:

1. The starting point of the ToC is the distinction between the three MOIP building blocks mentioned above. Each of them contains the *design principles* that constitute the essence of what an MOIP approach is.
2. The second stage of the ToC includes the *intermediate outcomes* of the MOIPs. For instance, it is expected that gathering a wide range of public and private actors from different disciplines, sectors and policy areas will allow the development of a shared agenda with strong legitimacy among these actors and beyond, as well as other features, notably a high level of directionality and intentionality.[2] At the end of the development process, this shared agenda should consist of a broader set of potential options that are directed towards more ambitious and long-term objectives. In combination with elements of the policy coordination building block, this shared agenda should also result in a higher budget for the mission, as well as stronger buy-in and commitment from the private actors

[2] These are the "design principles" mentioned in Larrue (2021).

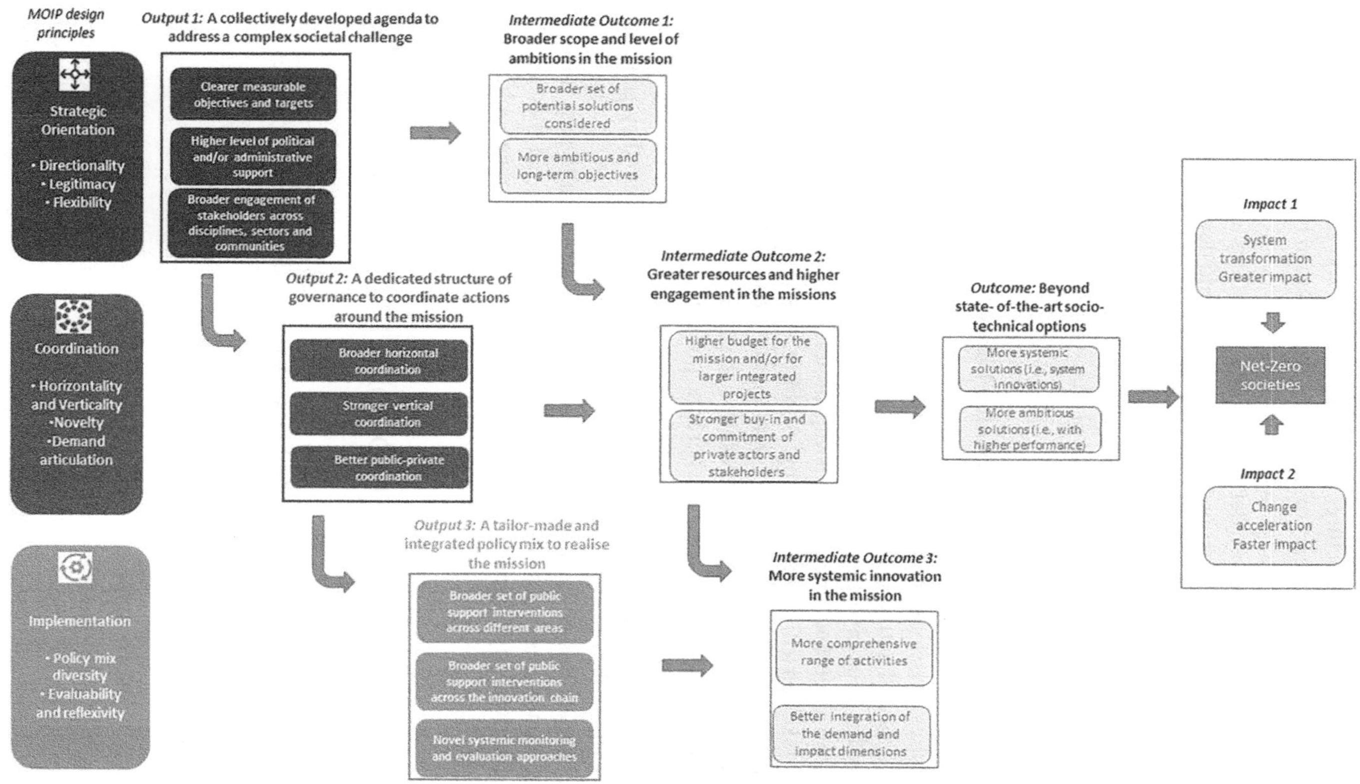

Source: Author's own elaboration.

Figure 6.4 *The theory of change for a net-zero mission: from MOIP design principles to net-zero achievements*

and stakeholders that are involved in the realisation of the mission and support for a more comprehensive range of activities (i.e., not only technological aspects but also elements related to the required skills, infrastructure, regulatory framework, user awareness, etc.).

3. Finally, in combination with the elements of the policy coordination building block, missions produce *outcomes*. A wide range of actors across all silos, strongly committed to achieving the mission by exploring a broader set of potential options and a wider range of supported activities, and benefiting from greater resources, could devise more ambitious solutions (i.e., with higher performance) that could be articulated in a more systemic way.
4. At the end of the causal chain, the *impact* of an MOIP is the realisation of its mission. Building on the abovementioned differentiation between acceleration and transformative missions, a distinction can be drawn between two primary forms of impact: system transformation and change acceleration.

6.3.2 The Double Additionality of Mission-Oriented Policy

In the vast majority of cases, missions consist of an additional layer of coordination on top of existing policy instruments and activities, which is aimed at achieving common objectives. There are only a few missions that involve new interventions, instruments, or funding streams. Instead, most missions leverage and integrate existing policies. In this context, the whole concept of additionality takes on a different meaning. Additionality is generally defined as the extent to which something happens as a result of an intervention and would not have occurred in the absence of the intervention. In the case of MOIPs, the counterfactual scenario is not the absence of policy intervention, but a scenario where the policies are less directional, coordinated, and integrated. The MOIP ToC aims to use this scenario as a point of reference for evaluating not only the missions' overall impact, but also the additional effects they generate beyond those of the existing policy instruments they incorporate (Figure 6.5).

To be considered legitimate this policy approach must therefore successfully pass a rigorous dual assessment: not only must missions achieve their stated objectives, but they should also do so in a more effective and efficient manner when compared to traditional policy and governance frameworks.

6.4 WHAT ARE MISSION-ORIENTED POLICIES DELIVERING TO HELP ACHIEVE NET ZERO?

This section presents the result of the OECD's ongoing work on mission-oriented policies for net zero.[3]

6.4.1 Net-zero Missions

The OECD has identified 29 MOIP initiatives that align with the MOIP definition *to a significant extent*. When using the ideal-type approach, it is impossible to draw a strict line between initiatives that are or are not MOIPs. Some MOIPs start with very incremental changes vis-à-vis the existing framework and a lot can be learned from these initiatives, provided that there

3 Preliminary results of the study have been presented in OECD (2023).

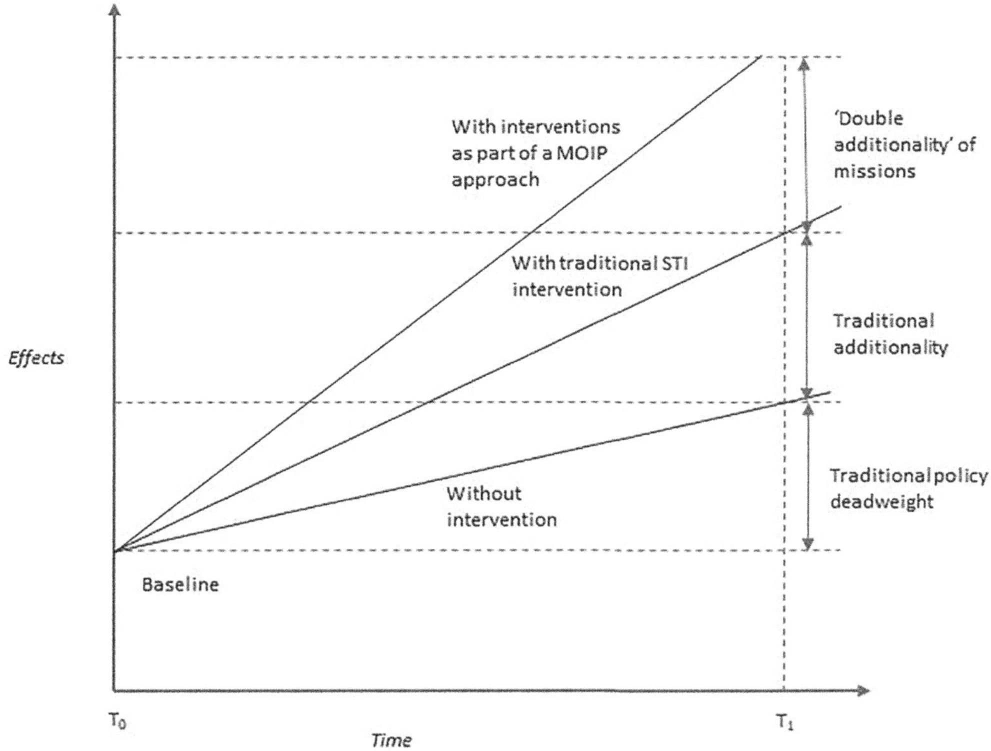

Source: Author's own elaboration.

Figure 6.5 Conceptual view of "double additionality" in MOIPs

is a genuine intent to adopt a mission-oriented approach. In the end, the acceptable distance from the ideal type depends on the analysts' judgement. However, based on the analysts' experience, it is possible to exclude initiatives that are clearly attempts at "mission-washing".

Among these initiatives, the fieldwork focusses on 74 missions that aim to reduce greenhouse gas emissions (see Appendix). Twenty in-depth case studies were conducted on net-zero missions, with each including at least two interviews.

The ToC was used to systematically analyse the effects of the missions.

6.4.2 Net-zero Mission Outputs

Analysing existing net-zero missions using the framework outlined above shows that net-zero missions produce many of their expected outputs and that, in most cases, these missions represent a marked improvement in terms of policy design when compared to traditional STI policy mixes.

Table 6.2a, Table 6.2b, and Table 6.2c (panel A) illustrate the results of the analysis of how existing net-zero MOIPs perform against the three outputs highlighted in the ToC. These results clearly show the following features: a codesigned agenda, a dedicated structure of governance and, finally, a tailor-made and integrated policy mix. In contrast to traditional policy mixes, net-zero missions are characterised by the following:

1. *Stronger orientation*, with clearer objectives and measurable targets relating to reductions in greenhouse gas (GHG) emissions, although few of them correspond to the expected "SMART" goal format:[4] These initiatives tend to have codeveloped strategic agendas that are directly associated with financial resources and modes of implementation, which is rarely the case with other strategic frameworks.
2. *Broader coordination* of policy plans across administrative silos: this brings together the authorities in charge of research and innovation policy and the "owners" of the challenges they tackle – for instance, the policy and regulatory bodies in charge of transportation or environmental affairs. However, to date, budgets are not commensurate with the transformative objectives of net-zero missions, and those budgets originate almost exclusively from STI public authorities.
3. *Higher integration* of various support instruments across the different stages of the innovation chain (from support for research to the strengthening of skills and, for some of

Table 6.2a Panel A. Missions – outputs: The collective development of a strategic agenda for addressing a complex societal challenge

Net-zero mission effects	Main results
Clearer measurable objectives and targets	Missions have enabled the setting of clearer goals than in traditional programmes, although few of them correspond to the expected "SMART" goal format. Only half of them have established clear targets. Strategic agendas allow "continuous directionality" and complement mission objectives in a context of high uncertainty and contestation. In stark contrast with traditional strategies and roadmaps, strategic agendas are directly associated with mission budgets, coordination structures, and modes of implementation.
Higher level of political/administrative support	Missions are political by nature due to normative goals related to societal impacts. Broad national missions attract more attention from politicians and higher administrative levels, which strengthens their legitimacy but can add pressure to obtain early results.
Broader engagement of stakeholders	Strategic agendas are developed by a wide range of actors from different communities, which increases their ownership of the mission and subsequent engagement.

Notes: Synthesis of the main results across the MOIP theory of change.
Source: OECD, 2023.

4 Specific, measurable, achievable, relevant, and time-bound.

Table 6.2b *Panel A. Missions – outputs: The alignment and holistic monitoring of the plans of various actors*

Net-zero mission effects	**Main results**
Broader horizontal coordination	Almost all of the net-zero missions that formed the subject of one of the case studies represent a significant expansion of the scope of coordination between different actors across the government structure. Leadership is assumed by the STI authorities that launched the mission. Cross-sector coordination is the main challenge for net-zero missions and, in the largest and most integrated ones, this generates significant transaction costs.
Stronger vertical coordination	Most net-zero missions are led by national (or EU) authorities and some of them include local authorities in the governance. Local authorities are involved in many mission activities to enable the demonstration of solutions and early transition to market. A small number of net-zero missions are being implemented at regional level, particularly in the context of the new generation of EU Smart Specialisation Strategies.
Better public-private coordination	Missions complement (and often benefit from) the existing public-private STI collaboration platforms. They add a well-targeted goal-oriented framework and ensure a more direct link to policy intervention.

Notes: Synthesis of the main results across the MOIP theory of change.
Source: OECD, 2023.

Table 6.2c *Panel A. Missions – outputs: The articulation and management of a portfolio of activities*

Net-zero mission effects	**Main results**
Broader set of public support interventions across different areas	All net-zero missions bundle different types of interventions together under a common strategic and governance framework; these interventions range from R&D grants to skill formation and market deployment. They allow the coordination of public support for different components of the systemic solutions.
Broader set of public support interventions across the innovation chain	Some net-zero missions map and connect the various support instruments across the different technology readiness levels in practical ways, with a view to providing more continuous support at different stages of the innovation chain.
Novel systemic monitoring and evaluation approaches	Many mission partners share the opinion that new evaluation methodologies and processes are needed to evaluate this approach, but very few MOIP evaluations have been undertaken to date. Meanwhile, those that have been conducted are not significantly different from traditional STI policy evaluations.

Notes: Synthesis of the main results across the MOIP theory of change.
Source: OECD, 2023.

them, market deployment through price-based mechanisms and public procurement). One of the areas where net-zero missions add most value is their result orientation, which leads mission partners to articulate societal needs and demands at different stages of the mission life cycle. Starting from the mission objectives, partners embed these demands and needs into every step of the mission activities, even in the more upstream research components (Figure 6.6). These linkages require new, proactive portfolio management practices that challenge prevailing skills, incentives structures, and processes in ministries and agencies.

6.4.3 Net-zero Mission Outcomes

Table 6.3a, Table 6.3b, and Table 6.3c (panel B) illustrate the results of the analysis of how existing net-zero MOIPs perform against the three outcome highlighted in the ToC. It shows that despite significant progress and conitnuous learning this policy approach has not yet delivered on all its promises.

Net-zero missions are broader in scale and scope than traditional innovation programmes, but they are still focussed on more or less narrowly defined forms of technological innovation. While mission managers often claim that social innovation is as important as technological innovation in their mission and far more prominent than in other STI initiatives, the social components are mainly limited to advocacy, information and communication campaigns, and various studies conducted at individual and societal levels to prepare for technology scale-up and market transition. Given the importance of combining both technological and social innovation, the transformative potential of the current set-up of these missions remains unclear. This is consistent with findings from other studies; for example, a study on the missions of the German High-Tech Strategy 2025 indicated that stakeholder involvement in the missions was still very much limited to the traditional R&I actors in the mission formulation process, and the missions did not fully realise their transformative potential as a consequence (Roth et al., 2022).

Even in the case of the broadest, most ambitious missions, analyses in terms of effective leadership and real commitments show that, in practice, they are also strictly funded with "STI money" and relate to programmes with an STI mandate. As vast numbers of stakeholders need to be mobilised in far more fields than just STI, the question must be asked as to whether research and innovation programmes are the best location for "hosting" such ambitious objectives.

Missions have codeveloped broad, systemic, shared strategic agenda to realise their objectives and set up dedicated, holistic governance structures that bring together a range of policymakers from different sectors with the goal of aligning their plans and monitoring their actions against the strategic agenda for a particular mission. This applies in particular to ecosystem-based missions that empower partners to define their own agendas; in such missions, a significant share of resources is dedicated to the formation of the ecosystem and various partnerships ranging from support for networking to coordination and orchestration. However, the influence of the strategic agenda on actual collective decision-making regarding budget allocation and policy implementation is less clear. Therefore, while most missions have been successful in setting legitimate and powerful objectives and targets, real collective action that builds on the shared strategic agenda is less frequent. Missions are generally less integrated at the implementation level than they are at the orientation and coordination levels.

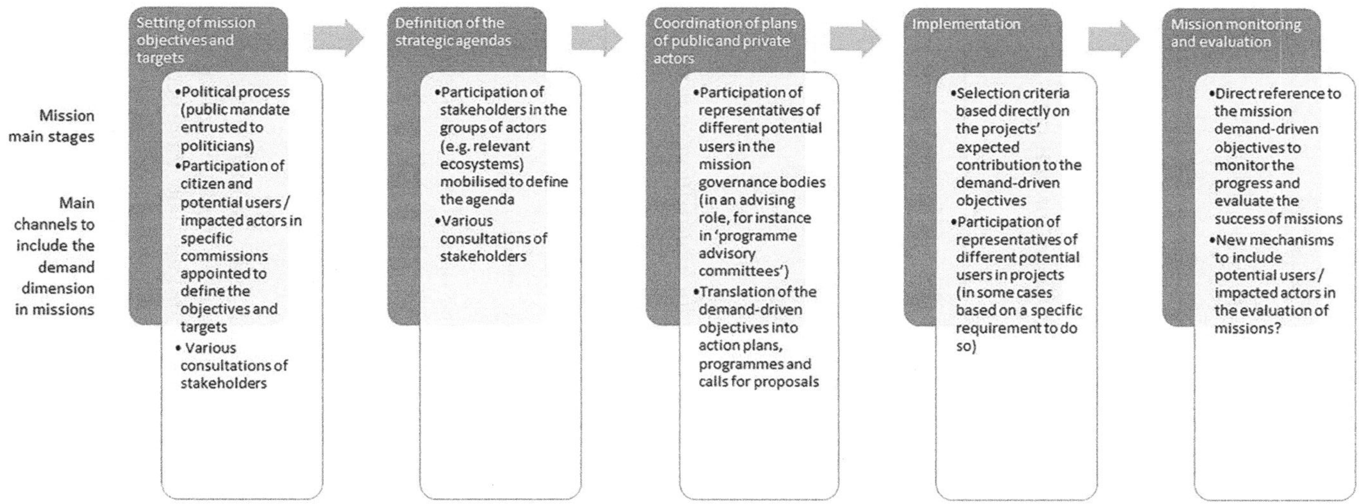

Source: Author's own elaboration.

Figure 6.6 *Integration of the demand dimension into net-zero missions at different stages of the mission lifecycle*

Table 6.3a *Panel B. Missions – outcomes: Broader scope and level of ambitions in missions*

Net-zero mission effects
Broader set of potential solutions considered
Most missions have set objectives that are geared towards achieving results by 2030 and 2050 The difference in missions' levels of ambition is difficult to assess

Notes: Synthesis of the main results across the MOIP theory of change.
Source: OECD, 2023.

Table 6.3b *Panel B. Missions – outcomes: Greater resources and higher engagement in net-zero missions*

Net-zero mission effects
Higher and longer-term funding
Strong engagement in development of the strategic agenda but insufficient information on private sector financial contributions and, therefore, the leverage effects of missions

Notes: Synthesis of the main results across the MOIP theory of change.
Source: OECD, 2023.

Table 6.3c *Panel B. Missions – outcomes: More systemic innovation in net-zero missions*

Net-zero mission effects
Better integration of the demand and impact dimensions
Net-zero missions allow for a wide and more consistent range of activities from basic research to deployment, capacity building, and communication and advocacy Proactive portfolio management practices are necessary in order to reap the systemic benefits of missions. However, they require significant resources and new skills, rules, and procedures in ministries and agencies

Notes: Synthesis of the main results across the MOIP theory of change.
Source: OECD, 2023.

Beyond the issue of missions' influence on underpinning public policies, their success will also depend on their leverage effect in mobilising the private sector. The latter will need to provide enormous amounts of resources to fund sustainability transitions and, more generally, to engage as key agents of change in these transitions. While policymakers and analysts are still currently focussing on public funding and cross-ministerial coordination in this phase, the ultimate test of missions will be the contribution and financial commitments of private businesses and investors. While many policy instruments that are aimed at helping the business sector to reduce GHG emissions are already in place, it remains unclear whether the added value of their being embedded in a mission approach (i.e., the "double additionality") will strengthen business engagement and investment more than their contribution to the development of objectives and roadmaps already does.

6.5 WHAT IS THE NEXT STAGE FOR MISSION IMPLEMENTATION AND ANALYSIS?

6.5.1 Missions Must Not Only Accelerate Technological Change but also Transform Socio-technical Systems

Overall, the analysis presented in this chapter enables a better understanding of the current MOIPs and highlights their achievements and limitations. According to the abovementioned typology developed by Kuittinen et al. (2018), the mission-oriented policies analysed in this study are closer in some respects to accelerator missions or umbrella missions than the transformative missions presented in Figure 6.3. Therefore, at this early stage of their existence where significant experimentation and learning are still ongoing, net-zero missions have yet to become the frameworks for systemic collective action that they are expected to be. This conclusion is consistent with the results of a recent study conducted by the Joint Research Centre (JRC) on transformative innovation policies, where the authors concluded that none of the 12 cases analysed in depth (most of which were MOIPs according to the OECD definition) stand out as being "fully" transformative despite their bold ambitions (Janssen et al., 2023). These results can be partially explained by the diversity of organisational models: the various initiatives and the different contexts they experiment in do not all match the expected features of transformative innovation policies or missions. As the authors rightly point out, there is no one, perfect pathway to transformation.

Another reason for these results lies in the MOIPs' current position in their "policy trajectory". Although they are very recent in most cases, existing experimental and pilot mission initiatives have already generated important learnings, and they continue to evolve. For example, the "Future Strategy" follows the missions of the German High-Tech Strategy 2025, which were criticised for their weak directionality and loose coordination. As was suggested by the advisory body, the High-Tech Forum, prior to its dissolution, the new German initiative is intended to be mission-like. This is why it has dedicated governance structures, for example. In Austria, the mission-oriented, topic-specific programmes "Building of Tomorrow", "Mobility of the Future" and "City of Tomorrow" have been taken over by (or embedded in) four national, directional, and cross-ministerial missions (Biegelbauer et al., 2020). Meanwhile, in Ireland, the funding and institutional dynamics of the Irish Recovery and Resilience Plan will allow the SFI Challenge Research programmes to expand and "deepen" their mission-oriented approach in the "National Challenge Fund". This initiative has greater directionality, which should allow for a more consistent and interrelated project portfolio. Likewise, the Dutch Mission-oriented Top Sectors and Innovation Policy (MTIP), a broad policy framework that includes 25 missions, will also be improved in terms of strategy and efficiency. In Japan, plans are in place to modify the design of the third generation of Cross-ministerial Strategic Innovation Promotion Programs significantly, in order to make them more targeted and integrated.

However, there are only a few of these kinds of initiatives, and more work must be done in terms of making missions "fit" to bring about the transformative changes needed to achieve the net-zero goal. This process will require effective monitoring and evaluation frameworks, so that adequate methodologies and processes can be developed both for assessing mission results and for capturing the added value and weaknesses of their mission-oriented features.

Transiting missions to their next stage and ensuring that they escape the "STI only" and "orientation" traps mentioned above will also depend on changes in their underlying environment. Missions often "stall" not because they are poorly designed but because the mandates, administrative and legal rules, accounting structures, and governance models of the relevant ministries and agencies cannot easily adapt to the mission approach (Aagaard et al., 2022). Changes at this level are far beyond the reach of STI authorities alone and will require significant political support and active engagement.

Net-zero missions will be able to realise their transformative potential if they build on their early results in terms of output and learning from good practices and succeed in finding a way out of two common traps:

1. The "*STI trap*": Despite displaying some systemic features, most net-zero missions continue to focus on supporting research and innovation; they are also led by STI authorities, and they draw almost exclusively on STI funds.
2. The "*orientation trap*": So far, most net-zero missions' achievements have involved defining strategic agendas and setting up governance structures. Evidence of joint implementation is rarer and tends to be on a small scale.

The ability of STI authorities in their current role as "mission champions" to attract stronger commitment from the sector-specific authorities that "own" the societal challenges (and as such, many of the programmes and regulations related to these challenges) will be key to mission success. This raises the question of where the missions that aim to transform socio-technical systems and achieve net zero should be incorporated into government structures. There are three main options here, namely STI public authorities (research and/or innovation), the "challenge owners" (sector-specific authorities), and centre of government (president, prime minister, or cabinet offices). Some countries with more extensive experience of this policy approach, such as Sweden and Norway, where missions fall mainly within the remit of the research and innovation agencies, have tried to "elevate" missions to a higher and broader level of governance, although the results have so far been mixed. The example of the French missions (the Acceleration Strategies), which are led by an autonomous agency attached to the prime minister's office, with strong support from the president and a dedicated budget covering a broad systemic portfolio of actions over different generations of socio-technical solutions, offers a different (centralised) model for institutionalising missions. Novel options for steering, governing, and possibly managing large systemic missions have yet to be designed and experimented on at different levels across government structures. Other potential options for incorporating missions include a dedicated agency, a virtual cross-agency platform, a large public-private partnership/platform, or an autonomous organisation with foundation status.

6.5.2 Knowledge Gaps regarding Net-zero Missions

The interest in mission-oriented policy is still very recent, but the five last years have been rich in terms of learning about their implementation. However, there is still a great need for more knowledge when it comes to bringing net-zero missions to the next stage, i.e., going beyond the two traps mentioned above. The OECD's work on net-zero missions, enriched by a thorough review of the literature on missions, points to the following knowledge gaps:

1. How can the missions be made truly systemic, so that they expand beyond the realm of STI actors, policies, and funding? This issue encompasses questions such as the following: Where should missions be incorporated within national innovation systems (see above)? What kind of incentives could promote this form of joint endeavour involving STI and sector-specific public authorities? What supporting mechanisms and processes (including in terms of budget and evaluation) are required?
2. What should "policy integration" within missions include, and how? It is essential to provide the practical tools for understanding what needs to be integrated in the specific case of each challenge. The term policy integration, which includes a variety of items (budget, infrastructure, policy instruments, etc.) across different established policies, needs to be "debunked". While it is clear that a broad, cross-sector collaborative approach is critical to achieving net zero, little is known when it comes to the question of what the boundaries and content of policy integration should be for different challenges at different moments in time.
3. What changes must be implemented in public sector governance if mission-oriented policies are to be successful and how can these changes be promoted? The mission-oriented policy approach challenges public sector structures (e.g., governance and roles and responsibilities within different organisations), processes (e.g., budgetary and human resource management), skills (e.g., for portfolio project management), and mindsets (e.g., system thinking, impact, and future orientation). Many organisations have adopted this policy approach without having reformed any of these structures, which limits the transformative potential of missions.
4. What linkages are needed between net-zero missions and the broad and high-level national governance and policy framework for net zero? Net-zero missions cannot be all-encompassing; in most cases, they are focused on particular sources of emissions (e.g., agriculture, transport, etc.), technological domains (e.g., hydrogen or carbon capture, storage and utilisation (CCSU)) or geographical areas (e.g., cities). As such, there is a need for coordination between the missions themselves and with all other activities relevant to the achievement of each country's net-zero commitment. Some countries – especially those that have translated their net-zero targets into law – have set up a dedicated governance structure to lead the progress reporting and review processes and coordinate and/or advise on actions at a national and holistic level. A better understanding of how these frameworks operate and how they provide and use information from and to the net-zero missions (especially the broad, systemic missions) would be essential to understanding how the nested structure of systemic approaches contributes to the achievement of net-zero objectives in a consistent way.

REFERENCES

Aagaard, K., Norn, M. T., & Stage, A. K. (2022). How mission-driven policies challenge traditional research funding systems [version 1; peer review: awaiting peer review]. *F1000Research*. https://doi.org/10.12688/f1000research.123367.1

Biegelbauer, P., Hartmann, C., Polt, W., Wang, A., & Weber, M. (2020). *Mission-oriented innovation policies in Austria: A case study for the OECD.* Austrian Institute for Technology and Joanneum Research. https://www.joanneum.at/fileadmin/POLICIES/Publikationen/Endbericht_BMK_OECD_MOP_Final.pdf

Cervantes, M., Criscuolo, C., Dechezleprêtre, A., & Pilat, P. (2023). Driving low-carbon innovations for climate neutrality. *Science, Technology and Industry Policy Papers.* OECD Publishing. https://doi.org/10.1787/8e6ae16b-en

Geels, F., & Schot, J. (2007). Typology of sociotechnical transition pathways. *Research Policy, 36*(3), 399–417. https://doi.org/10.1016/J.RESPOL.2007.01.003

Hekkert, J., Wesseling, J. H., & Negro, S. O. (2020). Mission-oriented innovation systems. *Environmental Innovation and Societal Transitions, 34,* 76–79. https://doi.org/10.1016/j.eist.2019.11.011

IEA (2022). *Energy technology RD&D budgets.* https://www.iea.org/data-and-statistics/data-product/energy-technology-rd-and-d-budget-database-2

IPCC (2018). *Special report: Global warming of 1.5°c – Summary for policymakers.* https://doi.org/10.1017/9781009157940.001

IPCC (2022). *Climate change 2022 – Mitigation of climate change – Summary for policy makers,* https://www.ipcc.ch/report/ar6/wg3/.

Janssen, M.J., Torrens, J., Wesseling, J.H., & Wanzenböck, I. (2021). The promises and premises of mission-oriented innovation policy – A reflection and ways forward. *Science and Public Policy, 48*(3), 438–444. https://doi.org/10.1093/scipol/scaa072

Janssen, M., Wanzenböck, I., Fünfschilling, L., & Pontikakis, D. (2023). *Capacities for transformative innovation in public administrations and governance systems: Evidence from pioneering policy practice.* https://doi.org/10.2760/220273

Jeudy-Hugo, S., Lo Re, L., & Falduto, C. (2021). *Understanding countries' net-zero emissions targets.* OECD/IEA Climate Change Expert Group Papers (2021/03). https://doi.org/10.1787/8d25a20c-en

Kuittinen, H., Polt, W., & Weber M. (2018). Mission Europe? A revival of mission-oriented policy in the EU. In RFTE (Ed.), *RE:thinking Europe.* Austrian Council for Research and Technology Development, pp. 191–207.

Larrue, P. (2021). The design and implementation of mission-oriented innovation policies a new systemic policy approach to address societal challenges. *OECD Science, Technology and Industry Policy Papers.* https://doi.org/10.1787/3f6c76a4-en

Lenton, T.M., Rockström, J., Gaffney, O., & Rahmstorf, S. (2019). Climate tipping points – Too risky to bet against. *Nature, 575.* https://doi.org/10.1038/d41586–019–03595–0

Mazzucato, M. (2016). From market fixing to market-creating: A new framework for innovation policy. *Industry and Innovation.* https://doi.org/10.1080/13662716.2016.1146124

OECD (2019). *Governance as an SDG accelerator: Country experiences and tools.* OECD Publishing. https://doi.org/10.1787/0666b085-en

OECD (2020). *Science, technology and innovation: How co-ordination at home can help the global fight against COVID-19.* https://doi.org/10.1787/aa547c11-en.

OECD (2021). *Transport strategies for net-zero systems by design.* OECD Publishing. https://doi.org/10.1787/0a20f779-en

OECD (2022a). *Climate Tipping Points: Insights for Effective Policy Action,* OECD Publishing. https://doi.org/10.1787/abc5a69e-en

OECD (2022b). *Mapping Business Innovation Support (MABIS) – R&D tax incentives reporting, OECD R&D tax incentives database.* https://www.oecd.org/sti/rd-tax-stats-database.pdf

OECD (2023). Reaching net zero: Do mission-oriented policies deliver on their many promises? In OECD (Ed.), *OECD Science, Technology and Innovation outlook 2023: Enabling transitions in times of disruption.* OECD Publishing. https://doi.org/10.1787/fd790e2c-en

Roth, F., Wittmann, F., Hufnagl, M., & Linder, R. (2022). *Putting mission-oriented innovation policies to work: A case study of the German High-Tech Strategy 2025.* https://www.isi.fraunhofer.de/content/dam/isi/dokumente/cci/innovation-systems-policy-analysis/2022/discussionpaper_75_2022.pdf

Schot, J., & Steinmueller, W. (2018). Three frames for innovation policy: R&D, systems of innovation and transformative change. *Research Policy, 47*(9), 1554–1567. https://doi.org/10.1016/J.RESPOL.2018.08.011

UNEP (2021). *Emissions gap report 2021: The heat is on – A world of climate promises not yet delivered*. https://www.unep.org/emissions-gap-report-2021
Wittmann, F., Hufnagl, M., Lindner, R., Roth, F., & Edler, J. (2021). Governing varieties of mission-oriented innovation policies: A new typology. *Science and Public Policy*, *48*(5), 727–738. https://doi.org/10.1093/scipol/scab044
World Resources Institute (2020). *State of climate action: Assessing progress toward 2030 and 2050*. https://doi.org/10.46830/wrirpt.20.00001

Table A.1 List of missions in the OECD net-zero mission database

Country	MOIP initiatives	Net-zero missions
Australia	**CSIRO's Missions**	Climate Resilient Enterprises Drought Resilience Mission Ending Plastic Waste Mission Hydrogen Industry Mission Towards Net Zero Mission
Austria	**Building of Tomorrow (plus)**	Innovative, sustainable concepts both for new buildings and for renovating existing ones
Austria	**City of Tomorrow**	City of Tomorrow
Austria	**Mobility of the Future**	Fit4UrbanMission
Belgium	**Smart Specialisation Strategy of Wallonia**	Circular materials
Canada	**NRC Challenge program**	Materials for Clean Fuels Challenge
Denmark	**INNOmissions**	Carbon capture and storage or utilisation Climate and environment-friendly agriculture and food production Green fuels for transportation and industry (Power-to-X, etc.) Recycling and reduction of plastic waste
European Union	**Horizon Europe's missions**	Adaptation to climate change including societal transformation (ACC) Climate-neutral and smart cities Healthy oceans, seas, coastal, and inland waters Soil health and food
Finland	**Flagship programmes**	Atmosphere and Climate Competence Center flagship programme FinnCERES Flagship Competence Centre for the Materials Bioeconomy
Finland	**Growth Engines**	Accelerating the growth of the carbon capture market Growth Engine (Compensate platform company) Baltic Offshore wind Growth Engine (Gaia consulting orchestrator) From a socio-sized demo of renewable energy production to an international total solution provider Growth Engine (Flexens platform company) GreenE2 Growth engine (CLIC Innovation orchestrator)
France	**Stratégies Nationales d'Accélération**	Decarbonation de l'industrie Hydrogène décarboné

Country	MOIP initiatives	Net-zero missions
Germany	**High-Tech Strategy 2025**	Achieve substantial greenhouse gas neutrality in industry ("GGNII") Build up a battery cell production in Germany Create sustainable circular economies Develop safe, networked and clean mobility
Ireland	**National Challenge Fund**	2050 Challenge
Ireland	**Science Foundation Ireland's Innovative Prizes**	AI for Societal Good Challenge Plastics Challenge Zero Emissions Challenge
Japan	**Cross-ministerial Strategic Innovation Promotion Program (SIP)**	Technologies for smart bio-industry and agriculture (Smart Food System)
Japan	**Moonshot Research and Development Program**	Moonshot goal 4: Realisation of sustainable resource circulation to recover the global environment by 2050 Moonshot goal 5: Creation of the industry that enables sustainable global food supply by exploiting unused biological resources by 2050
Korea	**Advanced Research Programme (KARPA)**	Plasma-based CO_2 free waste organic material as a basic raw material
Korea	**The Alchemist**	Artificial eco-food Low-cost CO_2-free hydrogen production facilities
Lithuania	**Mission-oriented science and innovation programmes**	Smart and climate-neutral Lithuania
Netherlands	**Mission-driven Top-Sector Policy**	A carbon-free built environment by 2050 A net climate neutral agriculture and nature system in 2050 A sustainable and completely circular economy by 2050, with resource use halved by 2030 An entirely carbon-free electricity system by 2050 Carbon-neutral industry with reuse of raw materials and products by 2050 Zero-emission mobility of people and goods by 2050
Norway	**Pilot E**	"Green shipping" (amalgamated Pilot E missions) A zero-emissions hydrogen value chain Zero-emissions construction and facilities
Norway	**Pilot-T**	New Solutions in the Interface between Transport and ICT

Country	MOIP initiatives	Net-zero missions
Spain	**CDTI missions**	Promote a more sustainable agri-food sector adapted to the new conditions associated with climate change thanks to the relevant use of advanced biotechnological tools Promote the substitution, recovery, and valuation of mineral resources and strategic materials for the Ecological Transition Strengthen technological capacities for safe and sustainable energy autonomy (fusion, hydrogen, and renewables)
Sweden	**Challenge Driven Innovation Initiative**	Business model innovation for closed-loop furniture flows (Reuse of furniture to reduce climate impact) New technology for carbon dioxide-free asphalt (Swedish Magnetite Microwave Asphalt Road Technology; SMMART) The Swedish Innovation Platform for Textile Sorting
Sweden	**Innovation Partnership Programmes**	Climate neutral industry Innovation Partnership Programme Skills supply and lifelong learning Innovation Partnership Programme
Sweden	**Strategic Innovation Programmes**	Viable cities
Sweden	**Vinnova's pilot missions**	Healthy sustainable food: Ensuring that every student in Sweden eats healthy, sustainable, and tasty school food Healthy sustainable mobility (Street Moves 2) Healthy sustainable mobility (Street Moves 1)
United Kingdom	**Industrial Strategy Challenge Fund**	Driving the electric revolution challenge Faraday battery challenge Future flight challenge Industrial decarbonisation Prospering from the energy revolution challenge Smart sustainable plastic packaging (SSPP) challenge Transforming construction challenge Transforming food production challenge Transforming foundation industries challenge
United States	**Energy Earthshots**	Carbon Negative Shot Hydrogen Shot Long Duration Storage Shot

Source: OECD, 2023.

7. Transformative missions as part of sustainability research programmes: experiences from the German FONA framework programme

Rainer Walz, Susanne Bührer and Sarah Seus

7.1 INTRODUCTION

In recent years, science, technology, and innovation (STI) policies have increasingly been topics of discussion in the context of mission-oriented policies. Mission-oriented policies have various distinct characteristics, such as directionality when it comes to targeting a specific problem or challenge within a well-defined time frame, and using different instruments together in an actively coordinated manner. Within these general characteristics, various classification systems have been proposed to define different kinds of missions that involve varied challenges, goals, and governance requirements (Wittmann et al., 2020).

Sustainability addresses key global challenges, as spelt out in the UN's Sustainable Development Goals (SDGs). We need innovations and system transformations to deal with sustainability challenges, and there have been numerous calls – in Germany, for example – for the scientific sector to address this need (Wissenschaftsplattform Nachhaltigkeit, 2020; Rat für Nachhaltige Entwicklung, 2021; Wissenschaftsrat, 2021).

Within Germany, the research framework programme for sustainability (Forschung für Nachhaltigkeit, FONA) has been at the centre of a global-challenge-driven research policy. Building on the experience of energy and environmental research programmes, which emerged from the 1970s onwards, FONA was established in 2004 by the German Federal Ministry of Education and Research (BMBF). It has been renewed three times since then, with FONA IV the current version. FONA is designed to be a learning programme and has acquired certain characteristics of a mission-oriented programme – particularly in its latest version. This framework programme addresses major societal challenges and is similar to an umbrella mission as defined by Kuittinen et al. (2018). FONA addresses various mission-like goals. These goals were very broad in the early days of FONA but have since been narrowed down and become more specific over time. However, although FONA is increasingly taking on the characteristics of a mission policy, it is not a fully comprehensive mission-oriented policy. First, even though some milestones have been added, not all of its goals have been assigned clear time frames and specific success markers. Second, there is a strong need to improve coordination between the more supply-oriented instruments used within FONA and the numerous demand-side-oriented instruments necessary for achieving the goals; these are the responsibility of the various sector-specific ministries, such as the Ministry for Economic Affairs and Climate (BMWK), the Ministry for Digital Affairs and Transport (BMDV), and the Environment Ministry (BMUV). Despite these shortcomings, we will nevertheless refer to FONA in this chapter as a mission-oriented policy.

In terms of addressing global issues, FONA focusses mostly on various environmental challenges. However, from the beginning, FONA has also been associated with a more instrumental mission in the form of introducing sustainability science into the research system. Early on, FONA was associated with a paradigmatic shift from narrowly defined environmental research towards sustainability science (Korbun, 2004; Jahn, 2004). During the process of FONA's evolution, this has become more explicit regarding changing the research mode and making the research system more adaptive to these changes. In particular, FONA III called explicitly for a shift towards interdisciplinarity and transdisciplinarity in the mode of research and also explicitly positioned FONA within the context of promoting changes within the research system (BMBF, 2014). The most notable sign of this has been the establishment of the Sustainability In Science Initiative (SISI) and its biannual national conferences.

Given this background, this chapter focusses on how successful FONA has been in achieving its goals. This is examined through two distinct research questions. The first looks at the impacts of FONA, drawing on the results of a comprehensive programme evaluation (Bührer et al., 2020b). The second research question focusses on the transformation of the research system and considers how FONA could contribute to this. The rationale behind such a "transformative meta mission" is that sustainability science could be managed in a new way, enabling it to mobilise sustainability-oriented mission policies and targeted transitions. The analysis of this second research question also draws on the results of the evaluation by Bührer et al. (2020b). In this way, both research questions are linked. However, whereas the first research question takes a more conventional approach to evaluating FONA as a research programme, the second one frames FONA in a multi-level perspective (MLP) context, with sustainability research constituting a niche within the regime of research activities in the traditional research system.

The chapter is structured as follows: In Section 7.2, we introduce the specific conceptual background for the second research question. First, we explain the paradigm of sustainability research. Second, we demonstrate the link between research modes, researchers' motivations, and the research system. Finally, we show how sustainability science has been framed as a niche within the regime of the traditional research system. In Section 7.3, we introduce the characteristics of FONA and present the key results of an evaluation of the programme with regard to its scientific output, its impact on business and policymaking, and its impact on research modes. In Section 7.4, we interpret these results with regard to mission-oriented policies. This interpretation involves three perspectives, which are related to the two research questions mentioned above: (1) the first perspective takes up research question two on the transformation of the research system. It looks into FONA's success in acting as a transition management tool for the research system and asks what additional elements may be necessary to achieve further success; (2) the second perspective addresses research question one, particularly as regards the impact on achieving sustainability-related missions. It develops conclusions on how sustainability research programmes can be structured to align with sustainability mission-oriented policies; and (3) both research questions involve assessing the impacts of FONA, either on the development of the research system (research question two) or on achieving environmental and other sustainability goals (research question one). The third perspective, therefore, looks at the lessons that can be learnt from an assessment of FONA's impact on mission-oriented policies. Finally, in Section 7.5, we summarise the results and contextualise them in terms of future development.

7.2 CONCEPTUAL BACKGROUND

7.2.1 Sustainability Science as an Example of Mode 2 Research

Government R&D subsidies in the energy and environmental domain have a long history. Environmental problems, particularly involving air and water pollution and waste disposal, have triggered environmental research. Starting after the first oil price crisis in the early 1970s, energy research programmes were established to focus on reducing dependency on oil. These programmes originated from the German Ministry of Research and Technology and the ministries in charge of environmental protection and energy. The directionality of these programmes indicates that they were intended to be high-impact projects – the impact endpoint for these R&D subsidies was the programmes' effects on energy use and reducing environmental damage.

Debate around sustainability strategies in the 1990s emphasised the various dimensions of sustainability – environmental, economic, social – that are all interrelated. This underlines the importance of looking at societal and economic impacts in addition to environmental ones when funding sustainability-focussed research. However, in parallel to the well-known debate on sustainability strategies, a debate also emerged concerning the development of a sustainability science paradigm. This debate linked sustainability to the increasing complexity of societal problems and challenges, which has led to discussions around new forms of knowledge production (Nowotny et al., 2003; Etzkowitz & Leydesdorff, 2000). The common factor across these discussions is that they see a shift occurring in terms of the scientific paradigm. Traditional "Mode 1" science is characterised by research questions defined primarily in an academic context, orientation towards a specific discipline, quality control that is based exclusively on inherently scientific criteria, and the production of knowledge without considering the context of use. In contrast, the new "post-normal science", also called "Mode 2", is characterised by a much stronger need for interdisciplinary interaction between the research system and society, and also involves processes of reflection to a much greater extent (Funtowicz & Ravetz, 1993; Nowotny et al., 2003; Frederichs, 1999; Gibbons et al., 1994; Weingart, 1997). Some parallel is evident here with the taxonomy of Georghiou et al. (2018), who distinguish between more easily achievable Type 1 missions and Type 2 missions, which have complex underlying challenges.

Drawing on this development, a debate about a specific paradigm of sustainability science has emerged (Kates et al., 2001; Kates, 2011; Komiyama & Takeuchi, 2006). It calls for a specific mode of research in sustainability science and has emphasised the inter- and transdisciplinarity of research early on. Because sustainability research has important implications for directionality and because its impact is so vital, it is appropriate that the usability and transferability of research results should be a core component in the context of sustainability research. This implies changes in the research processes themselves. Sustainability science aims at transcending disciplinary boundaries by uniting different scientific disciplines around a problem (interdisciplinarity) and requires that actors outside the science system be included in the research process (transdisciplinarity). There are two lines of argument for the importance of inter- and transdisciplinary principles for sustainability research, which also link sustainability science to Mode 2:

1. First, sustainability research is about contributing to solving complex problems. Disciplinary boundaries, however, do not necessarily follow the structure of knowledge necessary for coming up with solutions. Kaufmann and Cleveland (1995) point towards the interdisciplinary nature of sustainability, i.e., linking environmental, economic, and social dimensions. According to Newell (2001) and Goschin and Zaman (2010), complexity is a key justification for taking an interdisciplinary approach. Furthermore, the different dimensions of sustainability are interwoven. This increases the complexity of sustainability challenges.
2. Secondly, various characteristics of the challenges call for a transdisciplinary approach. The plurality of values and norms involved increases the complexity and calls for a holistic view associated with transdisciplinarity and Mode 2 research (Hirsch Hadorn et al., 2006; Newig et al., 2019). According to Lang et al. (2012), a transdisciplinary approach is necessary to improve the impact of directional sustainability research. It allows non-scientific stakeholders to become involved in collaborative problem-framing, co-creating solution-oriented and transferable knowledge through collaborative research, and (re-) integrating and applying the co-created knowledge.
3. More recently, these arguments have also proved important within the debate around transformative science (Schneidewind et al., 2016; Scholz, 2017; Shelley-Egan, 2020; Grunwald et al., 2020), which focusses on the role of science in bringing about the necessary transformations towards sustainability. It is clear that the mode of research and the impact of research become intertwined. There are many different values and norms to be dealt with in the course of transformations, which further increases the complexity of transformation processes. This also places sustainability science much more firmly within the framework of the research field's societal responsibility. Indeed, the literature review by Daedlow et al. (2016) highlights that the mode of sustainability research is increasingly seen as more embedded, self-reflective, and co-productive, and there is an acknowledged need for more inter- and transdisciplinary focus, as well as problem and application orientation.

7.2.2 The Link between Mode of Research, Researchers' Motivations and Research Systems

Researchers deliberately choose a research mode; however, research is subject to various external conditions, such as funding or organisational strategies. From the perspective of innovation research, the interplay between researchers and framework conditions can be viewed as contributing to path dependency. Looking at this from an institutional perspective helps to provide an explanation: institutions mirror the rules of human society, with social norms and expectations shaping organisational behaviour. This means that the conventional norms and rules of Mode 1 are also transferred to research organisations. At the same time, individual actors follow routines (DiMaggio & Powell, 1991). They are not completely free in their actions, but act in accordance with the cultural and normative structure of the institution in which they are located. In the case of research organisations, this can explain how researchers, despite their freedom of research, tend to take research approaches in line with Mode 1.

Applying the perspective of self-determination theory (Gagné & Deci, 2005) leads to similar results. This theory states that individuals react on a continuum of extrinsic and intrinsic motivations. This approach has been taken up to explain the motivations of researchers. Four kinds of motivations have been identified:

1. "Ribbon" is related to researchers being recognised by their peers, based on receiving publications, citations, or prizes. It is also the basis of extrinsic rewards in academia, which take the forms of promotions and higher salaries (Lam, 2011). System pressures, such as those outlined above, also translate into "ribbon" by influencing extrinsic rewards.
2. "Gold" is another extrinsic motivation, which relates to financial benefits resulting from the application of research results (Lam, 2011).
3. "Puzzle" is an intrinsic motivation related to the process of being attracted by intellectual pursuits (Lam, 2011).
4. "Mission motivation" (Blind et al., 2022), or "pro-social objectives" (Perkmann et al., 2021), is based on insights from Iorio et al. (2017), Suominen (2021) and Van der Burgwal (2019) that point to another form of intrinsic motivation. It is related to the envisioned impact of the actor's own research. With regard to sustainability, this is particularly relevant for researchers feeling intrinsically attracted to achieving sustainability transformations.

Self-determination theory is also useful for understanding the conflicts with the existing research system that researchers experience when they are motivated by sustainability. As long as "ribbon" is associated with Mode 1 and not with the specific modes of sustainability research, mission-oriented motivation for fostering sustainability transformations could conflict with "ribbon". "Ribbon" also influences future career paths. This can go some way towards explaining the path dependency of institutions to follow a Mode 1 logic described above.

7.2.3 Interpretation of Sustainability Science from a Multi-level Perspective

Sustainability transformations require input and drive from the sustainability science sphere. But is the science system ready to deliver this needed input? Is it able to include sustainability science as an accepted paradigm that can flourish within the existing institutional setting? Posing these questions highlights the relationship between the traditional science system and sustainability and transformative science. In the introduction of this chapter, FONA's reason for changing the research mode for sustainability was characterised as a "transformative meta mission", in that sustainability science could be managed in a new way, enabling it to mobilise sustainability-oriented mission policies and targeted transitions. However, such a transformation must contend with numerous challenges and obstacles.

The multi-level perspectives (MLP) approach is a widely used heuristic for analysing the transitions of socio-technical regimes towards sustainability. However, it is also increasingly being applied to less technical domains such as social innovations, consumption patterns, and behavioural transitions towards sustainability (Hölsgens et al., 2018; Bodenheimer, 2019). Section 7.2.2 points to a path dependency in favour of Mode 1 and conflict around moving towards sustainability science. The MLP approach also highlights the importance of analysing transitions on a meso level, which involves taking a more aggregated and systemic perspective than a micro-level analysis of individual behaviour. Thus, we have interpreted the relationship between sustainability and transformative science and the existing research system in line with an MLP approach (Figure 7.1):

1. The existing traditional research system builds a regime. Instead of a socio-technical system, which serves to fulfil specific functions using technology, the function of the research system is to create new knowledge. The traditional research system has evolved in parallel with good scientific practices that must be followed, but also with the notion of

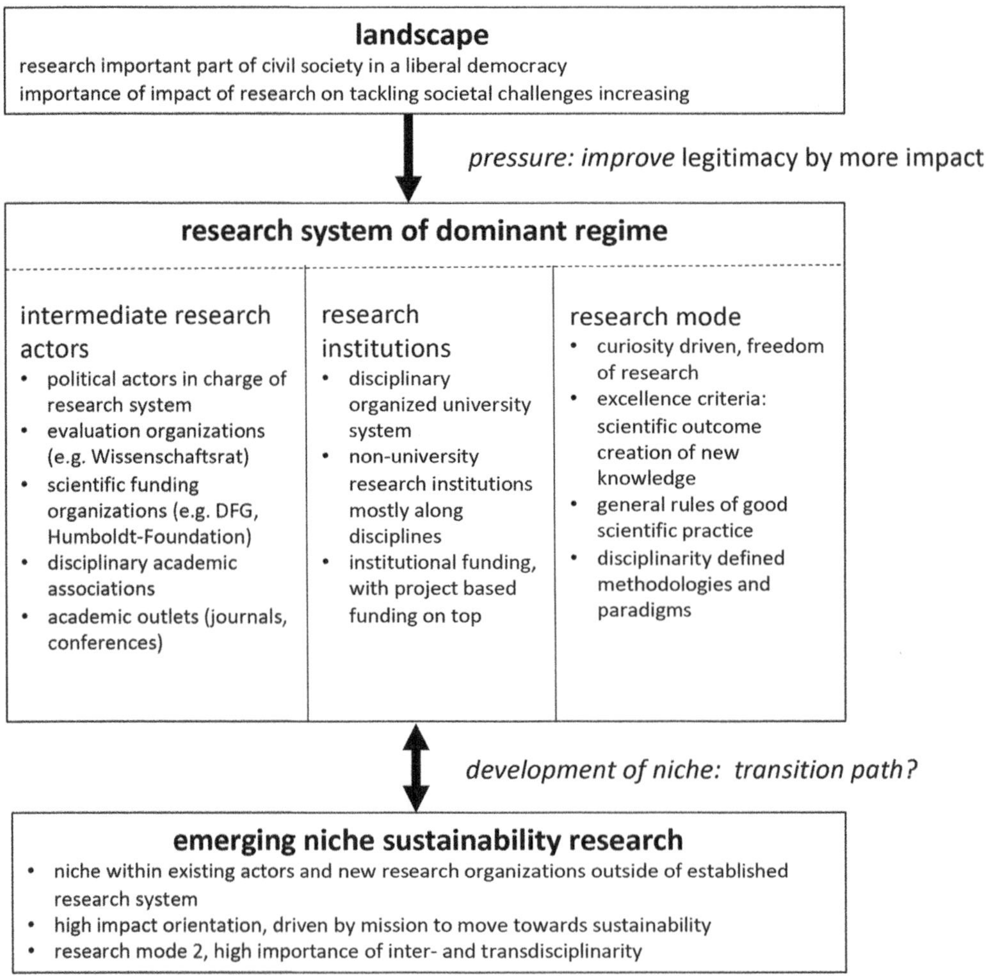

Source: Author's own elaboration.

Figure 7.1 *Characterisation of sustainability science as a niche within the research system*

freedom of research. It follows Mode 1, which rewards scientific advancement, self-regulates using peer reviews of output, and specialises in and within disciplines; this specialisation also contributes to the selection environment with its criteria of funding, careers, and status of science institutions. The mechanisms described by neo-institutionalism and self-determination theories both work towards aligning the prevailing system logic with the actions of individual researchers under the conditions of freedom of research.

2. At the landscape level, the ethical understanding of freedom of research is embedded within the concepts of liberal democracies, but also includes responsibilities for the researchers (Potthast, 2015; Vogt, 2020; Wäscher & Deplazes-Zemp, 2020). Various changes at the landscape level have been taking place over the years, which have put the traditional science system under increased pressure. First, the need to legitimise public funding of research has become more closely related to the impact of this research. Second, there is increased demand for all actors to show societal responsibility; this results in calls not only for businesses to demonstrate corporate and social responsibility, but increasingly also for the scientific sector to do the same. Third, as global challenges have become more important, the scientific sector is increasingly being expected to fulfil its societal responsibilities, which include contributing to solving sustainability problems. These developments put the traditional regime under pressure.
3. Sustainability and transformative science can be considered a niche. It emphasises the directionality of research, which results in directing research topics towards sustainability problems but could also potentially be interpreted as running counter to the idea of freedom of research. It also supports the notion of impact as a key justification for research, spreading the changes at the landscape level. Furthermore, it calls for a change in the mode of research, as shown in Section 7.2.1. However, it runs into potential conflict with the existing regime, which favours another mode of research.
4. Within the MLP approach, Geels and Schot (2007) have characterised various transition paths: transformation, reconfiguration, technological substitution, and de-alignment and re-alignment. One important characteristic influencing the transition path taken is the nature of interaction between niche innovation and the regime: "Niche innovations have a *competitive* relationship with the existing regime, when they aim to overturn or substantially modify it. Niche innovations have *symbiotic* relationships if they can be integrated as competence-enhancing add-ons in the existing regime to solve problems and improve performance" (Geels & Schot, 2007, p. 406). Symbiotic niche innovations can, for example, add to the existing regime without replacing its basic architecture (transition pathway "transformation") (Geels & Schot, 2007). From a theoretical perspective alone, it is not clear what type of interaction exists between sustainability research and the traditional research system. It can be assumed that the relationship will be more symbiotic if sustainability research achieves the traditional success factors such as publication output, because this would involve less demand for changing the basic architecture of the existing research system.

Transition management calls for the nurturing of niches. One form of STI policy is research programmes, which aim at developing sustainability innovations and transformations, and also emphasise the application of specific research modes. These programmes are legitimised by their perceived impact in terms of fostering sustainability transformations. However, when it comes to managing the transformation of the science system, such programmes can also be perceived as nurturing the niche of sustainability and transformative science. Thus, specific research programmes supporting sustainability and transformative science are not only part of directionality-driven STI policies but also necessary for transforming the science system.

So far, we only have limited evidence of the effects of large sustainability research programmes on the science system. Are they successful in helping to establish a niche? How does this niche fit with the structure of the established regime? Are there any indications that point towards one transition path configuration or another? What are the key issues that must be solved to further support this niche? In order to move towards answering such questions, we will use the empirical evidence provided by the German FONA programme. The evaluation with regard to the research mode taken and the impacts on transdisciplinary actors from business and politics shows whether FONA has been successful, not only with regard to the functional requirements of a sustainability research programme, but also with regard to fostering the key characteristics of a sustainability research niche. Furthermore, the evaluation of FONA with regard to publication output also indicates whether the niche is more likely to interact with the established regime in a more symbiotic or competitive way.

7.3 CASE STUDY OF FONA

7.3.1 Description of FONA

The FONA programme has recently undergone a thorough evaluation (Bührer et al., 2020b). This evaluation involved extensive bibliometric research, a survey of the projects' principal investigators (PIs), case studies, three focus groups (on interdisciplinarity, transdisciplinarity, and transfer of research), as well as interviews. As a complement to the evaluation, 50 interviews were carried out with key representatives from the research system regarding the impact of FONA, focussing particularly on its effects on the research mode and the legitimacy of sustainability research (Bührer et al., 2020a).

FONA was established in 2004 by the German BMBF. Between 2006 and 2018, FONA funded about 10,000 research projects with total funding of about 5 billion euros. The majority of this funding – about 70 per cent – was given to universities and non-university research organisations, while the remainder went to enterprises, local and regional administrations, and NGOs (Figure 7.2).

With FONA, the BMBF aimed to create an environmental research programme that was much more strongly oriented towards innovation and impact than conventional programmes. From a strategic perspective, this meant involving more actors in the innovation system than just research organisations and businesses. In this way, FONA called for a more systemic approach – an approach that was amplified by the High-Tech Strategy, the German framework research concept that the BMBF has been developing since 2006. Furthermore, Germany had introduced a sustainability strategy in 2002, which drove environmental research towards sustainability research. This amplified the need to look at impact categories beyond the environmental factor of sustainability. In addition, it also pointed to the need to base research on two research principles that sustainability research highlights: first, on interdisciplinarity, in order to account for the integration of ecological, economic, and social impacts and increased complexity; and second, on transdisciplinarity, in order to involve non-scientific stakeholders early on in the research process. This will in turn increase the opportunities for societal

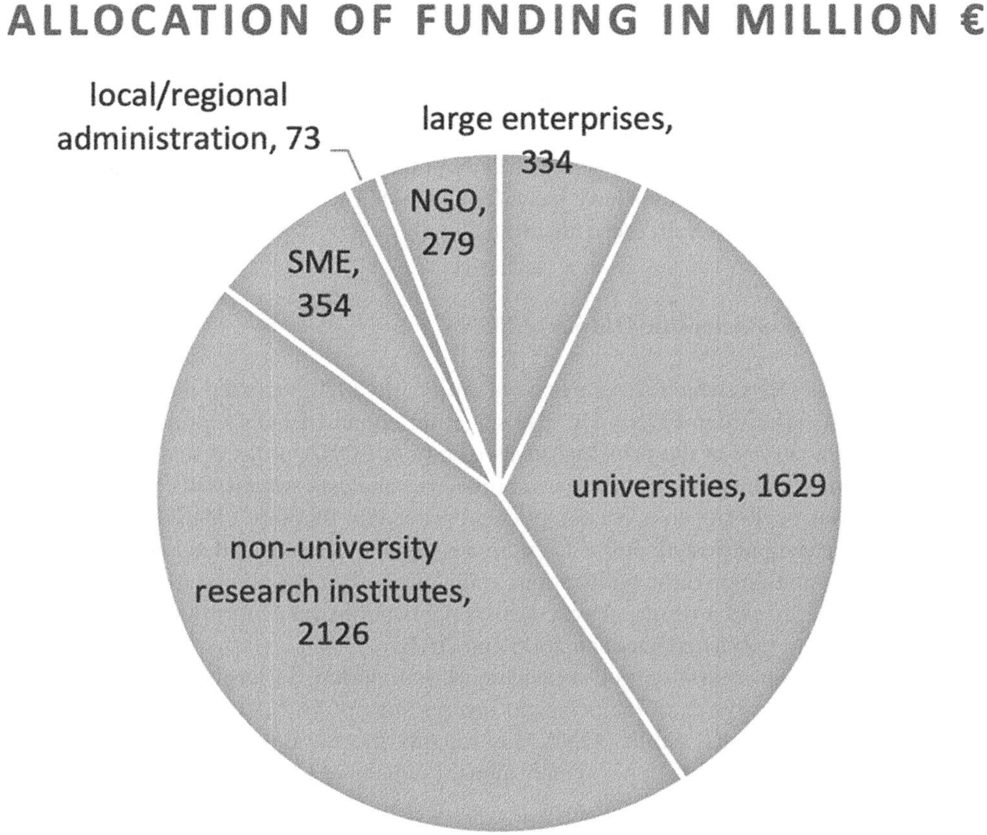

Source: Author's own elaboration. Data obtained in evaluation project (for project report, see Bührer et al., 2020b).

Figure 7.2 Allocation of FONA budget to institutions

impact, among other effects. As such, FONA's goals were originally and continue to be comprised of three different layers:

1. First, as is the case with every research programme, FONA aims to increase academic knowledge using conventional methods, e.g., publications.
2. Second, it aims to produce research results that have an impact on solving sustainability challenges.
3. Third, from a more process-oriented perspective, it aims to induce a change in the research mode towards increased inter- and trans-disciplinarity.

FONA has defined a wide range of sustainability challenges that must be addressed. On the one hand, the broad challenge of sustainability has been interpreted through various broad-based targets and principles. FONA II (BMBF, 2009), for example, identifies the following guiding principles for developing and characterising the different research initiatives undertaken within the FONA programme: international responsibility, research into earth systems, the climate and energy, sustainable economy and resource use, and societal developments. These guiding principles continued through to FONA III and FONA IV. However, increasingly, more specific research goals have also been added. FONA III (BMBF, 2014) defined "key initiatives" such as the green economy, the city of the future, and the energy transition. In FONA IV, three overarching goals have been defined, along with eight fields of action (FoA) that consist of 25 actions overall; these are aligned with various SDGs (Table 7.1a, Table 7.1b, Table 7.1c).

7.3.2 Evaluation of Scientific Output of FONA

The evaluation of FONA, which was carried out until early 2020, focussed mainly on the first phases of FONA rather than FONA IV. The scientific output of FONA projects was measured using both a survey of the principal investigators of FONA projects and a bibliometric analysis. The main database used for the bibliometric analysis was SCOPUS. Due to data restrictions, it was necessary to use a complicated procedure that first identified the publications of FONA principal investigators. Then, in a second step, various databanks were used to find more information on these publications and identify the ones related to FONA projects (Bührer et al., 2020b). As a result, almost 40,000 publications related to FONA were identified, which had been published between 2006 and 2017.

To evaluate the publication output, a benchmark is required. In terms of the quantity of publications, about two-thirds of the principal investigators of FONA projects evaluated that the number of publications within FONA was roughly the same as in other projects. The remainder was about equally split between responses indicating lower intensity and responses

Table 7.1a Structure of FONA IV: Goal 1: Achieve the climate goals

Field of action	Action
AFoA 1: Climate mitigation	Action 1: Reducing industrial process emissions and using CO2 as raw material A 2: Establishing green hydrogen in Germany A 3: Examining eco-friendly methods of removing CO_2 from the atmosphere
AFoA 2: Climate adaptation	A 4: Researching climate change-induced extreme events in Germany A 5: Understanding and preventing the health impacts of climate change A 6: Increasing the resilience of cities and regions
AFoA 3: Knowledge for effective climate policy	A 7: Improving global climate modelling A 8: Monitoring greenhouse gases for climate protection A 9: Understanding marine and polar regions – the climate engines of Earth

Note: AFoA = Field of action; A = Action.
Source: Author's own elaboration (c.f. BMBF, 2021).

Table 7.1b *Structure of FONA IV: Goal 2: Research, protect and use habitats, and natural resources*

Field of action	Action
AFoA 4: Conservation of biodiversity	A 10: Expanding development of biodiversity monitoring in Germany A 11: Understanding the systemic correlations of changes in biodiversity A 12: Conserving habitats and ecosystems
AFoA 5: Securing natural resources	A 13: Mitigating water crisis events worldwide A 14: Stopping the pollution of rivers and seas A 15: Maintaining healthy soils and sustainable use of land A 16: Expanding development of agri-food systems
AFoA 6: Circular economy	A 17: Increasing overall raw material productivity A 18: Bioeconomy: Utilising bio-based raw materials and avoiding waste A 19: Closing the plastics cycle A 20: Recycling phosphorus: Reusing waste streams, recovering resources

Note: AFoA = Field of action; A = Action.
Source: Author's own elaboration (c.f. BMBF, 2021).

Table 7.1c *Structure of FONA IV: Goal 3: Develop society and the economy – good living conditions throughout the country*

Field of action	Action
AFoA 7: Shaping society together – strengthening cohesion	A 21: Equal living conditions A 22: Supporting sustainable economic and financial systems
AFoA 8: Innovative regions	A 23: Shaping structural change in coal-mining areas A 24: Making urban, rural and regional transformation sustainable A 25: Ensuring sustainable urban and rural mobility

Note: AFoA = Field of action; A = Action.
Source: Author's own elaboration (c.f. BMBF, 2021).

indicating higher intensity (Bührer et al., 2020b). For the publication quality, benchmarks were established based on the citations of publications within SCOPUS. Publications were retrieved for three reference groups within Germany: total German publications in the sustainability field (DE), DFG-funded publications in the sustainability field (German DFG), and DFG-funded publications in the sustainability field – however, works were only included if they were authored by researchers who had also been active as principal investigators in

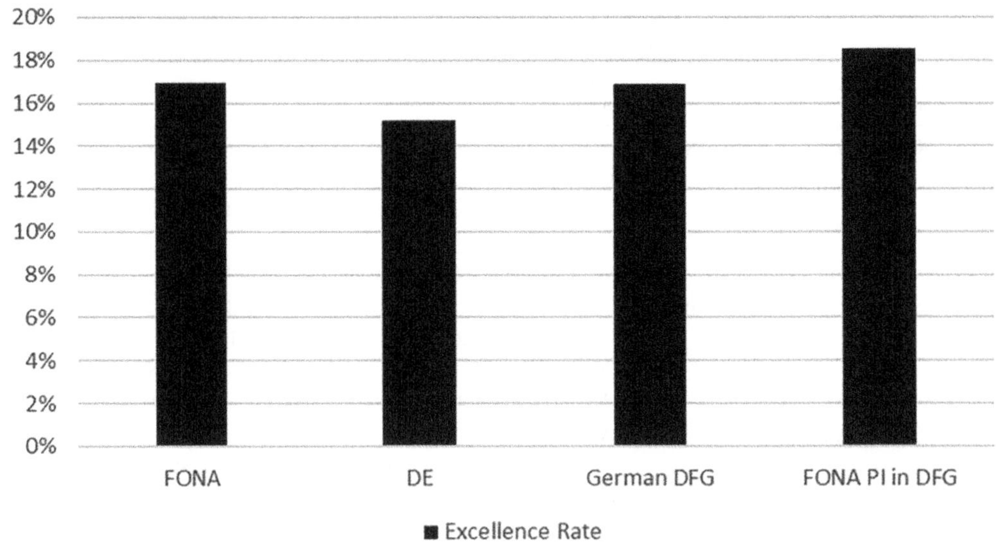

Source: Author's own elaboration. Data obtained in evaluation project (for project report, see Bührer et al., 2020b).

Figure 7.3 *Excellence rate of publications in the sustainability field for various reference groups*

FONA projects (FONA PI in DFG). The excellence rate[1] of the publications was calculated for each reference group (Figure 7.3).

The excellence rate indicates that FONA publications are of high quality – higher on average than sustainability publications from Germany. The indicators used for measuring the quality of publications from FONA seem to be in the same order of magnitude as those for sustainability publications funded within DFG projects. It is particularly interesting to compare the excellence rate of principal investigators in FONA projects with the excellence rate of FONA principal investigators who also authored publications based on projects funded by the DFG. Here, the latter have a somewhat higher excellence rate. As both excellence rates relate to the same group of persons, it seems plausible that these differences may relate to different situations around funding. This seems to indicate that the trade-off between the explicitly impact-oriented goals of FONA and traditional publication success is small or non-existent.

[1] The excellence rate denotes the share of highly cited publications from a group of researchers in the sustainability field. The number of publications belonging to the top 10 per cent of cited publications in the sustainability field is calculated for the respective groups and set in relation to the total number of their publications in the sustainability field listed in SCOPUS.

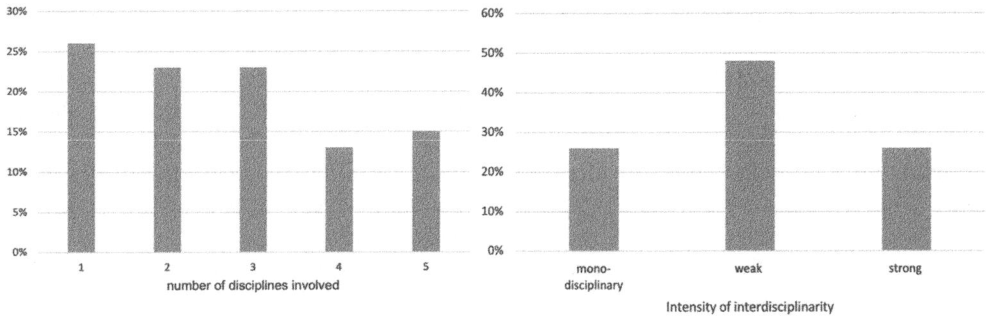

Source: Author's own elaboration. Data obtained in evaluation project (for project report, see Bührer et al., 2020b).

Figure 7.4 Composition of projects with regard to interdisciplinarity

7.3.3 Evaluation of Impact of FONA on Research Mode

One aim in evaluating FONA was to find out whether or not FONA projects succeed in following an inter- and transdisciplinary mode. The survey of the principal investigators therefore included various questions relating to inter- and transdisciplinarity. According to the survey of the principal investigators, about 75 per cent of FONA projects were in an interdisciplinary setting, with about half of all projects involving more than two disciplines (Figure 7.4, left side). About one quarter of all projects involved a high degree of interdisciplinarity between the natural and engineering sciences and the social sciences (Figure 7.4, right side). These numbers indicate that the call for increased interdisciplinarity did indeed have some influence on research processes.

However, there are also indications within FONA of the limits of an interdisciplinary research mode. A detailed analysis of the publications revealed that 24.1 per cent of deeply interdisciplinary projects have difficulties when it comes to finding suitable peer-reviewed journals. Even more telling is the detailed analysis of the interdisciplinary character of the FONA publications, which revealed that only 5.5 per cent of FONA publications can be characterised as deeply interdisciplinary (Bührer et al., 2020b). However, this is a much lower percentage compared to the 25 per cent of projects that were found in the survey to be deeply interdisciplinary. This means that even the deeply interdisciplinary projects resulted in a predominantly disciplinary output. The survey also indicated problems in regard to the acceptance of young interdisciplinary research groups in academic settings. All aspects relating to scientific work (reputation building, profile enhancement, how isolated the group is within the institution, appreciation for interdisciplinary/application-oriented research at the institution) are rated far less positively by those surveyed from deeply interdisciplinary cooperation than by the other young research group leaders. A significant difference can also be seen with regard to the establishment of young research group leaders in the German scientific system. Highly interdisciplinary alliances are at a disadvantage here: only 23.3 per cent of highly interdisciplinary young research group leaders evaluated their research group as a stepping stone for their future career within academia. This is a much lower percentage than the respondents from young research leader groups within FONA that have lower levels of interdisciplinarity, and highlights the disadvantage to the reputation of these groups.

The survey of FONA principal investigators also revealed a high proportion of transdisciplinary projects – 75 per cent. These were defined as projects involving a partner from outside the scientific sector. However, many science-business collaborations also involve research-intensive companies, which are more aligned with the same research process as research institutions. A second category was therefore created to indicate a deeper transdisciplinary setting. Research was considered to be deeply transdisciplinary if partners from outside academia and business were also involved in the research project. The survey indicated that 31 per cent of all FONA projects were operating in such a setting. A closer look at the role of the non-academic partners in the projects shows the strengths and weaknesses of the transdisciplinary mode within FONA. The participation of non-academic partners was a particularly strong indicator of collaboration between science and industry, and a somewhat less strong indicator for deeply transdisciplinary projects – among these, the roles of non-academics were predominantly to discuss research results (92 per cent) and transfer the results to practice (85 per cent). To a lesser extent, they also worked on defining research goals (74 per cent) and participated in research processes (62 per cent).

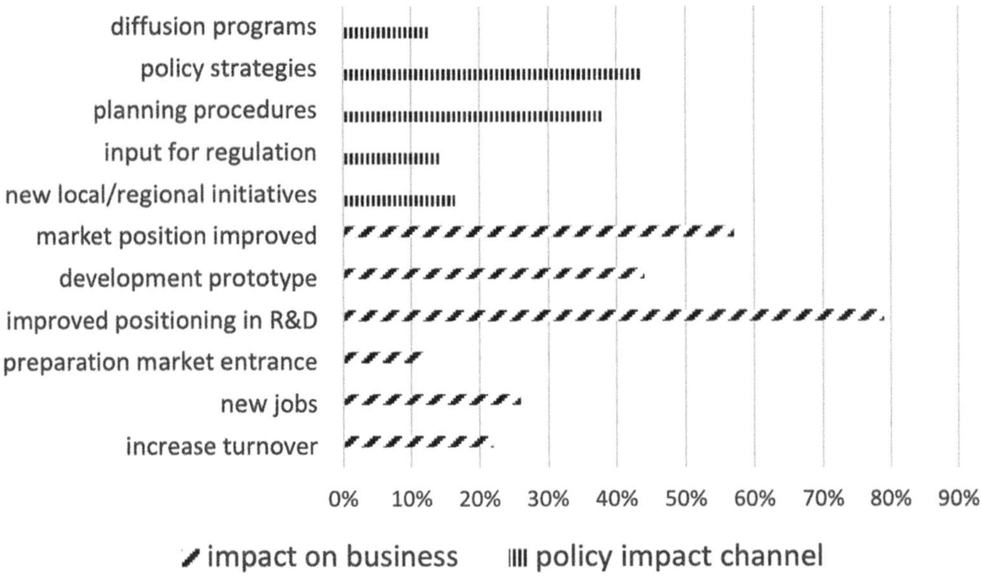

Source: Author's own elaboration. Data obtained in evaluation project (for project report, see Bührer et al., 2020).

Figure 7.5 *FONA projects' policy impact channels and impacts on businesses*

7.3.4 Evaluation of Impact of FONA on Policymaking, Business, and Sustainability

For a programme such as FONA, impact is a key success factor. Various indicators point towards FONA having an impact on policymaking and business. About 40 per cent of FONA projects developed strategies and planned procedures for implementing sustainable solutions (Figure 7.5). Supporting local or regional initiatives played a role in about 16 per cent of the projects. More than 10 per cent of the projects provided input for strengthening regulation or designing programmes to foster the diffusion of innovations. Technology development by the businesses involved in the projects is an important impact channel. About 44 per cent of the participating companies reported that they have already developed a prototype; an additional 24 per cent planned to do so based on the project results. Twelve per cent of the companies had prepared for market entry, with an additional 27 per cent planning to do so. This indicates that FONA led to technological improvements that boosted the market positions for these businesses.

The importance of these non-academic impacts and impact channels also corresponds to non-academic communication outlets. About 40 per cent of the FONA projects engaged in non-academic knowledge transfer activities, such as the publication of handbooks and brochures, or non-academic workshops.

The presented impact of FONA on business and policymaking is only an intermediate step on the impact pathway from research to the endpoint of improving sustainability – it cannot deliver a conclusive quantification of FONA's effects on sustainability. This results from the following underlying principles of impact pathways: newly generated knowledge feeds into different levels of action, from multilateral organisations to individual communities, businesses, and consumers. These are integrated into different rationales for taking action, financial constraints, and possibilities of control. The effect of this newly generated knowledge must be considered on a macro level that accounts for the relevant political and economic ecosystem within a network of influencing factors. The evaluation of FONA did not come up with a comprehensive assessment of the possible impacts at the macro level. Instead, it implemented qualitative case studies on specific sub-programmes within FONA. The following five case studies illustrate the heterogeneity of the impacts triggered by FONA and the different levels and actors involved in the projects:

1. The case study "Chemical processes and material use of CO2" exemplifies the creation of technical solutions. For example, projects in this sub-programme created a method for using CO_2 in polyols as a precursor for soft polyurethane foam, converted CO_2 into CO_2 synthetic fuel, and developed a technology platform for composite membranes, the commercial products of which have been used by customers.
2. FONA also included the funding of the Technical Support Unit of Working Group III of the IPCC, which supported the process of climate negotiations. The interviews conducted as part of the case study unanimously found that without an appropriate IPCC process, it is unlikely that the Treaty of Paris would have been adopted.
3. In the sub-programme "Transformation Energy Systems", empirical evidence showed that at the local level, the backing of political representatives is crucial for implementation.
4. The development and establishment of a tsunami warning system as part of FONA underlined that there were not only infrastructural-technical issues to be addressed, but that the governance perspective was of critical importance. The implementation of this system

involved actors along the entire process chain, from the national Indonesian disaster management authority to local actors.

5. Findings from the sub-programme "'Risk Analysis of Pharmaceutical Residues in Drinking Water" were received by the Federal Environment Agency and the Federal Ministry of Health and fed into European processes for the environmental impact assessment of pharmaceuticals.

In summary, the empirical evidence clearly points towards the importance of FONA's non-academic goals. Furthermore, there is evidence that FONA has not only achieved academic output but has also carried out activities that have had societal impact. However, the interpretation of the results with regard to an actual impact on societal transformation should be undertaken with caution, for two reasons. First, the evidence is partially based on a survey of the principal investigators. FONA is expected to have societal impacts, so this could have led to a confirmation or social-desirability bias in the investigators' responses. Secondly, there is data available for activities that are linked to inducing a societal impact, but no assessment of the impact on sustainability goals. We will come back to that issue in the next section on interpretation and outlook.

7.4 INTERPRETATION AND OUTLOOK

7.4.1 Interpretation of FONA as an Instrument to Change the Mode of Research

Considering the results of the FONA evaluation in terms of the conceptual framework developed in Section 7.2.3 highlights the fact that FONA strongly contributed to the creation and nurturing of a sustainability research niche. It provided funding for research aimed at sustainability, and explicitly called for a research mode that puts high value on inter- and transdisciplinary approaches. It also created visibility by organising the "FONA Forum", a platform that unites policymakers and sustainability researchers. Securing additional funding for research within the research system also legitimised this approach from the perspective of the established research system, particularly when linked to the traditional method of allocating "ribbon", i.e., peer-reviewed publications (Bührer et al., 2020a).

However, the evaluation also showed there has been some resistance from the established regime, evidenced by the low acceptance of strongly interdisciplinary research. There are still some incompatibilities with the regime, and in comparison to other, more traditional forms of academic career paths, FONA is not so often viewed as a stepping stone for academic careers. When it comes to gaining a positive reputation in academia through deeply interdisciplinary research, an issue emerges around explaining the paradox that only one-fifth of deeply interdisciplinary projects publish in deeply interdisciplinary mode. As long as there is more emphasis in the academic sphere on publishing work in disciplinary structured journals to build a personal reputation in the respective discipline, FONA researchers will be incentivised to adapt to the existing regime and to steer research results in a disciplinary direction to make them ready for publication in disciplinary journals. As a result, some form of "regime symbiotic sustainability science" is emerging, which is delivering research money and publications. However, an adaptation strategy of this kind involves a certain trade-off: on the one hand, there will be an emphasis on not only proposing but actually conducting truly

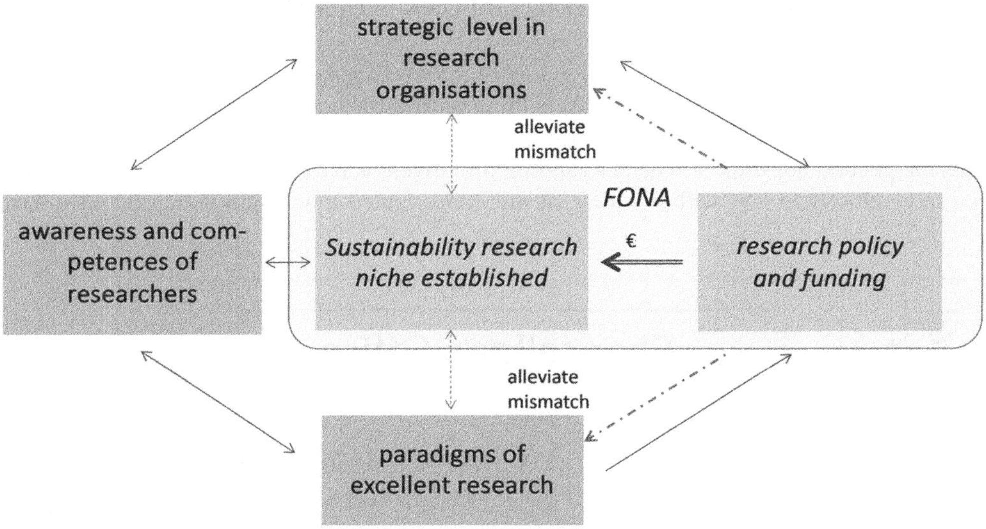

Source: Author's own elaboration.

Figure 7.6 Sustainability research fostered by FONA as a niche within the overall German research system

inter- and transdisciplinary research – but on the other hand, during the course of the project, the research will be steered towards disciplinary research results, for which recognition can more easily be achieved.

To summarise our interpretation, we believe that FONA has contributed to establishing a sustainability research niche (Figure 7.6). However, when it comes to this niche being fully integrated into the research regime, there are still various obstacles to overcome. On a strategic level, many research organisations define career paths as being built on traditionally earned "ribbon". The traditional form of research excellence, which defines "ribbon" in terms of the traditional forms of results of curiosity-driven disciplinary research, and inter- and transdisciplinary research aimed at societal impact, is not aligned with the above integration goal. Finally, there are also obstacles around the ability to perform inter- and transdisciplinary research (Helming et al., 2016). It therefore seems that additional measures are necessary to bring about more stringent changes in the research system, thus enabling more room for stronger forms of sustainability research to be included in the regime. These measures include (1) transferral of the criteria used in FONA's project-based funding to the institutional funding of non-university research organisations active in the field of sustainability; (2) institutional development towards universities adopting sustainability transformation as a focus; and (3) greater reflection on what all of this means for interpreting the quality of research and the excellence of research institutions. Within the SISI process, projects such as LeNa Shape and the "transformation of universities" research initiative have taken up these issues; however, the outcome of this remains to be seen.

7.4.2 Strengthening the Characteristics of FONA to Create Mission-Oriented Policy

As Section 7.3 showed, although not all of FONA's goals have been assigned clear timeframes and specific success markers, FONA IV is moving towards a mission-oriented architecture, with clearer targets across a hierarchical structure that has different levels of granularity. This poses the general question of whether funding initiatives such as FONA should be structured even more closely in line with the logic of mission-oriented innovation policy than we have

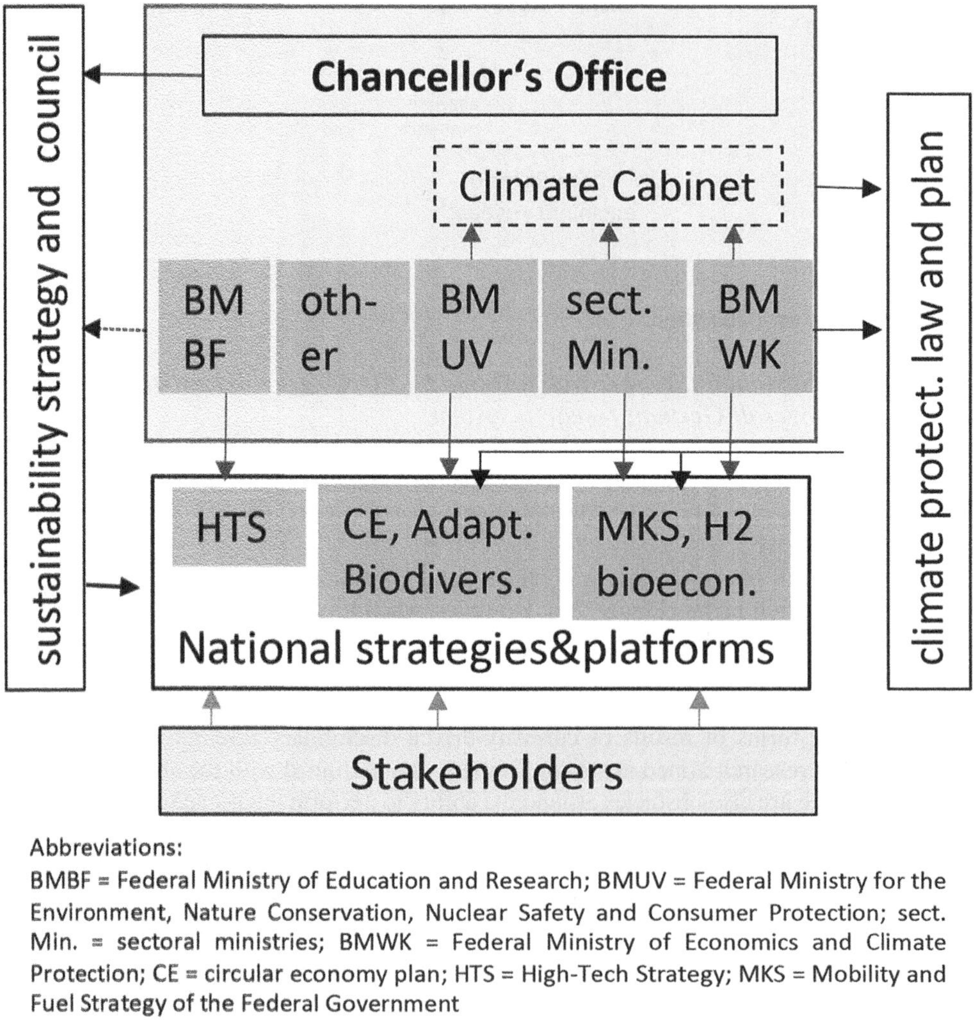

Source: Author's own elaboration.

Figure 7.7 *Governance structure of German key national strategies relating to energy and environmental transformations*

seen with FONA IV. This way, the initiative would be orientated more firmly towards measurable targets backed by indicators, have a clear timeframe, and be in alignment with key policy initiatives on the innovation demand side.

The German sustainability strategy defines targets with goals and timelines. From the perspective of innovation policy, demand-side policies are an important instrument to foster the generation of new innovation. When it comes to environmental policy, the same instrument is seen as an instrument to foster the diffusion of existing innovations (Rennings, 2000). The challenges associated with energy and environmental transitions affect multiple different ministries, with sector-specific ministries playing a very strong leading role in tackling them. This can also be seen in Germany, where various national strategies have been taking on the challenge of formulating policies and engaging stakeholders (Figure 7.7). These national strategies also correspond to the fields of action in FONA IV, as shown in Table 7.2:

In principle, its chosen environmental themes enable FONA to become a centrepiece of research to support German missions for energy and environmental transformation. Programmes such as FONA would thus be suitable as part of the supply side of innovation within the national strategies addressing sustainability. However, there is a need for further interdepartmental integration of policies between the ministry in charge of research (and FONA) and the sector-specific ministries. This need for integration runs both ways. On the one hand, measures to accelerate the diffusion of innovations also affect the generation of further innovations, in the sense of demand-oriented innovation policy in interaction with supply-side R&I policies. The challenge is to ensure that in the process of planning sector-specific

Table 7.2 Key policy initiatives corresponding to Fields of Action (FoA) of FONA IV

Field of action	Corresponding key policy initiative and key actors
AFoA 1 of FONA IV targets climate mitigation	Climate mitigation has been politically shaped by the climate cabinet – in which BMBF had not been included – and the resulting KlimaschutzGesetz (Climate Change Act, KSG), with the BMWK taking on a stronger role in the current government.
AFoA 2 of FONA IV targets climate adaptation	Climate adaption is politically shaped by the national adaptation strategy and the Interministerial Working Group "Adaptation Strategy," led by the BMUV.
FoA 4 of FONA IV targets the conservation of biological diversity	The national strategy on biological diversity is being led by the BMUV.
FoA 5 of FONA IV targets securing natural resources	One key policy initiative is the German water strategy, presented in early 2023 and led by the BMUV.
FoA 6 of FONA IV targets the circular economy	Currently, a national strategy on the circular economy (NKWS) is in the making; this process is being led by the BMUV. Another related strategy, the German strategy on raw materials, is also being led by BMWK.
FoA 8 of FONA IV targets mobility issues, among others	A key political initiative is the German mobility and fuel initiative (MKS); this is under the responsibility of the BMDV.

Source: Author's own elaboration.

demand-side measures, not only do the sector-specific departments contribute their sector-specific expertise but also their supply-side functional know-how regarding the generation of innovations induced by environmental and sector-specific policies. However, the knowledge required for this goes beyond the traditional field of activity of the sector-specific departments. On the other hand, defining specific mission-oriented research goals requires that the planning of the research funding instruments is integrated into the overall structure of the mission. Thus, specific mission-oriented research goals should be integrated into missions developed within the various national strategies to provide further support.

However, there are also arguments against making all of FONA part of a specific mission-oriented policy. Three lines of argument are particularly important here:

1. Missions are developed with a specific time frame and specific goals to be achieved. Such mission goals require a certain amount of political consensus around the policy goals, which sometimes falls short of the changes necessary for sustainability. If all FONA research concentrates solely on the stringency of the defined missions, the research being done may become too incremental, and there will be less ambitious research being carried out on sustainability topics beyond the consensus of today's mission stringency.
2. An early orientation towards societal needs and missions presupposes knowing what the societal needs of the future look like. If one focusses too much on current needs, then one fails to build up comprehensive knowledge of needs that will become relevant in the future. Addressing future challenges, therefore, also requires research to be carried out beyond the missions that have already been identified. In this context, research within FONA could also lay important foundations for identifying and developing future missions as part of an adaptive process involving policymakers, researchers, and civil society.
3. It is increasingly becoming accepted that sustainability requires the transformation of systems. The path to such transformation, however, is not linear and cannot be foreseen precisely. With regard to mission-oriented policy, this holds especially true for transformative missions, which tend to be more open-ended than accelerator missions (Wittmann et al., 2020). Sustainability transformations, therefore, also require knowledge beyond the foreseeable research goals defined in missions. From this perspective, sustainability transformation is linked to the concept of societal resilience (Walz et al., 2022). In terms of the societal resilience concept, however, the degree of directionality of knowledge is less specifically differentiated. This is because it is aimed very strongly at process characteristics so that new requirements can quickly be integrated into the system (Roth et al., 2021; Walz et al., 2022). Thus, to achieve societal resilience in the research arena, we should not focus so much on specifically foreseen research missions, but rather on creating a diverse and broad knowledge base that can be quickly adapted. In this way, areas can be linked to address necessary adaptations that were not foreseen when the missions were formulated.

Taken together, these perspectives imply that sustainability research programmes should not only focus on being part of specific missions that are defined today. Instead, research funding approaches such as FONA should be tailored to address the challenges on two levels. Parts of programmes such as FONA should be fully integrated with mission policy, which requires the specification of research mission goals that are integrated into the overall missions and of indicators for measuring their fulfilment. Another part, however, should explicitly be more radical, higher risk, and more broad in its focus. However, this second part of FONA should also be directional and follow a trans- and interdisciplinary research mode when it comes to sustainability science. This clearly differentiates this type of research from other programmes on sustainability that operate with a more traditional view of research excellence.

7.4.3 Consequences for Impact Assessment

Contributing to addressing sustainability challenges is a strong rationale for providing public funding to sustainability research and could be a strong argument for overcoming the established regime's resistance towards sustainability research. However, the evaluation of FONA shows that backing up this claim with empirical evidence is more difficult, especially at the macro level. There are two fundamental issues that make such an assessment difficult:

1. The impact on sustainability goals cannot be attributed to a single factor (attribution problem). This is because numerous (success) conditions must be in place for a successful contribution of research, such as a successful demand-side innovation policy that complements the supply side.
2. The impacts occur with a considerable time lag and therefore it is not possible for monitoring activities to identify this impact while research is still ongoing (ex ante problem). This requires that methodologies be developed for future-oriented assessments.
3. The need to deal with these conceptual issues is currently leading to a search for new methodological approaches. There are both qualitative approaches such as those outlined by Ghosh et al. (2021) and Matt et al. (2023), and more quantitatively-oriented approaches, such as in the context of the discussion of Horizon Europe's impact assessment strategy. Within the latter, Key Impact Pathways are tasked with assessing impact from a short-, medium-, and long-term perspective for each key impact dimension. To assess long-term effects for the Key Impact Pathway 7 focussing on economic growth, macroeconometric models are used. For the Key Impact "Delivering benefits and impact through research and innovation missions", the aim is to determine the output and outcome of research relevant to the mission in the short and medium term. The long-term aim is then to assess the mission goals achieved through research (see Nixon, 2022). How a solution will be found for ex ante measurement of this long-term perspective, however, is still unresolved in the EU approach, but one method is to adapt macroeconometric models.

Currently, another promising approach for addressing these challenges is being developed for the German FONA IV framework programme. The impact assessment will consist of three components (Figure 7.8):

1. First, a combination of proven project evaluation procedures with methodological developments that make use of external indicators shows how project results are taken on board by businesses and in public discourse: this leads to an enhanced impact-oriented monitoring system that captures the short- to medium-term effects of FONA measures (lower left-hand side of Figure 7.8).
2. Second, this impact monitoring will be used for qualitative analysis of how the research will enable the uptake of the generated knowledge. In this context, the foresight

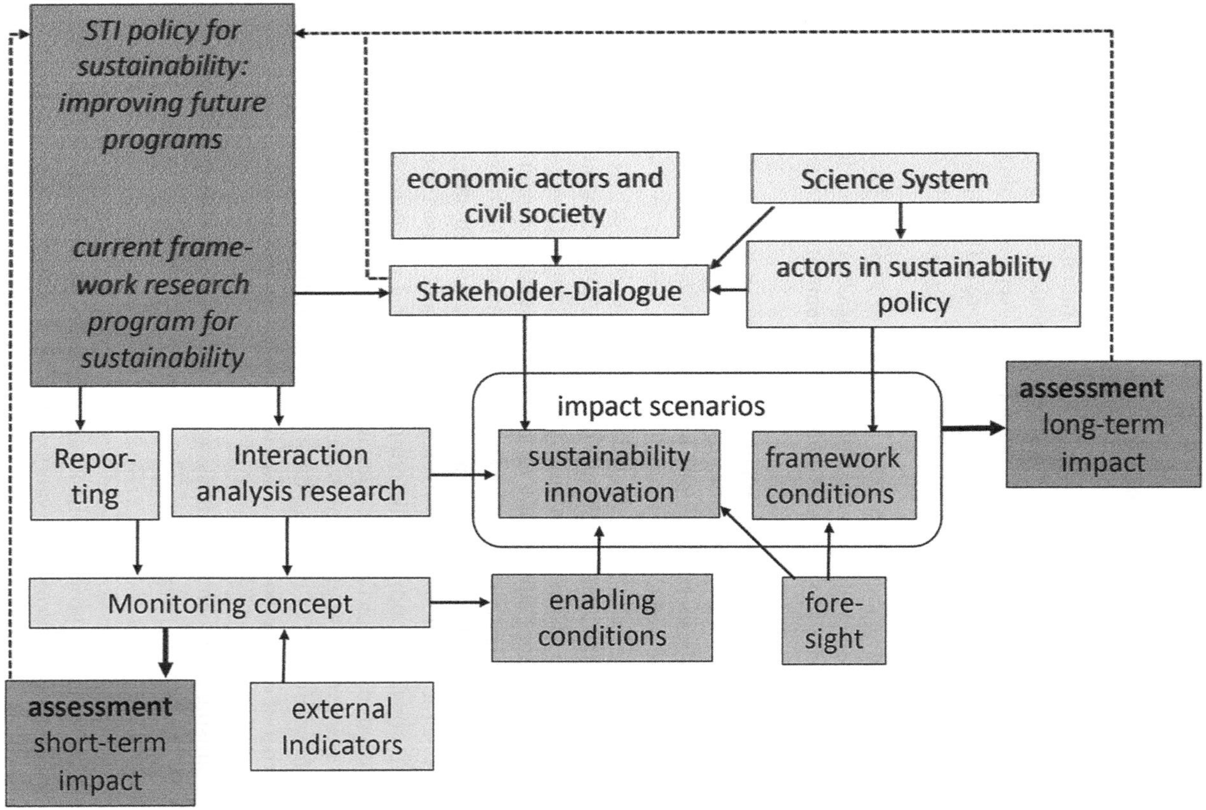

Source: Compiled by authors.

Figure 7.8 *Blueprint of enhanced impact assessment model for FONA*

methodology will be used to generate knowledge about possible future framework conditions for sustainability. These future conditions will be expressed in the form of context and impact scenarios, with a view to estimating the long-term contribution of research (lower right-hand side of Figure 7.8).

3. Third, there will be systematic interaction with the key stakeholders in sustainability strategies so that their views on framing the impact assessment scenarios can be considered. This approach intends to link qualitative-oriented evaluation methodologies with more quantitative-based methodologies, which are employed for scenario-based strategic impact assessments of sustainability strategies (upper right-hand side of Figure 7.8).

A mission-oriented STI policy brings the focus onto the achievement of specific goals. An impact assessment of a mission-oriented innovation policy therefore not only needs to look at the output and outcome of the research programmes, but also at the impact on the target. Thus, it seems likely that in the future, attribution issues and the ex ante problem will also play an increasingly important role in impact assessments of mission-oriented policies beyond sustainability. Furthermore, it is probable that transformative missions in particular will be affected by this challenge. "Transformer missions aim to achieve a comprehensive change affecting a socio-technical system as a whole and are therefore not limited to scientific progress and regulatory changes" (Wittmann et al., 2020, p. 14). Characteristics of transformations include the interweaving of several types of innovation and diverse interactions with the social and political environment. The impact assessment should no longer be focussed granularly on individual innovations, but rather on the overall system. This increases the number of causal factors that influence the success of an innovation, which in turn increases the complexity of the impact assessment. This means transformation missions in particular could benefit from advances in impact assessment methodologies, such as those envisaged for FONA IV.

7.4 SUMMARY AND CONCLUSION

It is increasingly being acknowledged that addressing environmental challenges requires sustainability transformations. Intentionally funding research that addresses sustainability challenges is important for fostering such transformations. Over the last 20 years, a specific sustainability science has been developed that not only emphasises directional research to address sustainability challenges, but also calls for a research mode characterised by inter- and transdisciplinary research. The research system itself has not yet adapted to this research mode and can be seen as part of the existing regime, while sustainability science is a niche characterised by a specific research mode. As such, in any discussion of transformation missions in sustainability research programmes, there are two different research questions that must be addressed: first, the impact of research results on the transformation towards sustainability; and second, the impact of sustainability research funding on the transformation of the science system towards the sustainability science research mode.

The previous analysis has taken FONA as an example to investigate both of these research questions. With regard to the first question, the empirical evidence from a thorough evaluation of FONA shows substantial success in terms of the outcome of research for sustainability; however, the impact on sustainability transformations is difficult to assess. With regard to the second research question, the experience shows some success in terms of establishing sustainability as a niche within the science system. However, this comes at the cost of compromising the sustainability research mode somewhat by shifting the focus more towards disciplinary-oriented scientific publication output.

In order to further improve the impact of research on sustainability transformations, three steps appear to be necessary. First, sustainability research criteria must become mainstream in the research system, e.g., with regard to institutional funding of research organisations. Second, there must be discussion around adapting how the quality of research is defined. This will help to make directionality-driven research and the corresponding sustainability research modes a more accepted part of the research system. Third, a methodology must be developed to better assess the impacts of sustainability research on future sustainability transformations. A promising approach is to link qualitatively-oriented evaluation methodologies with more quantitative-based methodologies, which are employed for scenario-based strategic impact assessments of sustainability strategies.

From the perspective of supporting mission-oriented policy, FONA can complement the move towards developing national strategies for various sustainability challenges. Currently, these strategies are driven mainly by ministries responsible for sector-specific policies. From an innovation perspective, these are demand-side innovation policies. As a key research funding programme, FONA could strengthen the knowledge base and innovation supply side of these policies. The thematic structure of FONA and the national strategies already show a high degree of complementarity. There are two avenues for improving FONA's mission orientation. First, mission goals for research within FONA must be systematically defined in quantitative terms and with a specific time frame, and include corresponding indicators to measure success. Second, the supply side of STI policies must be included early on in the formulation of the national strategies. This will require supplementing the sector-specific knowledge of the relevant ministries more strongly with knowledge about innovation processes and STI policies. However, there are also reasons why not all of FONA should follow such a mission-oriented approach. Sustainability transformation requires more than carrying out research towards a target with a given timeline – it also calls for the creation of variety to cater to the difficulties and uncertainties associated with the transformation pathway. Thus, there should also be space within FONA for sustainability science to look beyond missions and foster knowledge about the long term and create more radical innovations to move towards sustainability.

REFERENCES

Blind, K., Filipović, E., & Lazina, L. K. (2022). Motives to publish, to patent and to standardize: An explorative study based on individual engineers' assessments. *Technological Forecasting and Social Change, 175*, 121420. https://doi.org/10.1016/j.techfore.2021.121420

BMBF (2009). *Forschung für nachhaltige Entwicklungen. Rahmenprogramm des BMBF.*

BMBF (2014). *Forschung für nachhaltige Entwicklungen – FONA³. Rahmenprogramm des BMBF.*

BMBF (2021). *Research for sustainability. A strategy of the Federal Ministry of Education and Research.*

Bodenheimer, M. (2019). Behavioral transitions to social sustainability in global production networks. IRB-Verlag.

Bührer, S., Sarah S., Walz, R., Neumann, M., Astor, M., & Malik, F. (2020a). *Ergebnisse aus der Diskursfortsetzung zum zukünftigen FONA.* Abschlussbericht. Karlsruhe: Fraunhofer ISI.

Bührer, S., Walz, R., Seus, S., Astor, M., Stehnken, T., & Malik, F. (2020b). *Evaluation der BMBF-Rahmenprogramme Forschung für die Nachhaltigkeit FONA 1 (2005–2009) & Forschung für Nachhaltige Entwicklungen FONA 2 (2010–2014)*. Abschlussbericht.

Daedlow, K., Podhora, A., Winkelmann, M., Kopfmüller, J., Walz, R., Helming, K. (2016). Socially responsible research for sustainable transformation: An integrated assessment framework. *Current Opinion in Environmental Sustainability, 23*, 1–11.

DiMaggio, P. J., & Powell, W. W. (1991). Introduction. In W. W Powell & P. J. DiMaggio (Eds.), *The new institutionalism in organizational analysis* (pp. 1–40). University of Chicago Press.

Etzkowitz, H., & Leydesdorff, L. (2000). The dynamics of innovation: From National Systems and "Mode 2" to a Triple Helix of university–industry–government relations. *Research Policy, 29*, 109–123.

Frederichs, G. (1999). Der Wandel der Wissenschaft. *TATuP – Zeitschrift für Technikfolgenabschätzung in Theorie und Praxis, 8*(3/4), 16–25.

Funtowicz, S., & Ravetz, J. (1993). Science for the post-normal age. *Futures, 25*(7), 739–755.

Gagné, M., & Deci, E. L. (2005). Self-determination theory and work motivation. *Journal of Organizational Behavior, 26*(4), 331–362. https://doi.org/10.1002/job.322

Geels, F., & Schot, J. (2007). Typology of sociotechnical transition pathways. *Research Policy, 36*(3), 399–417.

Georghiou, L., Tataj, D., Celio, J., Giannini, S., Pavalkis, D., Verganti, R., & Renda, A. (2018). *Mission-oriented research and innovation policy. A RISE perspective*. European Commission – Directorate-General for Research and Innovation.

Ghosh, B., Kivimaa, P., Ramirez, M., Schot, J., & Torrens, J. (2021). Transformative outcomes: Assessing and reorienting experimentation with transformative innovation policy. *Science and Public Policy, 48*(5), 739–756.

Gibbons, M., Limoges, C., Nowotny, H., Schwartzman, S., Scott, P., & Trow, M. (1994). *The new production of knowledge: the dynamics of science and research in contemporary societies*. Sage.

Goschin, Z., & Zaman, G. (2010). Multidisciplinarity, interdisciplinarity and transdisciplinarity: Theoretical approaches and implications for the strategy of post-crisis sustainable development. *Theoretical and Applied Economics, 17*(12), 5–20.

Grunwald, A., Schäfer, M., & Bergmann, M. (2020). Neue Formate transdisziplinärer Forschung. Ausdifferenzierte Brücken zwischen Wissenschaft und Praxis. *Gaia, 29*(2), 106–114.

Helming, K., Ferretti, J., Daedlow, K., Kopfmüller, J., Winkelmann, M., Bertling, J., & Walz, R. (2016). Forschen für nachhaltige Entwicklung: Kriterien für gesellschaftlich verantwortliche Forschungsprozesse. *GAIA, 25*(3), 161–165.

Hirsch Hadorn, G., Bradley, D., Pohl, C., Rist, S., & Wiesmann, U. (2006). Implications of transdisciplinarity for sustainability research. *Ecological Economics, 60*(1), 119–128.

Hölsgens, R., Lübke, S., & Hasselkuss, M. (2018). Social innovations in the German energy transition an attempt to use the heuristics of the multi-level perspective of transitions to analyze the diffusion process of social innovations. *Energy, Sustainability and Society, 8*(8). http://doi.org/10.1186/s13705-018-0150-7

Iorio, R., Labory, S., & Rentocchini, F. (2017). The importance of pro-social behaviour for the breadth and depth of knowledge transfer activities: An analysis of Italian academic scientists. *Research Policy, 46*, 497–509.

Jahn, T. (2004). Von der Umwelt- zur Nachhaltigkeitsforschung. Neue Spieler in der Wissenschaft. *Ökologisches Wirtschaften, 6*, 12–14.

Kates, R. W. (2011). What kind of a science is sustainability science? *PNAS, 108*(49), 19449–19450.

Kates, R. W., Clark, W. C., Corell, R., Hall, J. M., Jaeger, C. C., Lowe, I., McCarthy, J. J., Schellnhuber, H.-J., Bolin, B., Dickson, N. M., Faucheux, S., Gallopin, G. C., Grübler, A., Huntley, B., Jaeger, J., Jodha, N. S., Kasperson, R. E., Mabogunje, A., Matson, P., Mooney, H., Moore, B. III., O'Riordan, T., & Svedin, U. (2001). Sustainability science. *Science, 292*(5517), 641–642.

Kaufmann, R. K., & Cleveland, C. J. (1995). Measuring sustainability: Needed – An interdisciplinary approach to an interdisciplinary concept. *Ecological Economics, 15*, 109–112.

Komiyama, H., & Takeuchi, K. (2006). Sustainability science: Building a new discipline. *Sustain Science, 1*, 1–6.

Korbun, T. (2004). Kurs Nachhaltigkeit? Perspektiven der Förderung von Nachhaltigkeitsforschung. *Ökologisches Wirtschaften*, *6*, 10–11.

Kuittinen, H., Polt, W., & Weber, K. M. (2018). Mission Europe: A revival of mission-oriented policy in the European Union. In RFTE (Ed.), *Re:* Thinking Europe*: Positions on* shaping *an idea* (pp. 191–207). Austrian Council for Research and Technology Development, Holzhausen.

Lam, A. (2011). What motivates academic scientists to engage in research commercialization: "Gold", "ribbon" or "puzzle"? *Research Policy*, *40*(10), 1354–1368. http://doi.org/10.1016/j.respol.2011.09.002

Lang, D. J., Wiek, A., Bergmann, M., Swilling, M., & Thomas, C. J. (2012). Transdisciplinary research in sustainability science: Practice, principles, and challenges. *Sustainability Science*, *7*, 25–43.

Matt, M., Robinson, D., Joly, P.-B., Van Dis, R., & Colinet, L. (2023). *ASIRPA$^{Real\text{-}Time}$ in the making or how to empower researchers to steer research towards desired societal goals. Research Evaluation*. https://doi.org/10.1093/reseval/rvad004

Newell, W. H. (2001). A theory of interdisciplinary studies. *Issues in Integrative Studies*, *19*, 1–25.

Newig, J., Jahn, S., Lang, D. J., Kahle, J., & Bergmann, M. (2019). Linking modes of research to their scientific and societal outcomes. Evidence from 81 sustainability-oriented research projects. *Environmental Science & Policy*, *101*, 147–155.

Nixon, J. (Ed.) (2022). *Study to support the monitoring and evaluation of the Framework Programme for research and innovation along Key Impact Pathways. Indicator methodology and metadata handbook*. Publications Office of the European Union.

Nowotny, H., Scott, P., & Gibbons, M. (2003). Mode 2 revisited: The new production of knowledge. *Minerva*, *41*, 179–194.

Perkmann, M., Salandra, R., Tartari, V., McKelvey, M., & Hughes, A. (2021). Academic engagement: A review of the literature 2011–2019. *Research Policy*, *50*(1), 104114. https://doi.org/10.2139/ssrn.3461621

Potthast, T. (2015). Ethics and sustainability science beyond Hume, Moore and Weber – Taking epistemic-moral hybrids seriously. In S. Meisch, J. Lundershausen, L. Bossert, & M. Rockoff (Eds.), *Ethics of science in the research for sustainable development* (pp. 129–152). Nomos.

Rat für Nachhaltige Entwicklung (2021). Empfehlungen zur Reform der Regierungsarbeit – Bessere Governance für die Nachhaltige Entwicklung. Rat für Nachhaltige Entwicklung (Stellungnahmen des RNE). https://www.nachhaltigkeitsrat.de/wp-content/uploads/2021/10/20211004_RNE-Stellungnahme_Reform-der-Regierungsarbeit-Bessere-Governance-fuer-die-Nachhaltige-Entwicklung.pdf

Rennings, K. (2000). Redefining innovation – Eco-innovation research and the contribution from ecological economics. *Ecological Economics*, *32*(2), 319–332.

Roth, F., Warnke, P., Niessen, P., & Edler, J. (2021). *Insights into systemic resilience from innovation research*. Policy Brief 03–2021. Fraunhofer ISI. https://www.isi.fraunhofer.de/content/dam/isi/dokumente/policy-briefs/policy_brief_resilience_en_web.pdf

Schneidewind, U., Singer-Brodowski, M., & Augenstein, K. (2016). Transformative science for sustainability transitions. In Brauch, H. G., Spring, U. O., Grin, J., & Scheffran, J. (Eds.), *Handbook on sustainability transition and sustainable peace* (pp. 123–136). Springer.

Scholz, R. (2017). The normative dimension in transdisciplinarity, transition management, and transformation sciences: New roles of science and universities in sustainable transitioning. *Sustainability*, *9*(6), 991.

Shelley-Egan, C., Gjefsen, M. D., & Nydal, R. (2020). Consolidating RRI and Open Science: Understanding the potential for transformative change. *Life Sciences, Society and Policy*, *16*(7). https://doi.org/10.1186/s40504-020-00103-5

Suominen, A., Kauppinen, H., & Hyytinen, K. (2021). "Gold", "Ribbon" or "Puzzle": What motivates researchers to work in Research and Technology Organizations. *Technological Forecasting and Social Change*, *170*, 120882. https://doi.org/10.1016/j.techfore.2021.120882

Van der Burgwal, L. H. M., Hendrikse, R., & Claassen, E. (2019). Aiming for impact. Differential effect of motivational drivers on effort and performance in knowledge valorisation. *Science and Public Policy*, *46*(5), 747–762. https://doi.org/10.1093/scipol/scz027

Vogt, M. (2020). *Ethik des wissens. Freiheit und verantwortung der wissenschaft in zeiten des klimawandels*. Oekom.

Walz, R., Bodenheimer, M., & Roth, F. (2022). Zwei jahre corona-pandemie: Lehren für gesellschaftliche resilienz und die nachhaltigkeitstransformation. Gutachten des fraunhofer ISI für den RNE. Rat für Nachhaltige Entwicklung.

Wäscher, S., & Deplazes-Zemp, A. (2020). Freedom of research in a democratic society. Is there a contradiction between socially desirable science and freedom of research? *EMBO Reports, 21*, e49928. https://doi.org/10.15252/embr.201949928

Weingart, P. (1997). *Neue formen der wissensproduktion: Fakt, fiktion und mode.* Institut für Wissenschafts- und Technikforschung, Univ. Bielefeld, IWT Paper 15.

Wissenschaftsplattform Nachhaltigkeit (2020). *Nachhaltig aus der corona-krise! Impulse aus der arbeit der Wissenschaftsplattform Nachhaltigkeit 2030 zur Stärkung deutscher Nachhaltigkeitspolitik. Hg. v. Wissenschaftsplattform Nachhaltigkeit.* https://www.wpn2030.de/wp-content/uploads/2020/08/Nachhaltig-aus-der-Corona-Krise-2.pdf

Wissenschaftsrat (2021). *Impulse aus der Covid-19-krise für die weiterentwicklung des wissenschaftssystems in Deutschland. Wissenschaftsrat (Drs. 8834–21).* https://www.wissenschaftsrat.de/download/2021/8834–21.pdf?__blob=publicationFile&v=15

Wittmann, F., Hufnagl, M., Lindner, R., Roth, F., & Edler, J. (2020). Developing a typology for mission-oriented innovation policies. *Discussion Papers Innovation Systems and Policy Analysis, 64*. Fraunhofer ISI. https://www.bing.com/ck/a?!&&p=69c7781db8b63ec7b9edd7f4ba27ba96b93ca47410deafb04f7ff8cc808be25dJmltdHM9MTczOTQwNDgwMA&ptn=3&ver=2&hsh=4&fclid=04b86492-6a66-6ebf-27ae-703b6b0d6fad&psq=wittmann+edler+mission+oriented&u=a1aHR0cHM6Ly93d3cuaXNpLmZyYXVuaG9mZXIuZGUvY29udGVudC9kYW0vaXNpL2RvY3VtZW50ZS9jY2kvaW5ub3ZhdGlvbi1zeXN0ZW1zLXBvbGljeS1hbmFseXNpcy8yMDIwL2Rpc2N1c3Npb25wYXBlcl82NF8yMDIwLnBkZg&ntb=1

8. Conversion of the green mission for transformation-oriented industrial innovation: green growth and 2050 carbon neutrality in South Korea

Soohong Eum

8.1 INTRODUCTION

South Korea has seen two iterations of the green mission: the first time as Green Growth, the second time as 2050 Carbon Neutrality. The country is facing the task of transforming its system on an unprecedented scale due to the twin challenges of economic recession and climate change, compounded by its heavy reliance on energy imports and its massive carbon emissions (Climate Watch, 2020). One of the most paramount tasks for the South Korean government is to mitigate the impact of changing global conditions and ensure economic growth by reducing carbon emissions to fulfil international obligations for combating climate change. To address these challenges, the government has prioritised green transformation throughout the industrial structure and identified green industries as a key driver for the transition to a low-carbon and environmentally sustainable economy. These industries have been recognised for their potential to contribute to global efforts to deal with climate change, ensure energy supply stability, and generate new job opportunities.

In 2008, the Lee Myung-bak administration launched Green Growth as the first national-level green mission. In response to the global challenges of climate change and the economic recession resulting from the financial crisis, it aimed to solve both economic and environmental problems. This was to be achieved through the development of green technologies and new and renewable energy sources that would mitigate the impact of climate change, reduce environmental damage, lower greenhouse gas emissions, and secure new sources of economic growth. More specifically, it offered a national vision to transform the prevailing paradigm of fossil fuel-based economic growth into a low-carbon, environmentally sustainable economy, industry, and lifestyle, driven by new growth engines centred on green industries (Presidential Council for Future and Vision, 2009). Green Growth held the top position among national agendas, as it focussed on achieving comprehensive national development that covered not only the economy but also society, the environment, and quality of life. It also restructured individual policies so that they would consistently support this objective and played a pivotal role in ensuring that government programmes were aligned and consistent with its goals and strategies.

The second version of the green mission, 2050 Carbon Neutrality, was introduced in 2020 by the Moon Jae-in administration. In light of the COVID-19 pandemic and the subsequent global recession, the government was compelled to redefine national goals to overcome the immediate crisis, establish a more resilient and sustainable economy, and align with international efforts to transition to a carbon-neutral system. To achieve these objectives, South Korea pledged to achieve carbon neutrality by 2050 and launched the Green New Deal as its core project. The goal of 2050 Carbon Neutrality was to achieve decarbonisation across the entire system, encompassing energy, industry, and society. It strived to foster green industries that eliminate carbon emissions and to establish a new economic and social structure that facilitates an inclusive transition. To support this vision, a policy framework was implemented, encompassing the carbon net-zero, investment in clean energy and infrastructure, and the creation of job opportunities.

These two green missions have been subject to both criticism and praise. Green Growth was a groundbreaking initiative that made "green" objectives an established part of the national policy agenda. South Korea has been recognised for its commitment and leadership in achieving a green economy, with the country adopting the world's greenest stimulus package and allocating the largest share to the environmental sector (Bernard et al., 2009; United Nations Environment Programme, 2009a). The declaration of achieving carbon neutrality and the Green New Deal were applauded as significant steps towards accelerating the energy transition (International Energy Agency, 2020). Green Growth has been criticised for neglecting environmental concerns and focussing solely on economic growth, with civil engineering work carried out to create economic stimulus, which destroyed the country's land and ecosystem. 2050 Carbon Neutrality has been criticised for failing to provide a clear path towards transition, and the Green New Deal has been labelled as a disappointing green stimulus package (Y. Lee, 2020).

The purpose of this chapter is to compare the intentions and content of Green Growth and 2050 Carbon Neutrality and examine the policy measures that have been shaped and implemented to support these goals. The focus is not on assessing the "greenness" of the missions, but on enhancing the understanding of the South Korean government's approach to addressing societal challenges and achieving sustainable economic growth. To this end, the chapter will compare the two missions in terms of three aspects: first, the problems that the government was required to address due to the circumstances surrounding the missions; second, how the programmes were organised to achieve their goals within the framework of the missions; and third, what changed between the two missions.

For this purpose, and following the approach developed by the Organisation for Economic Co-operation and Development (OECD), this chapter applies a framework that breaks down mission-oriented innovation policy into three building blocks: strategic orientation, policy coordination, and policy implementation (Larrue, 2021). In regard to strategic orientation, this chapter examines how the actors in the two missions understood the challenges they aimed to address, the underlying intentions behind the direction and goals that were set, and the ways of ensuring that the missions were seen as legitimate. In terms of policy coordination, the chapter focusses on the decision-making governance within the missions, particularly highlighting the close relationship between the government and business sectors, as well as the engagement of additional stakeholders. Lastly, concerning policy implementation, it investigates how the above characteristics of the missions were expressed in specific policies and implemented in the areas of New Deals and the energy transition.

The remainder of this chapter is structured as follows. Section 8.2 provides background information on two missions and explains how the relevant actors' understanding of the challenges to be addressed contributed to setting the missions' objectives and directions. Section 8.3 presents the Green New Deal projects implemented under the mission to revitalise the national economy, including their nature and organisational structure. Section 8.4 compares the energy policies aimed at the long-term transition in terms of actors and representative projects. Finally, Section 8.5 discusses the challenges that green missions would face in South Korea and provides suggestions for addressing them.

8.2 EMERGENCE OF TWO GREEN MISSIONS IN CRISES

8.2.1 Parallel Backgrounds for Green Missions

Green Growth, declared as a new national vision by President Lee Myung-bak in August 2008, was launched to overcome the global recession caused by the subprime mortgage crisis. When Lee was elected president in December 2007, South Korea was in the midst of this global recession. As the aftermath of the downturn in the financial sector spread to the real economy, economic activity contracted and demand decreased, which had a profound impact on the South Korean economy (H. Lee & Rhee, 2012). With the slogan, "revitalising the economy", Lee was elected based on broad expectations that he would help the country overcome the difficult economic conditions and achieve growth again. The new administration therefore made it a top priority to mitigate socioeconomic shocks and stimulate the economy.

The Kyoto Protocol set a global target for greenhouse gas emissions reduction, compelling South Korean industries to explore innovative approaches in response. South Korea's rapid economic growth was based on export-oriented industrialisation strategies, relying heavily on manufacturing and heavy chemical industries. However, a large portion of the flagship industries such as steel, automobiles, and shipbuilding emitted substantial amounts of greenhouse gases and had high energy requirements and carbon intensity (National Assembly Budget Office, 2022). Furthermore, changes in the climate policies of export destination countries, including the implementation of trade barriers related to greenhouse gas emissions, raised concerns about potentially weakening South Korean companies' price competitiveness and negatively impacting the South Korean economy, which depends heavily on foreign trade rather than the domestic market.

Lee presented the concept of Green Growth as a "comprehensive national vision" that encompasses not only energy and environmental concerns but also job creation, business competitiveness, and lifestyle transformation, and expressed the need to change the situation by forming a green market and generating new business opportunities. To achieve this, he proposed a green policy package that covered industries, energy, and infrastructure to boost the economy and reorganise the entire industrial value chain into an environmentally friendly, low-carbon model (Korean Government Ministries Concerned, 2009a; Office of Prime Minister, 2008; Presidential Committee on Green Growth, 2009a).

The government made efforts to achieve the Green Growth vision by enacting the Framework Act on Low Carbon, Green Growth and several national-level strategies. These defined the goals that the nation should pursue under the mission and established foundations for planning and implementing green programmes. All government ministries and organisations put their best efforts toward achieving this vision in their respective areas. However, after the end of Lee's presidency in 2013, the political momentum for Green Growth as a mission waned and the vision disappeared.

The green mission was reignited in 2020 under the Moon Jae-in administration in response to the COVID-19 pandemic. Faced with the possibility of a recession and widespread unemployment due to the global socioeconomic impact of the pandemic and lockdown measures, the government had to quickly take action to mitigate the negative effects on the country, stimulate the economy, and build resilience against future crises. Moon responded by launching the Korean New Deal, which aimed to create new jobs and revive economic growth through large-scale national projects during the crisis (Korean Government Ministries Concerned, 2020a).

Green was chosen as one of the two pillars of the New Deal. The demand for the Green New Deal was first raised by politicians and civil society during the April 2020 National Assembly elections. The ruling Democratic Party included a Green New Deal manifesto in their campaign, which proposed green stimulus packages aimed at building decarbonisation infrastructure, investing in the transition to green industries, and creating green jobs (Democratic Party, 2020; Kim Sungwhan Assembly Office, 2020). Both the Justice Party and the Green Party, which had close partnerships with civil society organisations, also proposed implementing the Green New Deal.

International trends have bolstered the legitimacy of green policy measures. The European Green Deal was launched as the EU's new growth strategy in December 2019. Although the Green New Deal was not formally adopted as an official agenda, it was proposed following the 2018 election in the United States, and Joe Biden announced a plan to invest in green infrastructure and energy during his presidential campaign. These domestic and international trends have provided reasons for the South Korean government to use policy tools to actively stimulate the economy through green programmes, especially in response to the pandemic.

There has also been a growing demand for South Korea – a country with a significant carbon footprint – to take more decisive action in response to climate change, given the global climate regime's call for enhanced mitigation measures. South Korea has received criticism for its passive approach to reducing emissions. However, as a signatory of the Paris Agreement, it was obligated to submit its Nationally Determined Contribution and Long-term Low Greenhouse Gas Emission Development Strategies to the United Nations Framework Convention on Climate Change by the end of 2020 (Climate Analytics, 2020; OECD, 2017). To this end, the National Assembly adopted a resolution in September 2020 calling for the government to align its 2030 greenhouse gas reduction targets with the recommended levels set by the Intergovernmental Panel on Climate Change (National Assembly, 2020). In this way, the South Korean government has been urged to set more ambitious targets and demonstrate its commitment to fulfilling its responsibilities as a member of the international community.

The declaration of the 2050 Carbon Neutrality vision in December 2020 concretised efforts to achieve the aforementioned objectives by combining and strengthening the two interconnected policy concepts of the Green New Deal and carbon neutrality. The long-term vision for transforming the entire system was redefined, encompassing industry, energy production, and energy consumption, as well as lifestyle. In pursuit of this vision, the government set the 2050 Carbon Neutrality scenario, which outlined a transitional path towards carbon neutrality (Korean Government Ministries Concerned, 2020b). These processes placed the green mission back at the top of the national political agenda, ready to steer national programmes in a new direction, and positioned the Green New Deal as a key strategy for achieving these goals.

8.2.2 Adapting Industry to the New Market Environment

The two missions shared the goals of transforming the economic base to reduce its carbon emissions and make it more eco-friendly and of entering the global green market. Lee believed that winning the race to secure a leading position in newly emerging green industry sectors would be crucial for the national economy and competitiveness. He identified fostering green industries as an area of national interest in response to the changing market environment. Consequently, the government prioritised intensive investments in promising sectors to facilitate the transformation of the industrial structure and enable early entry into the global green market.

As the global climate regime becomes a major challenge for South Korea, the 2050 Carbon Neutrality initiative also focussed on the economy and industry. The government expects enormous economic losses for South Korea if it maintains a carbon-intensive structure and if stricter global environmental regulations are implemented. As a result, it has become imperative to restructure all aspects of the country, including the economy, industry, and society, to transform the existing system. The rationale for this mission was further bolstered by the prospect that each country's efforts to shift towards a green economy, coupled with active investments in response to the pandemic, will accelerate structural changes and intensify competition, allowing countries to achieve a leading position in the newly established global value chain (Korean Government Ministries Concerned, 2020b). To adapt to these changes, the mission goals include preemptively reorganising the industrial structure and fostering areas related to carbon neutrality as drivers of future growth (Korean Government Ministries Concerned, 2021a).

These understandings led to similar goals being set for the two missions in terms of decarbonising the economy and industry, fostering new industries, and transforming society. Their first pillar focussed on reducing greenhouse gas emissions from economic and industrial activities by supplying clean energy, decreasing energy consumption, and increasing efficiency ("Climate Change Adaptation and Energy Independence" and "Low Carbonisation of the Economic Structure"). The second pillar aimed to drive the development and industrialisation of green technologies while realigning existing innovation ecosystems to meet the goal of carbon neutrality ("Create New Growth Engine" and "Creation of a Low-carbon Industrial Ecosystem"). The third pillar sought to create green living spaces and ensure inclusivity by encompassing all members of society as well as industries in the process of transitioning towards a carbon-neutral society ("Improve Quality of Life and Enhance National Standing" and "Just Transition towards a Carbon-neutral Society") (Korean Government Ministries Concerned, 2020b; Presidential Committee on Green Growth, 2009b).

Although the two missions shared a framework for green goals, they presented different visions and intentions. The Lee administration aspired to position the country as a global green powerhouse and designed Green Growth from the perspective of economic growth and industrial competitiveness. As revealed through the notion of '"growth", the Lee administration prioritised economic growth and viewed sustainable development as an abstract moral declaration that ignored the demands to realise industrial modernisation (Jin, 2013; Presidential Committee on Green Growth, 2009a). Green Growth regarded green factors as drivers of national competitiveness and economic growth, perceiving environmental programmes as extensions or complements to national development strategies. This perspective embodied the government's developmental environmentalism (S.-Y. Kim & Thurbon, 2015),

Table 8.1 The objectives of the two green missions

Plan	Vision	Policy directions and tasks
National Strategy for Green Growth and Five-Year Plan for Green Growth (Lee Myung-bak administration, 2009)	To rank among the world's top seven green powers by 2020 and the top five green powers by 2050	Direction 1. Climate Change Adaption and Energy Independence • Task 1. Effectively mitigate GHG emissions • Task 2. Drive oil phase-out and strengthen energy independence • Task 3. Improve capacity to adapt to climate change Direction 2. Create New Growth Engines • Task 4. Advance green technology development and create growth engines • Task 5. Make industry greener and foster new green industries • Task 6. Advance industrial structure • Task 7. Create base for green economy Direction 3. Improve Quality of Life and Enhance National Standing • Task 8. Create green homeland and transport • Task 9. Drive a green lifestyle revolution • Task 10. Become a model nation globally for green growth
Carbon Neutral Green Growth Strategy (Moon Jae- in administration, 2020)	From "adaptive greenhouse gas emissions reduction" to "proactive climate response"	Direction 1. Low Carbonisation of the Economic Structure • Task 1. Accelerating the energy transition • Task 2. Innovation in relation to high-carbon industrial structure • Task 3. Transition towards mobility of the future • Task 4. Decarbonisation of cities and land Direction 2. Creation of a Low-carbon Industrial Ecosystem • Task 5. Fostering promising new industries • Task 6. Establishing foundations of an innovation ecosystem • Task 7. Revitalising of the circular economy Direction 3. Just Transition towards a Carbon-neutral Society • Task 8. Protection of vulnerable industries and classes • Task 9. Realisation of regional carbon neutrality • Task 10. Raising public awareness of a carbon-neutral society

Source: Author's own elaboration

or authoritarian environmentalism (Han, 2015). Consequently, the primary objective of Green Growth was to develop, commercialise, produce, and export green technology, products, and services.

In contrast, the Moon administration incorporated climate goals as objectives alongside the pursuit of economic growth. This mission was given legitimacy by the fact that transforming the socioeconomic system towards carbon neutrality was inevitable. If South Korea continued its lukewarm attitude towards the international community's efforts to respond to climate change, it would have a negative impact on the national economy. Unlike the Kyoto Protocol, which did not mandate that South Korea establish targets for reducing greenhouse gas emissions, the Paris Agreement imposed this requirement on all participating countries. Consequently, South Korea committed to carbon neutrality as a response to climate change. While carbon neutrality was an unavoidable objective, the government presented an ambitious vision that aimed to achieve greater resilience against climate change, sustainable economic growth, and an improved quality of life through the transition to a carbon-neutral society (Korean Government Ministries Concerned, 2021b). In this respect, unlike Green Growth, 2050 Carbon Neutrality pursued changes that encompassed all areas of the country, including industry, the economy, and society.

The remainder of the chapter explores how distinct interpretations of the societal challenges relating to the two missions, as well as the resulting divergences in vision, influenced how policies were formulated and executed. A key argument is that the implementation of both missions was characterised by a combination of short-term New Deals and long-term policies aimed at transforming the energy system. By illustrating a mix of policies with specific directions and targets, it provides a better understanding of the rationale behind the way in which the South Korean government has aligned policy interventions with its long-term vision so as to accomplish the mission.

8.3 GREEN NEW DEALS FOR ECONOMIC STIMULUS AND PROCESS CHANGES

8.3.1 Economic Stimulus and Job Creation through Large-scale Infrastructure Projects

Since the launches of both missions were triggered by global crises, their implementation commenced with the introduction of New Deals as stimulus packages to mitigate shock and swiftly revive economic activity. The government defined its role as a facilitator that invests state funds to create a market and trigger private sector engagement. As New Deals sought legitimacy as an urgent response to the recession, the government selected projects that could create jobs in a short period and show tangible results in terms of stimulating the economy.

The Lee administration's Green New Deal, launched in 2009, included large-scale projects worth 50 trillion Korean Won (approximately 36 billion EUR), equivalent to 3 per cent of the country's GDP, to stimulate the economy and create 956,000 jobs. The primary focus of these projects, such as the Four Major Rivers Restoration Project, was to allocate a significant portion of the government's budget towards public construction and civil engineering. This allocation aimed to function as a social safety net, mitigating the impact of widespread unemployment by providing job opportunities to a substantial number of unskilled and socially vulnerable groups (Korean Government Ministries Concerned, 2009b). To achieve a stimulation effect at an early stage, many projects were front-loaded, and about 20 per cent of the budget was allocated in the first half of 2009 (United Nations Environment Programme, 2009b).

The Moon administration launched the Korean New Deal in 2020 in response to the COVID-19 pandemic. The New Deal had three main goals: first, to create jobs that support the economic transition; second, to invest heavily in essential infrastructure to restore normal economic growth; and third, to establish a foundation for adapting to and leading structural changes. Additionally, the New Deal was designed to alleviate problems such as unemployment and income inequality that may arise during the structural transition process and provide a strong safety net. To achieve these goals, the government allocated a total of 160 trillion Korean Won (115 billion euros) to create 1.9 million jobs by 2025 (Korean Government Ministries Concerned, 2020a).

The objective of the Green New Deal, which served as one of the two primary pillars of the New Deal, was to facilitate a shift towards eco-friendly and low-carbon systems by establishing infrastructure to tackle climate and environmental crises, proliferating low-carbon energy, and laying the groundwork for green industry innovation. The Green New Deal was allocated a total budget of 73.4 trillion Korean Won (52.8 billion euros), making it the largest environmental budget ever – it constituted approximately 1 per cent of the country's GDP. The Green New Deal was primarily divided into three categories (cf. Table 8.2a, Table 8.2b, Table 8.2c): "Green Transition of Infrastructures", which focussed on constructing eco-friendly infrastructure for everyday life; "Low-carbon and Decentralised Energy", which involved investing in research and development (R&D) and facilities for new and renewable energies; and "Innovation in the Green Industry", which aimed to identify key green industry areas and enhance infrastructure.

The 2020 Green New Deal also consisted of infrastructure projects aimed at boosting the economy; the government explained that this was an upgrade from the previous Green Growth strategy (Cho, 2020). However, it also differed from the previous version of the Green New Deal in the following ways: first, whereas the 2009 New Deal focussed on public construction and civil engineering, the Green New Deal allocated more than half of the budget to electric and hydrogen vehicles and renewable energy projects. These included expanding renewable energy facilities to increase power generation capacity, supplying electric and hydrogen vehicles, and expanding the infrastructure of charging and refuelling stations to support the adoption of these vehicles.

Second, the Green New Deal remained aligned with the objectives of the mission. The government established sector-specific greenhouse gas emission reduction targets that it aimed to achieve with the Green New Deal. Additionally, a year after its announcement, the New Deal 1.0 (cf. Table 8.2) was rebranded as the New Deal 2.0, and the "Building a Carbon Neutral Promotion Foundation" category was added to align with the objective of achieving carbon neutrality. The concept of carbon neutrality was also applied to the existing projects, and their scope and scale were expanded. Notably, plans were added to expand infrastructure for hydrogen production, distribution, storage, and utilisation, and to diversify R&D fields to promote the hydrogen economy.

The Green New Deals pursued by the two administrations shared the same objectives of boosting the economy and generating employment by investing heavily in infrastructure projects in line with the "Green Keynesianism" approach (Jacobs, 2013). This involved using

Table 8.2a *2020 Green New Deal projects: Green transition of infrastructures*

Projects	Targeted outputs
Turning public facilities into zero-energy buildings	Green remodelling of public rental housing, energy-saving cultural facilities, green-smart schools
Restoring the terrestrial, marine, and urban ecosystems	Smart green cities, urban green spaces, restoration of national parks, damaged urban areas, and tidal flats
Building a management system for clean and safe water	Smart water supply network, smart sewage treatment, improvements to old facilities, flood sensors

Table 8.2b *2020 Green New Deal projects: Low-carbon and decentralised energy*

Projects	Targeted outputs
Building a smart grid for more efficient energy management	Smart apartments power grid, micro power grids for islands, energy diagnoses for old buildings
Promoting renewable energy use and supporting a fair transition	Renewable energy generation capacity, core hydrogen technology, cooling and heating technology using river water
Expanding the supply of electric and hydrogen vehicles	Electric and hydrogen vehicles, early scrapping of old diesel vehicles, conversion of old diesel trucks to LPG

Table 8.2c *2020 Green New Deal projects: Innovation in the green industry*

Projects	Targeted outputs
Promoting prospective businesses to lead the green industry and establishing low-carbon and green industrial complexes	Supporting promising companies, green convergence cluster, smart energy platform, clean factories
Laying the foundation for green innovation through the R&D and financial sectors	CCUS demonstration and commercialisation, remanufacturing of old power equipment, recovery of rare metals, securing financing for future environmental industry

Table 8.2d *2020 Green New Deal projects: Laying the foundations for carbon neutrality (added in New Deal 2.0)*

Projects	Targeted outputs
Laying the foundations for greenhouse gas reduction, including systems and experts	N/A
Revitalisation of the circular economy and expansion of carbon sinks	N/A
Raise national awareness of carbon neutrality and support for climate change adaptation	N/A

Source: Author's own elaboration

fiscal stimulus packages to maintain aggregate demand in the economy by compensating for the losses incurred in private sector demand while tackling environmental challenges. Through substantial investments in the construction of green infrastructure, the Green New Deals aimed to provide significant economic stimulus, minimise job losses, and address economic inequality. They had clearly defined goals and targets set within a short time frame and sufficient amounts of committed resources, and as a result, they succeeded in turning around GDP growth rates that had previously plummeted (World Bank, 2022).

Both Green New Deals have been criticised for failing to align short-term actions with long-term goals. In the case of the 2009 New Deal, criticism was directed at the fact that it focussed solely on revitalising civil construction, while neglecting environmental concerns. This led to controversy over the Four Major Rivers Restoration Project, which received the largest budget allocation and was a key presidential campaign promise. Despite this project being presented as a climate change adaptation measure aimed at preventing floods and droughts, critics deemed it unnecessary and claimed it caused environmental destruction. Despite greenhouse gas regulation being an important background context for Green Growth, Lee failed to meet his goal of reducing greenhouse gas emissions by 30 per cent by 2020 compared to the emissions that would be expected if "business-as-usual" practices continued, as the CO_2 emissions increased by 17.7 per cent during his presidency (OECD, 2022).

The 2020 New Deal also faced criticism for focussing on short-term economic stimulus rather than transforming the socioeconomic system and responding to the long-term climate crisis. Critics have argued that it constituted a repackaging of existing programmes bundled together rather than a comprehensive plan to tackle climate change (National Assembly Budget Office, 2021; People Power Party, 2020). Furthermore, it has been pointed out that the Green New Deal lacked a clear intention to transform carbon-emitting industries, as the government's support plan for fossil-fuel-based industries did not provide adequate requirements for transitioning to low-carbon alternatives (Choi & Kim, 2020; Justice Party, 2020). For these reasons, opposition parties and NGOs have criticised the Green New Deal as a green-washed stimulus package that did not go beyond the previous Green Growth vision (J.-H. Lee & Woo, 2020).

8.3.2 Coordinating Bodies for Expedited Implementation

Despite the above shortcomings, the government successfully responded swiftly to global crises through fiscal expansion and implementing stimulus packages via Green New Deals. This agility was made possible by the president's authority to establish and execute missions. The president had the power to appoint ministers, direct government organisations, national committees, and several advisory and executive agencies, and influence the direction and goals of most state affairs. Additionally, the presidential initiatives gained political momentum, as the ruling party was overwhelmingly victorious in general elections held during the crisis.

The most significant aspect of the governance structure was the creation of a dedicated organisation to oversee and coordinate projects. East Asian countries that achieved rapid economic growth after World War II, known as "developmental states", implemented strategic industrial policies within their states through "pilot agencies" that directed national economic development (Onis, 1991). The pilot agencies, leveraging their expertise and "plan rationality" rooted in an elite bureaucracy, possessed substantial authority to inspect, plan, and coordinate policies from a higher hierarchical position than other ministries. They established the long-term national growth trajectory, which then cascaded into detailed plans for ministries, affiliated organisations, and the public and private sectors. Additionally, they effectively obtained information and cooperation from these entities and guided their activities through cooperative relationships with companies (Evans, 1995; Weiss & Hobson, 1995).

Both administrations created organisations to perform the above functions. The Lee administration established the Presidential Committee for Green Growth and the Green Growth Planning Office, which served as dual "control towers" for the mission (The Government of the Republic of Korea, 2013). The Green Growth Planning Office under the Presidential Office provided political support by delivering the goals of the mission pursued by the president to ministries, supervising the establishment of national-level plans, and connecting related government organisations and the private sector. The Presidential Committee for Green Growth was established as a central point for the president, prime minister, ministers, and private experts to discuss related issues and coordinate and implement appropriate measures. This committee was authorised to review and decide on plans for all policy domains related to climate change, energy, and green technology, and all major policies related to Green Growth were approved by members of the relevant ministries that made up the committee (Jin, 2013). The Green Growth Planning Team supported the tasks of the committee; this team was led by the Green Growth Planning Officer and comprised officials dispatched from ministries and affiliated organisations. Additionally, to ensure policies were consistently executed and coordinated, the new position of "Chief Green Officer" was created within each ministry to facilitate communication between ministries and committees.

The Green New Deal of the Moon administration relied on the Ministerial Meetings on the Korean New Deal as a key mechanism for governance. These meetings were chaired by the Deputy Prime Minister of Economy and attended by almost all ministers. In these meetings, policy measures to address the pandemic and stimulate the economy were reported, discussed, and coordinated, and the roles and responsibilities of each ministry were clarified. To support this process, a working-level task force was installed, headed by the Vice Minister of Economy and Finance, to monitor the progress and outcomes of the projects (Korean Government Ministries Concerned, 2020c).

However, 2050 Carbon Neutrality differed from Green Growth in terms of the actors involved. The policies resolved at the ministerial meetings were subsequently submitted to the Joint Committee between the government and the ruling party. To ensure a swift implementation of the New Deal, they cooperated to streamline political processes such as securing the budget and enacting necessary legislation. The plans finalised through this committee were announced at the Strategy Meeting on the Korean New Deal, which was chaired by the President, thereby endowing them with political authority.

Furthermore, a diverse array of stakeholders participated in the process of establishing the 2050 Carbon Neutrality Commission to deliberate on policies and monitor their implementation. Unlike Green Growth, which involved mainly bureaucrats and experts taking the lead to implement the president's directives, the Moon administration emphasised social consensus and public debate when it came to carbon neutrality (The Government of the Republic of Korea, 2020). Moon and his party have enjoyed a close relationship with civil society, and activists and former civic organisation leaders have been appointed to key positions in the government and the National Assembly. In order to set the national goal of achieving carbon neutrality by 2050, the Ministry of Environment established the Low-Carbon Society Vision Forum in March 2019; participants included climate and energy experts, civil society representatives, and industry stakeholders. Following a year of discussions, the forum submitted its official recommendations. These recommendations were subsequently discussed through consultations involving ministries, experts, industry representatives, and civil society from March to October 2020 (2050 Low Carbon Society Vision Forum, 2020). In addition, 17 provincial municipalities declared their own carbon-neutral goals in July 2020. These engagements helped to ensure that the goal of achieving carbon neutrality by 2050 received broad support and was based on a consensus among stakeholders.

The 2050 Carbon Neutrality Commission was established as a control tower in May 2021 to create a strategy for the transition to carbon neutrality across all sectors. The commission adopted participatory governance policies to facilitate the development of carbon neutrality scenarios. First, consultative bodies – consisting of diverse social groups – were created to discuss the draft prepared by experts and to gather feedback. Second, a national policy participation group was formed. This allowed the general public to participate and aimed to achieve a social consensus on the issues through learning, deliberation, and discussion regarding the scenarios (2050 Carbon Neutrality Commission, 2021). The scenarios that were formulated in the course of this procedure proposed the establishment of social dialogues to ensure stakeholder participation throughout the entire carbon neutrality process, so that all members of society could participate and share responsibilities; ideally, this would result in a fair and just transition. Furthermore, the option to include coal in the energy mix was rejected based on stakeholder opinions (Korean Government Ministries Concerned, 2021b).

However, the legitimacy of the carbon neutrality goal presented in the mission had certain limitations, despite efforts to involve diverse stakeholders. Although the Moon administration sought to develop an inclusive agenda through a democratic process and broad participation, the scenario draft was prepared by experts nominated by ministries; this meant that the people's concerns about climate change and demands for transformation were overlooked from the outset. Although the government intended to ensure procedural validity through discussions, the approach primarily involved one-way education; instead of fostering deliberation, information was delivered to participants, and engagement was not iterative but rather a one-time gathering of opinions (Chae & Youm, 2021). In addition, there was a limited time frame of only five months from the establishment of the Commission to the announcement of the scenario, which was a relatively short period for conducting comprehensive social discussions and building consensus with stakeholders (Korean Government Ministries Concerned, 2022a).

There were discrepancies evident between the mission's vision and its actual policies. While the long-term goal of carbon neutrality had been put in place, the government continued to invest in coal-fired power plants. Although a policy was announced to stop new coal-fired power plant construction and phase out ageing power plants, seven power plants that were already under construction were not cancelled, and overseas investment in coal-fired power plants continued despite the government's commitment to halting new investments (Korean Federation for Environmental Movement, 2020; Yong Hyein Assembly Office, 2021). As a result, opposition parties and environmental groups criticised the directionality and effectiveness of the carbon neutrality plan.

In summary, both the New Deals, which aimed to revitalise the economy, prioritised the construction of infrastructure to achieve results in a short period of time and ensured efficient implementation by establishing pilot agencies. However, in contrast to Green Growth, 2050 Carbon Neutrality attempted to reflect societal considerations, as it set up projects aligned with the mission's direction and encouraged stakeholder participation through a more horizontal governance arrangement. The following section explores these distinctions using examples of policies for a long-term energy transition.

8.4 EVOLUTION OF PUBLIC-PRIVATE PARTNERSHIPS FOR A GREEN ENERGY TRANSITION

8.4.1 Using Green Growth Alternative Energy Policies to Create Business Opportunities

While the Green New Deal of 2009 did not prioritise green energy as an economic stimulus, it remained a crucial component of Green Growth. Energy has been a critical issue in South Korea for a considerable period, given the necessity of a stable power supply to support the manufacturing industry and the export strategy (The Government of the Republic of Korea, 2020). In the wake of events such as the oil crisis in the 1970s, the government acknowledged energy security as a national interest and set goals focussed on diversifying energy sources and managing associated risks.

The global financial crisis resulted in a surge in oil prices and brought energy independence to the forefront of the policy agenda, because Korea relied heavily on imports for approximately 97 per cent of its energy sources. Green Growth prioritised energy sovereignty by preparing for potential changes and disruptions in the energy supply chain, decreasing dependence on imports, and developing alternative energy sources capable of replacing fossil fuels. To achieve these objectives, the government set targets for increasing the proportion of renewable and nuclear energy (Korean Government Ministries Concerned, 2008).

The government recognised the green energy industry as a new growth engine and took proactive measures to position itself in the global energy market by nurturing the sector in the long term. It also formulated strategies for developing green technologies as the country's key energy source; the South Korean industry needed to quickly establish competitiveness in core technologies to narrow the technological gap with advanced countries in the field of renewable energy and secure a global market share (Ministry of Knowledge Economy, 2008; Presidential Committee on Green Growth, 2009b). As a latecomer country in economic growth, South Korea has been learning, imitating, and improving on technologies from developed countries to catch up with the global frontier: South Korea succeeded in transitioning from the catch-up phase to the keeping-up and upgrading phase based on innovation and creation (L. Kim, 1997). However, the path-dependent strategy that followed the trajectory of an established dominant design necessitated a transition to the post-catch-up stage. It was imperative for South Korea to shift from being a fast follower to becoming a first mover, creating new paths and setting standards to compete in the emerging sector (Choung et al., 2016). The government emphasised that technology development needs to be industry- and business-oriented right from the basic research stage. To secure upstream core technology, by 2012 it had doubled its investment in green technology R&D compared to 2008 and had begun to prioritise 27 key green technologies based on technological capabilities, commercialisation potential, market environment, and contribution to economic growth (Chon, 2017).

The National Science and Technology Council served as a mechanism for ensuring consistency in green R&D policies and programmes across various ministries. As a "control tower" in the field of science and technology, the council oversaw the processes of setting R&D strategies, allocating and adjusting budgets, and evaluating performance to prevent duplication of efforts. Ministries shared roles and enhanced cohesion by discussing and aligning their policies in this governance arrangement. The Green Technology Committee under the Council planned a government-wide green technology development strategy and evaluated the feasibility of investment plans submitted by ministries to increase coordination effectiveness. The Korea Institute of Energy Technology Evaluation and Planning was established under the Ministry of Knowledge to plan R&D programmes in the energy field and distribute funding. The aim of dividing roles among government organisations in this way was to achieve systematic management of green energy R&D.

Another strategy of the government was to create a domestic foundation to link R&D outcomes to commercialisation and industrialisation for export (Ministry of Knowledge Economy, 2008). The government encouraged companies to expand their portfolios into the green energy sector and embrace high-risk innovations by eliminating problems such as lack of physical and knowledge infrastructure, slow formation of new networks, unfavourable institutions, and by generating new business opportunities. It also strengthened communication channels with businesses and fostered public-private partnerships. Experts from the private sector were appointed as members of committees under the National Science and Technology Council to facilitate communication and collaboration.

This was done because the private sector's capacity and expertise surpassed that of bureaucrats. In the past, the government was able to mobilise necessary resources and order companies to invest in new industries, such as the heavy chemical industry in the 1970s and the electronics and automobile industries in the 1980s, due to the lack of technological capacity in the private sector and asymmetric authority relationships (Choung et al., 2016; Kalinowski, 2021). However, as companies grew larger and acquired expertise in high-tech sectors, attempting to regulate and guide their activities through government orders became impractical. To adapt to this evolving landscape, the Lee administration institutionalised private-sector participation in the decision-making process, designed policies favourable to businesses, and shifted its role to that of a facilitator that encourages companies to engage in the policy process with a "business-friendly" slogan (Presidential Committee on Green Growth, 2009a).

8.4.1.1 Case: Korea Micro Energy Grid (K-MEG) project

The Korea Micro Energy Grid (K-MEG) project was a representative case where the government encouraged companies to participate in promising sectors that would become future flagship industries and facilitated the formation of consortia. The Ministry of Knowledge Economy funded the Future Industry-Leading Technology Development programme to boost the competitiveness of flagship industries and organise large-scale projects in prioritised areas to lead the global market within three to five years. The consortia, which were selected through competition to carry out projects in five fields, including micro energy grids, electric vehicles, and thin-film solar cells, were designated by the Office of Strategic R&D Planning under the ministry. The goal was to accelerate commercialisation by combining government contributions and matching funds from participating organisations, based on the domestic market foundations shaped by the government (Ministry of Knowledge Economy, 2010).

From July 2011 to June 2014, the Green Tomorrow Consortium carried out the K-MEG project, which received government funding amounting to 53.8 billion Korean Won (38.8 million euros), along with matching funds of 30.3 billion Korean Won (21.8 million euros).

Its goal was to develop an integrated energy management system that uses a combination of diverse energy sources and IT solutions to efficiently produce and utilise energy, thus enabling buildings and cities to achieve energy self-sufficiency. The primary focus of the project was to develop energy-saving technologies, as well as decentralised and renewable power sources, and demonstrate their effectiveness through the implementation of an integrated system.

The Korea Institute of Energy Technology Evaluation and Planning was designated as the project administrator to manage the project, which involved a consortium comprising 11 large companies, 33 SMEs, 12 government-funded research institutes and universities, and 5 foreign partners, covering the entire value chain from R&D to manufacturing, construction and system integration, operation, and management (Samsung C&T, 2014). To effectively implement the project and achieve commercial success, large companies were designated as leaders for five tasks, with Samsung C&T acting as the primary leader to manage and coordinate them (Ministry of Knowledge Economy, 2011).

The project's objectives included developing energy solutions and entering the global market. It was, therefore, crucial to establish demonstration sites and testbeds to verify the business model and facilitate scalability to guarantee the competitiveness of the new infant industry. The reliability and operability of the solution were tested in several domestic and international buildings and regions. For instance, Banwol-Sihwa National Industrial Complex had a heat grid installed to recycle waste heat, while Guro Digital Complex and Sejong City had systems to optimise cooling, heating, and energy operations in buildings. Overseas tests were conducted at solar power plants in Indonesia and wind farms in Australia. These efforts led to the consortium's success in developing, commercialising, and exporting micro energy grids (Kwon, 2011; Samsung C&T, 2014).

South Korean companies have successfully taken advantage of the government's promotion of green industries and integrated them into their core business portfolios. They have emerged as major manufacturers and drivers of global green industries, contributing to a fivefold increase in the turnover of the domestic environmental industry from 2004 to 2020 (Ministry of Environment, 2022). South Korea's export-oriented industrialisation strategy was complemented by the consumption-oriented strategies of key export destinations, including Europe and the United States. Supportive policies in these countries, including subsidies, instilled confidence among companies that the green industry will be a major part of their future business (Kalinowski, 2021).

8.4.2 A Two-track Strategy towards Energy Transition for Carbon Neutrality

The Moon administration's energy policy emphasised the environment, quality of life, and safety, while previous administrations focussed solely on energy supply security and industrialisation. During his presidential campaign, Moon pledged to decrease the proportion of nuclear power generation and raise the proportion of renewable energy generation in the energy mix. The administration identified "transition to safe and clean energy through a nuclear phase-out policy" as a national priority and developed a roadmap to cancel the construction of new nuclear power plants and disapprove the life extension of existing ones (Korean Government Ministries Concerned, 2017; National Planning Committee, 2017). Additionally, the administration aimed to nurture the renewable energy industry and develop future energy sources to increase the proportion of renewable energy generation from 7 per cent in 2017 to 20 per cent by 2030 (Ministry of Trade, Industry and Energy, 2017a).

The primacy of industrialisation continued with the introduction of 2050 Carbon Neutrality. To effectively carry out the industrialisation policy for renewable energy, a new vice-ministerial position was established in the Ministry of Trade, Industry and Energy, which is responsible for energy policy. The government proposed orienting R&D towards the mission to promote innovation in key areas. This involved strengthening the connection between R&D policies, technology development, commercialisation, and dissemination to consistently orient these towards carbon neutrality (Korean Government Ministries Concerned, 2022b). Private-sector participation was also encouraged throughout the entire R&D process, from technology development to demonstration, commercialisation, dissemination, and diffusion, to align with the target of carbon neutrality. To achieve this, the government introduced flagship R&D projects in which government-funded research institutes and public and private companies participate in different areas; this activity has large spillover effects on the industry (Korean Government Ministries Concerned, 2019a).

The Presidential Advisory Council on Science and Technology, which has taken over the functions of the National Science and Technology Council, also set strategic priorities for areas that contribute to carbon neutrality and enhance the competitiveness of export industries. The council established the Special Committee on Carbon Neutral Technology, consisting of senior government officials, industry executives, and researchers from universities and industry to encourage active involvement from the private sector (Special Committee on Carbon Neutral Technology, 2021). The government aimed to establish a role-sharing framework with the private sector, whereby the public sector proactively invests in frontier technologies to tackle societal challenges and lay the foundations for mitigating business uncertainty. On the other hand, the private sector takes the lead in terms of strategising and executing carbon-neutral R&D, making continuous efforts to build innovation ecosystems in newly emerging areas (Ministry of Science and ICT, 2020).

8.4.2.1 Hydrogen economy: Transferring the role of supplier from the government to the private sector

To achieve 2050 Carbon Neutrality, hydrogen was identified as a potential source for both the energy transition and industrialisation. With the phase-out of nuclear power and the reduction of greenhouse gas emissions from fossil fuels, South Korea needed to secure a breakthrough energy source that could act as a replacement and transform the energy consumption landscape. The government recognised hydrogen as a crucial catalyst for the energy transition, aiming to shift the energy paradigm away from carbon-based resources (Korean Government Ministries Concerned, 2019b; The Government of the Republic of Korea, 2020). In order to establish a comprehensive hydrogen economy that encompasses the production of clean hydrogen energy as well as its storage, transportation, and use in various applications such as hydrogen vehicles, the government took several significant steps. They released the Hydrogen Economy Revitalization Roadmap in January 2019, enacted the Hydrogen Economy Promotion and Hydrogen Safety Management Act in February 2020, established the Hydrogen Economy Committee, and proposed that the hydrogen industry be nurtured as a promising export sector.

To achieve a transition to a hydrogen economy in line with the mission's objectives, it has been essential for companies to expand their business portfolios into the hydrogen field. R&D programmes on hydrogen have been in place since hydrogen energy was first identified as one of the key green technologies under Green Growth. However, corporate participation has been relatively slow due to concerns about the small size and uncertainty of the global market

(J. Kim & Jang, 2021). The government proposed involving the private sector in the entire process, including planning R&D projects, to shift from a government-led growth engine policy to a government-company team system, and prioritised R&D in technologies that companies are willing to adopt. This was particularly important in areas where the private sector's crowding-in effect was expected to be significant (Korean Government Ministries Concerned, 2021c). The government made the decision to proactively invest in hydrogen R&D and create a domestic market to reduce uncertainty and allow companies to explore new business opportunities and enter the global market. As part of this strategy, the construction of hydrogen vehicle infrastructure was implemented through the Green New Deal.

While Green Growth appointed experts on green industries and technologies to government committees to encourage their participation, 2050 Carbon Neutrality institutionalised mechanisms for empowering the private sector and accelerating corporate participation. The Hydrogen Economy Committee, under the Prime Minister's Office, appointed business professionals as committee members to encourage private sector involvement in hydrogen initiatives. The Carbon Neutral Industrial Transformation Promotion Committee was established under the Ministry of Trade, Industry and Energy to lead discussions on carbon-neutral innovation strategies and strengthen government-private sector partnerships for industrialisation. The government established private organisations together with businesses to assume responsibilities relating to the promotion, distribution, and safety of the hydrogen industry (Ministry of Trade, Industry and Energy, 2021a).

The objective was to encourage innovation under private-sector leadership, and to boost large-scale collaboration between firms across the hydrogen industry's value chain by supporting the formation of consortia. The government acknowledged that mission-oriented carbon-neutral R&D requires integrating multiple sectors and technologies, including energy, transport, industry, and the environment (Korean Government Ministries Concerned, 2022b). HyNet and Kohygen, special-purpose companies established by public and private enterprises and municipalities, collaborated under the coordination of the Ministry of Trade, Industry and Energy to install hydrogen vehicle refuelling stations. The objective of these consortia was to reduce the cost of constructing these stations and achieve synergistic effects in the hydrogen business through cooperation between companies responsible for automobiles, hydrogen supply, and charging facilities.

The H2 STAR project was a consortium-led initiative involving companies across the entire value chain of the hydrogen industry, aimed at testing and demonstrating the generation of clean hydrogen from overseas, its transportation to South Korea, and its storage and utilisation at various locations. The project sought to leverage South Korean technology to generate renewable energy in other countries, electrolyse it into hydrogen, and transport it back to South Korea for applications in power generation, transport, and industry. The project is particularly relevant for achieving carbon neutrality by 2050, as it involved establishing a stable supply chain for producing "green" and "blue" hydrogen using renewable energy and carbon dioxide capture and storage technologies, rather than "grey" hydrogen produced from fossil fuel processing (Ministry of Trade, Industry and Energy, 2021b).

Based on these efforts, the leadership of the hydrogen economy is shifting from the government to the private sector, and carbon-intensive sectors are expanding their business portfolios to include hydrogen supply (J. Kim & Jang, 2021). Major conglomerates have announced their commitment to promoting various global hydrogen investment projects through a partnership between the Korea H2 Business Summit, an association of CEOs from large corporations, and the Hydrogen Economy Committee. Samsung Engineering's H2biscus project produces blue ammonia, green methanol, ammonia, and hydrogen from natural gas sourced from Malaysia, before importing and utilising them in industrial applications. GS Energy is collaborating with the United Arab Emirates to produce and utilise blue ammonia, commercialise hydrogen carriers, build hydrogen ports, and expand charging stations and pipeline networks. SK E&S plans to construct hydrogen liquefaction plants to distribute liquefied hydrogen and enter the global market for hydrogen facilities. Korea Electric Power Corporation uses green hydrogen and ammonia to generate carbon-free power. Through the process outlined above – from R&D investment to creating infrastructure and laying foundations, to networking and forming alliances between companies, right up to complementing private-led innovation with institutional initiatives – the South Korean government's intention to transition to a hydrogen economy has been set in motion.

8.4.2.2 Using demand-side projects to expand solar energy deployment

In a similar way to previous administrations, the Moon administration aimed to industrialise renewable energy through its energy policy. However, there was a clear shift toward rapidly increasing the share of renewable energy and promoting a decentralised and participatory energy system (Ministry of Trade, Industry and Energy, 2017a). The Third Energy Master Plan outlined specific goals for the energy transition and for promoting future energy sectors to enhance the competitiveness of the energy industry. This plan also encouraged the involvement of local governments and consumers in energy production to increase the amount of renewable energy being generated and distributed (Ministry of Trade, Industry and Energy, 2019).

To achieve the energy transition goals set by 2050 Carbon Neutrality, solar power generation facilities have been identified as a crucial component. South Korea has achieved a leading position in polysilicon production and silicon solar cell capacity thanks to its advanced solar cell and module technologies; however, the domestic consumption of solar energy has remained low (IEA PVPS, 2019). This was because most solar products were exported to the global market or manufactured by offshore facilities. The government allocated over 95 per cent of new power generation facilities to renewable sources, with a particular emphasis on increasing the installed capacity of solar power (Ministry of Trade, Industry and Energy, 2017b). The Green New Deal installed solar panels on public facilities and provided subsidies for setting up solar panels. Additionally, the Energy Technology Development Action Plan proposed building a large-scale photovoltaic power generation complex.

Solar energy, unlike large-scale, centralised facilities such as nuclear power plants, involves dispersed production and consumption. Hence, it is essential for the public to participate both as power producers and consumers, although consideration must be given to environmental and regional conditions when installing facilities. The government implemented various measures to encourage public involvement and investment in solar power generation. These included public participation projects in which residents, cooperatives, and municipalities led the installation of renewable energy facilities. The feed-in tariff system purchased energy from power generation companies at a fixed price to guarantee stable profits for small-scale power generation. Cooperatives and citizen funds leading power generation projects were incentivised, with a portion of the profits being returned to the respective regions (Ministry of Trade, Industry and Energy, 2017b).

In addition, the energy transition policy encouraged the participation of civil society in policymaking and decision-making processes. For instance, in the case of the nuclear phase-out policy, which was included in the presidential election pledge, a public debate committee comprising various stakeholders and a citizen participation group decided whether to resume the construction of nuclear power plants (Public Deliberation Committee on Shin-Kori Nuclear Reactors No. 5 and 6, 2017). The importance of this participatory model was particularly emphasised in terms of distributing renewable energy. The Renewable Energy 3020 Implementation Plan envisioned a transition to a participatory energy system and stressed the importance of residents' acceptance when constructing large-scale power generation facilities (Ministry of Trade, Industry and Energy, 2017b). This shift was prompted by criticism that previous policies had focussed mainly on areas with export potential while neglecting local and domestic demand and prioritising corporate interests over those of the public (Yun, 2009a, 2009b). Green Growth identified areas to receive intensive investment and focussed on developing core technologies to supply the industry. However, it was criticised for neglecting to link innovation activities with demand, and for not reflecting the expectations and requirements of consumers. For this reason, the existing solar power distribution projects were considered ineffective, and citizens were encouraged to participate in the production and consumption of renewable energy to enhance its acceptability.

The government proposed several policy measures to achieve the goal of a participatory energy system (Korean Government Ministries Concerned, 2019a). First, it recommended expanding opportunities for users to participate in proposing, developing, building demonstrators of, and commercialising technology, and conducting energy technology impact assessments before developing any new technology. Second, living labs were set up that allowed residents to experiment with and learn about the energy that they used. This facilitated the development and testing of local energy facilities, helped to establish optimal energy mixes, and enabled participatory technology development processes. Third, there are now plans for dedicated projects aimed at addressing issues and challenges in particular local contexts. For instance, environmental concerns such as deforestation caused by the construction of solar power plants have often led to disagreements with residents. In such cases, the local participation approach can be used to support reflexive governance (Park, 2020). The following examples were attempts to articulate the demand for energy supply policy by involving participation from locals in the region.

A living lab project related to small-scale solar power was conducted in the village of Seong-Dae-Gol in Seoul, initiated by a consortium of local community members, research organisations, and NGOs. This project carried out experiments to build an energy system at the village level, whereby residents' participation and feedback were used to solve technical problems, and the profits from the energy generation were shared in the village. Furthermore, based on the outcomes of this project, a cooperative for energy efficiency in housing was established, and the living lab was scaled up to implement a local energy grid. Efforts were also made to collaborate with other regions to transfer the knowledge and experience gained through this project (Seong et al., 2020).

In South Jeolla Province, a project was carried out under the leadership of the local government to define problems, increase municipalities' capacities for solving issues, and improve the acceptance of rural solar power. This project identified and addressed issues associated with installing solar power facilities in rural areas and established a resident-led energy generation platform to create a service model that could increase income and generate job opportunities

(KISTEP, 2020). It also developed guidelines for environmental impact assessments, based on a case study of residents' acceptance, to resolve conflicts that may arise from the introduction of energy facilities. The project concluded by articulating the need for follow-up projects, such as solar structures and stand-alone smart grids based on local characteristics (Jang, 2020).

The 2050 Carbon Neutrality energy transition policy marked a shift in the normative direction of the energy system. It established an institutional framework that enables diverse stakeholders to participate in the R&D process as both energy producers and consumers, and successfully facilitated the supply of renewable energy to dispersed regions. However, it also had its limitations. First, as the primary objective was to expand renewable energy production, the 2050 Carbon Neutrality projects focussed more on developing a business model for public investment in facilities rather than on socio-technical innovation. Since there was already a predominant design for solar panels and modules, and these were being mass-produced by South Korean companies, the aim was to install them in as many regions as possible without conflict. Therefore, instead of involving residents in the innovation process as end users and allowing them to interact with developers through experimentation and learning, the main activities focussed on minimising barriers and customising the technology to suit local conditions, thereby facilitating its acceptance.

Second, while living lab experiments have been conducted, they have not had a significant impact on energy policy, and there is currently no mechanism for ministries to evaluate, revise, or adjust policies based on the findings from these projects. Instead, the results are still mainly implemented at the site level or used as a reference for resolving local disputes. Although efforts are underway to apply the results of projects in Seong-Dae-Gol and collaborate with other regions, the projects are yet to be scaled up to a national level.

8.4.3 Comparison of the Two Missions' Energy Policies

Both missions were driven by the objective of transforming existing systems through the development of alternative energies. To achieve this, they focussed on effectively implementing R&D programmes, establishing a "control tower" in the field of science and technology, and setting strategic priorities. Additionally, both missions aimed to ensure a stable energy supply by commercialising green energy technologies and nurturing them as export industries. Through close relationships with companies, the government played a facilitating role in encouraging the private sector's entry into emerging areas and establishing a solid foundation for their endeavours. Companies are now seizing newly formed opportunities and expanding their business portfolios in that area.

However, there were notable differences between the two missions. While Green Growth primarily approached the energy transition from the perspective of energy security, 2050 Carbon Neutrality positioned responding to climate change as its primary objective. This shift in focus led to a change in direction and a wider range of policy portfolios, as it recognised the importance of creating a market for new industries as well as gaining consumer acceptance and addressing needs during the energy system transformation. The mix of policies was employed to secure industrial competitiveness through the development of advanced technologies and to distribute mature technologies to different regions, thereby supporting this endeavour.

8.5 PROSPECTS AND CHALLENGES OF THE GREEN MISSION IN KOREA

This chapter compared Green Growth and 2050 Carbon Neutrality by examining the intentions of the missions and characterising their implementation through examples of New Deals and green energy policies. Their components and key features are summarised in Table 8.3. There were parallels between the two missions due to South Korea's underlying policies and the circumstances that the missions faced at the time. Green missions were introduced in response to external pressures stemming from economic crises and climate policies that necessitated the reduction of greenhouse gas emissions and a shift away from carbon-intensive industries. These encompassed two primary aspects: Green New Deals, which concentrated on short-term economic stimulus through infrastructure construction projects, and green energy policies, which aimed at achieving long-term energy transition objectives.

The two missions also exhibited some differences in terms of their goals. While Green Growth prioritised economic growth over environmental concerns, 2050 Carbon Neutrality made reducing greenhouse gas emissions a key objective. Compared to the top-down implementation of Green Growth, which allocated most of its budget towards civil works, the New Deal of 2020 was more aligned with its mission and policies: it focussed on constructing infrastructure to provide a foundation for green industries and encouraged a broader range of stakeholders to participate. With regard to energy policy, 2050 Carbon Neutrality officially set the energy transition as a policy goal and adopted a mix of policy measures that included promoting bottom-up projects by residents to create a renewable energy supply. In this respect, South Korean green missions have been gradually evolving and becoming more consistent in addressing societal challenges and directing policy towards long-term goals. However, the green mission is faced with a challenge when it comes to securing its own sustainability. This is due to newly elected presidents who tend to quickly execute policy initiatives in order to pursue and achieve their own agendas within a fixed term in office (Linz, 1990). Particularly in South Korea, successors have removed their predecessors' political symbols and replaced them with their own agendas (Blaxekjær, 2016). This also applies to the green mission, which was a key part of the agenda for previous administrations. Lee's successor, Park Geun-hye's administration, erased all traces of green from the government and rendered Green Growth obsolete, despite belonging to the same political party.

The Yoon Seok-yeol administration, which succeeded Moon in 2022, is also removing the green mission. Yoon, who belongs to the opposing political party to Moon, criticised the carbon-neutral goal as lacking scientific basis and burdening the public, cut the New Deal's budget, and suspended the operation of the 2050 Carbon Neutrality Commission by reducing its size and function (Lim, 2022; Seo, 2022). Moreover, the plan for becoming the "most powerful nuclear power country" has undermined efforts to create an energy mix centred on renewable energy. These sudden discontinuations of policies highlight the risk of significant political agendas disappearing with a change in leadership.

However, it is too early to say that the green mission has completely disappeared in South Korea, despite such roll-backs. First, the legal and institutional framework established by the mission has not been abolished, as legislation falls under the responsibility of the National Assembly rather than the executive branch led by the president. Moreover, in the face of objections from the opposing party, the administration is not attempting to repeal existing laws. Even though Green Growth as a mission came to an end, the Framework Act on Low Carbon

Table 8.3a *Comparison of Green Growth and 2050 carbon neutrality: Emergence of mission*

Aspects	Green Growth	2050 Carbon Neutrality
Background	Global financial crisis and greenhouse gas emission regulations	COVID-19, mandatory submission of carbon reduction targets
Goals	Economic growth through green industrialisation	Economic growth and green industrial transformation for net zero

Table 8.3b *Comparison of Green Growth and 2050 carbon neutrality: Green New Deal (short-term)*

Aspects	Green Growth	2050 Carbon Neutrality
Target	Economic stimulus and job creation	Economic stimulus and job creation
Projects	Civil engineering work	Green industry infrastructure
Coordination	Top-down by bureaucrats and experts	Mix of top-down and stakeholder engagement

Table 8.3c *Comparison of Green Growth and 2050 carbon neutrality: Energy transition (long-term)*

Aspects	Green Growth	2050 Carbon Neutrality
Objectives	Energy independence, industrialisation of green energy	Energy transition, industrialisation, and dissemination of green energy
Approaches	Invite the private sector to government-led initiatives	Empowerment in the private sector
Case studies	K-MEG: a consortium of companies organised by the government	Hydrogen: handover of industrialisation initiatives from government to the private sector Soar energy: distributed energy system where residents participate

Source: Author's own elaboration

and Green Growth and the Green Growth Committee were not abolished, and their legacy was converted into the Framework Act on Carbon Neutrality and Green Growth and the 2050 Carbon Neutrality Commission by the Moon administration. This could continue to exist as a basis to support the green agenda when it emerges again.

Second, climate change and carbon neutrality are internationally mandated external elements that require countries to pursue policies in the same direction. Despite the significance and variety of national policy styles, the strengthening of global regulations on greenhouse gas emissions serves as a driving force for government organisations to continue detailed programmes in fields such as renewable energy and electric vehicles, and to support green industries, even if they diverge from the umbrella of the Green Growth mission. Yoon criticised the carbon-neutral goal but mentioned that it is a promise to the international community and should be implemented (Lim, 2022). Societal objectives constructed by various stakeholders at the global level and endorsed by many countries remain influential despite the fact that their implementation fluctuates at the national level.

Third, the climate responsibility currently imposed on South Korean companies by the country's export-oriented industrial structure serves as a driving force for persisting with green policies. As seen above, South Korean conglomerates are vying for leadership in the global market by self-organising energy projects – and the private sector expects them to succeed, as they previously have for solar power and secondary batteries. In addition, environmental regulations in export destination countries mean that products and services must be carbon-neutral and sustainable, and companies must fulfil these conditions to manage business risks. These global demands are factors that influence South Korean policy, shaping market opportunities and regulating business activities outside of the mission.

The final task of this chapter is to consider how the experience of South Korean green missions can inform the analysis of other mission-oriented policies that address societal challenges. One of the key lessons from the South Korean case is that the development of mission-oriented policies should consider not only the nature of the problem being addressed, but also the path dependency of the local political and governance context. While all countries were facing the same climate change and energy issues, South Korea understood the issues and established a vision through legacy-driven developmentalist policy logics, particularly through Green Growth. Policies were then implemented based on established governance arrangements. These factors indicate that missions aimed at addressing societal challenges can only be implemented by gradually reforming existing policy frameworks, even in East Asian countries with a history of developmental states or in countries that adopt a top-down approach based on authority.

Another lesson that can be learnt from the South Korean case is that there is potential for transforming the policy framework away from the above legacy. The green missions described in this chapter were perceived primarily as a response to external pressures rather than internally driven agendas. Their legitimacy was derived predominantly from their economic impacts rather than by addressing societal issues, and consequently, the design and implementation of policies were also oriented towards business interests. This highlighted that the bottom-up approach is only applicable to some Western countries and may encounter challenges when applied in diverse contexts in other nations (Angel & Rock, 2009; Karo, 2018). However, as found in 2050 Carbon Neutrality, there have been efforts to integrate societal goals into the mission's long-term objectives, align policies accordingly, and change the governance system to involve stakeholders. In particular, the engagement of political organisations and civil society, which has advocated for a response to climate change, could provide legitimacy and support to the mission.

Transformation in general, and energy transition in particular, entail making changes to the interconnected elements such as stakeholders, institutions, and norms that make up the system. It also necessitates considering the influence of path dependence and the existing legacy. The analysis of the South Korean case in this chapter offers an approach to comprehending missions and exploring potential avenues for change to address societal challenges. To effectively accomplish the mission, it emphasises the importance of not only active government intervention but also fostering a sense of ownership within society, which both demands and supports the transformation and normative turn of the system. In this regard, this chapter provides support for the idea that aligning long-term visions and policies with societal challenges can be better achieved through such engagement.

REFERENCES

2050 Carbon Neutrality Commission (2021). *The 1st plenary meeting of the 2050 Carbon Neutrality Commission.*

2050 Low Carbon Society Vision Forum (2020). *2050 Long-term low-carbon development strategy review.*

Angel, D., & Rock, M. T. (2009). Environmental rationalities and the development state in East Asia: Prospects for a sustainability transition. *Technological Forecasting and Social Change*, *76*(2), 229–240.

Bernard, S., Asokan, S., Warrell, H., & Lemer, J. (2009). The greenest bail-out? *Financial Times.*

Blaxekjær, L. (2016). Korea as green middle power: Green growth strategic action in the field of global environmental governance. *International Relations of the Asia-Pacific*, *16*(3), 443–476.

Chae, J. H., & Youm, J. (2021). *Green New Deal and social conflict: Consensus formation and conflict management strategies.* Korea Institute of Public Administration.

Cho, H. (2020). Blue House, "Green New Deal is an upgrade of the past government's Green Growth". *Media Today.*

Choi, W., & Kim, J. (2020). Economic stimulus was urgent, but... "Superficial" Green New Deal without climate crisis strategy. *Hankyoreh.*

Chon, S.-B. (2017). Science and technology policy governance analysis: From the Kim Dae-jung administration to the Park Geun-hye government. *Korean Governance Review*, *24*(3), 171–200.

Choung, J.-Y., Hwang, H.-R., & Choi, J. K. (2016). Post catch-up system transition failure: The case of ICT technology development in Korea. *Asian Journal of Technology Innovation*, *24*(supl), 78–102.

Climate Analytics. (2020). *Transitioning towards a zero-carbon society: Science-based emissions reduction pathways for South Korea under the Paris Agreement.*

Climate Watch. (2020). Climate Watch historical country greenhouse gas emissions data. https://www.climatewatchdata.org/ghg-emissions?end_year=2019&regions=KOR&start_year=1990

Democratic Party. (2020). *Green New Deal – general election pledges in the green energy sector.*

Evans, P. B. (1995). *Embedded autonomy: States and industrial transformation.* Princeton University Press.

Han, H. (2015). Authoritarian environmentalism under democracy: Korea's river restoration project. *Environmental Politics*, *24*(5), 810–829.

IEA PVPS. (2019). *National survey report of PV power applications in Korea 2019.*

International Energy Agency. (2020). Korea 2020 energy policy review. https://www.iea.org/reports/korea-2020

Jacobs, M. (2013). Green growth. In F. Robert (Ed.), *The handbook of global climate and environment policy* (pp. 197–214). John Wiley & Sons Ltd.

Jang, W. (2020). *A study on solutions for local residents' acceptance of agricultural solar power and offshore wind power.* Ulsan SESCOOP.

Jin, S. (2013). The path dependency of low carbon green growth policy in MB administration. *Korean Public Administration Quarterly*, *25*(4), 1049–1073.

Justice Party. (2020). *Korean Green New Deal, is it the Low Carbon Green Growth Season 2*?

Kalinowski, T. (2021). The politics of climate change in a neo-developmental state: The case of South Korea. *International Political Science Review*, *42*(1), 48–63.

Karo, E. (2018). Mission-oriented innovation policies and bureaucracies in East Asia. *Industrial and Corporate Change*, *27*(5), 867–881.

Kim, J., & Jang, S. (2021). *Strategy for Early settlement of market-led hydrogen economy.* Korea Energy Economics Institute.

Kim, L. (1997). *Imitation to innovation: The dynamics of Korea's technological learning.* Harvard Business School Press.
Kim, S.-Y., & Thurbon, E. (2015). Developmental Environmentalism. *Politics & Society, 43*(2), 213–240.
Kim Sungwhan Assembly Office. (2020). *Post-Corona era and Green New Deal.*
KISTEP. (2020). *A project to establish the foundation for social problem solving in the pan-Ministry.* Korea Institute of S&T Evaluation and Planning.
Korean Federation for Environmental Movement. (2020). *Coal power plant elimination policy proposal for climate crisis response.*
Korean Government Ministries Concerned. (2008). *The first national basic energy plan: 2008–2030.*
Korean Government Ministries Concerned. (2009a). *Comprehensive plan for the green technology R&D.*
Korean Government Ministries Concerned. (2009b). *Promotion plan for the Green New Deal Program to create job opportunities.*
Korean Government Ministries Concerned. (2017). *Follow-up measures for public deliberation on Shin-Kori Nuclear Reactors No. 5 and 6 and roadmap for energy transition (nuclear phase-out).*
Korean Government Ministries Concerned. (2019a). *The fourth energy technology development plan (2019–2028).*
Korean Government Ministries Concerned. (2019b). *Hydrogen economy revitalization roadmap.*
Korean Government Ministries Concerned. (2020a). *The Korean New Deal: National strategy for a great transformation.*
Korean Government Ministries Concerned. (2020b). *2050 carbon neutral strategy of the Republic of Korea: Towards a sustainable and green society.*
Korean Government Ministries Concerned. (2020c). *Comprehensive plan for the Korean New Deal.*
Korean Government Ministries Concerned. (2021a). *Korean New Deal 2.0.*
Korean Government Ministries Concerned. (2021b). *2050 Carbon neutrality scenario.*
Korean Government Ministries Concerned. (2021c). *R&D investment strategy for carbon neutrality.* Ministerial Meeting on Science, Technology.
Korean Government Ministries Concerned. (2022a). *Carbon neutral green growth strategy.*
Korean Government Ministries Concerned. (2022b). *Mission-oriented carbon-neutral R&D promotion plan.*
Kwon, Y. (2011). *Business-oriented Korea's flagship R&D program – Korea Micro Energy Grid (K-MEG) project.* Samsung C&T.
Larrue, P. (2021). *The design and implementation of mission-oriented innovation policies: A new systemic policy approach to address societal challenges.* OECD Publishing.
Lee, H., & Rhee, C. (2012). Lessons from the 1997 and the 2008 Crises in Korea. *Asian Economic Policy Review, 7*(1), 47–64.
Lee, J.-H., & Woo, J. (2020). Green New Deal policy of South Korea: Policy innovation for a sustainability transition. *Sustainability, 12*(23), 10191.
Lee, Y. (2020). South Korea's reforms should not be called a Green New Deal. *Heinrich-Böll-Stiftung Blog.* https://hk.boell.org/en/2020/09/14/south-koreas-reforms-should-not-be-called-green-new-deal
Lim, K. (2022). President Yoon, the Moon government's carbon reduction goal is "no scientific basis ... the burden of the people". *Pressian.*
Linz, J. J. (1990). The perils of presidentialism. *Journal of Democracy, 1*(1), 51–69.
Ministry of Environment. (2022). *Environmental industry ratio (relative to GDP).* https://www.index.go.kr/unify/idx-info.do?idxCd=4280&clasCd=7
Ministry of Knowledge Economy. (2008). *Green energy industry development strategy.*
Ministry of Knowledge Economy. (2010). *Briefing session on the Program for Five Future-Leading, Early Result-Creating Projects* [Press release].
Ministry of Knowledge Economy. (2011). *Selected contractor for Future-Leading Technology Development Program.*
Ministry of Science and ICT. (2020). *Science and technology future strategy 2045.* Presidential Advisory Council on Science and Technology.
Ministry of Trade, Industry and Energy. (2017a). *The 8th basic plan of electricity supply and demand.*
Ministry of Trade, Industry and Energy. (2017b). *The renewable energy 3020 implementation plan.*
Ministry of Trade, Industry and Energy. (2019). *The third energy master plan.*

Ministry of Trade, Industry and Energy. (2021a). *2021 plan of Hydrogen Economy Agency – Hydrogen Industry Promotion, Distribution, and Safety Agency.*
Ministry of Trade, Industry and Energy. (2021b). *Hydrogen economy achievements and vision of a leading country for hydrogen.*
National Assembly. (2020). *Resolution calling for an emergency response to the climate crisis.*
National Assembly Budget Office. (2021). *Comprehensive analysis of the 2022 Budget Bill – II.*
National Assembly Budget Office. (2022). *Economic overview of the Republic of Korea 2022.*
National Planning Committee. (2017). *Moon Jae-in government's five-year plan for state administration.*
OECD. (2017). *OECD environmental performance reviews: Korea 2017.* OECD Environmental Performance Reviews. OECD Publishing.
OECD. (2022). Air and GHG emissions. https://data.oecd.org/air/air-and-ghg-emissions.htm
Office of Prime Minister. (2008). *Green Growth promotion strategy.*
Onis, Z. (1991). The logic of the developmental state. *Comparative Politics, 24*(1), 109–126.
Park, J.-H. (2020). Technology innovation policy of renewable energy in Korea from the perspective of transformative innovation policy. *Journal of Korea Technology Innovation Society, 23*(2), 234–257.
People Power Party. (2020). *100 problem projects in 5 areas related to next year's budget bill.*
Presidential Committee on Green Growth. (2009a). *National strategy for the green growth.*
Presidential Committee on Green Growth. (2009b). *Five-year plan for the green growth.*
Presidential Council for Future and Vision. (2009). *The way to the green growth.* Joongang Books.
Public Deliberation Committee on Shin-Kori Nuclear Reactors No. 5 and 6. (2017). *Policy recommendation on Shin-Kori Nuclear Reactors No. 5 and 6.*
Samsung C&T. (2014). *K-MEG business model demonstration and commercialization final report.*
Seo, J. (2022). Reducing the organization and members of the 2nd Carbon Neutrality Commission by half… Yang Jungsuk pointed out the government's negligence in carbon neutrality. *Financial News.*
Seong, J., Song, W., & Kim, S. (2020). *A research on the spread of living-lab guideline for social problem-solving R&D activation.* Korea Institute of S&T Evaluation and Planning.
Special Committee on Carbon Neutral Technology. (2021). *Carbon-neutral key technologies.*
The Government of the Republic of Korea. (2013). *Lee Myung-bak administration white paper on government affairs, Volume 6: Green growth and future growth engines.*
The Government of the Republic of Korea. (2020). *2050 carbon neutral strategy of the Republic of Korea: Towards a sustainable and green society.*
United Nations Environment Programme. (2009a). *Overview of the Republic of Korea's green growth national vision.*
United Nations Environment Programme. (2009b). *Overview of the Republic of Korea's green growth national vision – An interim report.* https://wedocs.unep.org/20.500.11822/31513
Weiss, L., & Hobson, J. (1995). *States and economic development: A comparative historical analysis.* Polity Press.
World Bank. (2022). *GDP growth (annual %) – Korea, Rep.* https://data.worldbank.org/indicator/NY.GDP.MKTP.KD.ZG?locations=KR
Yong Hyein Assembly Office. (2021). *Export-Import Bank of the Moon Jae-in administration to increase support for renewable energy, bets on coal-fired power as much as solar power* [Press release].
Yun, S.-J. (2009a). Problems of green growth and degrading energy policy. *Environment and Life, 60*, 17–50.
Yun, S.-J. (2009b). The ideological basis and the reality of "low carbon green growth". *ECO, 13*(1), 219–266.

9. Mission-oriented innovation policy for transformative innovation in Japan: historical development and lessons from the case of automated driving

Kazuhito Oyamada and Tateo Arimoto

9.1 MISSION-ORIENTED INNOVATION POLICY INITIATIVES IN JAPAN

Transformative innovation, which aims to transform the social system itself so as to create a sustainable society, requires efforts that go beyond the traditional science, technology, and innovation (STI) policy framework (European Commission & Mazzucato, 2018; Schot & Steinmueller, 2018). While various policy approaches to transformative innovation have been envisioned, mission-oriented innovation policy (MOIP) is one specific policy approach that is being addressed in many countries. According to the study by the Organisation for Economic Co-operation and Development (OECD), an MOIP is defined as "a coordinated package of policy and regulatory measures tailored specifically to mobilise science, technology and innovation in order to address well-defined objectives related to a societal challenge, in a defined timeframe" (Larrue, 2021a, p. 8).

This chapter provides an overview of the specific ways in which Japan's science, technology, and innovation policies have been taking steps to address each of these elements of MOIPs. Currently, MOIPs involve a diverse range of policy approaches reflecting the nature of each country's policy framework and socioeconomic system. Among them, the OECD study characterises Japan's MOIPs as being led by centre-of-government (Larrue, 2021a; 2021b). This characteristic of Japan's MOIPs is largely due to the influence of the reform of the STI policy system that has taken place since the mid-1990s, particularly the strengthening of the "control tower" function in the Cabinet Office and the establishment of a large research and development (R&D) programme as a means to achieve this.

On the other hand, even if the legal basis is in place for the centre-of-government strategic leadership, it is not easy to coordinate the budgets and policy measures of the various ministries in an integrated manner towards achieving the mission. In addition, while large R&D programmes are the main policy instrument in Japan's MOIPs, the funding agencies that administer them usually do not have sufficient authority or resources to carry out the MOIP requirements of strategic direction, policy coordination, and policy implementation. Therefore, a mechanism to complement them is essential. This chapter discusses the development of an automated driving system and its social implementation as an example of such an effort. In this context, SIP-adus, an automated driving system R&D project under the Cross-Ministerial Strategic Innovation Promotion Program (SIP), a mission-oriented R&D programme of the Cabinet Office, and the public-private ITS Initiatives/Roadmap, a government-wide coordination framework, function in a mutually complementary manner, and the above requirements of the MOIP have been realised.

Like other countries, Japan's MOIP is also in the process of evolutionary development, and various initiatives beyond the conventional STI policy framework are beginning to emerge. In the last part of this chapter, we will introduce recent efforts in Japan and discuss issues around the development of such MOIP, especially in the implementation phase.

9.2 DEVELOPMENT OF JAPAN'S SCIENCE, TECHNOLOGY, AND INNOVATION POLICY AND MISSION-ORIENTED POLICIES

In the history of STI policy in Japan, the question of how to achieve interministerial coordination has been a long-standing issue. This section reviews the history of Japan's STI policy system from the post-war period to the present, particularly in terms of improving interministerial coordination.

9.2.1 Construction and Development of Japan's Science and Technology Policy System after World War II

The basic structure of Japan's current science and technology (S&T) policy system was created during the post-World War II reconstruction process. The central roles were played by the Ministry of Education, which was responsible for basic research at universities (academic research) and higher education policy, the Ministry of International Trade and Industry (MITI), which was in charge of industrial policy, and the Science and Technology Agency (STA), which was responsible for big scientific projects such as space development and nuclear power, and emerging technologies such as life science. The Science and Technology Council (STC), established in the Prime Minister's Office, and the STA, which served as its secretariat, were responsible for the overall coordination of R&D activities in each ministry. Okuma (2009) identified three main policy areas for the STA: formulating a plan for the promotion of S&T for Japan as a whole, conducting systematic R&D based on that plan, and supporting and utilising researchers. In particular, the STA uses funding called the Special Coordination Funds for Promoting Science and Technology to support cross-ministry and cross-field themes, R&D on emerging technologies such as internet-related technology and its implementation in society, new initiatives and projects such as the Human Frontier Science Program, and experimental efforts to reform the R&D system. However, when the STA was established in 1956, the jurisdiction of each ministry in charge of a particular field had already been defined, and it had limited scope for interministerial coordination (Okuma, 2009).

As part of the so-called industrial policy to strengthen industrial competitiveness, which was a national strategic goal at the time, the METI and the STA conducted large-scale, collaborative R&D projects between industry, academia, and government. These projects took the form of commissioned development projects based on public research institutes and consortiums of industry, academia, and government actors. Such industrial policies played an important role in Japan's post-war recovery and in helping it catch up with the US and European countries. These policies are said to have contributed greatly to strengthening Japan's industrial competitiveness, especially in fields such as semiconductors, supercomputers, and other high-tech equipment (Goto, 2000; Johnson, 1982; Sawai, 2011). However, as Japan's economy grew rapidly in the 1980s, trade friction with the United States intensified, and movement towards further liberalisation of the economy grew stronger; criticism of government-led industrial policies also increased, and large-scale projects were scaled back (Nakayama, 1995; Sawai, 2011).

For big projects such as space development and nuclear power generation, special corporations such as the Japan Atomic Energy Research Institute (JAERI) and the National Space Development Agency of Japan (NASDA) were established as mission agencies under the Science and Technology Agency. Initially, these agencies mainly introduced technologies from the US and other Western countries. However, they eventually acquired their own R&D capabilities and achieved results such as carrying out their own launches in the space development field and developing a private nuclear power industry (Okuma, 2009).

9.2.2 Institutional Reform of Science, Technology and Innovation Policy

Various institutional reforms of STI policies were implemented during Japan's period of prolonged economic stagnation after the 1990s. The institutions created during this period have defined the basic structure of Japan's STI policies up to the present day.

9.2.2.1 The basic Act on science and technology and the science and technology basic plans

The severe recession that followed the bursting of the bubble economy in the 1990s hit Japanese companies hard. Many companies eliminated or downsized their central research centres, creating great concern about the scientific and technological foundations of Japanese industry. This led to a strong demand from industry for government support to maintain and strengthen Japan's industrial competitiveness. Meanwhile, in the academic sector, problems such as ageing research facilities at universities and lack of employment opportunities for doctoral degree holders became apparent, and strengthening the research base became an issue. In addition, global environmental issues such as climate change emerged on the political agenda, and there was a need to respond to these issues as well. In response to these demands from industry, academia, and society, the Basic Act on Science and Technology was enacted in 1995, creating the first legal framework for the government to address S&T policy as a whole.

Based on the Act, the government has established a five-year plan: the Science and Technology Basic Plan. While the first basic plan focussed on strengthening the R&D system, the second and third basic plans emphasised R&D in priority fields. The second basic plan (fiscal year (FY) 2001–2005) set the goal of establishing a new relationship between science and technology and society with regard to solving social issues. This clearly designated "science and technology for society" as one of the pillars of Japan's science and technology policy, in addition to the previous "science and technology for knowledge" approach. Since then, the Research Institute of Science and Technology for Society (RISTEX) and other research and development programmes have been implemented to solve social issues (see Box 9.1).

In the fourth basic plan (FY 2011–2015) there was a major shift from the previous approach of focussing on specific priority areas to an approach that focusses on social issues to be solved for sustainable growth and social development in the future.[1] Another feature of the

[1] Three policy goals were set: recovery and reconstruction from the Great East Japan Earthquake that occurred in March 2011 and the Fukushima nuclear power plant accident; green innovation

fourth basic plan was the explicit inclusion of "innovation" as a policy target. This led to an expansion of the policy framework from the traditional "science and technology policy" to a "science, technology and innovation policy".[2] The fifth basic plan (FY 2016–2020) presented Japan's vision of future society as Society 5.0, where cyberspace and physical space are highly integrated.

BOX 9.1 OVERVIEW OF THE INSTITUTE OF SCIENCE AND TECHNOLOGY FOR SOCIETY IN JAPAN

The Japan Science and Technology Agency's Research Institute of Science and Technology for Society (RISTEX) was established in 2001 in response to the growing interest in using science and technology to address societal issues in Japan and abroad. Initially, RISTEX also conducted in-house research, but following a reorganisation in 2005, it became a funding organisation that conducts publicly solicited R&D programmes to support transdisciplinary research projects aimed at addressing societal issues. The term "science and technology for society (STfS)" is defined as "S&T for creating new social systems by integrating knowledge from natural sciences and social sciences and humanities (SSH)", and as S&T that "regards society itself as the object of R&D and seeks to resolve problems that either currently exist in society or that are anticipated to occur in the future" (RISTEX, 2023, p. 8). In order to achieve this, RISTEX is working on the following cycle (Figure 9.1).

Over the past 20 years, RISTEX has conducted numerous transdisciplinary research projects on themes such as ageing society, information technology, safety and security, environment, and science, technology and society (STS), as well as ethical, legal, and social issues (ELSI). As a result, it has produced solutions and knowledge. The programme also provides support for the dissemination and social implementation of the results and solutions generated. However, continuous communication and proactive activities with the ministries, agencies, local governments, and communities involved in solving actual problems are essential for disseminating and scaling up these results and solutions. This is a challenge. For this reason, RISTEX encourages the participation of stakeholders in a series of processes, from the establishment of R&D focus areas to the implementation and evaluation of R&D projects (RISTEX, 2023).

---

aimed at stable energy utilisation and low carbon emissions; and life innovation aimed at developing innovative preventive, diagnostic, and therapeutic methods and improving quality of life.

[2] The Fourth Science and Technology Basic Plan defines "Science, Technology and Innovation" as the "creation of intellectual/cultural values based on new knowledge obtained from scientific discovery, invention, etc., and the innovation to develop such knowledge into economic, social, or public value" (Government of Japan, 2011, p. 7).

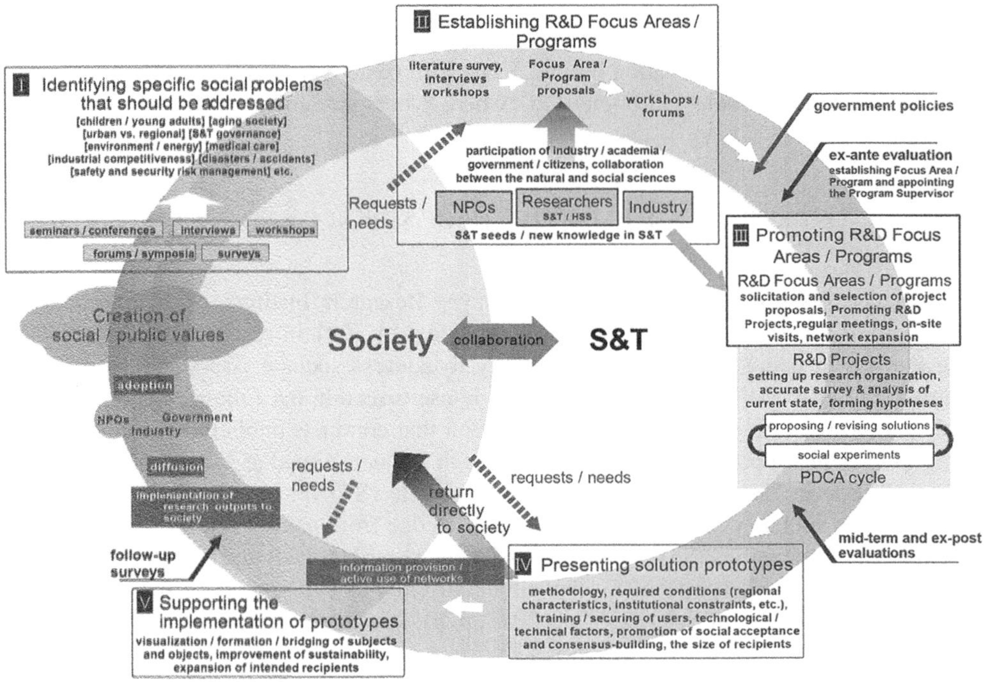

Source: RISTEX (2023) Method for Promotion of R&D https://www.jst.go.jp/ristex/en/funding/funding-outline/index.html

Figure 9.1 *RISTEX's process for identifying and implementing R&D focus areas to address societal problems*

9.2.2.2 Expansion of STI policy framework and acceleration of social transformation

In the 2010s, various institutional reforms were implemented in response to the growing awareness of social issues and the new challenge of achieving innovation, including social transformation towards a more sustainable society.

Firstly, the Basic Act on Science and Technology, which had been the legal foundation of Japan's S&T policy, was revised to become the Basic Act on Science, Technology and Innovation (to be enforced in 2021). The new act adds "creation of innovation"[3] to its scope of activity, along with the conventional goal of "advancing science and technology". The promotion of the humanities and social sciences was also included in the scope of this act.

The sixth Science, Technology and Innovation Basic Plan (FY 2021–2025), formulated based on the revised Basic Act, inherits the concept of "Society 5.0" presented in the fifth

[3] The Basic Act on Science, Technology and Innovation defines "creation of innovation" as "the creation and dissemination of new value through scientific discovery or invention, the development of new products or services, and other creative activities, thereby creating major changes in the economy and society" (Basic Act on Science, Technology and Innovation, Article 2, Paragraph 1). https://laws.e-gov.go.jp/law/407AC1000000130/

Basic Plan as Japan's vision for the future, and presents "a sustainable and resilient society that ensures the safety and security of the people" and "a society in which each individual can realise diverse happiness (well-being)" as goals to be achieved within that society (Government of Japan, 2021b). It also states that the concept of innovation is changing to transformative innovation[4] and that mission-oriented R&D and social implementation, such as SIP and Moonshot R&D programmes, will be strategically promoted.

9.2.2.3 Accelerating the efforts towards social transformation as a whole-government and STI policy

Thus, while social transformation was explicitly stated as a policy goal in STI policy, the government as a whole has also proposed strategies and goals in this area. In October 2020, the government declared itself carbon-neutral, with a goal of achieving zero overall greenhouse gas emissions by 2050. In light of this, the METI and the New Energy and Industrial Technology Development Organization (NEDO) established a 2 trillion yen (1.4 billion USD[5]) Green Innovation Fund under a supplementary budget for FY 2020. This fund will support projects from R&D and demonstration to social implementation over a period of 10 years, mainly targeting companies, based on ambitious and specific goals shared by the public and private sectors. In addition, the government has stated that it will realise green transformation (GX) investments of 150 trillion yen (1.1 trillion USD) over the 10 years from 2022 through collaboration between the public and private sector. This includes the creation of GX Transition Bonds to attract private investment and the GX League, a public-private consortium for market creation and rulemaking that also aims to create a carbon credit exchange scheme to help achieve 2050 carbon neutrality.[6]

In September 2021, the Digital Agency was launched as an organisation for promoting digital transformation (DX), as it has been noted that Japan is falling behind in this area. The mission of the Digital Agency includes not only the realisation of DX for the government, but also for Japanese society as a whole.

In Japan, there is also a need for transformation in terms of creating sustainable local communities in response to structural changes in the socioeconomic system, such as a shrinking and ageing population and changing industrial structure. As for "regional revitalisation", which has been an important policy issue in Japan for many years, the "Digital Garden City Nation Initiative" is being promoted from FY 2020, with the aim of bringing about transformation in rural areas through DX-based solutions to social issues. Although the measures under this initiative at this time mainly include the development and utilisation of human resources and support for entrepreneurship and resettlement, there are other public-private partnerships targeting local regions under various initiatives, such as the Smart City Public-Private Partnership Platform[7] and the SDGs Future City programme.[8] Through these efforts,

---

[4] The plan defines transformative innovation as "innovation aimed at social change in order to respond to complex and wide-ranging social issues such as global environmental issues" (Government of Japan, 2021a, p. 10).

[5] Calculated at 1 USD = 140 yen. The same rate shall apply hereinafter.

[6] The GX League Basic Concept, announced in February 2022, was endorsed by 440 companies, representing about 40 per cent of Japan's total CO_2 emissions.

[7] https://www.mlit.go.jp/scpf/about/index.html

[8] https://www.chisou.go.jp/tiiki/kankyo/index.html

local governments, private companies, NPOs, and others are increasingly participating on a proactive basis.[9]

As these efforts towards social transformation progress, the realisation of transformative innovation has become a major policy goal in STI policy. In addition to the science and technology-related ministries, universities, and research institutes that have played a central role in STI policy in the past, new actors such as sectoral ministries, local governments, and the civil sector are expected to participate. The concept of innovation and the policy goal of bringing this concept to life will play a role in connecting these actors. However, the current situation is complicated by the presence of various parallel initiatives and numerous headquarters and coordinating organisations, which makes collaboration processes more complex.

9.2.3 Strengthening the Cabinet Office's Cross-sector Coordination

Japan's STI policy has expanded over the past two decades; previously, it was a framework for research and development in specific technological areas and for strengthening collaboration between industry and academia. Now it has become a system for solving social problems, creating value through innovation, and transforming society to make it more sustainable and resilient. In addition, the sixth basic plan includes reference to demand-side policy measures, such as regulations and the promotion of practical applications, for the creation of innovation. However, with the existing policy system, in which R&D has been the main focus, the coordination of ministries and agencies, including measures other than R&D, will face difficulties in the institutional and budgetary domains, including the legal basis, as well as in actual implementation. In order to overcome these challenges, various efforts have been made to strengthen interministerial coordination and to orient and organise the efforts of each ministry and agency to assist with "strengthening the control tower function" in the Cabinet Office.

9.2.3.1 Establishment of the Council for Science and Technology Policy

The Cabinet Office, established as a result of the Central Government Reform in 2001, supports the Cabinet Secretariat, and its tasks are defined as planning and formulating important policies for the Cabinet and coordinating interministerial collaboration between the various ministries and agencies. The Council for Science and Technology Policy (CSTP) was established as one such collaboration on important policy issues (Shiroyama, 2018).

The CSTP is composed of the Prime Minister, the Chief Cabinet Secretary, the Minister of State for Science and Technology Policy, the Minister of Internal Affairs and Communications, the Minister of Finance, the Minister of Education, Culture, Sports, Science and Technology, the Minister of Economy, Trade and Industry, the President of the Science Council and executive members. The executive members are selected from academia and industry and appointed with the consent of the Parliament.[10] The Plenary Meeting, chaired by the Prime Minister, is to be held once a month in principle. In the course of these Plenary Meetings, expert panels are

---

[9] For example, the Smart City Institute of Japan (SCI-Japan), a public-private partnership, has developed the Liveable Well-Being City Index (LWC Index) as an indicator to measure well-being in the region. The index will be used in the Digital Rural City State Initiative. https://www.sci-japan.or.jp/LWCI/index.html

[10] Although the official English name in Japan is "the National Diet", in this article we will use "the Parliament", which is commonly used.

established to study individual policy issues. Shiroyama (2018) states that the fact researchers and people from the private sector with expertise participated in these expert panels has helped to build up a base of knowledge that formed the foundation of the coordination system.

The means of overall coordination in the CSTP include budget, planning, and evaluation (Ito, 2006). The budget was evaluated based on the following year's "Plan for Allocation of Budget, Human Resources, and Other Resources Related to Science and Technology" and interviews with ministries and agencies; however, its influence was limited (Shiroyama, 2018, p. 163). With regard to planning, the government was required to go through the CSTP's deliberations when formulating the S&T Basic Plan. However, the Ministry of Education, Culture, Sports, Science and Technology (MEXT), which was created through the merger of the STA and the Ministry of Education, still had jurisdiction over the planning and formulation of basic policies relating to science and technology; this had been the jurisdiction of the STA (Shiroyama, 2018, p. 163). For this reason, there have been long-standing issues around how to strengthen the Cabinet Office's strategic direction and cross-sectoral coordination.

During the implementation phase of the fourth S&T Basic Plan (FY 2011–2015), in which STI policies became issue-oriented, the CSTP tried to direct the ministries' policies by prioritising S&T-related budgets through the Action Plan for the Implementation of Important Science and Technology Policy Measures. The Action Plan was discussed by the Science, Technology and Innovation Strategy Councils. Consisting of representatives from industry, government, and academia, these councils were established to achieve three policy goals: reconstruction after the earthquake and nuclear power plant accident, green innovation, and life innovation. The Strategy Councils will develop strategies and initiatives to achieve specific goals; these will be promoted by the government outside of the framework of ministries and agencies. The councils will also examine the system reforms (regulatory and institutional reforms, measures to promote deployment, etc.) necessary to realise innovation; these reforms will then be compiled into a draft action plan. The Expert Panel for the Promotion of Science, Technology and Innovation Policy is tasked with formulating an action plan based on these drafts. However, in order to create value and return the results to society, it has been deemed necessary to take an integrated approach that includes not only R&D, but also the perspectives of business and society. In order to achieve this, strengthening the control tower function for the promotion of STI policy was required. The Study Group of Experts for Promoting Science, Technology and Innovation Policies, established in 2011 by the then Minister of State for Science and Technology under the Democratic Party of Japan (DPJ) administration, recommended the establishment of a Strategic Headquarters for Science, Technology and Innovation, which was to act as a control tower to promote the planning and implementation of government-wide policies, as well as a Science, Technology and Innovation Advisor. This advisor was to assist the Prime Minister and Ministers with understanding and implementing STI, and provide support to them by carrying out secretariat duties (Expert Group for the Promotion of Science, Technology and Innovation Policy, 2011). Although the recommendations of this study group did not materialise under the DPJ administration, they will be realised as a system in stages through subsequent Liberal Democratic Party (LDP) administrations, as described in the next section.[11]

[11] While this paper focusses on strengthening the control tower function, the Science, Technology and Innovation Advisor role proposed in this report in response to the Great East Japan Earthquake and the Fukushima nuclear power plant accident in March 2011, which made

9.2.3.2 Reorganisation into the Council for Science, Technology and Innovation

As mentioned above, since the 2010s, the policy framework has expanded from science and technology policy to science, technology, and innovation policy. In line with this expansion, it has become necessary to coordinate and manage a large number of diverse policies and initiatives – these include not only traditional R&D conducted by science and technology-related ministries and support for social implementation of the results of this R&D but also non-R&D policies of the sectoral ministries. The CSTP was to serve as a control tower for science and technology policy, but its role needed to be expanded in response to the broadening of the policy framework. To this end, the Cabinet Office Establishment Act was amended in 2014 to include "the comprehensive development of the environment to promote the creation of innovation through the practical application of the results of research and development", and the CSTP was reorganised to become the Council for Science, Technology and Innovation (CSTI).[12] The following section outlines the control tower functions of the CSTI.

9.2.3.3 Direction and interministerial coordination through an annual strategy on STI policy

Based on the CSTI's stronger governance powers as the control tower for STI policy, as described above, it will need to actually coordinate, direct, and concretely promote the various measures and initiatives of each ministry and agency. One of the means used to achieve this is an annual strategy on STI policy. The S&T/STI Basic Plan specifies the basic policies, strategic goals, and total amount of investment for the government's STI policies for the five-year period. However, it does not state the amount of investment for individual policies and measures, except for those that have already been decided. Therefore, budgets for new policies and measures will be allocated through the annual budgeting process, and this annual strategy is the policy tool by which CSTI directs and coordinates the budgets of each ministry through this budgeting process.

Since 2015, CSTI has developed a Comprehensive Strategy for Science, Technology and Innovation, which prioritises the policies of the various ministries. In addition, starting in 2018, this focus will be channelled into an "Integrated Innovation Strategy" for all government ministries, which will assign specific measures for each ministry to address over the course of one year in relation to STI policies. In order to promote the Integrated Innovation Strategy, the Cabinet has established the "Integrated Innovation Strategy Promotion Council".[13] The Council includes all ministries and agencies closely related to STI, such as information and telecommunications, intellectual property, medical and healthcare, space development, maritime policy, and geospatial information. Under the Council, "Expert Panels for the Promotion

---

bridging science and policy and strengthening its mediating function a major political issue, was realised as Science and Technology Advisor to the Minister of Foreign Affairs in 2015, Science and Technology Advisor to the Cabinet Secretariat in 2022, and Science and Technology Advisor to the Minister of Education, Culture, Sports, Science and Technology in 2023.

[12] At the same time, "Affairs related to the formulation and promotion of the Science and Technology Basic Plan" and "Affairs related to the coordination of cost estimation policies of relevant administrative agencies for science and technology" were transferred from the MEXT to the Cabinet Office. As a result, the authority related to the Science and Technology Basic Plan has been centralised in the CSTI (Cabinet Office, 2014).

[13] https://www8.cao.go.jp/cstp/tougosenryaku/kaigi.html

of Strengthening Innovation Policy" have also been established to formulate innovation strategies for strategic key technology areas such as artificial intelligence, biotechnology, and quantum technology.[14] The Cabinet Office is supposed to administer the affairs of the Integrated Innovation Strategy Promotion Council, and the secretariat of CSTI is in effect in charge of this.

The Integrated Innovation Strategy, which is decided and published around June each year, specifies the STI policy initiatives of each ministry and agency based on the government's priorities at that time. These policies are aligned with the "Basic Policies for Economic and Fiscal Management and Reform", which constitutes the budget policy of the entire government for each fiscal year and is released prior to the Strategy. In the subsequent budget elaboration for the following fiscal year, each ministry requests a budget based on these strategy documents, and the government as a whole, including sectoral ministries, is expected to manage and coordinate various measures.

9.2.3.4 Agenda-setting and progress management through regular meetings

In addition to coordination between ministries and agencies through the budgeting process as described above, a mechanism is required to respond to changes in the environment surrounding STI in Japan and abroad, and to monitor and manage the implementation of measures in each ministry to achieve strategic goals. Regarding Japan's recent policy system, Takashi Shiraishi, a political scientist who served as an executive member of the CSTP, compared the top-down approach based on the political leadership of the prime minister to the bottom-up approach based on expert analysis and opinion-gathering by the responsible departments of each ministry and agency. As a result, he pointed out that Japan has a strong capacity for individual policy formulation at the division level, which is the fundamental unit of policy planning and implementation at each ministry; however, he also noted that it is difficult to make strategic efforts to achieve major policy goals by aggregating individual policies (Shiraishi, 2013). Under these circumstances, the challenge is how to realise a system like the one described above that can flexibly respond to new issues, as well as provide direction, monitor progress, and manage initiatives for the government as a whole.

At its inception, the plan was for the CSTP to hold plenary sessions chaired by the prime minister about once a month. However, the prime minister's priorities and political situation at the time would ultimately affect the frequency of the meetings. Depending on the administration at the time, the meetings were held once every few months during certain years (Shiraishi, 2013). A reduction in meeting frequency causes problems in terms of agile responses and the overall direction, progress, and management of the government as a whole. For this reason, the current CSTI holds a meeting between the Minister of State for Science and Technology Policy, three other government ministers, and CSTI executive members[15] every Thursday morning. At this meeting, in addition to the topics presented by researchers and private-sector actors as part of the survey, each ministry gives progress reports and trend

---

[14] As of June 2023, six expert councils have been established: Fusion Energy Innovation Strategy; AI Strategy; Safety and Security; Biotechnology Strategy; Quantum Technology Innovation; and Materials Strategy. https://www8.cao.go.jp/cstp/tougosenryaku/kaigi.html

[15] If none of the ministers or the other three political officers participate, the meeting will be held as the "Roundtable Meeting of Expert Members". https://www8.cao.go.jp/cstp/gaiyo/yusikisha/index2022.html

reports on the policy goals set by CSTI. The results of the discussions at the meetings are reflected in the consideration and implementation of measures by each ministry and agency, thereby strengthening the top-down policymaking and coordination function of the political leadership.

9.2.3.5 Implementing large R&D programmes

In the 2010s, in addition to annual strategies being used to orient each ministry's policies, large-scale R&D programmes were implemented under the leadership of CSTI, as a means of strengthening the control tower function. These included the Impulsing Paradigm Change through Disruptive Technologies Program (ImPACT) (FY 2014–2018) and the Moonshot-type Research and Development Program (FY 2019–present). These primarily time-limited large-scale R&D programmes have been financed through supplementary budgets. These programmes were intended to promote R&D that would contribute to achieving the government's priority goals of enhancing industrial competitiveness and solving social issues. This was to be done by introducing new methods and approaches that had not been implemented in the R&D of individual ministries and agencies, with the expectation that this would have a spillover effect into R&D system reform.

In addition, the Cross-Ministerial Strategic Innovation Promotion Program (SIP) has been in place since FY 2014 as an initiative to promote interministerial cooperative R&D and help to solve social issues that cannot be handled by a single ministry. The details of SIP will be discussed in the next section. These large-scale R&D programmes, led by the Cabinet Office/CSTI, are defined as "mission-oriented R&D programmes" in the sixth STI Basic Plan (Government of Japan, 2021a).

9.2.3.6 Reinforcement of secretariat roles

Thus, the overall coordination roles of the CSTI have expanded beyond those of the CSTP. Therefore, it also became necessary to strengthen the secretariat role to support these functions. To this end, the Secretariat of Science, Technology and Innovation Policy was established in the Cabinet Office in 2021 as an organisation that coordinates science, technology, and innovation-related policies across the board, reorganising the Bureau for STI, which had been responsible for the secretariat responsibilities in the past. The Secretariat was designed as a higher-level organisation than the secretariats of the related Headquarters for Health and Medical Care, Intellectual Property, Space Development, and Ocean Policy, and is to coordinate the policies of each ministry and agency, including these related Headquarters (Cabinet Office, 2021a, p. 4).[16]

[16] Kobayashi (2017) points out that the establishment of multiple headquarters related to STI policy in the government creates policy enclaves, thereby narrowing the room for CSTI to conduct actual overall coordination. The establishment of the Secretariat for Science, Technology and Innovation Policy can be seen as an attempt to take such a situation into account and to organise the relationships among the headquarters' secretariats that are responsible for actual policy coordination.

9.3 CASE STUDY OF MISSION-ORIENTED INNOVATION IN JAPAN: AUTOMATED DRIVING SYSTEM

In the previous section, we reviewed how strategic direction, policy coordination, and policy implementation – all important measures for mission-oriented innovation policy – have been discussed and promoted in recent reforms of Japan's STI Policy, mainly in terms of orienting various policies across multiple ministries and agencies and coordinating them comprehensively. In Japan, efforts to strengthen industrial competitiveness and address societal challenges have so far been centred on mission-oriented R&D programmes (Larrue, 2021a). On the other hand, in recent years, the creation of innovation has been clearly set as a policy target, and the use of demand-side policy instruments such as laws and regulations and procurement by sectoral ministries has been recognised as means of implementing STI policy. Furthermore, implementing mission-oriented innovation policies to bring about transformative innovation requires a whole-of-government approach that includes the demand-side policy instruments of sectoral ministries (Larrue, 2021b). As seen so far, Japan's STI policy has also reinforced the interministerial coordination function as part of efforts to strengthen the control tower function of the Cabinet Office. However, whether or not such coordination works effectively can vary from situation to situation. In addition, bringing about transformative innovation requires a multi-layered and comprehensive approach that includes not only sectoral ministries and agencies, but also the voluntary efforts of various stakeholders such as the private sector and civil society (CRDS, 2022). In this context, comprehensive efforts that go beyond the framework of conventional STI policies are needed.

This section discusses automated driving systems as a case study of mission-oriented innovation being used to create transformative innovation in Japan. In addition to the SIP, an interministerial cooperative R&D programme by the Cabinet Office, called the Public-Private ITS Initiatives/Roadmap, was put in place. This policy tool was intended to help implement automated driving systems throughout society, and it functioned in a mutually complementary manner. Other interrelated factors included the automotive industry's sense of urgency due to changes in the international competition surrounding the industry, and a shared recognition of the need to transform Japan's regional social systems in response to the demographic changes of declining birthrates, and an ageing and shrinking population. It can be said that the interaction of these factors created the necessary impetus for realising transformative innovation.

9.3.1 SIP and SIP-adus

The Cross-Ministerial Strategic Innovation Promotion Program (SIP) is an interministerial cooperative R&D programme that was launched in FY 2014. As part of this programme, CSTI allocates budgets across ministries and promotes initiatives from basic research to practical application and commercialisation of the results, right through interministerial collaboration and industry-academia-government cooperation.

One of the characteristics of the SIP is that the CSTI makes top-down decisions in three areas: setting the targets that are essential to society and important for Japan's economic and industrial competitiveness; selecting the programme directors (PD), who are responsible for the overall measures taken to achieve these targets; and setting the budget. The PDs appointed by CSTI are responsible not only for implementing the R&D programmes but also for coordinating with the ministries, especially in terms of policies related to the social implementation of R&D results. For this purpose, a Promotion Committee has been established, chaired by the PD, with the participation of relevant ministries and agencies. In addition, the R&D institutions and funding agencies of closely related ministries are responsible for R&D implementation and related administrative work as research promotion organisations. The overall programme is promoted by a governing board composed of executive members of CSTI; this board evaluates and advises the programme as a whole (Cabinet Office, 2021b).

The SIP conducted 11 programmes (158 billion yen, 1.1 billion USD) in the first phase (FY 2014–2018), and 12 programmes (144.5 billion yen, 1.0 billion USD) in the second phase (FY 2018–FY 2022). While the specifics of R&D, the promotion structure, and the nature of interministerial collaboration vary considerably from programme to programme, SIP-Automated Driving for Universal Services (SIP-adus) – which focusses on R&D and the social implementation of automated driving systems – has been adopted in both the first and second phases of SIP.[17]

9.3.2 SIP-adus for Industry-Academia-Government Cooperation and Interministerial Collaborative R&D and Social Implementation

SIP-adus was launched in the first phase of the SIP, which began in FY 2014, under the title "Automated Driving Systems". The PD and sub-PDs responsible for the overall promotion of the programme were appointed from the major automobile companies Toyota, Honda, and Nissan. These companies worked together with industry, academia, and the government to develop technologies in the pre-competitive and collaborative areas related to automated driving, with the Tokyo Olympic and Paralympic Games – which were scheduled to be held in the summer of 2020 – highlighted as a milestone. The project promoted technological development through industry-academia-government and interministerial collaboration. Through this process, it became clear that it would be possible to overcome the three barriers of technology, the legal system, and social acceptability. Therefore, in the second phase of the SIP, the title of the project was changed to "Automated Driving (Extension of Systems and Services)". In addition to developing basic technologies through collaboration, plans were developed to perform demonstration experiments in the Tokyo waterfront area and rural areas and to foster social acceptance of automated driving systems. The sectoral ministries working in areas related to social implementation – such as the Cabinet Secretariat's IT Strategy Office (now the Digital Agency), the National Police Agency (NPA), the Ministry of Internal Affairs and Communications (MIC), METI, and the Ministry of Land, Infrastructure, Transport and Tourism (MLIT) – also participated as part of the promotion committee to advance efforts (Sugiyama & Kimura, 2023).

9.3.3 Public-Private ITS Initiatives/Roadmap as a Tool for Achieving Government-wide Coordination

The SIP-adus promotes R&D related to automated driving systems and the social implementation of these technologies through industry-government-academia and interministerial collaboration. In addition, it is necessary to coordinate the efforts of various government ministries and the private sector. It is also important to monitor the progress of these numerous

---

[17] SIP-adus web page: https://en.sip-adus.go.jp/

efforts and provide feedback to R&D and social implementation efforts as appropriate. The Public-Private ITS Initiatives/Roadmap (hereafter referred to as the Roadmap) was the tool used to achieve government-wide coordination to implement Japan's automated driving system (Arimoto, 2023).

The Roadmap involved four ministries and agencies, including the sectoral ministries and agencies, and was headquartered at the Cabinet Secretariat. The Roadmap is based on the understanding and analysis of information on R&D, laws and regulations, related policies, measures such as networks with automobile companies, local governments, business companies, NPOs, etc., and initiatives including social experiments conducted at the level of various ministries and local governments. In order to implement automated driving systems in society, a roadmap was prepared based on this information, and then published.

The Roadmap lays out plans for more than a decade into the future and serves to guide and direct the efforts of related ministries and stakeholders by setting a timeframe for achieving the vision of an automated driving society. The efforts of each party are reflected in the Roadmap in the form of specific action plans. The Roadmap is revised annually based on progress, which creates an institutional learning and feedback process.

Under this structure, the SIP-adus carries out pre-competitive technology development and social experiments that would be difficult for individual automobile companies to conduct on their own. The SIP-adus utilises the laws, regulations, and networks of the METI, the MIC, MLIT, and NPA, all of which are participating in the public-private ITS Initiatives/Roadmap framework, to implement R&D results and conduct numerous social experiments based on actual conditions in a variety of regions. Participating companies have also mobilised resources by providing equipment and personnel in kind (Kuzumaki, 2023) (see Figure 9.2).

9.3.4 Implications from the Case of Automated Driving

The background to the implementation of such a MOIP approach is, first of all, a widespread sense of urgency around maintaining the competitive position of the automotive sector – an important pillar of Japanese industry – in the international market in the future. Mobility, including automated driving, is currently at a major turning point, with large, established tech companies and emerging companies joining the industry one after another. This sense of urgency has led to the proactive participation of the private sector, including the Japan Automobile Manufacturers Association.

Another major driving force here is the fact that a wide range of people now need local social systems to be transformed due to the ageing and declining population. These demographic changes are having a tangible impact on local communities throughout Japan. In particular, it is becoming increasingly difficult to maintain the social infrastructure and transport systems built during the period of population expansion. It has been recognised that local governments and regional communities need to transform their traffic and transport systems and infrastructure by utilising automated driving and other technologies.

This sense of emergency and urgency is widespread among government leaders, relevant ministries and agencies, industry, local governments, academia, and other concerned parties. This pressure has provided major impetus for comprehensive efforts that transcend the boundaries of ministries and industry and has enabled the proactive participation of a wide range of stakeholders and the promotion of voluntary efforts. In order to support various

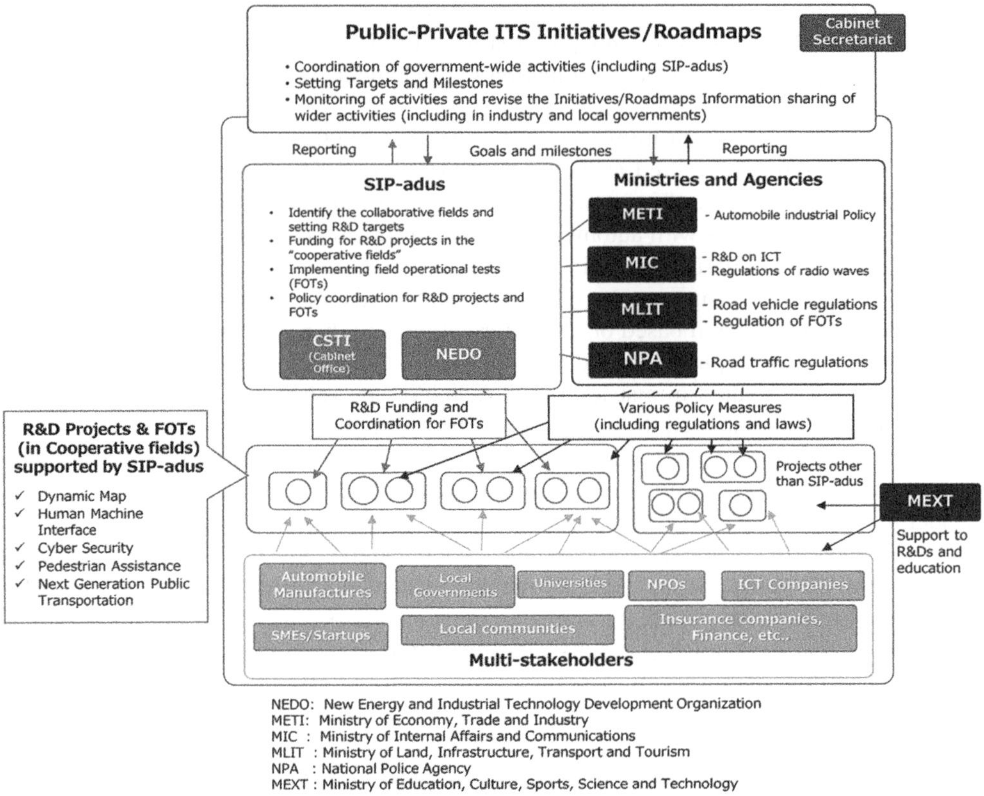

Source: Translated and modified from CRDS, 2022.

Figure 9.2 *Public-Private ITS Initiatives/Roadmap as a tool for implementing a whole-of-government approach*

mission-oriented initiatives in the future, it will be important to establish similar mechanisms in relation to other social issues.

From a technological perspective, there have also been limitations around automated driving. SIP-adus itself has changed its goals and targets between its first and second phases due to the development of the technology itself and changes in the competitive environment and society. In terms of mobility services, approaches and technologies other than automated driving must also be considered. In addition, the points of interaction with other areas, such as smart cities and community well-being, are also expanding. In order for automated driving systems to solve societal challenges and lead to social change, it is necessary to design the system from the perspective of a desirable society and with a future vision in mind. Establishing missions and themes with appropriate granularity and scope in light of these perspectives remains an issue for MOIP efforts.[18]

It has also been pointed out that the "all-Japan" approach to collaboration often emphasised in MOIP initiatives such as the SIP may not be compatible with the current industrial structure, where different interests intersect at various industry layers – especially in the case of systemised technologies and services such as automated driving (Yoshioka-Kobayashi, 2022). In this case, the question will be what legitimacy and logic will be used to organise cooperative and competitive relationships and policy interventions.[19]

9.4 FUTURE ISSUES AND CHALLENGES FOR TRANSFORMATIVE AND MISSION-ORIENTED INNOVATION POLICY IN JAPAN

Since the enactment of the Basic Act on Science and Technology in 1995, addressing societal challenges has been one of the key pillars of Japan's STI policy. In the almost three decades since then, various reform efforts and trials have been made at the strategic and operational levels. In view of the realisation of transformative innovation, which aims to transform the social system, there are challenges that need to be further addressed. This section discusses these challenges, dividing them into two categories: challenges related to strategic direction and interministerial coordination at the strategic level, and challenges at the operational (implementation) level.

9.4.1 Challenges in Interministerial Coordination at Strategic Level

9.4.1.1 Policy coordination based on the autonomy of each ministry

In MOIP, it is important to orient and coordinate the many policy measures across multiple ministries towards achieving common goals. In Japan, the interministerial coordination of measures has been a policy issue for many years. Currently, the CSTI (and the Integrated Innovation Strategy Promotion Council, which has the CSTI at its core), the Integrated Innovation Strategy, and sectoral strategies are being used to unite and direct the policies of each ministry towards the achievement of policy priorities. These strategies serve as an incentive to the ministries to increase the feasibility of obtaining budgets. In addition, the SIP, an interministerial collaborative R&D programme, is intended to promote the social implementation of R&D results by encouraging the participation of relevant personnel from each ministry and agency in the programme, thereby strengthening links with the measures

---

[18] In the theme-setting process for the third phase of the SIP, which began in 2023, the CSTI identified 10 areas corresponding to the 10 social visions in "Society 5.0", the vision for Japan's future society presented in the sixth STI Basic Plan, and five areas of fundamental technologies as candidate themes, including AI, data, and quantum technology. Based on this, 17 themes were selected through a call for opinions and consultations with various ministries. Among them is "Building a Smart Mobility Platform", as it relates to automated driving.

[19] For example, in automated driving, dynamic mapping, which adds various information and data to a map, is a technology in the cooperative domain at the layer of automobile companies, but in the layer of the mapping industry, it is a technology in the competitive domain.

implemented by sectoral ministries and agencies. However, since the actual implementation of the measures is the responsibility of each ministry or its affiliated organisation, the progress of specific collaborations within these frameworks is left to the autonomy of the participating ministries and organisations.

9.4.1.2 Silo problems within each ministry

There are still many issues regarding the use of demand-side policy measures such as laws and regulations and procurement. Efforts are being made to utilise demand-side policy instruments (Suzuki et al., 2021). However, these policy measures are under the jurisdiction of the departments in charge of projects related to societal challenges, not the departments in charge of R&D, even within sectoral ministries and agencies. As Japan's STI policy to date has centred on a technology-push approach when it comes to R&D programmes and disseminating their results, the Cabinet Office's counterpart for interministerial coordination in each ministry has been R&D or a similar department. Given the challenges of the decentralised Japanese government's planning and programming structure pointed out by Shiraishi (2013), overcoming such intra-ministerial silos is also a challenge.

9.4.1.3 Resource and capacity issues of the control tower organisation

In addition, coordinating the various policy instruments of each ministry, both horizontally and vertically, incurs significant administrative costs in terms of information, personnel, etc. In the SIP, the staff in charge of the Cabinet Office and the R&D promotion organisation (e.g., funding agencies and research institutes) are responsible for this task. The staff of the Cabinet Office are seconded from various ministries and private companies and are usually replaced after a few years. Although the staff of the R&D promotion organisation have knowledge of the technology in question and expertise in managing and operating R&D programmes and projects, many of them do not have connections with the sectoral ministries and other stakeholders or the authority to coordinate these actors, which is necessary to achieve the mission. These aspects of human resources and organisational authority therefore need to be strengthened, improved, or supplemented.

9.4.2 Challenges at the Operational Level

9.4.2.1 Design and management of project portfolios

The first challenge at the operational level is how to design and manage a portfolio of policy measures and projects to achieve the mission. Japan's STI policy has so far taken a technology-push approach, with R&D programmes as the mainstay. While this approach may work well in certain technology areas, bringing about transformative innovation requires designing a portfolio that includes demand-side policy measures and the voluntary efforts of a large number of stakeholders. The following are required in this regard: coordination between numerous ministries, agencies, and other actors; the collection and analysis of information on the efforts of each party; forecasting and studies to map out a path towards accomplishing the mission; analyses of bottlenecks that must be resolved in terms of science and technology; research on the feasibility of the measures being undertaken; and continuous monitoring of the efforts of each party using data and information. It will also be necessary to spend sufficient time in advance on institutional design and preparation.[20]

9.4.2.2 Involvement of numerous stakeholders at multiple levels

In order to achieve the mission, it is necessary for the many parties involved to participate with a sense of ownership and to make various efforts, including voluntary efforts. In the industrial policy of the 1970s and 1980s, research consortia played a role in encouraging such participation, but the participants were limited to major companies, universities, and public research institutions. Currently, a more diverse range of stakeholders is required to participate. One approach taken in Japan in recent years has been the widespread establishment of public-private partnership platforms and councils, and consortia of actors from industry, government, academia, and the private sector.[21]

9.4.2.3 Direction and mobilisation of numerous activities and resources

Another significant challenge in achieving the mission is ascertaining how best to orient and mobilise the activities of a large number of stakeholders and resources such as funds and personnel. As demonstrated in the case of automated driving, a sense of emergency and urgency among the stakeholders may motivate their participation. In addition, setting a more specific "local mission" with the community as the axis and in a way that is in line with local conditions may enable participants to have a sense of ownership.

9.4.2.4 Monitoring and evaluation

In order to bring about long-term transformation, initiatives in mission-oriented innovation policies, which require long-term and sustained efforts, need to involve multi-phase goals and clear time frames (CRDS, 2022). In order to promote numerous initiatives, including ambitious and experimental ones, continuous learning will be required to constantly monitor progress, including processes of trial and error, and to make improvements and revisions as appropriate based on such information and the results of analyses. In Japan in particular, it is necessary to reform the evaluation system – which has so far focussed on post-event summative evaluation – towards formative evaluation. This will better reflect the results in management during project implementation and reflect the evaluation results in the institutional design of policy measures and projects at the next stage, with an awareness of the linkage with the policy formation process.[22]

---

[20] The policy formation process in Japan is usually based on a one-year budgeting cycle. In recent years, large-scale research and development programmes have been established with supplementary budget funds, but these are usually designed within a very short period of time (a few months in the longest case).

[21] Typical examples are the GX League and the Smart City Public-Private Partnership Platform, which have already been mentioned in the text. Other examples include the projects of the SIP and Moonshot R&D programmes, where there is a movement to create industry-government-academia consortia with the aim of promoting social implementation of R&D results. However, in the case of consortia related to such large-scale, time-limited R&D programmes, some face issues of continuity after the projects are completed.

[22] As a recent example, an institutional verification of ImPACT was conducted one year before the end of the programme in order to use it as a reference for the institutional design of the Moonshot R&D programme, which was the successor to ImPACT.

9.4.2.5 Capacity building

A variety of organisational and professional capabilities are required to address the issues described so far. For a long time in Japan, government functions such as personnel have been downsized, and this includes the incorporation of national research institutes. In addition, national research institutions and national universities have also been incorporated and subsequently subjected to recurring budget reductions, which makes it difficult to develop human resources and build research capacity from a long-term perspective. In funding agencies, the creation of new R&D programmes is basically a scrap-and-build process involving the reorganisation of existing programmes, making it difficult to accumulate expertise within the organisation. In light of this current situation, it is necessary to rethink how organisational capacity is built up and maintained to achieve its mission.

9.4.3 Future Prospects

In recent years, STI policy initiatives that bring the realisation of transformative innovation to the forefront have expanded. In 2021, METI launched "Mission-Oriented Economic and Industrial Policy" as a new axis of economic and industrial policy. Based on subsequent studies, in 2023, eight missions were proposed as mission-oriented industrial policies, including GX, DX, and economic security (METI, 2023). METI also advocates agile governance for the realisation of Society 5.0 as a long-term strategic goal for Japan as a whole (METI, 2022).

In the third phase of the SIP (SIP3), which started in 2023, the R&D management tasks that were previously handled by the PD have been transferred to project managers (PMs) and research promotion organisations under the umbrella of the PD. This has created a system in which the PD can concentrate on managing overall efforts towards social implementation, including interministerial collaboration. In addition, SIP3 will formulate a social implementation plan based on five perspectives (technology, governance, business, social acceptability, and human resources) and monitor progress. The implementation plan drawn up from these five perspectives is intended to function as a common language among the various stakeholders participating in the project (Cabinet Office, 2023).

As described above, Japan's STI policy has positioned the realisation of transformative innovation aimed at future societal transformation as a priority policy goal, and the STI policy framework has been expanded. Institutional reforms have also been implemented to strengthen the overall coordination capacity of the Cabinet Office and CSTI to achieve this goal. However, there are still many challenges to be addressed in promoting the initiatives through a whole-of-government approach. In addition to the challenges at the strategic level, many challenges also remain at the operational level. These include ways of thinking and accepted practices that are oriented towards a technology-push approach (institutional inertia), dispersed and reduced organisational capacity due to the outsourcing of government functions, and a lack of resources related to operations. However, it is increasingly being recognised that bringing about social transformation and transformative innovation will be essential to realising the government's overall long-term strategic goals; this type of transformation is therefore expected to be extremely important in the next (seventh) STI Basic Plan. In order to promote MOIP to realise long-term social transformation, it will be necessary to address the removal of these constraints.

REFERENCES

Arimoto, T. (2023). SIP-adus and mission oriented STI policy. In SIP-adus Final Results Report Editional Committee & Society of Automotive Engineers of Japan, Inc (Eds.), *SIP 2nd phase: Automated driving for universal services – Final results report (2018–2022)* (pp. 268–270). New Energy and Industrial Technology Development Organization.

Cabinet Office. (2014). *Enforcement of the Act for Partial Revision of the Cabinet Office Establishment Act*. https://www8.cao.go.jp/cstp/stsonota/settihou/settihou.html

Cabinet Office. (2021a). *Summary of the act for partial revision of the basic act on science and technology, etc*. https://www8.cao.go.jp/cstp/cst/kihonhou/kaisei_gaiyo.pdf

Cabinet Office. (2021b). *SIP (Cross-ministerial strategic innovation promotion program) 2021 brochure*. https://www8.cao.go.jp/cstp/panhu/sip2021/sip2021.html

Cabinet Office. (2023). *Outline of the Cross-ministerial strategic innovation promotion program (SIP)*. https://www8.cao.go.jp/cstp/gaiyo/sip/sipgaiyou.pdf

Center for Research and Development Strategy (CRDS). (2022). *Promotion of mission-oriented science, technology and innovation policy, and research and development funding in Japan*. CRDS. https://www.jst.go.jp/crds/report/CRDS-FY2022-SP-01.html

European Commission, Directorate-General for Research and Innovation & Mazzucato M. (2018). *Mission-oriented research & innovation in the European Union: A problem-solving approach to fuel innovation-led growth*. Publications Office. https://data.europa.eu/doi/10.2777/360325

Expert Group for the Promotion of Science, Technology and Innovation Policy. (2011). *Final report*. https://www8.cao.go.jp/cstp/stsonota/kenkyukai/index.html

Goto, A. (2000). *Innovation and Japan's economy*. Iwanami Shoten.

Government of Japan. (2011). *The fourth science and technology basic plan*. https://www8.cao.go.jp/cstp/kihonkeikaku/4honbun.pdf

Government of Japan. (2021a). *The sixth science, technology and innovation basic plan* [tentative translation]. https://www8.cao.go.jp/cstp/english/sti_basic_plan.pdf

Government of Japan. (2021b). *Outline of the science, technology and innovation basic plan*. https://www8.cao.go.jp/cstp/english/outline_plan.pdf

Ito, M. (2006). The council for science and technology policy as a "specific general coordinating mechanism". *Journal of Public Policy Studies*, *6*, 43–55.

Johnson, C. (1982). *MITI and the Japanese miracle: The growth of industrial policy. 1925–1975*. Stanford University Press.

Kobayashi, S. (2017). The transformation of the council for science, technology, and innovation and instrumentalization of policy. *Kagaku*, *88*(1), 100–107.

Kuzumaki, S. (2023). Looking back upon SIP-adus history. In SIP-adus Final Results Report Editional Committee & Society of Automotive Engineers of Japan, Inc (Eds.), *SIP 2nd phase: Automated driving for universal services – Final results report (2018–2022)* (pp. 264–267). New Energy and Industrial Technology Development Organization.

Larrue, P. (2021a). *Mission-oriented innovation policy in Japan: Challenges, opportunities and future options*. OECD Science, Technology and Industry Policy Papers No. 106. OECD Publishing. https://doi.org/10.1787/a93ac4d4-en

Larrue, P. (2021b). *The design and implementation of mission-oriented innovation policies: A new systemic policy approach to address societal challenges*. OECD Science, Technology and Industry Policy Papers No. 100. OECD Publishing. https://doi.org/10.1787/3f6c76a4-en

Ministry of Economy, Trade and Industry (METI). (2022). *Agile governance update – How governments, businesses and civil society can create a better world by reimagining governance*. https://www.meti.go.jp/press/2022/08/20220808001/20220808001-b.pdf

Ministry of Economy, Trade and Industry (METI). (2023). *Second report of the Committee on new direction of economic and industrial policies*. https://www.meti.go.jp/shingikai/sankoshin/shin_kijiku/pdf/20230627_1e.pdf

Nakamura, M., Arimoto, T., Yamada, H., & Maruyama, R. (2021). Transforming Science, Technology, and Innovation (STI) for a sustainable and resilient society. *Science & Diplomacy*, (Fall). https://

www.sciencediplomacy.org/article/2021/transforming-science-technology-and-innovation-sti-for-sustainable-and-resilient

Nakayama, S. (1995). *Postwar history of science and technology*. Iwanami Shoten.

Okuma, K. (2009). History of the development of the science and technology agency policy. In Watanabe Memorial Foundation for the Advancement of New Technology (Eds.), *History of science and technology agency* (pp. 3–86). The Science News Ltd.

Research Institute of Science and Technology for Society (RISTEX). (2023). *RISTEX past, present, and future: 20 years of RISTEX/S&T for society*. RISTEX. https://www.jst.go.jp/ristex/en/aboutus/files/ristex_20th_booklet.pdf

Sawai, M. (2011). *History of Japan's trade and industry policy (9): Industrial technology policy*. Research Institute of Economy, Trade and Industry.

Schot, J., & Steinmueller, W. E. (2018). Three frames for innovation policy: R&D, systems of innovation and transformative change. *Research Policy*, *47*(9), 1554–1567.

Shiraishi, T. (2013). *The Abe administration's proposal to fundamentally strengthen the council for science and technology policy*. Nippon.com. https://www.nippon.com/ja/column/f00018/

Shiroyama, H. (2018). *Science and technology, and politics*. Minerva Shobo.

Sugiyama, K., & Kimura, H. (2023). The second phase of SIP-automated driving for universal services (overview). In SIP-adus Final Results Report Editional Committee & Society of Automotive Engineers of Japan, Inc (Eds.), *SIP 2nd phase: Automated driving for universal services – Final results report (2018–2022)* (pp. 11–15). New Energy and Industrial Technology Development Organization.

Suzuki, J., Yasuda, S., & Goto, A. (2021). *Transformation of the National Innovation System of Japan*. Yuhikaku Publishing Co., Ltd.

Yoshioka-Kobayashi, T. (2022). *Drawbacks to an "All Japan" approach to innovation policy*. The Tokyo Foundation for Policy Research. https://www.tokyofoundation.org/research/detail.php?id=885

10. The spatial and scalar implications of missions: challenges and opportunities for policy

Elvira Uyarra, Iris Wanzenböck and Kieron Flanagan

10.1 INTRODUCTION

In recent years, debates about innovation policy have highlighted a broader scope for action and a widening of the range of policy goals such policies are expected to (or might be expected to) address (Borrás, 2009; Flanagan et al., 2011). Scholars and analysts have both detected and advocated a shift from generic and primarily R&D-based innovation support measures towards a new (or third) "generation" of innovation policy – variously referred to as challenge-led, mission-oriented (Foray, 2018; Mazzucato, 2018b), or transformative innovation policies (Schot & Steinmueller, 2018). This new generation of innovation policy thinking is a response to major societal challenges such as climate change, migration, or food and energy security – the implication being that traditional innovation policies were either inadequate in responding to or else uninterested in such challenges. A more targeted and challenge-oriented innovation policy should, it is argued, help to deliver desired, and not just more, innovations (Boon & Edler, 2018; Kuhlmann & Rip, 2018). This implies a more active role of the state in funding risk-taking activities and in creating, not just correcting, markets. This "normative turn" in innovation policy has also been observed in the design and implementation of regional policies, with a greater emphasis on the socio-ecological dimension of innovation (Coenen & Morgan, 2020; Uyarra et al., 2019; Bugge et al., 2021), particularly in the context of the European Green Deal and the Innovation Strategies for Sustainability (S4) (Hassink et al., 2021; Morgan & Marques, 2019). Whilst there is much agreement that bolder, more customised, and directional policies are needed to tackle the societal challenges of our time, there is less consensus about how such policies should be implemented in practice.

As mission-oriented policy approaches gain prominence, questions emerge about how missions are best defined, implemented, and evaluated/studied. Mission-oriented innovation policies do not operate in a vacuum but within existing institutional and multi-level governance structures (Fisher et al., 2018). They will therefore vary according to context, including geographical context, yet the spatial and scalar dimensions of transformative missions have been neglected. The local level is key, given that it is where the new responses to the great challenges of society are experienced. Local administrations have competencies in relevant areas and can generate and test new solutions that can later be reproduced on a more global scale. Debates about transformative missions are also surprisingly divorced from parallel efforts to support regional innovation and the increasing importance of "missions" and challenges in such endeavours (McCann & Soete, 2020; Miedzinski et al., 2021; Pontikakis et al., 2022). Janssen et al. (2021) argue that missions risk "becoming a default answer that is applied uncritically to address societal issues, ignoring the accumulated experience from pre-existing and analogous goal-oriented policies and the complex social and political issues they bring to the surface" (p. 439).

This chapter seeks to address this shortcoming by discussing the spatial and scalar dimensions of mission-oriented policy frameworks. We draw from scholarly contributions in innovation policy studies, regional studies, and economic geography, to suggest how scale and place-based qualities may shape and be shaped by problem-solution representations in relation to wicked societal challenges. In trying to capture differences in how problems and solutions may be framed geographically, we discuss four stylised opportunity spaces, including governance strategies and instruments aligned with the prevalent geographical problem-solution constellations.

In the following section, we will first introduce the concept of transformative missions and review previous attempts to advance the definition and articulation of mission-oriented policies. We then discuss the importance of adopting greater spatial sensibility for the implementation of missions. In Section 10.4, we articulate these by proposing a tentative map of policy opportunities for transformative missions. Section 10.5 gives an overview of our research and conclusions. In this chapter, we use the term "regional level" to mean the subnational level of policymaking (including administrative regions and metropolitan areas) in contrast to the national or supranational level. While we acknowledge the diversity of regional settings in terms of industrial profile, institutional capacity and political competences, we do not discuss these in detail in the text.

10.2 UNDERSTANDING TRANSFORMATIVE MISSIONS

"Third generation" innovation policy approaches emphasise societal problems as key drivers for intervention and advocate a departure from generic innovation support measures – often based on R&D support – towards policies that seek more transformative system changes through not just technical solutions but also social or behavioural changes (Wesseling & Meijerhof, 2021). System-wide transformation requires attention to directionality, experimentation, demand articulation, policy learning and coordination (Weber & Rohracher, 2012). In a mission-oriented innovation policy, these ambitions materialise through ambitious, actionable, measurable, and time-bound goals (Larrue, 2021; Mazzucato, 2018a).

Wanzenböck et al. (2020, p. 3) define mission-oriented innovation policy (MOIP) as a directional policy that starts from the perspective of a societal problem and focusses on the formulation and implementation of a goal-oriented strategy by acknowledging the degree of wickedness of the underlying challenge, and the active role of policy in ensuring coordinated action and the legitimacy of both problems and innovative solutions across multiple actors.

In attempts to classify and compare different mission policy approaches, the innovation policy literature has drawn on the distinction between societal challenge-oriented "transformer" missions that target transformative change, and "accelerator"-type "moon-shot" missions (Fisher et al., 2018; Larrue, 2021; Wittmann et al., 2021). According to Fisher et al. (2018), accelerator missions are aimed at narrow, well-defined, and generally scientific or technological goals, while transformer missions target wicked societal problems requiring system transformation. However, they note that both are ideal types and that most mission-oriented policies may consist of a mix between the two. In a similar vein, Edler et al. (2021) define "transformative" missions as politically defined search areas for solutions to societal problems and differentiate them from other more technocratic types of missions that are more suitable in instances where "clear goals, targets and pathways can be defined but [are] possibly less appropriate in areas of wicked transformative change" (2021, p. 2).

Implicit to the idea of transformative missions is an acknowledgement that societal challenges are often "wicked problems" and as such are contested, uncertain, and highly complex (and also temporary in principle, in the sense that a challenge can be addressed). The focus on challenges of the new mission orientation is often contrasted with the top-down technological missions of the past (Nelson, 1977), which focussed on achieving "stretch" technological goals rather than on social innovation or on issues related to social acceptance, market conditions, and alternative technological and non-technological solutions (Edler et al., 2021; Wanzenböck & Frenken, 2020). However, whilst the stretch technological goals of "moonshots" (if not the means of achieving them) are clear and understood by all participants, neither the understanding of a wicked societal problem and/or the possible means of addressing it, nor the legitimacy or likely impact of those means, can be taken for granted (Wanzenböck et al., 2020). The risk that missions tackling such wicked problems are mistakenly framed as accelerator missions is ever present.

Wanzenböck et al. (2020) argue that the design of mission-oriented policies should start with a recognition of the heterogeneity of "underlying problem structures". Specifically, they suggest that mission design must consider the different "wickedness" dimensions (contestation, complexity, and uncertainty) of both problems and solutions. Contestation or legitimacy may exist around the feasibility of a solution or may arise because of conflicting claims, values, and framings on the part of diverse actors. Complexity may stem from technological interrelatedness and the need to restructure the broader socio-technical system, but also from the multi-scalar and multi-dimensional nature of societal problems to be addressed by policy: which actors, at which levels of governance, are responsible for the problem, and who decides about possible solutions? Uncertainty may result from limited knowledge about the availability or feasibility of potential solutions, their potential side-effects or unintended consequences, or from limited or fragmented knowledge about the problem. Wanzenböck et al. (2020) identify a need for more in-depth understanding of geographical and institutional conditions, including multi-level governance arrangements that may influence the alignment of problems and solutions, better insights into how legitimacy is built in policy processes, and more knowledge about the organisational capacity of public actors.

Mission-oriented policy crosses traditional public policy boundaries, addressing problems situated in domains such as health, mobility, energy, or food and of interest at multiple levels of governance. The articulation of missions will be a highly politicised matter because it involves prioritising some societal challenges over others and requires both sense-making and legitimacy-building processes (see e.g., Edler et al., 2021). Many problems compete for legitimacy and prioritisation, and there is a risk of capture by dominant groups (Wesseling & Meijerhof, 2021). Janssen et al. (2021) argue that missions tend to be situated at the interface between, on the one hand, the socioeconomic system associated with the relevant social domain relating to the challenge in question (in terms of the infrastructure, technologies, behaviour, and values for production and consumption) and, on the other hand, the science, technology, and innovation system that could be mobilised to address the challenge. Missions are thus embedded in, but also in tension with, those two systems. Moreover, science, technology, and innovation systems may often be poorly aligned with newly prioritised challenges (Cappellano et al., 2022; Janssen et al., 2021).

The complex and politicised nature of making and implementing mission-oriented innovation policies implies significant capacity on the part of state actors, who may struggle to maintain directionality and ensure the funding, articulation, and evolution of the required instrument mix (Larrue, 2021). While for clear-cut "moon-shot" accelerator missions it is possible to envisage a top-down approach (for instance, through the creation of a single agency to coordinate how the mission is addressed), transformative missions are more likely to require the mobilisation, alignment, and coordination of a variety of distributed actors (Craens et al., 2022; Edler et al., 2021) in more fluid networked arrangements that are likely to cut across different policy areas and levels of decision-making.

The design of transformative missions is therefore very unlikely to be done on a tabula rasa, but rather will be built on existing problem framings (Janssen et al., 2021) and institutional arrangements. Similarly, the development of new practices and technologies, as well as the phasing out of old ones, is likely to be shaped by existing innovation system structures, including knowledge bases and industrial specialisations (Bugge et al., 2021). New missions are defined and then implemented into a landscape of existing policy rationales and mixes, much of which will have become institutionalised, with their own epistemic, stakeholder, and "client" communities and their own momentum (Laranja et al., 2008). Legacies of earlier policy approaches and system structures may enable or hamper new mission approaches. Despite the stress on novelty, the existence of policy path dependencies and the need for missions to command widespread legitimacy and political alignment mean that new policy approaches are more likely to sit alongside existing policies, whether complementary or (potentially) in tension, than to displace them completely (Hassink et al., 2021; Scordato et al., 2022). In particular, transformative missions have implications not just for the science, technology, and innovation system, but also for the infrastructure, knowledge, technologies, behaviours, and values related to production and consumption that are relevant to the challenge in question (Janssen et al., 2021). The pursuit of a mission has the potential to mobilise or modify existing regional resources, actors, or institutions and, depending on the regional context, to shape (i.e., generate or constrain) new economic opportunities (Bugge et al., 2021).

Therefore, transformative mission approaches will vary not just according to the specific challenge or problem but also according to the existing configuration of innovation systems and institutional and political structures. New transformative missions may turn out to be in tension with existing policies, goals, and framings, including those at other levels of governance. Yet these spatial and scalar dimensions of transformative missions are often overlooked. It is not clear from the literature how subnational policy actors might best contribute to broader missions, or whether certain kinds of missions might be best pursued at the local or regional level. Equally unclear is how different types of regions might use mission-oriented approaches to advance their own economic development goals. Finally, the possibility of unintended distributional effects from mission policies across geographies constitutes a further blind spot in current debates.

10.3 CONSIDERING SCALE AND PLACE IN MISSION-ORIENTED INNOVATION POLICY

The question of how to "organise, coordinate and implement challenge-oriented innovation policies across different levels" has been neglected in current policy debates (Wanzenböck &

Frenken, 2020, p. 51). It is often implicitly assumed that societal challenges are best dealt with at the national or supranational level, where it is also assumed that the right levers, structures, resources, and technocratic competencies will be found, where critical mass is felt to be most achievable and economies of scale most readily exploited. This "big science for big problems" approach not only devalues the importance of cities and regions in the implementation phase of innovation policy, often in a multi-level governance setting, but also de facto excludes the majority of people and places (Heslop et al., 2019), potentially creating a profound challenge to societal and democratic legitimacy (Coenen & Morgan, 2020; Wanzenböck & Frenken, 2020). What is not clear is who chooses the mission and how, and what an inclusive and democratic process of mission selection might look like (Storper et al., 2022). Another risk of a space-blind approach is greater regional inequality: the top-down, technology-centred mission policies of the past had clear differential impacts on regions and cities, concentrating government and private R&D in a small number of locations (e.g., Sternberg, 1996), with potentially limited links and spillovers to other local economies.

There is a geography inherent to both societal challenges and the knowledge that might be required to tackle and resolve them. This is often left unacknowledged by mission-oriented approaches, which are then unable to "explain why solutions can be more successful in one place compared to another, and why some solutions spread beyond their place of origin and scale up, while others remain trapped by local context" (Coenen et al., 2015, p. 491). On the problem side, societal challenges are not only fuzzy and contested but also local and situated. Wanzenböck and Frenken (2020) invoke the principle of subsidiarity to suggest that societal issues should be addressed by those who are more likely to be affected by them. They argue that "challenges do not present themselves the same for every region or nation, as underlying problems affect places in different ways and to different extents". Despite the widespread use of terms such as "grand" and "global", Wanzenböck and Frenken argue that "challenges remain contextual" (p. 56). Place-dependent conditions and the contested nature of societal challenges mean that problems are likely to be felt, understood, and acted upon differently in different places. Moreover, problem representations might be shaped by what Rutten (2017) refers to as "distance dynamics", whereby definitions of needs and problems may be reflective of the specific economic, social, and environmental conditions of one place or may, in contrast, address needs and challenges common to multiple settings.

On the solution side, the know-how necessary to address societal challenges is partly tacit and situated, and interventions "require discovering the tacit elements of technology and adapting them to the local environment" (Hausmann & Rodrik, 2003). This helps explain why solutions and transformation pathways developed in one place, co-produced with local actors and communities, are not always easily transferred to another place (Boschma et al., 2017). Knowledge is spatially embedded in people and organisations and is distributed among many actors – including users on the demand side. Because contemporary grand challenges have an uncertain and contested nature, and because they demand the involvement of multiple, diverse stakeholders (Coenen et al., 2015), solutions will be shaped by geographical proximity. Solutions will also be shaped by pre-existing innovation system structures, such as the knowledge infrastructure of universities and research laboratories or incumbent firms and industry specialisation patterns (Bugge et al., 2021). For example, Tanner (2014) found that the fuel cell industry emerged in regions with competencies in related technological fields. Therefore, just as for problem representations, solution representations will be shaped by local place qualities (or "place dynamics" (Rutten, 2017)). This includes the knowledge bases and expertise of firms and the knowledge of the community at large, as well as other characteristics of place such as the culture and demand sophistication of users. Some authors have begun to explore the place-based conditions that might enable or hinder transformative system change, considering features such as regional innovation systems, the variety of regional assets (natural, industrial, and institutional) and firm- and system-level agency (Isaksen et al., 2018; Jakobsen et al., 2022; Trippl et al., 2020) able to mobilise (and transform) place-based, but also extra-regional, assets.

The geography of problems, then, can be different from the geography of solutions. Anchoring missions in local contexts may bring benefits not just in terms of local knowledge and new economic opportunities, but also through greater legitimacy for challenge-based policy action. Flanagan and Uyarra (2022, p. 544) note that "all regions have their own problems and challenges, some of which may provide a basis for legitimation, positioning, knowledge and market building" for new technologies. Orienting innovation policy missions according to place-based needs and problems, rather than to technological strengths, may enhance the opportunities of less-favoured regions, which are often poorly equipped with knowledge assets (Flanagan et al., 2023). However, analysing regional innovation systems from the perspective of problems or challenges may require a broader understanding of innovation, and of relevant innovation actors, and is likely to require greater attention to the application and consumption side. The latter is important because "economic value is increasingly created in places where new uses are invented and less in places where advanced technologies are produced" (Jeannerat & Crevoisier, 2022, p. 5).

Place is not only relevant to the orientation of missions but is also key to the effective implementation of mission-oriented innovation policies (Brown, 2020). Most discussions on innovation policy implicitly assume the existence of an idealised "policy-maker" (Morgan & Marques, 2019). But complex innovation policy mixes are characterised by multi-level, multi-actor governance (Flanagan et al., 2011). Many societal challenges are addressed or shaped by supra-national policies, such as targets, standards, and regulations that concern environmental issues, and much of the decision-making that influences economic development, education, industry, and finance remains at the national level (Capasso et al., 2019). Nevertheless, regions and cities are key sites for policy action around societal problems. They are responsible for a large share of public spending and have a wide range of policy competencies in areas such as transport, the built environment, housing, and energy that are likely to be key to system transformation. Increasingly identified as "agents on the ground", cities and regions are essential for experimenting with and implementing new policies and for providing context for the emergence of system innovations and new institutional arrangements (Bulkeley et al., 2010; Raven et al., 2019). This requires institutional capacity, which may be weak in some places (Morgan & Marques, 2019). It also creates a profound challenge of multi-actor, multi-domain, and multi-level coordination. While the mission policy literature increasingly acknowledges that a "bottom-up" perspective is required (Mazzucato, 2018a), it is hard to determine from the current literature how bottom-up and top-down approaches might be effectively combined in various mission settings – beyond some hand-waving references to the need for better coordination.

In the end, even for top-down approaches, policies must be implemented in real places, "on the ground". Regions can be active sites of policy learning, experimentation, and testing of solutions – "living labs" for and demonstrators of new solutions to societal problems (Bulkeley et al., 2016; Voytenko et al., 2016). Regions have the potential to become "lead markets" for other places facing similar challenges (Losacker & Liefner, 2020). Small-scale policy programmes or experiments can bring temporary local innovation systems into being, involving various actors, to advance societal objectives and pursue regional industrial transformation (Frenken, 2017; Suitner et al., 2023).

A key question is to what extent place-based solutions can diffuse and contribute to system transformation more widely (Coenen et al., 2015; Uyarra et al., 2020). As shown by Bours et al. (2022) for the case of plastic pollution, more robust, systemic, or multi-local dynamics do not necessarily emerge quickly out of local "small wins". They require institutional support and governance arrangements that facilitate coalition-building and networks at a wider geographical scale. The literature on experimentation in cities and regions (e.g. Bulkeley et al., 2010; Fünfschilling et al., 2019; Sengers et al., 2021; Lam et al., 2020) has addressed the problem of local government actions within a system of translocal and multilevel governance ; . In this view, national or supra-national entities can provide incentives through broad goals, regulation, coordination, and financial resources to address a shared problem such as sustainability. At the same time, cities or regions can decide upon their actions and experiment with new solutions tailored to their local conditions, politics, and values (Loorbach et al., 2020).

Experiences and practices with new solutions can travel through networks across scales (from lower to higher levels and vice versa) or horizontally between cities or regions through translocal networks. The connections between experts, solution providers, other stakeholders, or policies offer structures that help "diffuse contextualised solutions across territories and sectors" (Frenken, 2017, p. 45), leading to impacts beyond the individual experiments. The literature distinguishes several types of diffusion or scaling-up processes: embedding into local structures (e.g., institutional embedding); translation or replication in other spheres, institutional contexts, or localities; and scaling in terms of incorporating more domains and practices, actors and/or resources (e.g., von Wirth et al., 2019). Lam et al. (2020) similarly discuss several amplification processes seeking to: (a) accelerate the impact of individual initiatives (within amplification), (b) extend the impact to more people or places, and (c) change institutional structures, values, and mindsets beyond the initiative. Therefore, it is not so much a question of diffusing concrete solutions, technologies, or outcomes that emerged from a local innovation system, but more one of diffusing the lessons and experiences gained from particular problem-solving activities. In fact, most solutions are likely to stay local given the embedding and context-dependent assembling of actors, networks, infrastructures, technologies, and institutions. The difficulty for societal challenge-oriented innovation policy lies in finding ways to extract and codify place-based practices and achievements, e.g., in the form of new templates, artefacts, or indicators, and to let these travel through horizontal and vertical connections.

Moreover, a number of scholars have started to consider place and scale as a strategic tool that can open up opportunities for transformative innovation policies. In this sense, "scale" is not fixed but is rather a temporary structure that conditions actors and activities (Madsen, 2022). A constructivist understanding of policy scale sees narratives and framings about problems (and their solutions), as well as the resulting missions and policy targets, as the outcome of framing processes (Schön & Rein, 1994; Van Hulst & Yanow, 2016). Framings draw boundaries around a societal or political issue, define what counts and what does not, and become a guide to action for some while excluding others (Flanagan et al., 2023). This implies that policy designs building on specific problem and scale frames can be performative, influencing the innovation trajectories and the search for solutions within and across regions.

Consequently, not just solutions but also problems can be scaled up and down, and then deliberately used to disrupt system stability and help drive transformative change. Flanagan et al. (2023) note how problems can be scaled up to facilitate the embedding of local actors, networks, and solutions into global market structures, shaping the likely solution space and the direction of innovation (Ansell & Torfing, 2015). Nonetheless, current mission-oriented approaches and frameworks typically do not consider the role of local problem and solution conditions or the different place and distance dynamics intentionally or unintentionally engendered by these policies (Bugge et al., 2021; Uyarra et al., 2020).

10.4 A TENTATIVE MAP OF THE OPPORTUNITY SPACE FOR TRANSFORMATIVE MISSIONS

In previous sections, we have noted the spatial, scalar, and place dimensions that can shape societal problem representations, how potential solutions are articulated, and how transformative innovation policies are governed. In this section, we link problem-solution views on mission policies (Wanzenböck et al., 2020; Bugge et al., 2021) with work on territorial knowledge dynamics and the aforementioned notions of place and distance dynamics (Binz & Truffer, 2017; Rutten, 2017; Uyarra et al., 2020). Our aim is to derive a tentative map of the potential opportunity space for transformative mission orientation that better takes into account place, scale, and distance dynamics.

To recap, Wanzenböck et al. (2020) remind us that societal challenges are characterised by complexity, uncertainty, and contestation. "Wickedness" or divergence will be greater if, for instance, the division of responsibilities to tackle a problem across regions or countries in a multi-level governance system is not clear. It will also be greater in situations of institutional complexity, where state actors take multiple roles in transformations (Borrás & Edler, 2020), where many stakeholders are involved in policy formulation and implementation, or where actors play multiple roles as innovation producers, consumers, facilitators, and regulators, for instance, in the case of the built-environment or the healthcare sector (Fisher et al., 2018). Moreover, even where there is some agreement or shared vision around the nature of a problem or challenge, a solution may still not exist or remain uncertain and contested (the case of a problem looking for a solution). Proponents may argue for the feasibility of innovative solutions, but those solutions may lack broad-based societal legitimacy, only partially address the problem, or work better in certain (e.g., urban versus rural) environments and contexts than others. Missions may stagnate where the problem and solution representations do not match or where the challenge/mission is not sufficiently aligned with the available actors, competencies, networks, and institutions of the innovation system (Wanzenböck et al., 2020).

Shifting from a "wicked" towards a more aligned problem-solution constellation can follow a "problem-led", "solution-led", or "hybrid" pathway in which the issues of contestation, complexity, or uncertainty are addressed on the problem and solution side (Wanzenböck et al., 2020). As we will argue in this section, scaling down or embedding a problem and/or solution locally, or alternatively, bridging the distance to frame a challenge in a broader (or more global) sense, can be strategies to support problem-solution alignment and the activation of new innovation dynamics and, therefore, mission progress. At its simplest, a societal problem may be framed locally or globally, depending on whether it is perceived as unique – relevant to one locality – or shared by many places. Similarly, solutions may be sought locally or globally, depending on the nature of the innovation (technological versus non-technological), the knowledge and expertise available or needed locally, or on local aspirations to build such innovation competencies. The innovation dynamics and search for solutions may, in turn, be more "place-based" (i.e., local) or more "footloose" (i.e., global).

It is important to note that we use the local/global distinction rather than impose strict administrative boundaries or levels. Our aim is to capture stylised differences in how problems and solutions may be framed geographically. The tentative map of policy opportunities for transformative missions below (Figure 10.1) is offered as a reflective tool to foreground spatial assumptions and consequences when scrutinising transformative mission thinking on the one hand and to stimulate thinking about strategies for a more place- and context-sensitive mission governance on the other. We discuss four stylised opportunity spaces, including governance strategies and instruments aligned with the prevalent geographical problem-solution constellations. We are aware that problem and solution representations are rarely isolated or independent considerations. Instead, they may be constructed simultaneously and unconsciously in practice: how problems are framed in geographical terms may determine the solutions that

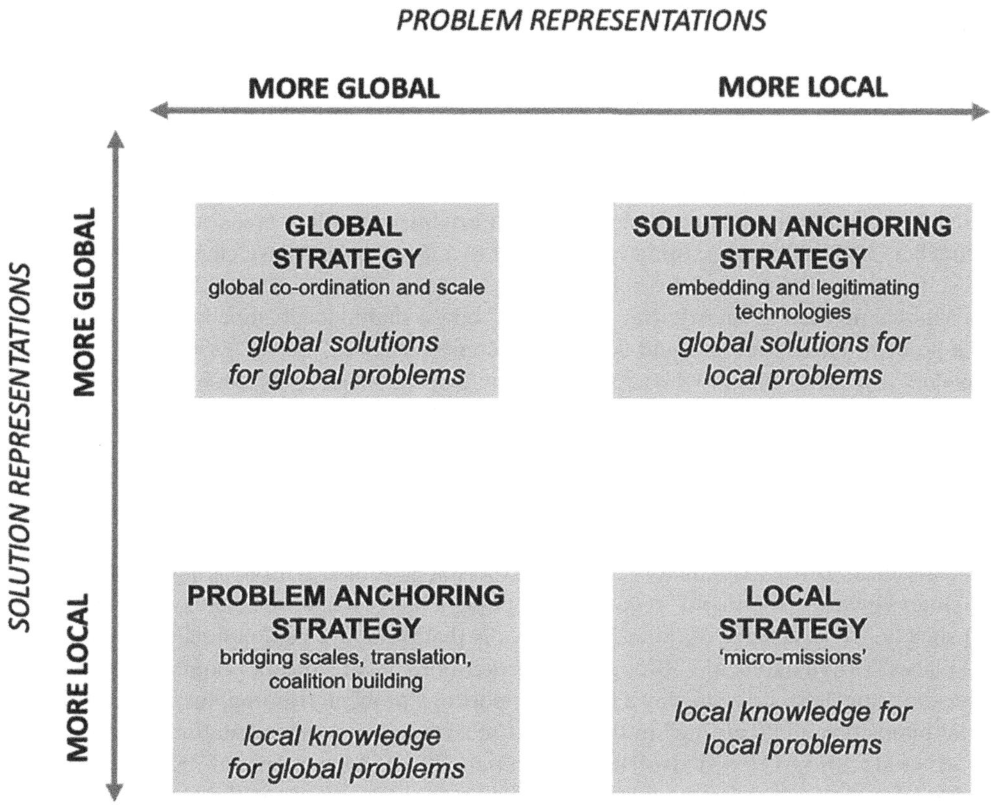

Source: Authors' own elaboration.

Figure 10.1 *Policy opportunities for transformative missions*

are sought and the place(s) where we search for them. Therefore, the geography of problems and the geography of solutions are highly interrelated. We are not proposing a fixed typology but rather a notional continuum of possibilities. It is also not intended to reflect or determine the transformative potential or nature of a specific governance structure, strategy, or instrument – specific governance strategies may not be intrinsically transformative, and the potential for transformative impacts will differ from situation to situation and context to context.

10.4.1 Local Knowledge for Local Problems

Local problem/solution representations that are place-specific and with limited distance dynamics may relate to wicked issues or missions that are highly complex, contested, or uncertain. In such situations, local networks along with institutional and cultural embedding can facilitate the search for solutions. A "local" strategy may be adequate to manage the complexity and contestation that can arise in such situations, benefiting from both place-based knowledge of the problem and potential solutions. In this context, Henderson et al. (2023) discuss the advantage of smaller scale "micro missions", namely those addressing specific place-based issues at a sub-national scale, in terms of bringing together local actors, supporting legitimacy, and balancing diverse interests and needs. Such approaches allow for an "experimentation logic" in which there is simultaneous learning about the "supply side" (knowledge and innovation development) and "demand side" (legitimacy and implementation) of a solution to a localised problem.

Transformative and system-oriented initiatives involving social or institutional innovations are easier to establish and support on a smaller scale, or temporarily, without facing the boundary and coordination challenges between established policy areas or levels (Weber & Rohracher, 2012). Moreover, initiatives relevant to a mission may have emerged "bottom-up" and self-organise their search for solutions. For instance, Bours et al. (2022) show in the case of a national mission of "plastic-free waterways" how a shared local embedding can help initiate a problem-based dialogue and connect different initiatives "on the ground", i.e., solution providers and stakeholders who are affected by or care about a common goal. Such local, system-oriented approaches can activate capacities to self-organise and undertake action – alone or in networks or cooperatives.

In recent years, several concepts and instruments have been discussed in the innovation and transition literatures which can be related to a local experimentation strategy. These range from niche experiments (e.g., Kemp et al., 1998), to urban experiments and living labs (e.g., Voytenko et al., 2016), to regulatory or policy experiments aimed at influencing the place-based dynamics between "top-down" regulatory or policy changes and the behaviour of businesses and society. An advantage of these approaches is that they "are geographically embedded in real places" (Voytenko et al., 2016, p. 47). Furthermore, local "anchor" organisations such as universities and hospitals can play a role in supporting "problem framing, institutional setting, social interaction and learning" in the context of such urban experimentation and living labs (Evans et al., 2015, p. 2; see also Henderson et al., 2023). An example of regions adopting a "living lab" approach to test, validate, and refine complex solutions is the Network of Living Labs of Health of Galicia (LABSAÚDE), which aims to respond to the needs of the Galician health system through the creation and implementation of innovative projects in a real-world environment where multiple agents (e.g., users, practitioners) interact.

However, local experimentation needs to be managed appropriately to avoid the downsides of "localness", such as too narrow a framing of the problem or solution, a poor uptake of solutions, or lock-in. It may be important here to focus not only on local embedding but also on complementing this with a translocal or global bridging strategy to access extra-regional networks or enable broad-based solution uptake (Gluckler, 2007).

System-based experimental approaches are increasingly discussed in light of an experimentalist governance logic (Sabel et al., 2012), as they attempt to combine interim goal-setting and contextual learning with monitoring and the systematic comparison of alternatives. Morgan (2018) argues that monitoring organised at higher governance levels, such as the EU, can support local strategies by ensuring both horizontal cross-regional learning and vertical coordination to diffuse solutions across contexts.

10.4.2 Global Solutions for Global Problems

In a situation where a broad understanding of, and consensus about, both technological priorities/possibilities and societal problems exist and where place-based knowledge or needs are not critical to the societal challenge at hand, a global strategy that involves bridging distance and making use of global knowledge creation dynamics could be appropriate (Wanzenböck & Frenken, 2020). More akin to "top-down" accelerator missions (Wittman et al., 2021), such a governance strategy consists of "global" coordination and scale; that is, seeking critical mass to tackle problems with a relatively clear definition and objective ("mission") designed according to a scientific "breakthrough logic".

The focus in this situation would be on building specific knowledge, capacity, and competencies; geographical proximity would be less relevant due to the ongoing innovation dynamics, codified search routines, and established connections of professionals over distance (Rutten, 2017). Recent large-scale vaccine developments and the European mission to fight cancer, both of which focus on transnational collaboration for scientific advancements, are good illustrations of this kind of approach. Limited value or attention to context-specific knowledge, local concerns, or serendipitous encounters could be a downside of a global amplification or scaling strategy that is focussed on scientific and technological advancement. Local or "mobile" brokers such as scientists, NGOs, consultants, or firms who are engaged in both local and international knowledge communities are needed in order to achieve regional engagement or local societal impact (Atta-Owusu & Fitjar, 2022; Gluckler, 2007). Non-core regions without strong local knowledge clusters or organisations can be disadvantaged by such global mission designs.

10.4.3 Local Knowledge for Global Problems

In other instances, problems may be framed as more general or global to stimulate local searches for solutions in multiple places. In this situation, the problem space may be well understood or relatively uncontested. At the same time, the application and diffusion of solutions may still be hampered by uncertainty, ambiguous regulatory frameworks, or weak decisions over technological choices. Broadening the geographical scale of the problem – e.g., by mobilising translocal networks, such as global networks of cities (Fastenrath & Coenen, 2021) – may help "bridge scales" and widen the scope of a market, upscaling potential solutions to new places or fields of application (Dewald & Truffer, 2012; Uyarra et al., 2017). Thus, a strategy could be based on translating and building local coalitions to tackle a challenge that, in its broad definition, is shared by many places, but where the creation and implementation of a solution are dominated by localised processes, routines, or conditions.

A place-based approach for anchoring global problems is relevant for exploiting local qualities such as the knowledge bases and expertise of firms or other knowledge actors, and the demand sophistication of local users and citizens. Existing regional innovation system configurations may be mobilised to co-develop solutions to societal problems. This could, for instance, be an appropriate strategy for many societal issues related to public health (particularly around care), education, or mobility, where problems may be both local and global and require complex solutions that combine technologies and knowledge from different sectors and industries but are implemented in local contexts. Regions could use these situations to encourage interactions between diverse and potentially co-located networks of industries and knowledge bases (Gluckler, 2007; Janssen & Frenken, 2019). Local bridging and recombination of unconnected actors and industries could create new opportunities for regional path creation, fostering new economic trajectories (Uyarra et al., 2020). The challenge here is to adequately articulate the problem in a local context and to mobilise actors and networks around a common vision.

Multi-local problem-solving, collaborative platforms (Ansell & Gash, 2018) can be an instrument to bring together localised and ongoing projects, collaborations, and networks. Such multi-local ambitions have been established in areas such as food, childhood obesity (Ansell & Gash, 2018; Borys et al., 2012), or sustainable mobility (van der Kam & Bekkers, 2020). In a mission context, these "networks of networks" can have different governance purposes but still work together on framing and creating broader commitment for a global problem, sharing and translating experience with localised innovations, or promoting joint implementations, specific certification schemes, or new standards to support multi-local solution diffusion.

An example of such collaborative platforms is the "shared agendas" for sustainability and social change adopted in Catalonia (Spain). These agendas aim to articulate, through a participatory governance model, the collective action of diverse actors to face a common societal challenge (usually related to the SDGs) within a territory and the problems that arise from facing the challenge (Fernández & Romagosa, 2020). These projects are based on cross-sectoral collaboration and the generation of shared knowledge between public administrations, academia, companies, and civil society, in order to understand and manage complex problems from a holistic perspective and take long-term effects into account. The efforts seek to change the model of production and consumption in a territory towards a green and circular economy model by responding to the specific needs and problems of groups in that territory. The aim is to replicate or scale successful solutions by connecting them to more global strategies and agendas, such as the EU's research and innovation missions (Boni et al., 2023).

10.4.4 Global Solutions for Local Problems

Finally, local problems may be mobilised to support the development of particular innovations or technologies. Innovations (technological or non-technological) that have the potential to tackle a certain societal problem may initially lack a local context. For instance, technological innovations might need to be accompanied by institutional or social innovations (Vale & Carvalho, 2013) to anchor technologies, embed them in the local institutional context, and increase their legitimacy and public acceptance. Anchoring thus implies "re-contextualising and diffusing external knowledge in place" (Vale & Carvalho, 2013, p. 1021) so that regional actors can access it and use it to transform regional structures (Binz et al., 2016).

Global solutions for local problems can only be achieved if legitimacy is created at the local level. For instance, Uyarra and Flanagan (2022) discussed how the government of the Galicia region in Spain mobilised all regional departments to articulate which problems and needs could demonstrate the application of UAV- (uncrewed aerial vehicles) based service solutions to public problems such as fire prevention, control of fisheries, and coastal surveillance and monitoring. Here, regional problems were leveraged to support external knowledge anchoring and legitimation for contested UAV technology.

As this example illustrates, peripheral places may make promising test-beds for solutions that can then diffuse more widely across organisations and locations. This is particularly important for regions which may be less suited to lead in the overall development of new technologies, but which can instead take the lead in specific applications (Janssen & Frenken, 2019). The development of generic technologies may boost existing industries but also open up opportunities for diversification by driving convergence among the knowledge bases of unrelated industries (Corradini & De Propris, 2017; Montresor & Quatraro, 2017).

As noted by Losacker and Liefner (2020), regions may have demand or regulatory advantages for the testing of complex technologies, which makes them ideal as regional (rather than national) lead markets. They argue that lag or follower markets "anticipate the benefits of the innovation designs first adopted in the lead market" (2020, p. 123). Demand advantages may arise from specific demand conditions, such as the risk of flooding or forest fires as in the case of Galicia. Looking at the case of waste management in Shanghai, Losacker and Liefner (2020) concluded that the large market, the legal conditions, and the interplay of these factors with a technological advantage explained why Shanghai outperformed Shenzhen and Beijing as potential lead markets for waste-related technologies. Uyarra and Gee (2013) similarly identified the purposeful articulation of demand for waste management, active lobbying at multiple scales, and technology legitimation as key mechanisms that helped the adoption of more sustainable waste management technologies in the Greater Manchester area. Binz et al. (2016) studied the development of on-site water recycling technology in Chinese city regions and found that Beijing had developed a sizeable industry (despite having the least favourable technological conditions) as a result of active mobilisation and anchoring processes of external resources through market formation, investment mobilisation, technology legitimation, and knowledge creation.

In these examples, the focus is not on developing new technical innovations, but instead on the social impacts of innovations and how they can be stimulated and used to support challenges affecting specific places. However, one danger is that discourse may be captured by special interests who intend to legitimise particular technologies with reference to their potential to solve societal challenges, but without sufficiently considering those challenges. This "solutionist" bias may arise if too much emphasis is placed on promising solutions at the expense of understanding the problem and its causes and consequences more deeply (Montero, 2020).

10.5 CONCLUDING REMARKS

In the last two decades, there has been a lively debate about the need for more directional or mission-oriented innovation policies to address so-called wicked societal challenges, requiring not just technical solutions but also social or behavioural changes. In common with the broader debate about innovation policy, the debate about mission orientation has focussed mostly on the definition and design of missions rather than on their practical implementation in real political and local contexts. It remains unclear how MOIPs will align in practice with multi-scalar governance, regional innovation systems, and place-based problems and challenges. This lack of spatial sensitivity leads to a neglect of cities and regions as key loci for innovation and experimentation and as a key element in multi-scalar transformative innovation policy.

Clearly, the need for societal change driving a mission has not been "out there" all along, nor is it identically manifested in all places; rather, it is contextual, socially constructed, follows various geographical dynamics, and involves different viewpoints and interests in relation to the kinds of innovation that might be needed. Given this fundamentally normative character, questions of how, where, at what point in time, and by whom problem discourses are created and societal needs articulated are highly relevant to attempts to make transformative mission-oriented innovation policy work in practice. Problem-driven innovation activities are, at least to some extent, directed by problem discourses and build on contextual narratives which specify the need for innovation and the nature of the change that is being sought. This means that both problems and solutions need to be considered, and that there is both a geography of problems and a geography of solutions.

This chapter has sought to illuminate these issues by integrating a spatial and scalar dimension into how we think about transformative mission-oriented innovation policy. It brings together contributions from innovation policy studies, regional studies, and economic geography, and argues that problem/solution representations in the context of mission-oriented policies not only have different degrees of "wickedness" but will also be more or less geographically anchored. The latter is shaped by considerations of place qualities, and how these are framed in relation to a single location or multiple places. On this basis, we propose a tentative mapping of the possible opportunity space for mission-oriented innovation policy to illustrate how cities and regions might articulate "missions" to advance societal challenges that consider not just their relevance to their territory and community but also the extent to which local (pre-)conditions may support or be supported by the search for solutions. We used this to explore four specific idealised opportunity spaces that are relevant for different types of "missions" and different types of places and that might be supported by diverse strategies and policy instruments. The aim is to foreground spatial and scalar implications and to encourage reflection on the potential for transformative mission-oriented policies. This is an initial, tentative exploration of these implications and inevitably has limitations – most notably that we neglect the fact that societal challenges are not only contested and complex but also subject to transformation over time and that the systems of innovation that emerge to solve them are not fixed but transient. The solution of one problem often leads to new and different problems (Metcalfe & Ramlogan, 2008). Since problems evolve, so do the content and the patterns of systemic interconnections between actors in the system.

In this chapter, we have also tried to reopen the question of what it means to be transformative. Transformative missions consider transformation in the sense of tackling a pressing societal challenge, but mission-oriented policies might also (directly or indirectly) transform places and communities. Top-down accelerator missions may be place-blind in theory, but no policy is spatially neutral in practice. Any MOIP will inevitably influence the agglomeration of economic activity, potentially aggravating economic and societal disparities across territories. Thus, place-blind transformative policies may exclude a large proportion of communities and places if not made more place-sensitive. Equally, "localness" may lead to fragmentation and poor take-up of otherwise potentially transformative solutions. Missions may also suffer from "solutionism", namely a greater emphasis on promising solutions rather than the goals and values that the policy seeks to address. Considering place and scale could help ameliorate this risk by better fitting solutions to problems on the ground and by building legitimacy for those solutions according to the local context and the prevalent innovation dynamics.

ACKNOWLEDGEMENTS

The authors would like to thank Jakob Edler, Mireille Matt, Wolfgang Polt, and Matthias Weber for their useful and helpful comments on our manuscript. Elvira Uyarra acknowledges support from the Research Council of Norway (Project RegReSir, grant number 316539). Iris Wanzenböck acknowledges support from the Research Council of Norway (Project INTRANSIT, grant number 295021).

REFERENCES

Ansell, C., & Gash, A. (2018). Collaborative platforms as a governance strategy. *Journal of Public Administration Research and Theory*, *28*(1), 16–32. https://doi.org/10.1093/jopart/mux030

Ansell, C., & Torfing, J. (2015). How does collaborative governance scale? *Policy & Politics*, *43*(3), 315–329. https://doi.org/10.1332/030557315X14353344872935

Atta-Owusu, K., & Fitjar, R. D. (2022). Engaging for the love of place? The role of place attachment in academics' regional engagement efforts. *Regional Studies*, *56*(12), 1–12. https://doi.org/10.1080/00343404.2022.2034778

Binz, C., & Truffer, B. (2017). Global innovation systems: A conceptual framework for innovation dynamics in transnational contexts. *Research Policy*, *46*(7), 1284–1298. https://doi.org/10.1016/j.respol.2017.05.012

Binz, C., Truffer, B., & Coenen, L. (2016). Path creation as a process of resource alignment and anchoring: Industry formation for on-site water recycling in Beijing. *Economic Geography*, *92*(2), 172–200. https://doi.org/10.1080/00130095.2015.1103177

Boni, A., Velasco, D., Acebillo, M., Sort, M. L., Gironès, X., Molas, J., & Fernández, T. (2023). Evaluación formativa y políticas regionales de innovación transformadoras. El caso de la Agenda Compartida del Bages. *Ekonomiaz: Revista vasca de economía*, *104*, 100–125.

Boon, W., & Edler, J. (2018). Demand, challenges, and innovation. Making sense of new trends in innovation policy. *Science and Public Policy*, *45*(4), 435–447. https://doi.org/10.1093/scipol/scy014

Borrás, S. (2009). *The widening and deepening of innovation policy: What conditions provide for effective governance?* https://smartech.gatech.edu/handle/1853/39803

Borrás, S., & Edler, J. (2020). The roles of the state in the governance of socio-technical systems' transformation. *Research Policy*, *49*(5), 103971. https://doi.org/10.1016/j.respol.2020.103971

Borys, J.-M., Le Bodo, Y., Jebb, S. A., Seidell, J. C., Summerbell, C., Richard, D., De Henauw, S., Moreno, L. A., Romon, M., Visscher, T. L. S., Raffin, S., Swinburn, B., & Group, The E. S. (2012). EPODE approach for childhood obesity prevention: Methods, progress and international development. *Obesity Reviews*, *13*(4), 299–315. https://doi.org/10.1111/j.1467–789X.2011.00950.x

Boschma, R., Coenen, L., Frenken, K., & Truffer, B. (2017). Towards a theory of regional diversification: Combining insights from evolutionary economic geography and transition studies. *Regional Studies, 51*(1), 31–45. https://doi.org/10.1080/00343404.2016.1258460

Bours, S. A., Wanzenböck, I., & Frenken, K. (2022). Small wins for grand challenges. A bottom-up governance approach to regional innovation policy. *European Planning Studies, 30*(11), 2245–2272. https://doi.org/10.1080/09654313.2021.1980502

Brown, R. (2020). Mission-oriented or mission adrift? A critical examination of mission-oriented innovation policies. *European Planning Studies, 29*(4), 1–23. https://doi.org/10.1080/09654313.2020.1779189

Bugge, M. M., Andersen, A. D., & Steen, M. (2021). The role of regional innovation systems in mission-oriented innovation policy: Exploring the problem-solution space in electrification of maritime transport. *European Planning Studies, 30*(11), 1–22. https://doi.org/10.1080/09654313.2021.1988907

Bulkeley, H., Broto, V. C., Hodson, M., & Marvin, S. (2010). *Cities and low carbon transitions*. Taylor & Francis.

Bulkeley, H., Coenen, L., Frantzeskaki, N., Hartmann, C., Kronsell, A., Mai, L., Marvin, S., McCormick, K., van Steenbergen, F., & Voytenko Palgan, Y. (2016). Urban living labs: Governing urban sustainability transitions. *Current Opinion in Environmental Sustainability, 22*, 13–17. https://doi.org/10.1016/j.cosust.2017.02.003

Capasso, M., Hansen, T., Heiberg, J., Klitkou, A., & Steen, M. (2019). Green growth – A synthesis of scientific findings. *Technological Forecasting and Social Change, 146*, 390–402. https://doi.org/10.1016/j.techfore.2019.06.013

Cappellano, F., Makkonen, T., Dotti, N. F., Morisson, A., & Rizzo, A. (2022). Where innovation meets directionality: An index to measure regional readiness to deal with societal challenges. *European Planning Studies, 30*(8), 1549–1576. https://doi.org/10.1080/09654313.2021.1976114

Coenen, L., Hansen, T., & Rekers, J. V. (2015). Innovation policy for grand challenges. An economic geography perspective. *Geography Compass, 9*(9), 483–496. https://doi.org/10.1111/gec3.12231

Coenen, L., & Morgan, K. (2020). Evolving geographies of innovation: Existing paradigms, critiques and possible alternatives. *Norsk Geografisk Tidsskrift – Norwegian Journal of Geography, 74*(1), 13–24. https://doi.org/10.1080/00291951.2019.1692065

Corradini, C., & De Propris, L. (2017). Beyond local search: Bridging platforms and inter-sectoral technological integration. *Research Policy, 46*(1), 196–206. https://doi.org/10.1016/j.respol.2016.09.017

Craens, J., Frenken, K., & Meelen, T. (2022). *Mission-oriented innovation policy: The case of the Swedish "Vision Zero" approach to traffic safety.* In Innovations in Transport (pp. 343-358). Edward Elgar Publishing.

Dewald, U., & Truffer, B. (2012). The local sources of market formation: explaining regional growth differentials in German photovoltaic markets. *European Planning Studies, 20*(3), 397–420. https://doi.org/10.1080/09654313.2012.651803

Edler, J., Matt, M., Polt, W., & Weber, K. (2021). *Transformative mission-oriented STI policy: Theoretical and conceptual rationales, intervention logics and challenges of an emerging type of STI policies*. EU-SPRI.

Evans, J., Jones, R., Karvonen, A., Millard, L., & Wendler, J. (2015). Living labs and co-production: University campuses as platforms for sustainability science. *Current Opinion in Environmental Sustainability, 16*, 1–6. https://doi.org/10.1016/j.cosust.2015.06.005

Fastenrath, S., & Coenen, L. (2021). Future-proof cities through governance experiments? Insights from the Resilient Melbourne Strategy (RMS). *Regional Studies, 55*(1), 138–149. https://doi.org/10.1080/00343404.2020.1744551

Fernández, T., & Romagosa, M. (2020). *L'articulació d'agendes compartides per a la sostenibilitat i el canvi social* (p. 48). Generalitat de Catalunya.

Fisher, R., Chicot, J., Domini, A., Polt, W., Turk, A., Unger, M., Kuittinen, H., Arrilucea, E., Van Der Zee, F., Goetheer, A., Lehenkari, J., Pelkonen, A., & Kristensen, F. S. (2018). *Mission-oriented research and innovation: Inventory and characterisation of initiatives: Final report.* European Commission EC. https://doi.org/10.2777/697082

Flanagan, K., Uyarra, E., & Laranja, M. (2011). Reconceptualising the "policy mix" for innovation. *Research Policy, 40*(5), 702–713.https://doi.org/10.1016/j.respol.2011.02.005

Flanagan, K., Uyarra, E., & Wanzenböck, I. (2023). Towards a problem-oriented regional industrial policy: possibilities for public intervention in framing, valuation and market formation. *Regional Studies*, *57*(6), 998–1010. https://doi.org/10.1080/00343404.2021.2016680
Foray, D. (2018). Smart specialization strategies as a case of mission-oriented policy: A case study on the emergence of new policy practices. *Industrial and Corporate Change*, *27*(5), 817–832. https://doi.org/10.1093/icc/dty030
Frenken, K. (2017). A complexity-theoretic perspective on innovation policy. *Complexity, Governance & Networks*, *3*(1), 35–47. https://doi.org/10.1080/10438590500141453
Fuenfschilling, L., Frantzeskaki, N., & Coenen, L. (2019). Urban experimentation & sustainability transitions. *European Planning Studies*, *27*(2), 219–228. https://doi.org/10.1080/09654313.2018.1532977
Gluckler, J. (2007). Economic geography and the evolution of networks. *Journal of Economic Geography*, *7*(5), 619–634. https://doi.org/10.1093/jeg/lbm023
Hassink, R., Gong, H., Fröhlich, K., & Herr, A. (2021). Exploring the scope of regions in challenge-oriented innovation policy: The case of Schleswig-Holstein, Germany. *European Planning Studies*, *30*(11), 1–19. https://doi.org/10.1080/09654313.2021.2017857
Hausmann, R., & Rodrik, D. (2003). Economic development as self-discovery. *Journal of Development Economics*, *72*(2), 603–633. https://doi.org/10.1016/S0304–3878(03)00124-X
Henderson, D., Morgan, K., & Delbridge, R. (2023). Putting missions in their place: Micro-missions and the role of universities in delivering challenge-led innovation. *Regional Studies*, *0*(0), 1–12. https://doi.org/10.1080/00343404.2023.2176840
Heslop, J., Morgan, K., & Tomaney, J. (2019). Debating the foundational economy. Renewal: A journal of social democracy, 27 (2): 5–12.
Isaksen, A., Tödtling, F., & Trippl, M. (2018). Innovation policies for regional structural change: Combining actor-based and system-based strategies. In A. Isaksen, R. Martin, & M. Trippl (Eds.), *New avenues for regional innovation systems: Theoretical advances, empirical cases and policy lessons* (pp. 221–238). Springer International Publishing. https://doi.org/10.1007/978–3-319–71661–9_11
Jakobsen, S.-E., Uyarra, E., Njøs, R., & Fløysand, A. (2022). Policy action for green restructuring in specialized industrial regions. *European Urban and Regional Studies*, *29*(3), 312–331. https://doi.org/10.1177/09697764211049116
Janssen, M. J., & Frenken, K. (2019). Cross-specialisation policy: Rationales and options for linking unrelated industries. *Cambridge Journal of Regions, Economy and Society*, *12*(2), 195–212. https://doi.org/10.1093/cjres/rsz001
Janssen, M. J., Torrens, J., Wesseling, J. H., & Wanzenböck, I. (2021). The promises and premises of mission-oriented innovation policy: A reflection and ways forward. *Science and Public Policy*, *48*(3), 438–444. https://doi.org/10.1093/scipol/scaa072
Jeannerat, H., & Crevoisier, O. (2022). From competitiveness to territorial value: Transformative territorial innovation policies and anchoring milieus. *European Planning Studies*, *30*(11), 2157–2177. https://doi.org/10.1080/09654313.2022.2042208
Kemp, R., Schot, J., & Hoogma, R. (1998). Regime shifts to sustainability through processes of niche formation: The approach of strategic niche management. *Technology Analysis & Strategic Management*, *10*(2), 175–198. https://doi.org/10.1080/09537329808524310
Kuhlmann, S., & Rip, A. (2018). Next-generation innovation policy and grand challenges. *Science and Public Policy*, *45*(4), 448–454. https://doi.org/10.1093/scipol/scy011
Lam, D. P. M., Martín-López, B., Wiek, A., Bennett, E. M., Frantzeskaki, N., Horcea-Milcu, A. I., & Lang, D. J. (2020). Scaling the impact of sustainability initiatives: A typology of amplification processes. *Urban Transformations*, *2*(1), 3. https://doi.org/10.1186/s42854–020–00007–9
Laranja, M., Uyarra, E., & Flanagan, K. (2008). Policies for science, technology and innovation: Translating rationales into regional policies in a multi-level setting. *Research Policy*, *37*(5), 823–835. https://doi.org/10.1016/j.respol.2008.03.006
Larrue, P. (2021). *The design and implementation of mission-oriented innovation policies: A new systemic policy approach to address societal challenges*. OECD. https://doi.org/10.1787/3f6c76a4-en
Loorbach, D., Wittmayer, J., Avelino, F., Von Wirth, T., & Frantzeskaki, N. (2020). Transformative innovation and translocal diffusion. *Environmental Innovation and Societal Transitions*, *35*, 251–260. https://doi.org/10.1016/j.eist.2020.01.009

Losacker, S., & Liefner, I. (2020). Regional lead markets for environmental innovation. *Environmental Innovation and Societal Transitions*, *37*, 120–139. https://doi.org/10.1016/j.eist.2020.08.003

Madsen, S. H. J. (2022). A constructivist approach to the spatial organization of transformative innovation policy. *Environmental Innovation and Societal Transitions*, *42*, 340–351. https://doi.org/10.1016/j.eist.2022.01.007

Mazzucato, M. (2018a). *Mission-oriented research & innovation in the European Union a problem-solving approach to fuel innovation-led growth*. European Commission.

Mazzucato, M. (2018b). Mission-oriented innovation policies: Challenges and opportunities. *Industrial and Corporate Change*, *27*(5), 803–815. https://doi.org/10.1093/icc/dty034

McCann, P., & Soete, L. (2020). *Place-based innovation for sustainability*. JRC Working Papers (No. JRC121271). Joint Research Centre (Seville site). https://ideas.repec.org/p/ipt/iptwpa/jrc121271.html

Metcalfe, S., & Ramlogan, R. (2008). Innovation systems and the competitive process in developing economies. *The Quarterly Review of Economics and Finance*, *48*(2), 433–446. https://doi.org/10.1016/j.qref.2006.12.021

Miedzinski, M., Stancova, K. C., Matusiak, M., & Coenen, L. (2021). *Addressing sustainability challenges and Sustainable Development Goals via Smart Specialisation. Towards a theoretical and conceptual framework*. JRC Research Reports (No. JRC126448; JRC Research Reports). Joint Research Centre (Seville site). https://ideas.repec.org/p/ipt/iptwpa/jrc126448.html

Montero, S. (2020). Leveraging Bogotá: Sustainable development, global philanthropy and the rise of urban solutionism. *Urban Studies*, *57*(11), 2263–2281. https://doi.org/10.1177/0042098018798555

Montresor, S., & Quatraro, F. (2017). Regional branching and key enabling technologies: Evidence from European patent data. *Economic Geography*, *93*(4), 367–396. https://doi.org/10.1080/00130095.2017.1326810

Morgan, K. (2018). *Experimental governance and territorial development*. OECD.

Morgan, K., & Marques, P. (2019). The public animateur: Mission-led innovation and the "smart state" in Europe. *Cambridge Journal of Regions, Economy and Society*, *12(*2), 179–193. https://doi.org/10.1093/cjres/rsz002

Nelson, R. (1977). *The moon and the ghetto*. Norton and Company.

Pontikakis, D., Gonzalez, V. I., Bianchi, G., Ranga, L., Marques, S. A., Reimeris, R., Mifsud, S., Morgan, K., Madrid, G. C., & Stierna, K. (2022, May 16). *Partnerships for regional innovation playbook*. JRC Publications Repository. https://doi.org/10.2760/775610

Raven, R., Sengers, F., Spaeth, P., Xie, L., Cheshmehzangi, A., & de Jong, M. (2019). Urban experimentation and institutional arrangements. *European Planning Studies*, *27*(2), 258–281. https://doi.org/10.1080/09654313.2017.1393047

Rutten, R. (2017). Beyond proximities: The socio-spatial dynamics of knowledge creation. *Progress in Human Geography*, *41*(2), 159–177. https://doi.org/10.1177/0309132516629003

Sabel, C., Moore, M., & Zeitlin, J. (2012). Experimentalist governance. In *The Oxford handbook of governance*. Oxford University Press, pp. 169–183. https://doi.org/10.1093/oxfordhb/9780199560530.013.0012

Schön, D., & Rein, M. (1994). *Frame reflection: Toward the resolution of intractable policy controversies*. Basic Books.

Schot, J., & Steinmueller, W. E. (2018). Three frames for innovation policy: R&D, systems of innovation and transformative change. *Research Policy*, *47*(9), 1554–1567. https://doi.org/10.1016/j.respol.2018.08.011

Scordato, L., Bugge, M. M., Hansen, T., Tanner, A., & Wicken, O. (2022). Walking the talk? Innovation policy approaches to unleash the transformative potentials of the Nordic bioeconomy. *Science and Public Policy*, *49*(2), 324–346. https://doi.org/10.1093/scipol/scab083

Sengers, F., Turnheim, B., & Berkhout, F. (2021). Beyond experiments: Embedding outcomes in climate governance. *Environment and Planning C: Politics and Space*, *39*(6), 1148–1171.

Sternberg, R. G. (1996). Government R & D expenditure and space: Empirical evidence from five industrialized countries. *Research Policy*, *25*(5), 741–758. https://doi.org/10.1016/0048–7333(95)00860–8

Storper, M., Nicholas Ziegler, J., Botelho, A. J. J., & Ornston, D. (2022). On Mariana Mazzucato's mission economy: A moonshot guide to changing capitalism. Allen Lane. *Socio-Economic Review*, *20*(3), 1501–1511. https://doi.org/10.1093/ser/mwac042

Suitner, J., Haider, W., & Philipp, S. (2023). Social innovation for regional energy transition? An agency perspective on transformative change in non-core regions. *Regional Studies*, *57*(8), 1498–1510.

Tanner, A. N. (2014). Regional branching reconsidered: Emergence of the fuel cell industry in European regions. *Economic Geography*, *90*(4), 403–427.

Trippl, M., Baumgartinger-Seiringer, S., Frangenheim, A., Isaksen, A., & Rypestøl, J. O. (2020). Unravelling green regional industrial path development: Regional preconditions, asset modification and agency. *Geoforum*, *111*, 189–197. https://doi.org/10.1016/j.geoforum.2020.02.016

Uyarra, E., & Flanagan, K. (2022). Going beyond the line of sight: Institutional entrepreneurship and system agency in regional path creation. *Regional Studies*, *56*(4), 536–547. https://doi.org/10.1080/00343404.2021.1980522

Uyarra, E., Flanagan, K., Magro, E., & Zabala-Iturriagagoitia, J. M. (2017). Anchoring the innovation impacts of public procurement to place: The role of conversations. *Environment and Planning C: Politics and Space*, *35*(5), 828–848. https://doi.org/10.1177/2399654417694620

Uyarra, E., & Gee, S. (2013). Transforming urban waste into sustainable material and energy usage: The case of Greater Manchester (UK). *Journal of Cleaner Production*, *50*, 101–110. https://doi.org/10.1016/j.jclepro.2012.11.046

Uyarra, E., Ribeiro, B., & Dale-Clough, L. (2019). Exploring the normative turn in regional innovation policy: responsibility and the quest for public value. *European Planning Studies*, *27*(12), 2359–2375. https://doi.org/10.1080/09654313.2019.1609425

Uyarra, E., Zabala-Iturriagagoitia, J. M., Magro, E., & Flanagan, K. (2020). Public procurement, innovation and industrial policy: Rationales, roles, capabilities and implementation. *Research Policy*, *49*(1), 103844

Vale, M., & Carvalho, L. (2013). Knowledge networks and processes of anchoring in Portuguese biotechnology. *Regional Studies*, *47*(7), 1018–1033. https://doi.org/10.1080/00343404.2011.644237

van der Kam, M., & Bekkers, R. N. A. (2020). *Comparative analysis of standardized protocols for EV roaming: Report D6.1 for the evRoaming4EU project.* Netherlands Knowledge Platform for Public Charging Infrastructure (NKL). https://evroaming.org/the-ideal-ev-roaming-protocol/?utm_source=Mailing+Lijst&utm_medium=email&utm_campaign=Wrap+up+EV+Roaming+Webinar

Van Hulst, M., & Yanow, D. (2016). From policy "frames" to "framing" theorizing a more dynamic, political approach. *The American Review of Public Administration*, *46*(1), 92–112. https://doi.org/10.1177/0275074014533142

von Wirth, T., Fuenfschilling, L., Frantzeskaki, N., & Coenen, L. (2019). Impacts of urban living labs on sustainability transitions: Mechanisms and strategies for systemic change through experimentation. *European Planning Studies*, *27*(2), 229–257. https://doi.org/10.1080/09654313.2018.1504895

Voytenko, Y., McCormick, K., Evans, J., & Schliwa, G. (2016). Urban living labs for sustainability and low carbon cities in Europe: Towards a research agenda. *Journal of Cleaner Production*, *123*, 45–54. https://doi.org/10.1016/j.jclepro.2015.08.053

Wanzenböck, I., & Frenken, K. (2020). The subsidiarity principle in innovation policy for societal challenges. *Global Transitions*, *2*, 51–59. https://doi.org/10.1016/j.glt.2020.02.002

Wanzenböck, I., Wesseling, J. H., Frenken, K., Hekkert, M. P., & Weber, K. M. (2020). A framework for mission-oriented innovation policy: Alternative pathways through the problem–solution space. *Science and Public Policy*, *47*(4), 474–489. https://doi.org/10.1093/scipol/scaa027

Weber, K. M., & Rohracher, H. (2012). Legitimizing research, technology and innovation policies for transformative change: Combining insights from innovation systems and multi-level perspective in a comprehensive "failures" framework. *Research Policy*, *41*(6), 1037–1047. https://doi.org/10.1016/j.respol.2011.10.015

Wesseling, J., & Meijerhof, N. (2021). *Developing and applying the Mission-oriented Innovation Systems (MIS) approach* [Preprint]. SocArXiv. https://doi.org/10.31235/osf.io/xwg4e

Wittmann, F., Hufnagl, M., Lindner, R., Roth, F., & Edler, J. (2021). Governing varieties of mission-oriented innovation policies: A new typology. *Science and Public Policy*, *48*(5), 727–738. https://doi.org/10.1093/scipol/scab044

11. Transformative innovation policy in the making: how directionality has emerged for mobility as a service in Finland

Paula Kivimaa and Jonas Torrens

11.1 INTRODUCTION

Innovation policy is increasingly expected to address societal challenges and refocus from pursuing economic growth to enabling systemic shifts or transformations. Transformative innovation policy (TIP) and mission-oriented innovation policies (MOIP) have emerged as prominent rationales for a more engaged stance from both science and innovation funders (e.g., Mazzucato, 2018; Schot & Steinmueller, 2018). Both TIP and MOIP rationales emphasise "directionality", i.e., how policies and interventions can help orient societal efforts around innovation and align them with societal needs (c.f. Edler & Boon, 2018). A fundamental question that can inform TIP and MOIP implementation is understanding how directionality develops in practice.

The term "directionality" refers to a key insight of socio-technical innovation studies: innovations can develop along many distinct normative "directions", which are often obfuscated in public discourse that emphasises progress as unidirectional and self-evident (Stirling, 2009; Stirling et al., 2007). Any vision based on innovation excludes and downplays other alternative directions of change (Stirling, 2009). This implicit emphasis on direction and unquestioned progress of economic or industrial advancement has been typical in traditional, market-oriented innovation policy. Over the past two decades, the expectation that policies should be "technology-neutral" (for example, via a general carbon-price) and the associated fear of governments "picking winners" have been central to the rhetoric legitimising innovation policies (Azar & Sandén, 2011; Weber & Rohracher, 2012).

However, expecting innovation policy to proactively contribute to transformative change demands new rationales and sources of legitimacy. Calls for new challenge-oriented innovation policies bring forth "directionality" as one critical dimension where policy must intervene (Weber & Rohracher, 2012). However, given the uncertainty and surprises involved in transformative or sustainability transition processes, it is not evident whether directionality for innovation policy should be set as "top-down" by the public sector given "missions", as "bottom-up" via inclusive stakeholder processes (with some risk that many people cannot imagine out-of-the-box future solutions), or some combination of the two. Thus, in this chapter, we are interested in the following questions: Where can the processes of setting directionality in forming transformative and mission-oriented policies occur? What processes contribute to the formation of directionality? What can be the role of different actors within networks to influence directionality?

We will investigate the case of Mobility as a Service (MaaS) in Finland during 1995–2019. We aim to understand how directionality towards MaaS emerged in the Finnish context.

Presently, MaaS promises to transform transport systems by reducing the private ownership of vehicles through innovative, customised mobility service packages for consumers. Such a transition is envisaged to free up city space for other uses, reduce transport emissions, cater to ageing rural populations, and create new digital services. The case of MaaS in Finland is relevant for TIP and MOIP debates as an example of proactive governmental engagement with an emerging problem. This is because the case involves experimentation and consolidated vision-building efforts that are supported by an unconventional actor constellation and go beyond the traditional contours of the innovation system (Kivimaa & Rogge, 2022).

Based on this empirical case, we will show how directionality evolved in the process, heavily shaped by public policy – highlighting the importance of concrete cross-domain coordination for transformative and challenge-driven innovation policy. The chapter is organised as follows. The first section reviews the debate on directionality in TIP/MOIP. The second section explains briefly the methodological approach taken. We will then explore the key events concerning MaaS development in Finland, analysed via six elements found relevant for directionality. The final section discusses the significance of the MaaS case from the perspective of TIP and MOIP.

11.2 DIRECTIONALITY IN TRANSFORMATIVE AND MISSION-ORIENTED INNOVATION POLICY

In the European context, the past decade has seen a growing interest in rekindling innovation policy to address the need for long-term, transformative change in ways that exceed the capacities of innovation systems and the scope of traditional innovation policy rationales. In a seminal paper, Weber and Rohracher (2012) suggested that innovation policy should move beyond focussing solely on market or innovation system failures and address four transformational failures concerning directionality, demand articulation, policy coordination, and reflexivity. Addressing these failures requires expanding the scope of the rationales that legitimise innovation policy. In this section, we focus primarily on the "directionality question".

Transformative change is conceptualised by Weber and Rohracher (2012, p. 1042) as "intimately connected to the question of direction and requires the setting of collective priorities; priorities that require a strategic policy approach to be in place". It is not just a matter of correcting market and innovation system failures to incite innovation in particular sectors, but doing so to move "in the direction of desired long-term transformative change" (Weber & Rohracher, 2012). According to that perspective, building a collective vision is "only one option" for enabling collective coordination. One should not assume a consensus about a direction, as there are many intractable controversies around supposedly consensual issues. One should also not assume that the process is purely rational and strategic; the power and agency of those involved and the implicit but incongruent worldviews greatly influence vision formation.

Directionality matters in sustainability-related challenges because innovations can be appropriated and embedded in drastically different socio-technical configurations, leading to systems of provision that have divergent outcomes. There is a plurality of possible "pathways" for socio-technical change, which can be terminated prematurely or explored more intentionally, depending on the conduct of political and policy processes (Stirling et al., 2007). Electric vehicles, for instance, could be a substitute for individually owned gasoline and diesel vehicles or be part of a wider shift towards sharing modes that displace individual ownership. These different pathways represent distinct directions and framings of the problem and solution and require specific policy and institutional support. Crucially, they encompass dramatically different values and normativities. Each pathway is justified with distinct ways of appraising technology and its effects. In this context, "letting the market decide" easily reproduces values and outcomes that are compatible with the existing socio-technical regime, even when they are demonstrably sub-optimal vis-à-vis policy and societal goals (Geels, 2005). As the long-standing literature on path dependence clarifies, early decisions or features of these innovations can become obdurate. Therefore, early attention to how directionality emerges and is maintained is critical to understanding the pathways in which sustainability transitions unfold.

The concern over directionality covers at least three distinct debates with different levels of analysis. First, in the sustainability transitions literature, the *directionality of existing socio-technical systems* has mainly been ascribed to the alignment of multiple socio-technical dimensions, with an essential role for institutions. For example, public policies can play significant roles both in "directing the search" for innovation processes, as well as destabilising existing unsustainable socio-technical regimes to create space for new directions (Kivimaa & Kern, 2016). Institutions have been described as the "deep rules" of socio-technical regimes, which profoundly influence the stability of existing regimes and need to change if transformation is to be achieved (Fuenfschilling & Binz, 2018; Geels, 2004; Geels & Schot, 2007). North (1991, p. 97) has described institutions as "the humanly devised constraints that structure political, economic and social interaction. They consist of both informal constraints (sanctions, taboos, customs, traditions, and codes of conduct), and formal rules (constitutions, laws, property rights)". Streeck and Thelen (2005) view institutions as building blocks of social order that include rights and obligations for actors, which may change incrementally or abruptly when multiple actors collectively change their logic of action. In a general sense, institutions change slowly, and hence breaking the lock-ins of established systems is difficult and the emergence of new directionalities slow.

The *directionality of innovation trajectories* has often been the object of studies concerning niche development or technology appraisal. Early on, the sustainability transitions literature, and in particular, the strategic niche management approach (Geels & Schot, 2007; Kemp et al., 1998; Smith & Raven, 2012), was put forward as a way to modulate the development of alternatives in ways that could lead to the emergence of multiple alternative trajectories of socio-technical change. Here, niches act as protective spaces where radical alternatives can emerge through experimentation to become more robust before competing with mainstream systems. No singular factor determines which technological trajectories develop, though some patterns can be discerned. These patterns have inspired different typologies that concern "transitions pathways" (Geels & Schot, 2007) which centre on the interactions between niche innovations, the prevailing socio-technical regime, and an exogenous context (landscape).

Recent studies have given rise to a more complex understanding of this area. Pel et al. (2020), writing on the "driverless car" transition in the Netherlands, proposed developing "directionality-conscious" efforts to cultivate niches comprising three distinct angles. First, they highlight "socio-technical multiplicity", drawing attention to innovations as contingent outcomes "emerging out of a range of possible socio-technical configurations" – rather than predetermined or fixed – with distinct coalitions of actors pursuing competing configurations (Pel et al., 2020, p. 3). Second, they explain that divergent normative understandings of what is desirable are always at play: actors appraise technological options using divergent criteria.

Third, their study shows that directionality can also be studied through "process dynamics and junctures" to emphasise the "turns in the innovation trajectory". They also showed how a focus on "wider socio-technical and institutional constellations" instead of isolated technologies was necessary, as well as the need to attend to the varied efforts to synchronise governance that involve coordination and adjustment.

In turn, Andersson et al. (2021) proposed to examine the "development trajectories" of emerging technologies by studying the morphology of socio-technical configurations and their levels of diffusion over time. In the course of their project, they determined that multiple technical, social, and spatial parameters can be used to that effect. Meanwhile, Yang et al. (2021) investigated actors' strategies to shape directionality, exploring divergent paths in the scaling of solar photovoltaic energy in two Chinese provinces. Their study highlighted both incumbent and niche actors' institutional work and considered whether those actors are invested in maintaining or disrupting existing institutions. These contributions show that the directionality of innovation trajectories is contingent on the interplay between different innovation efforts, pre-existing institutional settings and policy efforts, and societal discourses surrounding innovations.

The third debate concerns the *directionality of innovation policy*, which provides different interpretations of how to proactively construe normative directions for socio-technical change to meet a given societal challenge. Two key rationales articulate how to attain transformative change, foregrounding different steering or coordination processes.

First, following closely from both Weber and Rohracher (2012) and Steward (2012), who emphasise transformations as resulting from socio-technical system change, the notion of transformative innovation policy has been gaining momentum. Transition-informed scholars have been demanding policies that accommodate new types of actor-networks willing to work on transformative change in an experimental setting (Schot & Steinmueller, 2018). According to the Transformative Innovation Policy Consortium (Schot et al., 2019), which popularised TIP, such policies can be discerned when they: (1) are directional (not assuming normative neutrality), (2) focus on addressing societal goals, (3) aim to have an impact at a socio-technical system level, (4) facilitate learning and reflexivity, (5) accept contestation rather than assume consensus, and (6) focus on inclusiveness (e.g., civil society and end users). Rather than a top-down assertion, this perspective foregrounds flexibility, experimentation, acceptance of failure, mobilisation, circulation, and the upscaling of local initiatives, which follows closely from the strategic niche management perspective (c.f. Ghosh et al., 2021).

Meanwhile, mission-oriented innovation policy, as popularised by Mazzucato (2018), places more emphasis on the role of the entrepreneurial state in co-creating and shaping markets and in providing directionality to economic growth. This approach centres attention on the formulation of missions that are ambitious and mobilising. Successful innovation policy, in this sense, "combines the need to set directions from above with the ability to enable bottom-up experimentation and learning", including a need for intermediary organisations to facilitate between science and industries (Robinson & Mazzucato, 2019, p. 3). Along these lines, Wanzenböck et al. (2020, p. 3) define MOIPs as:

> a directional policy that starts from the perspective of a societal problem and focuses on the formulation and implementation of a goal-oriented strategy by acknowledging the degree of wickedness of the underlying challenge, and the active role of policy in ensuring coordinated action and legitimacy of both problems and innovative solutions across multiple actors.

While academics have emphasised the contested and deliberative nature of such missions (see Janssen et al., 2021), the fast adoption of MOIP risks emboldening the government to the detriment of less acquiescent actors such as those in civil society.

In Mazzucato's pioneering work on missions, there is tension between using MOIPs to "actively set a direction of change" and acknowledging that bottom-up debates about these directions are necessary to ensure democratic legitimacy. The moon-shot era mission-oriented policies are oft-cited examples of "setting a clear direction for problems to be solved" (Mazzucato et al., 2020). The rationale is that by setting this direction, missions "stimulate the development of a range of different solutions to achieve the objective" (Mazzucato et al., 2020, p. 6). The concern is redressing the "blindness" of markets and highlighting the possibilities of an entrepreneurial state that plays a more proactive role in shaping or creating new markets to meet particular needs. Critically, directionality according to Mazzucato is a matter of *envisioning* and *picking*[1] directions (Mazzucato, 2015, p. 6), which demands "understanding how particular 'directions' and routes can be chosen and determining how to mobilise and manage activities that can lead to the achievement of dynamic social and technological challenges". These decisions are seen primarily as investment decisions – what to fund, at what level, with which conditions and which requirements. In MOIP, the government provides directions for change "which bottom-up solutions can then experiment around" (Mazzucato, 2015, p. 19). Janssen et al. (2021, p. 8) point toward the dilemma around how strong and narrow this direction should be, with the risk of "selecting out" necessary but unforeseen solutions or keeping an open-ended and exploratory mode. They argue that individual missions should be seen as a "negotiated outcome between different interests, concerns and imperatives", surrounded by politics and not only a technocratic selection exercise.

These rationales outline ideal-type policies that emphasise particular justifications, instruments, and ways to conceptualise policy coordination and impact. Whether and how these ideal types are translated, adapted, and appropriated in practice deserves more empirical attention (Janssen et al., 2021). In effect, the literature has started to show signs that these categories are not necessarily mutually exclusive and can overlap around "transformative missions". Hence, missions also feature in the TIP formulation, epitomised by Schot and Steinmueller (2018). Here, they are understood to be "implemented in an open and tentative way" and as an outcome rather than the starting point of policy. For instance, the Transformative Innovation Policy Consortium has emphasised experimental policy engagements through which a sense of directionality can be derived from multiple experimental interventions (Schot et al., 2019). As discussed above, this understanding can be traced back to the strategic niche management approach, in which experimentation in a protected space ("niche") is seen as necessary to allow for the development of an emerging socio-technical trajectory with the potential to challenge the prevalent socio-technical regime. This can be accomplished via constituting a nurturing process that includes the articulation of expectations and visions, broadening actor networks, and learning in multiple dimensions (e.g., Kemp et al., 1998; Smith & Raven, 2012). In an understanding of transformative innovation policy, these processes are combined

[1] The argument is framed as an opposition from the old criticism that governments should not be "picking winners", which perhaps explains the choice of words. However, the emphasis in Mazzucato's writing tends to be on assertive choosing, with less attention to how envisioning takes place.

with destabilising, unlocking, strengthening regime-niche interactions, and unlearning within established socio-technical systems (Ghosh et al., 2021).

Crucially, Kuhlmann and Rip (2018) mediate between these two positions, having argued for a nuanced debate on directionality and highlighting how multiple new actors are joining and therefore reconfiguring the public discussions around innovation in "new constellations of actors". They contrast the notion of vertical coordination (*orchestration*) by a central actor with that of horizontal coordination (*concertation*), whereby the basis for concerted action emerges by actors "mutually taking actions of others into account and pro-actively adjust[ing]" (Kuhlmann & Rip, 2018, p. 449). In this reading, open-ended missions are posited as a way forward, acknowledging the need for rekindling the contexts (socio-technical systems and innovation systems) of innovations and the engagement with new actor constellations (e.g., new sponsors and new performing actors).

Both TIP and MOIP also acknowledge intermediary actors have a role in brokering between new niches and established regimes (Kivimaa et al., 2019), contributing to achieving transformative outcomes (Ghosh et al., 2021). Systematic intermediaries may develop the capacities and scope for mediating the setting of new directionalities, operating at the interface of destabilising old unsustainable systems and creating or reconfiguring new ones.

Therefore, how "directionality" is construed is a pivotal issue in the debate between MOIP and TIP, with relevance for pressing policy agendas. What emerges from this discussion are two archetypal approaches to fostering directionality. The first, derived from transition studies and, in particular the strategic niche management (e.g., Pel et al., 2020; Yang et al., 2021) and the multi-level perspective informing TIP, emphasises engagement with multiple socio-technical configurations that can be appraised in diverse ways and are prone to contestation. According to this approach, policies can create the conditions for these processes and proactively establish experimental policy engagements through which different options are explored and developed. The latter approach, MOIP, emphasises the setting of missions to co-create and co-shape markets in order to influence investment decisions and gradually reorient the actions of private and public actors through a combination of top-down steering and bottom-up (entrepreneurial) experimentation.

However, in concrete cases, elements of both approaches may be present yet in conflict with each other as a process of identifying and framing a particular challenge or opportunity amenable to transformation unfolds. In the remainder of this chapter, we contrast these archetypal propositions with the case in Finland to explore how directionality emerged in the case of MaaS development. In analysing this case, we seek to identify which elements most influenced the emergence of new socio-technical arrangements and the overall direction of that change.

11.2.1 Research Approach and Method

The empirical case of MaaS in Finland draws on qualitative exploratory case study research with specific interest in the role of public governance in the unfolding of this innovation trajectory. The study comprises triangulated evidence from 22 research interviews conducted in 2017 and 2019, an event history workshop organised in 2017, and documentary evidence from government documents and news reports. The research was intentionally conducted in two phases, with time in between to monitor the development of directionality and the impacts of governance changes. The first phase explored the Finnish mobility transition more broadly, covering different niche innovations. MaaS arose in this context as an emerging niche with potentially large transformative effects on transport practices and routines. The second phase examined the development of MaaS specifically and how new directionality was constructed around it. The innovation history workshop, organised by the Transformative Innovation Policy Consortium (TIPC) in 2017, gave an initial timeline of events of the Finnish transport transition. The timeline was drafted by workshop organisers and feedback and elaboration were requested from its nine participants. Table 11.1 shows the empirical material the analysis is based on.

The interviews were recorded and transcribed verbatim. The first author created a 6,000-word case narrative based on the interview transcripts and supplementary written materials (policy documents, websites, news stories) that was sent to all the interviewees for fact-checking. In this paper, the 6,000-word case narrative and a more detailed coding of interview transcripts in NVivo were used to inductively identify topics relevant to the emergence of directionality around MaaS.

This process led to the identification of six elements in the Finnish MaaS development that can be seen as relevant to transformative and challenge-driven innovation policy by contributing to the process of building directionality. Some of these elements connect directly to sustainability transitions thinking and transformative innovation policy (e.g., Ghosh et al., 2021), emphasising the roles of *visioning* and *networking*. The other elements are more closely associated with changing innovation governance and include *policy experimentation*,

Table 11.1 Empirical material used in the case study

Phase I: May–June 2017	**Unstructured interviews & innovation history workshop for scoping analysis**
Interviews of 7 experts	Business: 2 (I1, I6) University: 1 (I2) Innovation agency: 2 (I3, I5) Research institute: 1 (I4) Ministry: 1 (I7)
Innovation history workshop involving 9 experts	Business: 2 University: 1 Innovation agency: 3 Research institute: 1 Ministry: 1 Transport agency: 1
Phase II: January–May 2019	**Semi-structured interviews for case study construction**
Interview of 17 experts in 16 interviews	Business: 3 (I16, I19, I20) Innovation intermediary: 2 (I15, I22) Ministry: 3 (I9, I10) Transport agency: 3 (I11, I12, I18) Network organisation: 3 (I13, I17, I21) Think tank: 1 (I8) Public transport provider: 1 (I14) Innovation agency: 1 (I23)

intermediation, and *institutional change*. The final element, *contestation and conflict*, emphasises that the processes shaping directionality are not just positive and smooth but experience tensions along the way. Analysing the case with these elements allows for a more discerning exploration of similarities and differences within TIP/MOIP conceptualisations.

11.3 MAAS DEVELOPMENT IN FINLAND

Finland is a Northern European country located beside the Baltic Sea, its land bordering Sweden, Norway, and Russia. The country has a large land area (338,000 square metres), but its 5.6 million inhabitants are primarily located in the south; 1.8 million people live in the Uusimaa Region, where the capital city Helsinki is located. Therefore, while the generally sparsely populated country creates more difficult conditions for efficient collective transport, new mobility services are relevant to southern areas and larger municipalities where the urban structure is fragmented.

Finland is the country where telecommunications company Nokia was founded, and it still possesses high levels of digital expertise. In the late 2000s, the declining business of the Nokia-centric telecommunications sector sparked interest in new industries to utilise similar skillsets. Along with private companies, the administrative sector of the Ministry of Transport and Communications (MTC) has been regarded internationally as a thought leader concerning future mobility and communication systems, new ways of working, and creating an "experimental culture" that stimulates innovation (Kivimaa & Rogge, 2022). MTC has collaborated with other ministries, such as the Ministry of Economic Affairs and Employment (in charge of innovation policy) and the Ministry of Finance. Political support received from different political party ministers has grounded the change trajectory at MTC. Due to this support, Finland has pioneered intelligent transport innovations such as MaaS and in 2017 was the first country globally to pass framework legislation for new transport services.

The development of MaaS in Finland during 2013–2019 shows the government's sustained and pioneering role. Since 2014, the Finnish public sector has taken a proactive role in conceptualising and facilitating the emergence of MaaS via a series of policy experiments and institutional changes (see Kivimaa & Rogge, 2022, for a detailed study of this interplay). The key actors included the MTC, the innovation agency Tekes (Business Finland, since 2017), and the public-private Intelligent Transport Network ITS Finland. The Innovation Fund Sitra also played a smaller role. MaaS Global, a company developed with support from public sector interventions, was the only fully operational MaaS operator company in Finland but went bankrupt in 2024. Another start-up, Kyyti, was a platform and solution provider for MaaS, operating until 2021. Banking, insurance, and telecommunications companies have also conducted MaaS-related activities. Kivimaa and Rogge (2022) drew on the same empirical material as this chapter and identified five phases in the early development of MaaS (see Table 11.2).

11.3.1 Visioning

Visioning has been very important for MaaS development, both in terms of broader visions for new fossil-free and intelligent transport systems and specific visions for operationalising MaaS. This topic was explicit in 13 out of 22 interviews. At first, the broader vision concerned

Table 11.2 Phases of MaaS development, summarising Kivimaa & Rogge (2022)

Phases in MaaS development	**Key events**
(1) Early developments towards an intelligent transport system (1995–2012)	In 2004, MTC launched Intelligent Transport Systems Finland (ITS Finland) as an open forum for collaboration between companies, researchers, and the public administration, becoming a non-profit association in 2006. Its former CEO, Sampo Hietanen, had an idea to compare transport to telecommunications, using an operator that connects to customers by offering different "packages". MTC's first Intelligent Transport Strategy (2009) identified a need to renew transport services, followed by a merger of previously separate agencies into a new multi-modal transport agency. A changing mindset towards transport is visible in 2011–2012 policy documents.
(2) Active vision formation for MaaS (2013–2014)	The New Transport Policy Club (NTPC) was set up for 2012–2014 to renew transport policy and respond to challenges created by emissions, automation, and servitisation. Sampo Hietanen presented his idea about MaaS in a club meeting in 2013, receiving interest from high-level civil servants. MTC launched the idea of MaaS in 2014 and established a Traffic Lab to support the creation of new intelligent transport services.
(3) New business model and market creation (2014–2015)	Innovation agency Tekes created, in collaboration with MTC, an experimental two-stage call to fund MaaS operators. ITS Finland queried companies about their interest in setting up an operator, with 24 actors (Siemens, Ericsson, Uber, HSL, TeliaSonera, Elisa, etc.) agreeing to fund the preparation of a business plan. Sampo Hietanen became the CEO of a new company, MaaS Global, the first to develop MaaS operator services.
(4) High-level policy support and entrepreneurial experiments (2015–2017)	A new government programme in 2015 highlighted digitalisation, new services, and experimental culture, specifically mentioning MaaS. In 2016, MTC executed an organisational change where the transport and communications departments merged to form new departments. This followed a shift towards more holistic thinking in the ministry. On the ground, entrepreneurial pilots were conducted by a telecommunications company, TeliaSonera, MaaS Global, and some research consortia.
(5) Execution of major institutional change and coupling of transport and innovation policy (2017–2019)	In 2017, new framework legislation, the Transport Service Act, was formed to proactively support new transport services, influenced by prior experimentation on MaaS and intelligent transport services. This was followed by further business developments. In 2019, previous transport and communication agencies were merged to form a new organisation, Traficom.

"intelligent transport systems" already in place during early development (phase 1), combining transport and communications policy domains. However, the more specific vision created for MaaS in 2013–2014 (phase 2) was gradually refined and made more concrete via experimental programmes taking place from 2014 onwards (phases 3–4).

The same ministry (MTC) is responsible for both transport and communications; its ministers and high-level civil servants put forward several visionary strategies such as the *Transport*

2030 strategy in 2007 and the first *Intelligent transport strategy* ever developed in 2009 during early development (phase 1). The former described climate change as a significant challenge and sought to halt the fragmentation of urban structures and the growth of private car use. The latter proposed to renew transport services in response to climate change, globalisation, and stringent finances. These were followed, in lockstep, by the *Transport Revolution programme* for "developing a new mind-set for urban and transport planning and policies and policy implementation" (MTC, 2011, p. 1) with contributions from the Finnish Innovation Fund Sitra, along with ministries. Notably, the programme stated that transport and logistics would be advanced as a service and include experimental projects. In 2012, a government high-level policy report to parliament mentioned a "serving transport system" (MTC, 2012).

Although these strategies were relatively ambiguous, they facilitated cooperation between distinct policy domains and the continuing search for directions on how to renew transport policy and rethink the entire transport system. This search was conducted with the help of the New Transport Policy Club, set up by the Minister of Transport and Communications, which ran from 2012–2014, involving active vision formation for MaaS (phase 2). In this Club, the outlines of the MaaS vision began taking shape. MaaS was put forward as a way to accomplish a vision of future mobility that is greenhouse gas emissions-free, affordable to everyone, uses monetary resources carefully, does not waste users' time, should not occupy urban space, and aims to end road fatalities (Suikkanen & Hietanen, 2017). It presented an idea of freedom of mobility based on public-private mobility solutions independent of car ownership. Sampo Hietanen, an entrepreneur who founded the first MaaS operator (MaaS Global), was the first to articulate this concept. He drew an analogy to the telecommunications sector: in telecoms, an operator connects customers and uses infrastructure, offering different "packages" of services to consumers (e.g., SMS, data, calls). Mobility could be sold as part of a bundle of services to meet diverse consumer needs.

In 2013, this idea was welcomed by the New Transport Policy Club. It formed the beginnings of a shared vision for MaaS. The MTC was particularly interested and assumed a decisive role in further development: "*Without the ministry, this would have probably never happened*" (Interview comment). Yet, the conceptualisation of MaaS was still vague at the time. Clarification was provided in May 2014 via an MSc thesis, "Mobility as a Service – A proposal for action for the public administration, Case Helsinki' (Heikkilä, 2014) that developed a vision for MaaS. The leading newspaper in Finland, *Helsingin Sanomat*, published a news article referring to the thesis and presented a vision of how the residents of Helsinki would not need to own cars in ten years' time. That thesis ended up being very important to MaaS research and gained international attention. In June 2014, the MTC explicitly launched the idea of mobility as a service, publishing a supporting animated explanation on YouTube.[2] These early instances emphasised that consumers should be able to choose between different packages comprising both transport and telecommunications services (internet connection and videoconferencing), with a single monthly subscription at a comparatively low cost. They underscored multiple perceived benefits for those individuals (more free time, less trouble with maintenance, freedom, more comfortable commuting, etc.) and the city (less pollution, less parking).

Institutionalising this vision took little time and happened during phase 4, "high-level policy support and entrepreneurial experimentation". By 2015, the government programme,

2 https://www.youtube.com/watch?v=ZQieTU7_5xo

Finland's most important policy document, openly mentioned MaaS: "Innovation and service platforms will be promoted in sectors where the public administration plays a role in terms of the functioning of the markets. Such sectors include mobility as a service..." (Prime Minister's Office, 2015, p. 28). On the business side, in 2016, Sampo Hietanen set up MaaS Global as the first MaaS operator company in Finland and engaged in experiments and pilot programmes in different European cities. The following year, during phase 5, the Ministry of Economic Affairs and Employment began coordinating the development of the National Growth Programme for the Transport Sector 2018–2022 that presented MaaS as one of six focus areas. The vision was also further developed into a MaaS programme coordinated by the national innovation agency Tekes, which grouped different actors together to discuss the business opportunities from MaaS. The innovation history workshop held in 2017 noted that the emergence of a shared vision among all key players was the most crucial success factor in the Finnish MaaS, which worked as a basis for updating legislation and reorganising the administration.

By 2019, the MaaS vision was widely circulated around the globe, drawing interest well beyond the Finnish context. In Europe, this was advanced by establishing an independent public-private organisation called MaaS Alliance in 2016, hosted by ERTICO ITS Europe and influenced by Finnish stakeholders from the public and private sectors, including MTC and MaaS Global. At the International Transport Forum 2019, MaaS was discussed as a way to redress the paradox of the limited capacity of mobility networks (as well as their congestion) with their excess capacity of individual vehicles and "the need to provide a compelling vision or reality" for the future of urban mobility. In such events, MaaS Global continued to be perceived as the pioneering operator, offering "a service that covers all your mobility needs".

11.3.2 Experimental Culture and Policy Experimentation

Enacting that vision took more than broad agreement among actors. Around the same time the MaaS vision was emerging, an experimental era in Finnish policymaking began. This covers phases 2–4. The experimental portfolio of transport (a series of stakeholder events) paved the way for the Traffic Lab during 2014–2015, while the government programme of 2015 also emphasised experimental culture:

> The administrative culture has perhaps traditionally been such that one needs to be 100 percent sure that it will work before something is done. Experiments sort of released that and gave a permission to fail. So, I think that has been significant. (Interview comment)

This culture of experimentation emphasised experiments conducted by public and private actors and attempts to test new ways of policymaking. An instance of the latter, the New Transport Policy Club, met two to four times per year for two-and-a-half years to renew transport policy and respond to the needs to reduce greenhouse gas emissions and digitalise the transport system. It addressed future direction as something explicitly different from the existing transport sector and departing from traditional transport infrastructure and cost-based policy. The Club was responsive to new pressures around environment, security, finance, and automation. The Club's creation was enabled by expanding from the "usual suspects" and choosing participants considered as visionary people from the public, private, and research sectors, within and beyond the transport sector. The Club also included the minister herself to create a shared vision and goals around the new direction for transport policy. The discussions covered how to renew regulations, reorganise transport services, and drive this change.

At the public-private interface, the Traffic Lab played a key role in stimulating experimentation. Established in 2014, it aimed to enable new market creation for intelligent transport services and included Tekes, the innovation agency. An experimental project was extended and institutionalised as a format for collaboration between the public sector, companies, and research:

> It was a new way of operating in collaboration so that there is facilitation, and the government authority comes closer to the company interface... to enable that from the perspective of mutual benefit, we can communicate, facilitate workshops and seminars and such that promote the advancement of this thing. (Interview comment)

The Lab developed into a forum for advancing a broad range of experiments to encourage innovation, create a better understanding of the future direction of transport, and to activate and grow the network around it. It acted as an umbrella for public-private co-experimentation to explore future markets and determine whether regulatory changes were required. Expectations for the Lab changed along the way, resulting in a variety of experiments on MaaS and beyond, which connected to climate change and digitalisation and also created insights on the future of transport.

Tekes ran an informal experimental innovation programme in 2014–2015, which became a venue for dialogue between different actors about the direction of MaaS and set a two-stage call to create MaaS operator business models. The first stage financed projects that co-designed real MaaS pilots with Tekes-funded pre-studies. Many companies, including MaaS Global, received funding that was regarded by our interviewees as instrumental. What made this experimental was the programme's quick setup, which bypassed Tekes' typical bureaucracy, and the intense dialogue with the Traffic Lab. Consequently, by the time a vision had started forming, the Finnish context already offered a favourable environment to support the iteration of that vision through concrete pilots (see Kivimaa & Rogge, 2022).

11.3.3 Networking

Networking was less explicit than visioning, but it has been deeply intertwined with the processes of visioning and experimentation. The development of MaaS built on a collaborative format in Finnish society, where different administrative domains frequently interact, both officially and unofficially, and dense collaboration networks partly pre-existed the vision. For example, the various transport strategies, during early development (phase 1), high-level policy support (phase 4), and institutional change (phase 5), were formulated in cooperation between the ministries in charge of transport and communications, the innovation agency Tekes/Business Finland, and the more arm's-length innovation foundation operating independently under the supervision of the Parliament, Sitra. Thus, the directionalities envisaged in public policy strategies are not those of isolated ministries, but are a shared interpretation of a broader range of actors. Both Tekes and Sitra can be regarded as intermediary actors who brought together views of different private sector actors (see 11.3.4). Prior to this, in 2004, intermediary actor ITS Finland was set up as an extensive public-private network of actors and companies interested in developing future intelligent transport systems, emphasising the close interactions between public administration and companies in the transport sector (phase 1). These networks were later relevant to operationalising MaaS, when ITS Finland offered companies the opportunity to contribute by investing only 5,000 euros to the setup of MaaS Global; 24 companies expressed an interest in investing (phase 5).

Further networking took place in the experimental arenas (phases 2–3): an experiment portfolio for transport was enlarged and institutionalised as the Traffic Lab with an expanding network; the New Transport Policy Club was instituted as a network of visionaries and change agents; and the Tekes MaaS programme collected different actors to shape the vision for MaaS.

Internationally, networking was furthered by the MaaS Alliance, established in 2016 (phase 4). Its founding members included the Finnish Ministry of Transport, European Road Transport Telematics Implementation Coordination (ERTICO – ITS Europe), the European Automobile Manufacturer's Association (ACEA), the World Transport Organisation (IRU), Fédération Internationale de l'Automobile (FIA), Cubic, MaaS Global, RACC Transport, Siemens, Xerox, and eight other representatives from public authorities, transport companies, and organisations. The broader international potential of MaaS was viewed as dependent upon collaboration across country boundaries:

> How MaaS succeeds or fails, it requires a much broader collaboration or larger number of actors than perhaps the traditional transport sector supply chain has had. This collaboration needs to be based on completely new kinds of values and a new kind of division of risk and profit than traditionally. MaaS Alliance facilitates this public-private collaboration. (Interview comment)

The networking activities on multiple fora have demonstrated active participation of actors in both public and private sectors in formulating future directionality via means of governance. What remains less clear is whether and how ordinary citizens or transport users have been included in these processes.

11.3.4 Intermediaries

In Finland, key intermediaries pre-existed the MaaS "mission" and were well-embedded in the aforementioned networks and the processes of building directionality, and were also backed by powerful actors. ITS Finland, for instance, was set up in 2004 as a non-profit public-private network to intermediate systemic change towards intelligent transport, with backing from the ministry (phase 1). Its former CEO championed MaaS, supporting the initial idea and later setting up MaaS Global (phases 2–5). ITS Finland has acted as an open forum for the collaboration of companies, researchers, and public administration to make intelligent transport options visible and facilitate networking between actors, thereby creating opportunities to explore distinct normative directions. It has been described as a "primus motor" in advancing the direction of MaaS.

An innovation agency was also an instrumental intermediary. Tekes (Business Finland since 2018) provided vital innovation funding and an experimental innovation programme for the development of MaaS (phases 2–4) and has also been involved in transport sector policy processes. Tekes showed flexibility when intermediating the vision and network-building processes for MaaS, compiling information, and allocating funding. Innovation funding implemented via Business Finland has been essential for MaaS-ecosystem development.

> All central companies have in one way or another... received at least seed money or initial funding... from Business Finland. (Interview comment)

In this process, the agency's role was not limited to funding but also included contributing to constructing directionality. It stimulated companies to consider possible market disruptions in a rather informal capacity. That role was evident in a joint campaign with the Ministry of Transport and Communications that hosted round tables on different topics linked to MaaS, inviting companies to discuss their vision, the needs of municipalities and public transport providers, service design, and logistics.

> Tekes organised, so that they then had a big role, even as a thought leader, to make sure it advanced. (Interview comment)

In addition, a specific intermediary, Traffic Lab, was set up as an umbrella organisation for public-private experimentation. Initially a temporary experiment, it was later institutionalised as a permanent part of transport governance. The Innovation Fund Sitra, otherwise identified as a critical systemic intermediary in many transition processes in Finland (Kivimaa, 2014; Kivimaa et al., 2020), played a surprisingly small role. In 2018, Sitra began funding a project that aims to develop a digital platform in three pilot areas to advance rural MaaS services with the idea of creating a national technology platform for rural MaaS. None of the above intermediary actors in this process played a purely instrumental "process intermediary" role (cf. Kivimaa et al., 2019) but contributed to shaping directionality with the pre-existing agency and systemic orientation they possessed.

11.3.5 Institutional Change

In our case, institutions tie in strongly with cross-sectoral collaboration and directionality. The relatively speedy and transformative development would not have been possible without combining innovation policy instruments (a targeted MaaS funding programme during 2015 and a growth motor funding instrument since 2018) with systemic changes in the administrative sector of the MTC (see Kivimaa & Rogge, 2022).

The administrative sector has undergone two sets of supportive organisational changes that embedded the principle of multimodality and created openings for exploring synergies between transport and communications. First, in 2010 during phase 1, different transport modes were merged under one agency. The aviation, road, rail, and marine transport agencies were fused into a new multimodal transport infrastructure agency, the Transport Agency and the Transport Safety Agency (Trafi). There was a shift of attention in the ministry from simply building roads to solving problems and developing intelligent transport systems. Second, during 2016–2019 (phases 4–5), transport and communications were integrated under the same departments and agencies, focussing on both in combination. In 2016, the ministry's organisational structure was changed from distinct departments for transport and communications to four new departments: ministerial governance, networks, data, and services – the latter three comprising all transport modes and communications.

In addition to changes in the organisational structure of the administration, the emerging policy mix removed a key barrier to MaaS expansion when a change in the framework law of the transport sector, the Transport Service Act of 2017, required transport operators to open their ticket interfaces to third parties. This has been labelled as groundbreaking and unique transport sector legislation globally. The objective was to proactively open space for new transport service innovations and the envisioned transformation.

These institutional changes have created openings to explore new directionalities (e.g., multimodality), break down sectoral siloes, and gather the institutional backing for concretising the vision. In other words, major institutional change contributed to directionality by breaking down or destabilising past directions and confirming the new emerging direction.

11.3.6 Contestation and Conflict

While there have been many synergies and common expectations in building directionality, this process has not been without tensions and contestation. In particular, the Transport Service Act and the changes it provided during phase 5 were strongly opposed by taxi drivers, who lost their local or regional monopoly positions.

In addition, certain public transport providers have resisted allowing MaaS operators entry into their ticket sales systems, viewing MaaS services as competitors to traditional public transport offerings. For example, in March to April 2019, the Transport and Communications Agency Traficom delivered a notice of penalty payment to the Helsinki metropolitan regional transport provider HSL for non-compliance with the Transport Service Act, which was later resolved. The inflexibility of public transport ticket pricing and concessions is also an issue in this context. More broadly, advancing Finnish export via MaaS is not an objective on the priority list of public transport providers.

Not everyone in the ministry has been on board with the MaaS thinking, and some have had a different perception of what MaaS means. In addition, different ministries have disagreed on which policy instruments are most suitable for advancing MaaS.

Directionality and vision-building have thus far mainly happened on the national level and by national-level actors, often with international markets in sight. However, the local city level, where day-to-day mobility gets realised, has been excluded. It also became clear that contestation increased when new directionalities began to influence actors' day-to-day activities, despite seeming harmless in earlier vision formation and experimental phases.

11.4 DISCUSSION AND CONCLUSIONS

This chapter studied the Finnish MaaS development, exploring how directionality emerged, what processes contributed to its formation, and the role of different actors.

The MaaS case shows how initially broad and unspecific visions, such as those of intelligent and fossil-fuel-free transport systems, have given rise to more specific and mobilising visions around Mobility-as-a-Service as a model for addressing multiple transformative pressures. MaaS can be used to address climate change, digitalisation, automation, and social equality (though the latter has been less addressed). In collectively pursuing these broad yet underspecified visions and focussing efforts on MaaS, the shared visions have gradually formed from the interaction between public and private actors, as well as from different formal and semiformal networks around selected policy processes, intermediary organisations, and experimental programmes – rather than a specific and clear mission setting (as indicated by MOIP). The inclusion of actors has been "deep" via mutual trust and "broad" in terms of a plurality of perspectives (cf. Ghosh et al. 2021), but at the same time partly excluded more vulnerable groups or voices. Both local governments and users seem to have been rather poorly represented in these processes, showing a lack of "inclusivity" in setting the new direction. The development has followed the critical dimensions of sustainability transitions – a backbone for TIP – in setting directionality: forming visions and expectations, networking and learning via experimentation. However, contrary to the traditionally described niche-led transition processes, this process has involved experimentation and networking at the policy level and niche-regime interactions from the very start. The latter indicates that a proactive stance and steering action from governance actors, akin to some of the ideas around MOIP, is beneficial for accelerating socio-technical transitions, but the processes of setting directionality involved a more nuanced and partly bottom-up development, akin to TIP. In essence, in this real-world process, we saw elements of both TIP and MOIP conceptualisations.

The case study, which had key policy events on the national scale but experienced drivers and barriers from local and EU scales, illustrated six elements that we noted contribute to the formation of directionality: visioning, experimentation, networking, intermediation, institutional change, and contestation and conflicts (Table 11.3). While the first five relate mainly to advances, the sixth indicates that conflicts and tensions influence the setting in which directionality occurs. In this case, directionality occurred in a later phase when institutional changes around the new directionality began to take place. Thus, we see directionality as a process contingent on different sub-processes and events, but not determined by a singular one.

The six elements we identified help understand the emergence of directionality, providing insights into implementing TIP and MOIP. The elements match sustainability transitions thinking but also bring forth a decisive governance aspect around policy experimentation and institutional change. The case highlights how gradual institutional change, intermediary organisations, and cross-sectoral collaboration can create conditions to sustain the process of "concertation" long enough for a transformative mission to emerge. The MaaS case also showcased bureaucratic hindrances of established innovation policy structures, emphasising a technology and business-oriented transformation framework over an environmental framework. MaaS also faces challenges posed by changing landscapes, such as the impact of COVID-19 on the demand for collective transport and a growing political appetite for the electrification of private vehicles, which diverts attention and resources away from new mobility services. Recent developments may temporarily compromise the "durability" and momentum of the MaaS mission.

The MaaS case stands out as a counterintuitive example of how transformative missions may emerge from a collaborative and experimental culture of working in relatively open-ended and exploratory ways and that is enabled by public governance but complemented by emergent bottom-up processes – though even then the process is not entirely without conflict and contestation. Such contestation may also be argued as an integral part of mission-building processes, as Janssen et al. (2023) describes missions as outcomes of contestation and negotiation on strategic arenas. What differs in our case from the typical understanding in sustainability transition studies is that the attention of public policymakers was proactively interested in this development early on and so high-level agreement was achieved before much of the experimentation took place. This resembles the MOIP notion of a proactive, direction-setting public sector. However, the MaaS development process has been less goal-centric and more open for joint deliberation of directionality than MOIP has been characterised as. Although

Table 11.3 Contribution of six elements on emergence and evolution of directionality

Element	Contribution to directionality for MaaS
Visioning	Directionality involves the creation of shared future visions (Weber & Rohracher, 2012). Several visionary strategies by MTC highlighted the need to renew transport services to address climate change, globalisation, and the growth of private car use. One strategy was titled "Transport Revolution" and aimed for a new mindset for transport planning/policy. Shared visions created in a policy committee (New Transport Policy Club) put together visionary thinkers across domains to rethink a new vision for transport policy – resulting in MaaS as one solution. The MaaS vision, which contributed a new direction, was endorsed by media and the government programme, resulting in a new European network established to promote MaaS.
Experimentation	Directionality is also about collective actions and proactive stimulation and innovation activities to contribute to a specific direction of technological development (Weber & Rohracher, 2012; Yap & Truffer, 2019). The policy experiments covered in this case study addressed the future direction of transport systems/policy *explicitly* as something that deviates from the status quo and needs major restructuring. They used experimental policy/governance designs to create insights that will contribute to this future direction.
Networking	Directionality resulted from interaction among actors (Yap & Truffer, 2019). Networking was deeply interconnected with the processes of visioning and experimentation. Visions were created in policy strategies and experimental engagement fora where different actors came together to form shared interpretations of the future direction of mobility systems and policy. However, the actors tended to include representation from citizens and the civil society, comprised mainly of national and local public governance, private enterprises, and intermediary organisations.
Intermediation	Intermediaries acted as brokers in actor networks for creating shared directionality by turning visions and expectations into material actions (Kivimaa et al., 2019). Here, pre-existing key intermediaries played a role in shaping directionality by possessing agency in systemic changes and/or established systems. They orchestrated (ITS Finland, Business Finland, Traffic Lab) or were embedded in (Sitra, Business Finland) networks and experiments advancing the MaaS development by assembling knowledge and encouraging disruptions to established directions.
Institutional change	Transitions studies posit that institutional rules create the deep structure of socio-technical regimes; changing institutions are crucial for new directionalities. In this case, major institutional changes contributed to directionality by breaking or destabilising past directions and confirming the new emerging direction.
Contestation & conflicts	System incumbents may attempt to hinder broader change or adopt niche innovations to optimise the existing system without changing the overall directionality (Kanger et al., 2020). In this case, opposition became stronger when the new directionality was confirmed by institutional change, meaning that it disrupted the status quo of past day-to-day operations.

recent research also explores missions as open-ended boundary objects (Janssen et al., 2023) our case highlights a relatively long "gestation period" of a vision, which was preceded by multiple organisational and institutional shifts.

The results also point to the dilemma between setting a strong and narrow or a more open-ended and broad direction. In the MaaS case, a broad direction was evident in the early attempts to formulate a vision for intelligent transport, which has been relatively successful. Over time, this direction became more specific through the experimentation and partial consensus developed in multiple fora, but this has left the other, less "packaged" ways of (re)organising MaaS to the side and has potentially contributed to tensions with public transport providers. The process has been enabled by horizontal coordination or concertation (Kuhlmann & Rip, 2018), and various actors from public and private domains have conducted joint actions without a central conductor. This coordination included many national-level actors but could have engaged city and regional actors more actively. In essence, more specific missions can be seen as outcomes, instead of starting points, for transformative policy processes and the coordination they engender, in line with the TIP notion of transformative missions.

Regarding actors, the MaaS case certainly shows how directionality is shaped in the interaction of and within the range of processes undertaken by different public and private actors, who need to be engaged early on in the process. Transforming innovation policies alters the setting for "traditional" innovation policy actors, changing approaches within innovation agencies and necessitating much more horizontal coordination with different sector-specific actors. Specialised innovation agencies and intermediaries may be needed for missions, as the existing innovation agencies may lack the substantial knowledge base required. However, existing innovation agencies have many pre-existing capacities to utilise, such as established networks of actors and administrative processes to set up necessary innovation programmes or funding instruments quickly. A key question is whether existing agencies, such as Business Finland, allow the flexibility (as in this case) to experiment with new funding formats and collaboration tools to support emergent transition and innovation processes in the context of TIP and MOIP.

Finally, critical points for implementing directionality around TIP and MOIP stand out from this analysis, which we present here as potential hypotheses for future research.

First, niche- and bottom-up-led mechanisms proposed by TIP and policy- or regime-led mechanisms advocated by MOIP are interconnected and co-evolving. Shared directionality for transformation is unlikely unless changes in one are complemented with changes in the other.

Second, long-term institutional change is vital to enable mission formulation and specification. It is often also an interim outcome of the unfolding process that is crucial for "confirming" the direction emerging in partly bottom-up or niche-led processes.

Third, cross-sector collaboration, with innovation agencies' involvement and public-private collaboration, can (and perhaps should) occur via both formal and informal routes and fora, gradually consolidating around an insipid "mission" rather than the opposite way around.

Fourth, a range of experimentation supported by institutional change, cross-sectoral collaboration, and intermediary actors' coordination is necessary to foster learning and reflexivity in formulating, implementing, and recalibrating the direction for change. Thus, the "question of directionality" is not simply one of "picking a direction" but instead a more complex process through which legitimate and mobilising visions are shaped.

In sum, we see benefits in both MOIP and TIP that support the formulation of directionalities in "real-life" settings. While TIP is more suitable for exploration and accelerated learning, MOIP enables the streamlining of actors around a particular vision or mission. In essence, the two policy approaches are complementary and can be used in different phases of setting directionality. Rather than rigidly following particular policy prescriptions, the reflexive process of engaging, mobilising, and shaping a joint effort appears to be more effective.

To conclude, combining thinking around TIP and MOIP and tracing how actors put them into practice helps make sense of these unfolding transitions and builds up the capacity for navigating transformations and strategically intervening in those processes. Innovation policy is being asked to play new roles in complex societal and environmental challenges. Developing the ability to navigate cross-sectoral, experimental, collaborative, and participatory processes is more pertinent than ever. More empirical cases such as the one examined are needed for a better understanding of how transformations can be induced. The six elements deriving from our analysis can be used as a framework to study directionality setting in different contexts.

ACKNOWLEDGEMENTS

Empirical data was collected in the Transformative Innovation Policy Consortium and the Smart Energy Transition project, 2015–2021, funded by the Strategic Research Council of Finland. Paula Kivimaa acknowledges funding from the Academy of Finland (decision number 322667) for writing this chapter. Laur Kanger carried out the first scoping interviews in 2017, and the workshop was carried out jointly by Laur Kanger and Paula Kivimaa. We would like to thank Laur for the use of this material. We also thank Mireille Matt and Matthias Weber for excellent constructive comments on earlier versions of this chapter.

REFERENCES

Andersson, J., Hellsmark, H., & Sandén, B. (2021). The outcomes of directionality: Towards a morphology of sociotechnical systems. *Environmental Innovation and Societal Transitions, 40*, 108–131. https://doi.org/10.1016/J.EIST.2021.06.008

Azar, C., & Sandén, B. A. (2011). The elusive quest for technology-neutral policies. *Environmental Innovation and Societal Transitions, 1*(1), 135–139. https://doi.org/10.1016/J.EIST.2011.03.003

Edler, J., & Boon, W. P. (2018). "The next generation of innovation policy: Directionality and the role of demand-oriented instruments" – Introduction to the special section. *Science and Public Policy, 45*(4), 433–434. https://doi.org/10.1093/SCIPOL/SCY026

Fuenfschilling, L., & Binz, C. (2018). Global socio-technical regimes. *Research Policy, 47*(4), 735–749. https://doi.org/10.1016/J.RESPOL.2018.02.003

Geels, F. W. (2004). From sectoral systems of innovation to socio-technical systems: Insights about dynamics and change from sociology and institutional theory. *Research Policy, 33*(6–7), 897–920. https://doi.org/10.1016/j.respol.2004.01.015

Geels, F. W. (2005). *Technological transitions and system innovations: A co-evolutionary and socio-technical analysis*. Edward Elgar. https://doi.org/10.4337/9781845424596

Geels, F. W., & Schot, J. (2007). Typology of sociotechnical transition pathways. *Research Policy, 36*(3), 399–417. https://doi.org/10.1016/j.respol.2007.01.003

Ghosh, B., Kivimaa, P., Ramirez, M., Schot, J., & Torrens, J. (2021). Transformative outcomes: Assessing and reorienting experimentation with transformative innovation policy. *Science and Public Policy, 48*(5), 739–756. https://doi.org/10.1093/scipol/scab045

Heikkilä, S. (2014). *Mobility as a Service – A* proposal *for* action *for the* public administration *Case Helsinki.* Aalto University. www.aalto.fi

Janssen, M. J., Torrens, J., Wesseling, J. H., & Wanzenböck, I. (2021). The promises and premises of mission-oriented innovation policy – A reflection and ways forward. *Science and Public Policy, 48*(3), 438–444. https://doi.org/10.1093/SCIPOL/SCAA072

Janssen, M.J., Wesseling, J., Torrens, J., Weber, K.M, Penna, C., & Klerkx, L. (2023). Missions as boundary objects for transformative change: Understanding coordination across policy, research, and stakeholder communities. *Science and Public Policy, 50*(3), 398–415. https://doi.org/10.1093/scipol/scac080

Kanger, L., Sovacool, B., & Noorkõiv, M. (2020). Six policy intervention points for sustainability transitions: A conceptual framework and a systematic literature review. *Research Policy, 49*(7), 104072. https://doi.org/10.1016/j.respol.2020.104072

Kemp, R., Schot, J., & Hoogma, R. (1998). Regime shifts to sustainability through processes of niche formation: The approach of strategic niche management. *Technology Analysis & Strategic Management, 10*(2), 175–198. https://doi.org/10.1080/09537329808524310

Kivimaa, P. (2014). Government-affiliated intermediary organisations as actors in system-level transitions. *Research Policy, 43*(8), 1370–1380. https://doi.org/10.1016/j.respol.2014.02.007

Kivimaa, P., Boon, W., Hyysalo, S., & Klerkx, L. (2019). Towards a typology of intermediaries in sustainability transitions: A systematic review and a research agenda. *Research Policy, 48*(4), 1062–1075. https://doi.org/10.1016/j.respol.2018.10.006

Kivimaa, P., & Kern, F. (2016). Creative destruction or mere niche support? Innovation policy mixes for sustainability transitions. *Research Policy, 45*(1), 205–217. https://doi.org/10.1016/j.respol.2015.09.008

Kivimaa, P., Primmer, E., & Lukkarinen, J. (2020). Intermediating policy for transitions towards net-zero energy buildings. *Environmental Innovation and Societal Transitions, 36*, 418–432. https://doi.org/10.1016/j.eist.2020.01.007

Kivimaa, P., & Rogge, K. S. (2022). Interplay of policy experimentation and institutional change in sustainability transitions: The case of mobility as a service in Finland. *Research Policy, 51*(1), 104412. https://doi.org/10.1016/J.RESPOL.2021.104412

Kuhlmann, S., & Rip, A. (2018). Next-generation innovation policy and Grand Challenges. *Science and Public Policy, 45*(4), 448–454. https://doi.org/10.1093/SCIPOL/SCY011

Mazzucato, M. (2015). *Building the entrepreneurial state: A new framework for envisioning and evaluating a mission-oriented public sector.* Levy Economics Institute of Bard College Working Paper, 824.

Mazzucato, M. (2018). Mission-oriented innovation policies: Challenges and opportunities. *Industrial and Corporate Change, 27*(5), 803–815. https://doi.org/10.1093/icc/dty034

Mazzucato, M., Kattel, R., & Ryan-Collins, J. (2020). Challenge-driven innovation policy: Towards a new policy toolkit. *Journal of Industry, Competition and Trade, 20*(2), 421–437. https://doi.org/10.1007/S10842–019–00329-W/TABLES/1

MTC. (2011). *Transport revolution programme.* Publications 28/2011. Ministry of Transport and Communications, Helsinki. http://urn.fi/URN:ISBN:978–952–243–253–7

MTC. (2012). *Competitiveness and well-being through responsible transport.* Government Report on Transport Policy submitted to the Parliament of Finland 2012. http://julkaisut.valtioneuvosto.fi/handle/10024/78027

North, D. (1991). Institutions. *Journal of Economic Perspectives, 5*, 97–112. DOI: 10.1257/jep.5.1.97

Pel, B., Raven, R., & van Est, R. (2020). Transitions governance with a sense of direction: Synchronization challenges in the case of the Dutch "Driverless Car" transition. *Technological Forecasting and Social Change, 160*, 120244. https://doi.org/10.1016/j.techfore.2020.120244

Prime Minister's Office (2015). *Programme of Prime Minister Sipilä's government.* http://urn.fi/URN:ISBN:978-952-287-185-5

Robinson, D. K. R., & Mazzucato, M. (2019). The evolution of mission-oriented policies: Exploring changing market creating policies in the US and European space sector. *Research Policy, 48*(4), 936–948. https://doi.org/10.1016/j.respol.2018.10.005

Schot, J., Kivimaa, P., & Torrens, J. (2019). *Transforming experimentation: Experimental policy engagements and their transformative outcomes*.TIP research report, March 2019. https://www.tipconsortium.net/wp-content/uploads/2019/07/Transforming-Experimentation.pdf

Schot, J., & Steinmueller, W. E. (2018). Three frames for innovation policy: R&D, systems of innovation and transformative change. *Research Policy*, *47*(9), 1554–1567. https://doi.org/10.1016/j.respol.2018.08.011

Smith, A., & Raven, R. (2012). What is protective space? Reconsidering niches in transitions to sustainability. *Research Policy*, *41*(6), 1025–1036. https://doi.org/10.1016/j.respol.2011.12.012

Steward, F. (2012). Transformative innovation policy to meet the challenge of climate change: Sociotechnical networks aligned with consumption and end-use as new transition arenas for a low-carbon society or green economy. *Technology Analysis & Strategic Management*, *24*(4), 331–343. https://www.tandfonline.com/doi/full/10.1080/09537325.2012.663959

Stirling, A. (2009). *Direction, distribution and diversity! Pluralising progress in innovation, sustainability and development*. STEPs Working Paper, 32.

Stirling, A., Smith, A., Leach, M., Pellizzoni, L., Levidow, L., Hendriks, C., Owens, S., von Tunzelmann, N., & Wynne, B. (2007). "Opening up" and "Closing down": Power, participation, and pluralism in the social appraisal of technology. *Science, Technology, and Human Values*, *33*(2), 262–294. https://doi.org/10.1177/0162243907311265

Streeck, W., & Thelen, K. (2005). *Introduction*. In W. Streeck & K. Thelen (Eds.), *Beyond continuity: Institutional change in advanced political economies* (pp. 1–39). Oxford University Press.

Suikkanen, H., & Hietanen, S. (2017). What if mobility became the symbol of the future once again? *Demos Helsinki Blog*. https://www.demoshelsinki.fi/en/2017/11/30/mobility-became-symbol-future/

Wanzenböck, I., Wesseling, J. H., Frenken, K., Hekkert, M. P., & Weber, K. M. (2020). A framework for mission-oriented innovation policy: Alternative pathways through the problem-solution space. *Science and Public Policy*, *47*(4), 474–489. https://doi.org/10.1093/scipol/scaa027

Weber, K. M., & Rohracher, H. (2012). Legitimizing research, technology and innovation policies for transformative change: Combining insights from innovation systems and multi-level perspective in a comprehensive "failures" framework. *Research Policy*, *41*(6), 1037–1047. https://doi.org/10.1016/j.respol.2011.10.015

Yang, K., Schot, J., & Truffer, B. (2021). Shaping the directionality of sustainability transitions: The diverging development patterns of solar photovoltaics in two Chinese provinces. *Regional Studies*, *56*(5), 751–769. https://doi.org/10.1080/00343404.2021.1903412

Yap, X-S., & Truffer, B. (2019). Shaping selection environments for industrial catch-up and sustainability transitions: A systemic perspective on endogenizing windows of opportunity. *Research Policy*, *48*(4), 1030–1047. https://doi.org/10.1016/j.respol.2018.10.002

12. The transformative potential of Horizon Europe R&I Partnerships in the areas of climate, energy, and mobility

Kathy Berger, Christoph Brodnik and Michael Dinges

12.1 INTRODUCTION

More than 50 per cent of global greenhouse gas (GHG) emissions are attributable to the energy and mobility sector (IPCC, 2014), which makes it one of the key sectors for climate neutrality. As such, it falls within the policy priorities set out by the European Union in the European Green Deal, which is a set of proposals aimed at transforming Europe and making it the first climate-neutral continent by 2050. Investment in "cutting-edge research and innovation" represents a vital part of the Green Deal. The importance of research and innovation in realising a transformation of this nature is reflected in the resurging debate on "missions". Such missions highlight the role that R&I and the associated policy instruments have to play in addressing persistent and wicked societal challenges (Wanzenböck et al., 2020) and point to a new role for R&I initiatives (Glennie, 2019). This policy agenda is the result of a shift in logic whereby innovation should no longer be pursued for the sake of economic growth alone but with the aim of addressing major societal challenges through a system-oriented R&I policy approach that delivers transformative change in coordination with other policy domains such as sector-specific, market, and regulatory policies.

Within the domain of European R&I policies, the European Partnerships in Horizon Europe are a policy instrument that has a transformative ambition (European Commission, 2021a). The European Partnerships bring together private and/or public (policy) partners to address global challenges in four key[1] areas through coordinated R&I initiatives. In the context of Horizon Europe, these partnerships, which have strong roots in previous framework programmes (cf. Weber et al., 2022), serve as a key tool for coordinating, combining, and augmenting resources from the European Commission, EU member states, and research and industry stakeholders, with a view to accelerating and more effectively tackling societal challenges such as climate change or those outlined in the Sustainable Development Goals (SDGs). For this purpose, the European Partnerships develop Strategic Research and Innovation Agendas (SRIA) and proposals for implementation, with the aim of not only developing a shared vision but also ensuring a commitment of EU member states' national resources and investments so as to achieve the jointly agreed objectives (cf. Dinges et al., 2020).

[1] The areas are (1) health; (2) digital industry and space; (3) climate, energy, and mobility; and (4) food, bioeconomy, natural resources, agriculture, and environment. In addition to these areas, the European Institute of Innovation and Technology (EIT)-Knowledge and Innovation Communities, the EU Open Science Cloud, and the Innovative SME Initiative are part of the partnership community.

The European Commission (2019) claims that European Partnerships are a powerful tool for providing solutions to global challenges and driving transformation towards achieving the Sustainable Development Goals.[2] However, it is less clear whether the partnerships can live up to this ambition and to what extent they address important aspects of transformative mission-oriented STI policy. To shed light on this issue, this chapter analyses the European Partnerships' Theory of Change (ToC)[3] through a transformative innovation policy lens.

Our empirical research questions are as follows:

1. What is the transformation potential of Horizon Europe partnerships in the cluster "climate, energy, and mobility"?
2. How can the concept of Transformative Outcomes be used to characterise the transformational potential of Horizon Europe partnerships?
3. What can we learn from this form of characterisation?

Empirically, the analysis conducted in this chapter is based on a review of the recently published European Partnership proposals from Horizon Europe.[4] We have conducted a qualitative text analysis that considers how these partnerships frame transformation by examining their Theory of Change, i.e., the planned objectives, activities, outputs, outcomes, and impacts. We then consider and highlight transformative elements from the perspective of sustainable transition theory. Conceptually, this chapter combines the programme theory (Theory of Change) approach (Rogers, 2008) with the Multi-Level Perspective on system transformation and the related concept of Transformative Outcomes (Ghosh et al., 2021). We consider these two conceptual building blocks to be useful for the following reasons:

1. The programme theory approach connects the underlying rationales of a programme (a specific challenge to be addressed) with an overall roadmap that outlines how specific activities are expected to produce immediate outputs in relation to outcomes/intermediate impacts and eventually result in the realisation of the objectives.[5] This helps us understand the partnerships' rationales, how they relate their outputs to specific outcomes and impacts, and how they believe they contribute to transformation.
2. The concept of transformative outcomes builds on the literature on sustainability transitions and the multi-level perspective. It also explicates a set of leverage points for innovation programmes and projects that aim to achieve transformative outcomes that can be understood as mechanisms for turning activities into transformative results (Penna et al., 2023; Ghosh et al., 2021; Schot et al., 2019). This concept helps us to better understand

---

[2] European Commission (2019), Orientations towards the first Strategic Plan for Horizon Europe. https://ec.europa.eu/info/sites/default/files/research_and_innovation/strategy_on_research_and_innovation/documents/ec_rtd_orientations-he-strategic-plan_122019.pdf

[3] We use the concept of Theory of Change to describe what is understood as "intervention logic" in the partnership proposals.

[4] The proposals can be found at the following link under the section "Partnership candidates and contact details": https://research-and-innovation.ec.europa.eu/funding/funding-opportunities/funding-programmes-and-open-calls/horizon-europe/european-partnerships-horizon-europe_en#the-european-partnership-portfolio

[5] See Amanatidou et al., 2016, in the context of European partnerships.

the extent to which partnerships contribute to system transformation in terms of facilitating the growth and establishment of innovative practices and the destabilisation of entrenched and unsustainable ones.

In this way, we will unpack the rationale of a "European" STI policy instrument by contrasting it with the transformative outcomes framework. The main focus of this chapter is to analyse and discuss the transformative potential of the partnerships by exploring how they address forces for increasing their transformative potential.

As a starting point, the next section will provide a brief introduction and overview of the existing literature on programme theory and Transformative Outcomes. Section 12.3 describes the Horizon Europe partnerships as an instrument and addresses the rationale and structure of the programme. Section 12.4 presents the data and methodology, while Section 12.5 addresses the first research question linking the partnerships' theory of change to each of the Transformative Outcomes. Finally, Section 12.6 discusses the findings and draws conclusions, ultimately placing the partnerships in a broader context and reflecting on the concept of Transformative Outcomes.

12.2 THEORIES OF CHANGE (TOCS) FOR TRANSFORMATION

In European innovation policymaking, developing a ToC is a common method for understanding the strategy and steps that a project, programme, or any other type of intervention is taking to achieve its desired impact (cf. European Commission, 2021b). A ToC thereby focusses on the process of change that occurs on the road to a particular goal, as a result of the cumulative effects of project or programme inputs, activities, outputs, and outcomes. It also specifies the impact that a project or programme aims to achieve and the assumptions that underpin the change process (Brown, 2020).

A ToC is often based on a conceptual model that expresses certain assumptions about how change unfolds and how impact can be achieved. Such models can form either an explicit or implicit part of a ToC. Moreover, a model can be theoretically derived or influenced by particular experiences or worldviews. For the purposes of this chapter, we will adopt a theoretically based perspective to analyse the ToCs of the partnerships through the lens of sustainability transition research. More specifically, we will draw on the sustainability transition research field and its concept of transformative outcomes (Ghosh et al., 2021), which we will discuss in more detail below.

The sustainability transition research field depicts transformative change as a fundamental shift in the logic and functioning of the socio-technical system (Markard et al., 2012). Based on the established multi-level perspective (MLP) theory, sustainability transition literature (Loorbach, 2014) proposes a model whereby transformative change occurs as an evolutionary interaction between stable *regimes* and emerging *niche* innovation alternatives under exogenous *landscape* pressure (Geels, 2011). The interactions between all three levels can have different impacts on socio-technical systems and unfold in different transition pathways (Geels & Schot, 2007).

The governance of sustainability transitions is difficult due to the complexity of systemic problems and the evolutionary nature of the change process, which cannot be directly controlled by any single set of actors (Kivimaa et al., 2019). Therefore, in order to facilitate transformative change, it is essential to carefully craft processes that will enable the creation and expansion of niches, in addition to unlocking incumbent and unsustainable regimes. These are the transformative processes, each of which is composed of a set of 12 Transformative Outcomes (Ghosh et al., 2021). The Transformative Outcomes are a burgeoning framework in Transformative Innovation Policy (TIP) research and practice. They are particularly valuable for reflecting on ongoing initiatives by STI actors (Molas-Gallart et al., 2021; Alvial Palavicino et al., 2022). The key principles guiding the formulation of these outcomes are their potential to: include and navigate tensions between a broad range of actors across multiple system dimensions; enable shifts to deeper rules (values and norms) in systems; and allow transformations to unfold at different geographical scales. The Transformative Outcomes are based on these principles and offer ways to reflect on and direct innovation initiatives along more transformative lines (Ghosh et al., 2021).

The twelve Transformative Outcomes are grouped into the following three overarching processes of transformative change: building and nurturing niches; expanding and mainstreaming niches; and opening up and unlocking regimes.

Building and nurturing niches: This refers to the process by which novel ideas and experiments grow, learn, and mature. It includes the following Transformative Outcomes, each of which can be broad (extending their scope) or deep (quality and directionality):

1. *Shielding*: This is the active and passive protection of the niche.
2. *Learning:* This refers particularly to second-order learning that challenges the beliefs and assumptions of actors and the formalisation of this knowledge.
3. *Networking*: This refers to the construction, extension, and strengthening of networks of actors, along with nurturing the variety of ideas and deepening the opportunities for resource mobilisation.
4. *Navigating expectations:* This is achieved by creating shared visions about the future and making them more robust, which legitimises the actions of actors in the niche.

Expanding and mainstreaming niches: This process refers to moving experiments beyond the niche and expanding their scope and scale. It is associated with the following four Transformative Outcomes:

1. *Upscaling* the niche beyond its existing boundaries by involving more users.
2. *Replicating* all or part of the niche in a different context, such as a new group or setting.
3. *Circulating* ideas through texts, products, and learning, and transferring ideas from one niche to another.
4. *Institutionalising* norms, rules, or expectations associated with the niche by developing shared narratives, definitions, standards, and interpretations from niche experimentations.

Opening up and unlocking regimes: This process refers to destabilising the practices, beliefs, and rules that characterise a dominant regime and thus creating space for alternatives to emerge. It includes the following Transformative Outcomes:

1. *De-aligning and destabilising:* Weakening the reproduction of the regime and creating opportunities for alternatives.

2. *Unlearning and deep learning in regimes:* Helping regime actors reassess the routines, rules, and expectations of the regime in comparison to new rules and routines associated with solving sustainability challenges.
3. *Strengthening regime-niche interactions:* Creating stronger links between regime and niche actors, thus facilitating exchange and improving resource flows between them.
4. *Changing perceptions of landscape pressures:* Challenging individual and collective perceptions about wider socio-political developments.

12.3 HORIZON EUROPE R&I PARTNERSHIPS

In Horizon Europe, "European Partnership" means an initiative where the EU, together with private and/or public partners (such as industry companies, research organisations, bodies with a public service mission, or civil society organisations), commit to jointly supporting the development and implementation of a programme of R&I activities (European Council, 2019). As a policy approach, the European Partnerships provide the main mechanism for linking European R&I policy with the EU member state level and developing close synergies with national and regional programmes. The aim of the partnerships is to bring together a broad range of innovation actors to work towards a common goal – and turn research into socioeconomic results.

The European Partnerships are framed around three main rationales that have developed over time: (1) alignment of the fragmented European Research Area; (2) strengthening the competitiveness of the EU; and (3) tackling societal challenges and contributing to transformation (see Weber et al., 2022). Rationales 1 and 2 have a long history that reaches back to the original concept of the European Research Area in 2000 and the Lisbon Agenda.[6] Rationale 1 led to the design of various public-public partnership instruments, including, in particular, the European Union Area Networks (ERA-Nets), as a means of more effectively coordinating European and national R&I policymaking. Rationale 2 led to the design of Joint Technology Initiatives (JTIs) in the Seventh Framework Programme (FP7) as the primary instrument for strategically pooling public and private investments in R&I. Tackling societal challenges only became a key ambition as a response to the Lund Declaration in 2009, which stressed the relevance of focussing research on the grand challenges of our times and aligning European and national strategies and instruments. As a response, 10 Joint Programming Initiatives (JPIs) have been formed. These initiatives are intended to build partnerships based on common visions and Strategic Research and Innovation Agendas (SRIA) in order to address major societal challenges. Thereafter, the intention is for JPIs to be implemented through a variety of measures including joint calls for proposals, fast-track activities, knowledge hubs, and task forces, among others.

In the programming process for Horizon Europe, the European Commission and EU member states aimed to introduce a more strategic and impact-driven approach to partnerships. The development of the partnership approach has also gone through a major reform. As expressed by the New ERA Communication (European Commission, 2021a), European Partnerships in

6 https://www.consilium.europa.eu/uedocs/cms_data/docs/pressdata/en/ec/00100-r1.en0.htm

key policy areas[7] are seen as an essential means of achieving the following goals: prioritising investments and reforms in research and innovation so as to complete the green and digital transitions; supporting Europe's economic recovery after the COVID-19 pandemic; and meeting a target of spending 5 per cent of national public funding on joint research and development programmes and European partnerships by 2030.

Three types of European Partnership instruments have been developed for Horizon Europe: co-funded partnerships, co-programmed partnerships, and institutionalised partnerships (see ERA-LEARN):

1. *Co-funded partnerships* have a (more or less centralised) blend of EU and national public and/or other R&I funding sources (e.g., ERA-Nets, European Joint Programme (EJP), Future and Engineering Technologies (FET) Flagships model in H2020). Co-funded partnerships involve EU countries, and the core members of their consortia are research funding bodies and other public authorities. They develop joint research agendas and implement them with joint calls for proposals and other collective activities.
2. *Co-programmed partnerships* are arrangements that involve the EU, member states, or associated countries, and/or other stakeholders. These partnerships are based on memoranda of understanding or contractual arrangements with partners (i.e., the Contractual Public Private Partnerships (cPPP) model in H2020). Co-programmed partnerships develop joint research agendas which are planned and implemented within the Horizon Europe work programmes.
3. *Institutionalised partnerships (based on Art. 185 or 187 of the Treaty on the Functioning of the EU (TFEU) and the EIT regulation for knowledge and innovation communities (KICs))* are implemented only when other parts of the Horizon Europe programme, including other forms of European Partnerships (co-funded or co-programmed), cannot achieve the objectives or generate the necessary expected impact. The preparation of such institutionalised partnerships requires new EU legislation and the setting up of specific legal structures (funding bodies) based on Art. 185 or 187 of the TFEU. As such, all institutionalised partnerships must be justified with an impact assessment prior to the preparation of the legislative proposals.

EIT KICs are also institutionalised partnerships. EIT KICs aim to address skills shortages and are established under Horizon 2020. Higher education institutions, research organisations, companies with a focus on entrepreneurs and start-ups, and other stakeholders such as end users are key partners in EIT KICs.

The Horizon Europe partnerships are spread across the clusters of Horizon Europe. The health cluster underlines the importance of the UN SDGs that call for universal health coverage for all persons of all ages by 2030, so as to leave no one behind and end preventable deaths. The digital and industry cluster follows on from the Industrial Leadership Pillar of Horizon 2020 and targets "Leadership in Enabling and Industrial Technologies (LEIT)" and particularly Key Enabling Technologies (KETs: nanotechnologies, advanced materials, advanced manufacturing and processing, and biotechnology). The food and natural resources cluster aims to reduce environmental degradation by halting and reversing the decline of biodiversity

[7] The communication mentions health, accessibility, digitalisation, industrial competitiveness, climate, energy, mobility, natural resources, and food systems.

on land, in inland waters, and at sea and improving management of natural resources through transformative changes of the economy and society in both urban and rural areas.

The climate, energy, and mobility cluster, which is the main cluster of interest in the following analysis, aims to make the energy and transport sectors more environmentally and climate-friendly, efficient, safe, and competitive. It focusses on various areas of intervention in the process. European Partnership candidates in this cluster deal with topics such as the transformation of Europe's rail system, air traffic management, clean aviation, clean hydrogen, built environments for people, zero-emission road transport, connected cooperative and automated driving, zero-emission waterborne transport, the industrial battery value chain, driving urban transitions to a sustainable future (DUT), and the clean energy transition.

12.4 METHODOLOGY AND APPROACH

12.4.1 Method and Sample

To analyse the transformative potential of the partnerships, we conducted a qualitative text analysis of the partnership proposals submitted to the European Commission in the course of the application process. Of the total 48 partnerships currently in preparation, we focused on all 11 partnerships in the area of climate, energy, and mobility (see Table 12.1). This focus was chosen due to its relevance to the ongoing research and debates on climate neutrality, as well as the availability of proposals at the time at which our analysis took place,[8] the feasibility of deeper analysis and discussion within this study, and the fact that all three different categories of partnerships are represented in this cluster.

12.4.2 Coding and Research Approach

To systematically analyse, contrast, and compare the relevant text segments within the proposals, we used MAXQDA.[9] Our research process (see Figure 12.1) has been divided into the following four different phases:

1. A literature review examining the theoretical foundations and understanding (see the section on theories of change for transformation).
2. The development of the analytical categories based on the transformative outcome (TO).
3. The development of a coding structure based on the partnership proposals, their ToC, and reasons for transition.
4. A mapping of the partnerships' ToC to specific transformative outcomes that form the basis for our analysis (see the Results section). The arrows in the figure, which move in a circular structure, indicate an iterative process and a back-and-forth between the different phases.

---

[8] Our initial screening of all proposals revealed that, of the 48 partnerships/topics planned by the European Commission, only 29 partnerships had been submitted by the time of the analysis in April 2021. The climate, energy, and mobility cluster was the only cluster for which all proposals were available at that time.

[9] MAXQDA Version 2020 from VERBI GmbH, for further info see: https://www.maxqda.com/new-maxqda-2020

Table 12.1 *Overview of partnership types, names, and visions in the climate, energy, and mobility cluster*

Type of partnership	Name of European partnership	Transformative vision
Institutionalised partnerships	Shift2Rail (Transforming Europe's rail system)	"The proposed Rail European Partnership will deliver technological and operational solutions that respond to a new Concept of Operations for Rail, through a System of Systems service-oriented approach, in which an integrated rail system, including freight, urban, suburban, regional and mainline, will realise its full potential."
Institutionalised partnerships	Single European Sky Air Traffic Management (ATM)	"Based on the identified problems, the overall objective of the proposed European Partnership is to develop and validate ATM technological solutions that support the achievement of the Digital European Sky making the European airspace the most efficient and environmentally friendly sky to fly in the world and support the competitiveness and recovery of the European aviation sector in a post-COVID crisis Europe."
Institutionalised partnerships	Clean Sky 2 (Clean Aviation)	"The grand challenge facing the aviation sector in this and the next decades is to develop and introduce safe, reliable and affordable low- to zero emissions air transport for citizens and ensure Europe's industrial leadership is maintained and strengthened throughout the transition to a climate-neutral Europe by 2050."
Institutionalised partnerships	Fuel Cells and Hydrogen (FCH 2)	"Our common vision for the partnership is that it would accelerate the development of clean hydrogen technologies to the point where market and policy mechanisms can take over."
Co-programmed partnerships	People-centric sustainable built environment (Built4People)	"The B4P partnership's common vision is [a] high quality, low carbon and highly energy and resource efficient built environment driving the societal and economic transition towards sustainability. It is reached through a user-centric, holistic, life cycle-based and circular approach to innovation that will contribute to achieving EU sustainability targets (including in related sectors such as energy and transport)."

Type of partnership	Name of European partnership	Transformative vision
Co-programmed partnerships	Towards zero-emission road transport (2ZERO)	"A climate-neutral and clean road transport system is possible by 2050. The 2Zero partnership will be strongly committed to [achieving] the use of 100% renewable energy carriers in road transport and share[s] the vision of Europe becoming the first climate-neutral continent by 2050. A high introduction rate of zero tailpipe emission powertrains (BEV and FCEV) is one of the key elements for climate neutrality within the transport sector."
Co-programmed partnerships	Connected and Automated Mobility (CCAM)	"The vision for the next 10 years is to make Europe a world leader in the development and deployment of connected and automated mobility and logistics services and systems."
Co-programmed partnerships	Zero-emission waterborne transport	"For Europe to lead and accelerate the transformation of the global waterborne transport sector into a zero-emission mode of transport which has eliminated all harmful environmental emissions (including GHG emissions, and air and water pollutants, including (underwater) noise) through innovative ship technologies and operations which underpin European growth and employment."
Co-programmed partnerships	Towards a competitive European industrial battery value chain	"Batteries Partnership vision is to establish by 2030 in Europe [a] best in the world innovation ecosystem to boost a competitive, sustainable and circular European battery value chain and to drive the transformation towards a carbon-neutral society."
Co-funded partnerships	Driving urban transitions to a sustainable future (DUT)	"The DUT partnership steps up the game to tackle urban challenges. We enable local authorities and municipalities, business and citizens to make global strategies into local action. We develop the skills and tools to make urban change happen and boost the urgently needed urban transformations."
Co-funded partnerships	Clean Energy Transition	"The Clean Energy Transition Partnership is a transformative Research and Innovation Programme across Europe boosting and accelerating energy transition in all its dimensions. It enables energy transition from regional to national and global level, co-transformed by industry, public organisations, research and citizens organisations to make Europe [the] frontrunner in energy innovation and implementation by becoming the first climate-neutral continent."

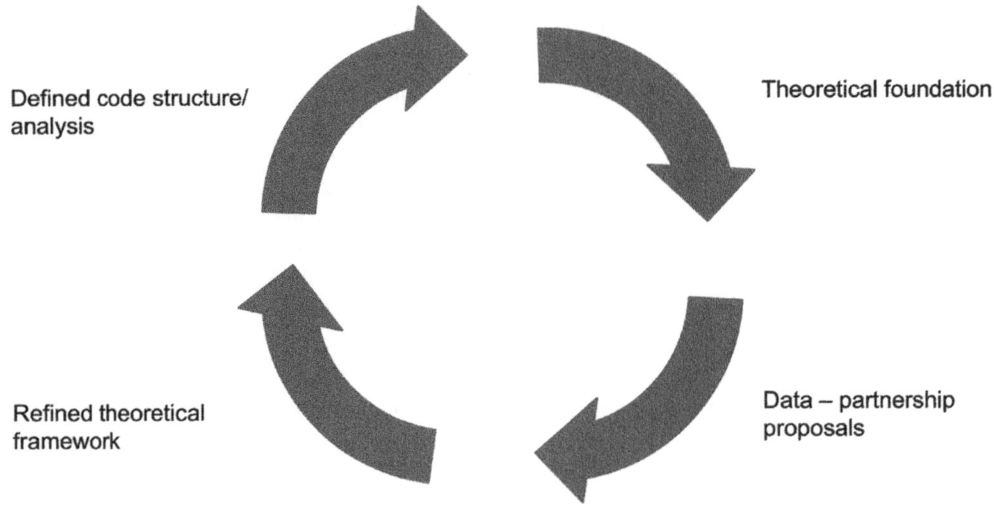

Source: Authors' own elaboration.

Figure 12.1 Overview of our (iterative) research process

The methodology for analysing the partnerships was guided by transition theory and influenced by the partnerships' ToC. In addition, we considered the overarching structures in which the partnerships are embedded, such as the overall policy objectives of the European Green Deal and the governance of the partnerships, including the actors involved. The analysis of the partnership proposals was guided by the development of two separate code systems: one for the transformative outcomes and another for mapping the partnerships' ToC, including the expected inputs, activities, outputs, outcomes, and impacts, as a self-determined basis for evaluations and the path to impact. These two separate code systems allowed us to quantitatively measure the relationships that emerged and determine whether specific steps within the ToC could be attributed to a particular TO. After the initial analysis of the proposals and mapping of the partnerships' ToC, we analysed the extent to which elements of the ToC can be attributed to the TO. To do this, we looked at the code segments in each TO code and developed subcodes that described the particular TO (see Table 12.2, column 3, "Analytical categories for detecting evidence of Transformative Outcomes"). These subcodes helped us to understand how each partnership's ToC addresses the TO. In general, we devised approximately three subcodes for each TO: these codes described a series of actions taken by the partnership. This code structure was then used to count the codes (see Figure 12.2), establish relationships between the code, the TO, and the ToC, and perform the qualitative text analysis of each sub-code.

Table 12.2 Operationalisation of Transformative Outcomes (analytical categories) and the supporting evidence in the partnership's ToC

Macro-process	Transformative Outcomes	Analytical categories for detecting evidence of Transformative Outcomes	Evidence in the partnership's ToC
Building and nurturing niches	*Shielding*	Creating positive conditions for innovation in niches but with a view to addressing real challenges as expressed by need owners and the application of a systemic/ holistic approach	Joint calls for R&I proposals Conduct R&I projects/ activities, labs, and experiments
Building and nurturing niches	*Learning*	Connect existing knowledge and activities Monitor activities Build capacities of various stakeholders	Building on experiences from previous partnerships/ networks Screening connections to other partnerships and identification of trends, needs, and lessons learned Defining, implementing (activity, objective, impact), and monitoring
Building and nurturing niches	*Networking*	Mobilising and engaging various innovation actors (including across niches), engaging actors beyond innovation actors (connecting niche- and regime-level actors) Building (international) cooperation Engaging countries across Europe	"Progressing together saves time" Creating a platform for dialogue and building transnational communities for specific topics

Macro-process	Transformative Outcomes	Analytical categories for detecting evidence of Transformative Outcomes	Evidence in the partnership's ToC
Building and nurturing niches	*Navigating expectations*	Alignment of strategies, programmes, and budgets across European Countries and across different stakeholder types	Roadmapping: SRIA (Strategic Research and Innovation Agenda) – creation of common understanding (agreed target visions, limits of current programme, identification of additional R&I needs) Jointly identifying issues and areas and activities where leverage effects of multilateral work can be achieved, fostering co-design of approaches and solutions Joint financial commitment (aligning public and private investments)
Expanding and mainstreaming niches	*Upscaling*	Make knowledge widely available and accessible Stimulate demand	"Bridge the gap between research and implementation" through combined cross-European demonstration Fostering high public acceptance Deploying technology and/or infrastructure by means such as market uptake, mass-scale production, integration into value chains, and modernisation of infrastructure
Expanding and mainstreaming niches	*Circulating*	Derived strategic knowledge Target-group-specific dissemination of strategic knowledge (policymakers, practitioners, problem owners) for exploitation Share knowledge with the scientific community to stimulate further R&I	Building a community of practice Exchange both vertically (EU member states, regions, and mission innovations) and horizontally (with other EU-programmes)

Macro-process	Transformative Outcomes	Analytical categories for detecting evidence of Transformative Outcomes	Evidence in the partnership's ToC
Expanding and mainstreaming niches	*Replicating*	Mobilise respective replication partners across Europe Facilitate adaptation in products, processes, and practices according to different country/market conditions across Europe	Deployment and implementation across regions and countries are "possible"; "replication orientation" – design actions in such a way that replication is key from the beginning Measurable expected outcome is to increase the number of regions implementing and deploying, e.g., a technology
Expanding and mainstreaming niches	*Institutionalising*	Provide evidence for implementation so as to improve policies, laws, regulations, and legislation with respect to systemic change Provide validated solutions/ evidence for practitioners	Inspiration for policies and support Strengthen the impact of R&I to develop, support, and implement EU policies
Regime destabilisation	De-aligning and destabilising regimes; unlearning and deep learning in regimes; niche-regime interactions; changing perceptions of landscape pressures	-	No evidence found

12.5 RESULTS

Our analysis of the partnership proposals shows that many partnerships are increasingly transformation-oriented in their overall ambitions and that they have adopted a narrative that actively communicates their desire to contribute to a green transition. This can be seen in their overall vision (see Table 12.1), which indicates that the partnerships contribute to a "climate-neutral continent by 2050" (mentioned in all partnerships), "achieving a zero-pollution ambition for a toxin-free environment" (Battery Value Chain), "global technological leadership in green ship technologies and solutions" (Waterborne Transport), and "reach net-zero GHG emissions" (mentioned in all proposals).

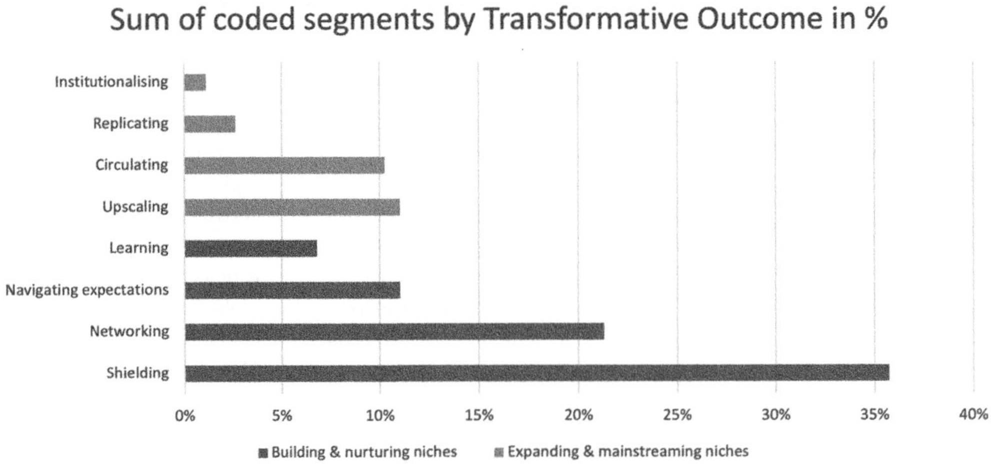

Source: Authors' own elaboration.

Figure 12.2 Sum of coded segments by Transformative Outcome in %

The proposals at the core of this agenda resemble a complex, transformative ToC that translates a joint vision of a partnership into operational mechanisms for delivering different types of impact. The aim is to contribute to tackling the grand challenges expressed by overarching UN and EU goals, such as the SDGs or the European Green Deal. These contributions range from the scientific and societal to the economic and technological, covering all of the major impact pathways identified by Horizon Europe.

On the one hand, the partnerships recognise that close collaboration among the various stakeholders in the quadruple helix is an important factor in addressing "systemic" and "cross-sectoral" issues (mentioned in all partnerships), by providing a "bridge between R&D activities and markets" (Built4People) and "scientific input for informed regulation" (2Zero). They recognise that technology often evolves in a complex, interdependent sociotechnical environment that must be viewed and assessed holistically in order to acquire a deeper understanding of the entire value chain that aims for transformative *systemic* change. Thus, the partnerships clearly indicate that they intend to go *beyond*[10] the preparation and implementation of joint calls for proposals by pointing to their far-reaching impact *beyond* time, space, sectors, and actors. On the other hand, it can be recognised that the partnerships have some specific factors to consider in their overall setup. For example, in the area of transport, the different transport modalities (airborne, water, road) are strictly separated, which of course has an impact on the actors involved in these partnerships. In other partnerships, the central focus is on key technological developments and emerging application environments, as in the hydrogen partnership and the battery value chain partnership, for example. In the remaining partnerships, overarching topics influenced by major societal challenges, such as the energy

[10] Across all the proposals, the word "beyond" was mentioned a total of 171 times in this context.

transition, the urban transition, and the process of creating a people-centric and sustainable built environment, form the main rationale for the partnership.

Although the partnerships clearly highlight their understanding of the complexity in which R&D is embedded, they simultaneously avoid mentioning the scope of their impact. On the one hand, they formulate it implicitly by using vague terms, such as "*support* replication, mainstreaming and scaling up" (DUT), "*inspire* policies for implementation" (CETP), and "replication *orientation* – design actions in a way that replication potential is key from the very beginning" (CETP). On the other hand, the partnership for batteries explicitly emphasises that although the intended outcomes were set, they were not reliant solely on achieving them through the partnership itself as "many of them are not under the control of the partnership". This language suggests that the partnerships are finding it difficult to clearly define and articulate their contributions to important mechanisms that go beyond the realm of traditional STI policy measures and are critical for generating wider impacts.

The analysis of the European Partnership proposals shows (see Figure 12.2) that about 75 per cent of the activities, outputs, and objectives described can be attributed to the macro-process of *building and nurturing niches.* Figure 12.2 shows the sum of coded segments by transformative outcome relative to all coded segments. Within the macro-process of *building and nurturing niches* (shown in Figure 12.2 in light grey), more than 35 per cent is attributed to *shielding.* This implies that the partnerships focus mainly on technology development and alignment of R&I activities across Europe rather than technology deployment at the regime level. Therefore, Figure 12.2 gives an indication of "how much" the partnership is doing in "which" areas. However, it would not be useful to interpret the data any further, as it is only a quantification of the qualitative coding information.[11]

Table 12.2 shows the qualitative results of our textual analysis. The first column provides information on our analytical categories as they relate to each TO. The second column summarises the evidence that we found on the partnerships' ToCs. Further detail on what the partnerships intend to do (distinguishing between activity, output, and outcome) in relation to the Transformative Outcomes is provided in the following sections.

12.5.1 Shielding

A number of different shielding activities can be detected within the partnerships. In the context of the grand challenges, the partnerships aim to address and justify the purpose of their existence by creating positive conditions for niche innovation by means of joint calls for proposals for research and innovation. Through these joint calls, they claim to better align and bridge science and practice by taking a holistic approach that engages multiple stakeholders, such as need owners and policymakers. As such, the funded projects are often described as an "experimental space'" (see also von Wirth et al., 2019) for testing new solutions, demonstrating their potential, and deploying them.

Projects are funded through joint calls in order to cover all technology-readiness levels (TRLs), engage multiple stakeholders, and demonstrate new solutions and their readiness

[11] Although we have tried to be as consistent as possible in coding, this information may be subject to "coding errors", e.g., overcoding of some segments due to splitting paragraphs and/or sentences into multiple coded segments. Nevertheless, the results show a trend within the data and give a first impression of what the partnerships claim to do.

for the market (e.g., cost-effectiveness and readiness of the service and/or product, including functionality, comfort, convenience, accessibility, health, and well-being for the end-user). This is primarily applicable for partnerships that aim to develop one or more technologies to make the sector itself less CO_2 intensive, e.g., aviation, ATM, waterborne transport, hydrogen, road transport, and buildings. While the aviation, air transport, road transport, automated mobility, and hydrogen partnerships focus for the most part on speeding up market readiness and upscaling specific technologies, partnerships in fields such as water and rail emphasise the need to go beyond technological solutions and focus on means of multimodal transport. We will return to this point in the discussion and challenge the arguments that they set out in their proposals by closely examining their governance structures and setup.

Other partnerships, such as "Driving Urban Transition" or the "Clean Energy Transition" aim to transform a whole system through both new technologies and the development of integrated urban planning approaches, such as "tools, methods, systems, and business models appropriate for transforming key areas".

12.5.2 Learning

Within this "experimental space", the partnerships provide opportunities for niche actors to participate in a process of reflexive learning, which is also referred to in the literature as "higher-order learning" or "learning by doing" (Ghosh et al., 2021). Within the partnerships, the partnership and the actors included in it learn through the coordinated R&I calls and their outcomes and the partnerships' activities, thus allowing them to build on existing knowledge from previous networks and partnerships by identifying trends, needs, and lessons learned. In addition, monitoring activities facilitate further learning from ongoing experiences. As the partnerships are still in their early stages, the structure of the monitoring processes and the content they will focus on is still unclear in most cases. Nevertheless, it can be noted that monitoring processes have been established at the programme level in order to assess project results and their impacts (e.g., Clean Aviation establishes indicators and metrics for climate (global warming) impacts, local air quality, noise emissions, and competitiveness) and to learn from the partnerships' funding practices. For example, the 2Zero Partnership argues that its governance structure and connections to other partnerships, national delegates, and the European Commission allow it to share and learn from ongoing experiences.

12.5.3 Networking

In addition, it is important that networks are established between a wide range of actors because diverse backgrounds, skills, and knowledge (Ghosh et al., 2021) are needed to develop breakthrough innovations. Most of the partnership activities involve the mobilisation and engagement of different innovation actors. In most partnerships, the key actors addressed are academia, government, and industry, which represents the usual configuration of R&I networks in recent decades, with limited scope for connecting niche- and regime-level actors. Society is often portrayed merely as the ultimate beneficiary of innovation, with limited attention given to actively involving civil society in the innovation process itself. Instead of engaging society as active participants, innovation in the Partnerships tends to be driven by the goal of improving and enhancing society's well-being from a distance. In that sense, society is often described as the end user of a product, whose current needs must be understood and addressed. In some partnerships, e.g., "Driving Urban Transition", stakeholder networking actions have been developed at the partnership network level. The urban transition stakeholder engagement platform "Agora" has the aim of "creating a space for multi-stakeholder exchange, discussions and co-creation for urban actors with diverse backgrounds (researchers, practitioners, public administrators, entrepreneurs, social innovators, etc.) to discuss current topics and priorities and identifying the most pressing urban challenges of today and the future" (JPI Urban Europe, 2020).

In addition to mobilising relevant actors, the partnerships focus on activities that create interconnections with other partnerships and other relevant areas. The European Rail Partnership and the Zero-Emission Road Transport Partnership aim to create synergies with the Clean Hydrogen and Battery partnerships in particular. The "Driving Urban Transition" (DUT) partnership sees the benefit of linking with the "Clean Energy Transition Partnership" (CETP) as, to a certain extent, they both face a common challenge. Furthermore, the partnerships aim to provide strategic access to various EU and international actors, broaden overall participation, and enable capacity building in the developing countries of the EU. In other words, they aim to lower the barriers to entry into joint European STI activities and overcome fragmentation.

12.5.4 Navigating Expectations

The process of navigating expectations within the partnerships is about creating a shared understanding of what the partnership is intended to accomplish, who it is intended for, who should be involved, and how it will be carried out. This is done by aligning strategies, programmes, and budgets among partners. Roadmapping and the development of a Strategic Research and Innovation Agenda (SRIA) are the key activities within this Transformative Outcome. The SRIA establishes joint visions, boundaries of current programmes, and additional R&I needs. This joint vision is then broken down into operational mechanisms that are intended to have different types of impact. Thus, developing the SRIA is an activity in which different actors and communities define not only the nature of the vision achieved but also new approaches for solving societal challenges. At the same time, some SRIAs also delineate potential areas of friction and conflicts of interest among different types of stakeholders. As a result, we have argued elsewhere that the SRIAs "constitute an integral part of new, transformation-oriented R&I policies" (Dinges et al., 2020) because the SRIA creation process in the partnerships involved consultations among the partnerships' stakeholders and joint sensemaking regarding the key challenges to be addressed. While some partnerships' development processes for SRIAs have been driven primarily by industrial associations, others have been more open and relied upon public consultations in addition to topic- and stakeholder-specific co-creation processes.

12.5.5 Upscaling

Upscaling is understood quite differently by the partnerships. However, the partnerships generally emphasise that "bridging the gap between research and implementation" (ATM) through clustered Europe-wide demonstration is key to upscaling. In particular, the demonstration of solutions (e.g., technologies, tools, methods, and approaches) in real-life environments should help to upscale some of their results through field testing and accompanying "confidence building". In summary, they define upscaling in three ways: (1) technological upscaling (e.g., mass production); (2) infrastructural upscaling (e.g., built environment [Built4People], ATM, and railway infrastructure [Shift2Rail]); and (3) promoting public (user) acceptance and deployment.

12.5.6 Circulating

Another key transformative outcome that the partnerships aim to achieve is the group-specific dissemination of strategic knowledge to different stakeholder groups for exploitation. Strategic knowledge derived from and created in the projects is disseminated to other niche spaces so that they can learn from it and use it in different spatial and cultural contexts (Ghosh et al., 2021). The partnerships also mention the importance of communicating the results of the calls both vertically to EU member states and regions and horizontally to other EU programmes where mutual benefit can be achieved.

12.5.7 Replicating and Institutionalising

As already highlighted in Figure 12.2, the partnerships make little reference to *how* they contribute to the replication and institutionalisation of R&D, but they clearly see their function as acting as a knowledge hub that develops around different actors with the aim of *inspiring* replication and institutionalisation. However, it is less clear which entry points they use for this. It can be noted that what the partnerships intend to deliver in terms of TO replication covers quite a broad bandwidth. For example, the 2Zero partnership aims for a clear replication outcome, with a "number of cities and regions implementing and deploying eMaaS (electric mobility as a service) solutions", whereas others, such as CETP, seem to see their contribution as "only" relating to the area of testing, demonstrating, and validating implementable solutions with a replication *orientation* in mind.

There is even less evidence for the TO of institutionalisation. Four out of eleven partnerships (DUT, CETP, Built4People, 2Zero) state that increased R&D should lead to actions beyond joint calls for a European Knowledge Community that will stimulate actions to provide a "bridge between research and policy" (DUT). For example, Built4People highlights that it wants to contribute to transnational policies, such as the EU Sustainable Finance Action Plan and the EU Green Taxonomy for Buildings by providing national implementation plans.

12.6 DISCUSSION AND CONCLUSION

Having elaborated on the partnership's ToC and their contributions to the specific TOs, this section places the results in the context of a broader discussion that answers our research question on the transformational potential of the partnerships.

The European Partnerships pool R&I funding across Europe, linking and combining competencies and capacities in a transnational way and enabling pan-European discourse. Through their inter- and transdisciplinary approach, they aim to tackle the wicked societal challenges by targeting system transformation, in order to contribute to the overarching goal of all partnerships: the decarbonisation of Europe by 2050. Their governance structure allows for an orchestration of intellectual and financial resources which collectively define the way forward and build on a shared understanding of what the central societal goal is. Together, they provide a platform where shared learning is possible and where results are aggregated, evaluated, and assessed against an overarching vision and mission that centres on transformative change. Given these characteristics, partnerships can help to overcome the fragmentation of R&I efforts by allowing a larger critical mass of innovation actors to be pooled. As such, they represent some of the core characteristics of transformative mission-oriented STI policy: the engagement of a broad range of actors (as is evident in the TO of networking and learning), a focus on system transformation, clearly defined goals and timeframes (as is evident from the ToC), a multi-level approach (as is evidenced by the partnerships' alignment of national and EU agendas), and directionality (see Chapter 1, this volume).

In analysing how the partnerships address the Transformative Outcomes, this research reveals that they mainly build and nurture niches through their activities, stimulating bottom-up transformation. This is perhaps not surprising as this was, and still is to a large extent, a key feature of innovation policy paradigms and practice (Schot & Steinmueller, 2018). Their particular advantage in doing so is that they integrate multiple regime actors in the way their governance is set up and in the way they operate. This is reflected most notably in the TO of networking. However, networking is not a necessary outcome of the partnership, but rather a means of achieving the desired results and impact more quickly and effectively. An openness to new actors (i.e., start-ups and emerging players in the field) is a key criterion for enabling purposeful networking; this is also reflected in the EC evaluation criteria for future partnerships. When openness is not ensured, there is a key risk that dominant regime actors will slow down the desired transformation processes by safeguarding their own short-term interests. In order to manage expectations in this regard, activities that promote and facilitate networking and co-creation between niche actors and regime actors are crucial.

It is evident that there are major differences within the partnerships in this area, firstly in terms of the activities they focus on and secondly in terms of the types of actors they involve. While all partnerships use a language of system transformation, using terms such as transformation, system, holistic understanding, inclusivity, openness, and transparency, it is less clear what this actually entails for them. As can be seen from the results section, most partnerships focus on activities that are aimed at promoting innovation in one or more technological areas and take an approach that is more likely to accelerate the development and distribution of a particular technology (see "accelerator mission", Wittmann et al., 2020) than to transform a system. For example, in the transport sector, there are six different partnerships in rail, water, road, and aviation. The partnerships for the ATM system and automated mobility do not cover a transport sector, but rather technologies that are intended to revolutionise air and road transport. Furthermore, if we look at the actors that make up the respective transport partnerships, we see that they mostly involve key actors from the value chain. These partnerships bring together large industrial players with long traditions and history, which is contrary to the views of the EC and its innovation model, which is based on openness and stakeholder inclusion. In order to challenge existing beliefs, views, and systems, it is necessary to engage with actors who may have opposing views and opinions. Therefore, we need to find ways to refine our understanding of and clarify terms such as directionality and openness by looking more closely at the actors involved in partnerships during evaluation.

In addition to the potential shortcomings of partnerships in addressing systemic transformation, they also face limitations in terms of how the respective system adopts innovations. As Jakob Edler and Jan Frageberg (2017) point out, innovation is not primarily about generating new ideas, but also about their exploitation and adoption by the regime (2017, p. 16). However, our analysis of their ToC suggests that these important outcomes are currently not strongly emphasised, pointing to a weaker strategic focus on expanding and mainstreaming niches. Most notably, the transformative outcomes of institutionalisation and replication are not well addressed by these partnerships.

Partnerships strive to change incredibly complex socio-technical systems at different levels (e.g., regional, national, and transnational) with different actors, policies, and challenges. The task of fundamentally changing socio-technical systems cannot be relegated to STI policy alone, let alone to partnerships, which are only one policy instrument among many. Instead, it may be better to consider their contributions to niche upscaling and mainstreaming and regime destabilisation in terms of finding synergies and interfaces with other actors and instruments that can cumulatively realise these processes. Seen in this way, the limitations of partnerships, and in particular, what they are *capable of* and should be achieving (i.e., their ambition), need to be examined in view of the broader mission governance arrangements that are currently emerging. In this context, it is worth examining the following governance angles more closely to find synergies and interfaces that, when taken as a whole, have the capacity to increase partnerships' potential to contribute to transformative change on different levels: (1) the EC and associated STI instruments level, and (2) the national and local level.

12.6.1 The EC and How It Can Create Additionality for Other Instruments (through Design)

The importance of coordination, streamlining, creating synergies, and coherence between the partnerships and the other instruments of the European Commission cannot be overemphasised. We see great value in aligning the partnerships with the following initiatives: the development of the work programmes in Horizon Europe (in particular the research and innovation activities); the European Innovation Council, which funds potential breakthrough and disruptive innovations and allocates more than 70 per cent of its funding to SMEs; and the Marie Skłodowska-Curie Actions and the European Research Council activities, which are aimed at strengthening European research capacities, pooling individual capacities for excellence, and accelerating innovation.

While the above initiatives emphasise R&I funding instruments, it is also important to create additionality through industrial policy and European Structural Funds policies, which can increase the leverage for transformation. Partnerships are aware of their contribution to the objectives of other overarching policies and their function as networks between R&I and other "external" stakeholders. In addition, partnerships have the ability to circulate, upscale, replicate, and institute new solutions. However, it is less clear how they can fulfil this function. Therefore, we emphasise how important it is for the EC to coordinate with other sector-specific and regional policies in order to create additionality by better dovetailing with existing instruments. Structural interventions that target the institutional framework conditions of industrial sectors and regions need to be coordinated with R&I policy to a much greater extent, depending on the topic-specific challenges (e.g., energy, mobility, health, etc.) that are addressed. Since the challenges and the national and institutional frameworks vary greatly between partnerships, it may be impossible to develop blueprints for the development of intersections between the different policy domains. This is because transformation differs according to the relevant area (e.g., health, climate, energy, etc.) and institutional (policy) setting. Therefore, some transformations may be largely driven by market dynamics, while other socio-technical transformations may require more state-led interventions (see introductory chapter, this volume).

Promoting complementarity by means of demand-driven innovation policies (Edler & Georghiou, 2007) in order to foster innovation is not a new endeavour on the EU agenda. Nevertheless, we call for the promotion of complementarity between supply-side *and* demand-side instruments such as public procurement, regulation, and regulation and standardisation. Adoption and greater integration of the partnership outcomes can bring about top-down transformation, destabilise the regime level, and create needed pressure from multiple directions.

12.6.2 The National and Local Levels

Another avenue for strengthening synergies pertains to the national and local levels. Specifically, we argue that partnerships can promote the implementation of transnational R&I programmes by encouraging streamlined national funding rules and further alignment of priority setting in specific fields. This would speed up the application procedures, encourage project collaboration across geographic regions, and support the inclusion of beneficiaries outside the public research sector (e.g., cities, municipalities, SMEs, and industry actors), which is a prerequisite for mission-oriented STI and important aspects of upscaling and replication. Similarly, guidelines and funding models could be standardised to give partnerships better opportunities to create synergies between STI actors and experts and pool resources and research capacity. To this end, we argue that partnerships can play an important role as an advocate for better alignment of strategy setting, research funding, and procedure implementation between countries. This role should be realised at the strategic level in the partnership's advisory and management groups as well as operationally when designing and implementing joint calls.

In addition to this, it is important to emphasise that EU partnerships (and missions) may only turn out to be successful if innovation actors not only acknowledge but also actively engage in the outlined transformation agendas on their own behalf. While participation in EU-funded calls is a core activity of many European research organisations and businesses, it is important that behavioural changes are reflected in adjustments to the organisations' strategies and operational procedures beyond participation in calls. Activities that promote and support change processes within organisations should be considered in this context.

12.6.3 Reflections on the Concept of Transformative Outcomes and Future Research

Having reviewed the transformative potential of partnerships and provided reference points to further illustrate their potential, we will now return to our second sub-question by reflecting on the concept of Transformative Outcomes and how the transformative potential of R&I instruments can be characterised. We will refine, delineate, and operationalise Transformative Outcomes by examining what they mean and what contributions the activities, outputs, and outcomes of partnerships make to the promotion of transformation. To this end, we will also seek to bridge the gap between researcher and practitioner terminology by identifying potential gaps and leverage points for policymaking. The concept of Transformative Outcomes is well-suited to analysing transformation processes within programmes and reflecting on the directionality of RTI interventions. It also allows us to identify gaps in programme design and approaches when analysing the European Partnerships.

However, further research is needed on the operationalisation of the concept and its use in projects. Firstly, Transformative Outcomes terminology is not well defined. Transformative Outcomes can be seen both as a *result* of an intervention (outcome) and as a *process* over which policymakers have control. Secondly, the linkages and complementarity of the Transformative Outcome concept could be further explored. This comprises the various conceptualisations of input-, output-, and behavioural additionality used in research, policy, and impact analysis in the context of RTI interventions. It also includes conceptualisations of organisational learning and knowledge spillover. Thirdly, a deeper understanding is required regarding the interplay of top-down and bottom-up mechanisms in transformation policies and their implications at the regime level. This is a particular point of weakness for the Transformative Outcomes approach, as it does not address how fundamental structures and processes such as regulation, standards, and norms can co-exist and be designed from the top down. This conceptualisation of Transformative Outcomes is very "innovation-heavy", as it focusses primarily on bottom-up induced system change processes, while somewhat neglecting the force of top-down policies that could trigger transformative change. In this section, we have shed light on this aspect by highlighting the importance of placing and assessing EU partnerships not only in a broader context of transnational and national R&I policy, but also in relation to the relevant sector-specific, regulatory, and fiscal policy areas.

ACKNOWLEDGEMENTS

We would like to express our sincere gratitude to Susanne Meyer, who supported us in the initial phase of our research process, from the idea stage to the operationalisation of the transformative outcomes and the analysis of the partnerships. She has put her knowledge in this area at our disposal, serving as a constant partner for discussion and reflection on the topic.

REFERENCES

Alvial Palavicino, C., Matti, C., & Witte, J. (2022). *Motion handbook: Developing a transformative theory of change.* Transformative Innovation Policy Consortium (TIPC), Utrecht University Centre for Global Challenges.

Amanatidou, E., Cunningham, P., Cox, D., Hunter, A., Dinges, M., Köngeter, A., Meyer, S. (2016). *Guide for P2P impact assessment.* ERA-LEARN 2020. https://repository.fteval.at/367/1/ERA-LEARN%202020%20D3.4a%20Guide%20revision_DEC2016_final.pdf

Brown, M. (2020). Unpacking the Theory of Change. *Stanford Social Innovation Review, 18*(4), 44–50.

Dinges, M., Meyer, S., & Brodnik, C. (2020). Key elements of evaluation frameworks for transformative R&I programmes in Europe. *Fteval Journal for Research and Technology Policy Evaluation, 51*, 26–40.

Edler, J., & Fagerberg, J. (2017). Innovation policy: what, why, and how. *Oxford Review of Economic Policy, 33*(1), 2–23. https://doi.org/10.1093/oxrep/grx001

Edler, J., & Georghiou, L. (2007). Public procurement and innovation – Resurrecting the demand side. *Research Policy, 36*(7), 949–963. https://doi.org/10.1016/j.respol.2007.03.003

European Commission (2019). *Orientations towards the first strategic plan for Horizon Europe.* https://ec.europa.eu/info/sites/default/files/research_and_innovation/strategy_on_research_and_innovation/documents/ec_rtd_orientations-he-strategic-plan_122019.pdf

European Commission (2021a). *European partnerships in Horizon Europe.* https://research-and-innovation.ec.europa.eu/funding/funding-opportunities/funding-programmes-and-open-calls/horizon-europe/european-partnerships-horizon-europe/partnerships-across-themes_en

European Commission (2021b). *Better regulation toolbox.* https://commission.europa.eu/system/files/2022–06/br_toolbox_-_nov_2021_-_chapter_6.pdf

European Council (2019). *Proposal for a Regulation of the European Parliament and of the Council establishing Horizon Europe – the Framework Programme for Research and Innovation, laying down its rules for participation and dissemination – Common understanding, Art. 8*, 2018/0224(COD). Council of the European Union.

Geels, F. W. (2011). The multi-level perspective on sustainability transitions: Responses to seven criticisms. *Environmental Innovation and Societal Transitions, 1*(1), 24–40. https://doi.org/10.1016/j.eist.2011.02.002

Geels, F. W., & Schot, J. (2007). Typology of sociotechnical transition pathways. *Research Policy, 36*(3), 399–417. https://doi.org/10.1016/j.respol.2007.01.003

Ghosh, B., Kivimaa, P., Ramirez, M., Schot, J., & Torrens, J. (2021). Transformative outcomes: Assessing and reorienting experimentation with transformative innovation policy. *Science and Public Policy, 48*(July), 739–756. https://doi.org/10.2139/ssrn.3686077

Glennie, A. (2019). *Agents of change making innovation agencies as innovative as those they support.* nesta. https://media.nesta.org.uk/documents/Agents_of_Change_v3.pdf

IPCC (2014). *Summary for policymakers.* In O. Edenhofer, R. Pichs-Madruga, Y. Sokona, E. Farahani, S. Kadner, K. Seyboth, A. Adler, I. Baum, S. Brunner, P. Eickemeier, B. Kriemann, J. Savolainen, S. Schlömer, C. von Stechow, T. Zwickel & J. C. Minx (Eds.), *Climate Change 2014: Mitigation of climate change. Contribution of Working Group III to the Fifth Assessment Report of the Intergovernmental Panel on Climate Change.* Cambridge University Press.

JPI Urban Europe (2020). *Preparation of the European partnership: Driving urban transitions. Report on the AGORA strategic dialogues.* https://jpi-urbaneurope.eu/wp-content/uploads/2020/10/DUT-AGORA-online-1.pdf

Kivimaa, P., Boon, W., Hyysalo, S., & Klerkx, L. (2019). Towards a typology of intermediaries in sustainability transitions: A systematic review and a research agenda. *Research Policy, 48*(4), 1062–1075. https://doi.org/10.1016/j.respol.2018.10.006

Loorbach, D. (2014). *To transition! Governance panarchy in the new transformation.* Erasmus University Rotterdam.

Markard, J., Raven, R., & Truffer, B. (2012). Sustainability transitions: An emerging field of research and its prospects. *Research Policy, 41*(6), 955–967. https://doi.org/10.1016/j.respol.2012.02.013

Molas-Gallart, J., Boni, A., Schot, J., & Giachi, S. (2021). *A formative approach to the evaluation of Transformative Innovation Policy.* TIPC Working Paper, TIPCWP 2020–01, p. 26.

Penna, C. C. R., Alvial-Palavicino, C., Ghosh, B., & Schot, J. (2023). Transformative innovation policy. In Encyclopedia of Social Innovation. Cheltenham, UK: Edward Elgar Publishing. https://doi.org/10.4337/9781800373358.ch73

Rogers, P. J. (2008). Using programme theory to evaluate complicated and complex aspects of interventions. *Evaluation, 14*(1), 29–48.

Schot, J., Kivimaa, P., & Torrens, J. (2019). *Transforming experimentation: Experimental policy, engagements and their Transformative Outcomes* (TIPC Research Report, p. 40). Transformative Innovation Policy Consortium. https://www.tipconsortium.net/wp-content/uploads/2019/07/Transforming-Experimentation.pdf

Schot, J., & Steinmueller, W. E. (2018). Three frames for innovation policy: R&D, systems of innovation and transformative change. *Research Policy, 47*(9), 1554–1567. https://doi.org/10.1016/j.respol.2018.08.011

Von der Leyen, U. (2019). *A Union that strives for more: My agenda for Europe.* https://ec.europa.eu/commission/sites/beta-political/files/political-guidelines-next-commission_en.pdf

Von Wirth, T., Fuenfschilling, L., Frantzeskaki, N., & Coenen, L. (2019). Impacts of urban living labs on sustainability transitions: Mechanisms and strategies for systemic change through experimentation. *European Planning Studies, 27*(2), 229–257. https://doi.org/10.1080/09654313.2018.1504895

Wanzenböck, I., Wesseling, J., Frenken, K., Hekkert, M. P., & Weber, M. (2020). A framework for mission-oriented innovation policy: Alternative pathways through the problem-solution space. *Science and Public Policy*, *47*(4) 1–29.

Weber, M., Biegelbauer, P., Dinges, M., Lamprecht, K. (2022). EU research, technological development and innovation policy. In H. Heinelt & S. Münche (Eds.), *The EU at a crossroads. EU Policies after the negotiations about the multiannual financial framework 2021–2027.* Edward Elgar Publishing.

Wittmann, F., Hufnagl, M., Lindner, R., Roth, F., Edler, J. (2020). *Developing a typology for mission-oriented innovation policies.* Fraunhofer ISI Discussion Papers – Innovation Systems and Policy Analysis, 64, Fraunhofer-Institut für System- und Innovationsforschung ISI. https://www.isi.fraunhofer.de/content/dam/isi/dokumente/cci/innovation-systems-policy-analysis/2020/discussionpaper_64_2020.pdf

PART III

EMPIRICAL COMPARISON OF DIFFERENT TYPES OF TRANSFORMATIONAL AREAS

13. Achilles' heels of acceleration? Assessing critical capacity for transformative policy mixes

Karoline S. Rogge and Qi Song

13.1 INTRODUCTION

Addressing the climate crisis calls for rapid and far-reaching sustainability transitions. Although several international transformative agreements, such as the Paris Climate Change Agreement, the 2030 Agenda for Sustainable Development, and the Sendai Framework for Disaster Risk Reduction, have been agreed upon in the past decade, recent scientific evidence has revealed there has been insufficient action and has repeatedly voiced the increasing urgency of addressing the climate crisis (Gills & Morgan, 2020; Ripple et al., 2021). Apart from the insufficiency of action so far, this urgency arises from three scientific characteristics of climate change: long atmospheric lifetimes of most greenhouse gases, the irreversibility of related damages, and the proximity of current concentrations to levels that can cause very serious impacts (Goulder, 2020). Viewing global climate change as a current crisis rather than a future risk underlines that more ambition, action, and collaboration towards rapid and fundamental transitions are immediately required to mobilise humanity (Zhenmin & Espinosa, 2019).

Meanwhile, scholars have been incorporating the urgency of addressing the climate crisis into their disciplinary and interdisciplinary research, focussing on timely and feasible ways to accelerate transformative change (Kern et al., 2016; Galvin, 2020; Gills & Morgan, 2021; McHugh et al., 2021). In particular, the interdisciplinary sustainability transitions literature (Markard et al., 2012) – rooted in innovation studies – argues that the increasing urgency of global sustainability challenges such as the climate crisis calls for an interventionist approach to accelerate socio-technical transitions towards more sustainable systems of production and consumption (Markard et al., 2020). To better address multiple failures in such transition processes (Weber & Rohracher, 2012), the policy mix perspective recommends designing and coordinating systemic policy interventions via policy strategies and instrument mixes (Rogge & Reichardt, 2016). In this context, two core policy mix aims – the stimulation of sustainable niche innovations on the one hand and the destabilisation of currently unsustainable regime arrangements on the other hand – have been highlighted (Kivimaa & Kern, 2016). This implies that transformative policy mixes call for a combination of transformative innovation and sector-specific policies. Therefore, in this chapter we focus on policy mixes for system innovation, which according to Kanger et al. (2020) should tackle not only creation and destruction but also address further policy intervention points for achieving transformative change, such as coordinating multi-system interactions.

However, since past policy mix research has had to focus empirically on the emergence phase of typically very slowly unfolding transition processes, we are still in the beginning stages of better understanding policy interventions for the accelerating phase of transitions. For this acceleration phase, Markard et al. (2020) have identified five challenges which bring with them various implications for policy intervention. It is in this phase where the transformative nature of policy mixes is put to the test, i.e., if they are actually fundamentally changing unsustainable regimes rather than mainly supporting experimentation and initial growth in sustainable niches. For this phase, as Rogge et al. (2020) point out, transformative policy mixes for accelerating sustainability transitions not only need the combination of different policy instruments targeting "creative destruction" (Kivimaa & Kern, 2016), but also need to tackle further aspects such as directionality, political resistance, institutional arrangements, and policy coordination. In this chapter, we argue that for a broader perspective, the role of policy actors, especially their skills and resources in the design of transformative policy mixes across the whole policy cycle, including formulation, adoption, and implementation of policy mix elements but also for their evaluation and revision, needs to be systematically investigated (Haddad et al., 2022; Siddiki & Curley, 2022). To achieve this, more dedicated work incorporating capacity considerations into policy mix research is a promising avenue.

13.2 CONCEPTUAL BACKGROUND

In this chapter, we propose an integrated framework for investigating critical capacities for transformative policy mixes in the acceleration phase of sustainability transitions. To do so, its three key conceptual components – policy capacities, policy mix intervention points, and acceleration challenges in sustainability transitions – are introduced in the following sections. By drawing on, tailoring, and combining these conceptual lenses to meet our research interests, we argue that the resulting integrated framework can serve as a useful heuristic for generating insights into which capacities are required by policy actors who are aiming for the acceleration of sustainability transitions.

13.2.1 Conceptual Component I: Policy Capacities

Innovation and transition scholars have proposed different capacity notions for analysing transformative innovation, such as the transformative capacity of new technologies (Augenstein, 2015; Dolata, 2009), urban transformative capacity (Wolfram, 2016; Wolfram et al., 2019), transformative capacity in climate governance (Hölscher et al., 2019), or the transformative capacity of socio-ecological systems (Olsson et al., 2010; Wilson et al., 2013). However, in line with Kern et al. (2019), we argue that unpacking the actor-centric capacity dynamics behind the design of transformative policy mixes calls for better bridging transition studies with policy studies. We are convinced that the notion of "policy capacity", originally from public administration and policy research (Bakvis, 2000; Painter & Pierre, 2005; Parsons, 2004), will help in advancing our understanding of critical capacity for transformative policy mixes, which is required for accelerating transitions. We follow the definition set by Wu et al. (2015), who defined policy capacity as "the set of skills and resources [...] necessary to perform policy functions" (2015, p. 166). Using this definition, we differentiate three types of essential capacities – analytical, operational, and political – for achieving policy success. Considering the comprehensiveness of this widely recognised conceptual approach in policy sciences, we argue that this definition can enable researchers to systematically investigate the various skills and resources of key actors involved in sustainability transitions.

Specifically, *analytical capacity* concerns the skills and resources to absorb, process, and utilise information and knowledge within and across policy fields. An example would be data collection and analysis for policy mix evaluation. In contrast, *operational capacity* addresses the skills and resources of relevant actors to allocate, mobilise, and distribute various available resources for performing policy functions into everyday routines or practices. For this, the availability of resources and the existence of coordination processes are a case in point. Finally, *political capacity* refers to the skills and resources needed to deal with political judgements, cross-actor negotiations or trade-offs, and the establishment of new governance structures. Such political capacity could, for instance, encompass stakeholder engagement, negotiations, and consensus-building.

Inspired by previous work (Howlett & Ramesh, 2016), we use this distinction of analytical, operational, and political capacity in our own integrated framework. However, as we want to focus on the role of policy actors – in the form of organisations – for accelerating transitions, we do not differentiate between individual, organisational, and system-level capacities, but instead concentrate on the organisational-level capacities of the actors governing sustainability transitions (Borrás & Edler, 2020). Also, while the term "policy capacity" suggests that the government is "the principal subject of inquiry", we follow Wu et al. (2015) in their understanding of policy capacity which recognises the role of non-government actors (e.g., NGOs and businesses) in influencing governments' capacities to perform policy functions (Wu et al., 2015).

13.2.2 Conceptual Component II: Policy Mix Intervention Points

As an intellectual fruit of cross-field fertilisation across policy sciences, environmental economics, and innovation studies (Rogge et al., 2017), the policy mix perspective lends itself to developing and analysing mission-oriented and transformative policy solutions for addressing grand challenges (Flanagan et al., 2011; Kern et al., 2019; Mazzucato, 2018). Here we follow Rogge and Reichardt (2016) who defined a policy mix as a combination of three building blocks: policy mix elements (constituted of policy strategies and instrument mixes), associated policy processes, and policy mix characteristics. We focus particularly on their broader understanding of policy mix elements, capturing both policy strategies (targets and roadmaps) and specific policy instruments (and their interactions), as this resonates well with mission innovation and transformative change. In addition, the attention to associated policy processes opens up the analysis to the role of established political institutions and governance structures in terms of shaping policy mixes (Kern & Rogge, 2018). For example, the attention to multiple stakeholders and their role in such policy processes can provide important insights into managing tensions induced by radical socio-technical reconfigurations that can be associated with sustainability missions. Finally, capturing the characteristics of mission-oriented policy mixes with concrete indicators can assist in dynamically assessing, learning about, and improving transformative policy solutions, which is key for achieving long-term mission-oriented innovation goals (Kattel & Mazzucato, 2018). Importantly, building on our focus on policy mixes for sustainability transitions, we highlight the importance of identifying the relevant policy mix for achieving missions (rather than mission-oriented innovation policy mixes), which can be identified through the bottom-up delineation approach introduced by

Ossenbrink et al. (2019). This implies that our focus on transformative policy mixes calls for the identification of policy strategies and various instruments influencing the organisational behaviour of actors that play an important role in achieving or hampering any given mission or intent; as such, our framework must encompass different policy fields and governance levels.

To better unpack the specific role of such policy mixes for sustainability transitions, Kanger et al. (2020) have summarised six policy intervention points (IP) identified through the lens of the Multi-Level Perspective (MLP, see Geels 2002, 2004). The first two points concern the niche level, in terms of supporting the emergence of various alternative niches and accelerating particularly promising niches to develop radical innovative solutions challenging incumbent unsustainable socio-technical regimes. Moreover, particularly in the acceleration phase, such established regimes should be directly destabilised through policy interventions, such as introducing carbon pricing or phasing out support for unsustainable technologies (Rosenbloom et al., 2020). The fourth intervention point covers the broader repercussions of such disruptive processes, stressing that these should be well managed through policy efforts in order to control unintended negative consequences arising from the sustainable reorientation of existing socio-technical systems. The fifth intervention point is the effective coordination of multi-regime interactions. For example, the diffusion of alternative niches in one socio-technical system, such as battery electric vehicles in the mobility system, might be hampered by the persistence of incumbent actors from adjacent socio-technical systems, such as those clinging to coal in the energy system. As such, multi-system policy efforts may be required to ensure deep decarbonisation through transport electrification. Lastly, Kanger et al. identified a sixth intervention point at the landscape level, of tilting these broader contextual and exogenous conditions to exert greater pressures on the regime, such as through international agreements arising from global negotiations on climate change mitigation.

In order to investigate the type of policy capacities required for the design of transformative policy mixes tackling these six different intervention points, we propose to assess the required policy capacity type for each policy mix intervention point (see the analytical matrix in Table 13.1).

As Table 13.1 shows, we use the research from Wu et al. (2015) for the distinction and definition of analytical, operational, and political capacity as outlined above. However, we implement two changes to the intervention points from Kanger et al. (2020). The first change we made is to add "provide direction" to the existing one "tilt the landscape" as the first intervention point to strengthen the explicit consideration of the directionality of socio-technical system change (Andersson et al., 2021), thereby acknowledging its articulation as an essential policy task to achieve transformative change (Braams et al., 2021). This extended intervention point is closely related to the notion of policy strategies in the policy mix conceptualisation by Rogge and Reichardt (2016). The second change we made is to use the name "coordinate multi-system interactions" instead of the original "provide coordination to multi-regime interaction", in order to broaden the scope of this intervention point. That is, we argue that policy mixes should not only target the coordination of different regimes, but also of multi-level niche-regime interactions in multi-system interactions and associated multi-actor interplays in transition governance (Markard et al., 2020; Rosenbloom, 2020).

Table 13.1 Associating policy intervention points with policy capacity types at the organisational level

Intervention points	*Analytical capacity*	*Operational capacity*	*Political capacity*
Provide direction and tilt the landscape (IP1)	Analytical capacity for providing direction and tilting the landscape	Operational capacity for providing direction and tilting the landscape	Political capacity for providing direction and tilting the landscape
Stimulate niches (IP2)	Analytical capacity for stimulating niches	Operational capacity for stimulating niches	Political capacity for stimulating niches
Accelerate niches (IP3)	Analytical capacity for accelerating niches	Operational capacity for accelerating niches	Political capacity for accelerating niches
Destabilise regimes (IP4)	Analytical capacity for destabilising regimes	Operational capacity for destabilising regimes	Political capacity for destabilising regimes
Address broader repercussions (IP5)	Analytical capacity for addressing broader repercussions	Operational capacity for addressing broader repercussions	Political capacity for addressing broader repercussions
Coordinate multi-system interactions (IP6)	Analytical capacity for coordinating multi-system interactions	Operational capacity for coordinating multi-system interactions	Political capacity for coordinating multi-system interactions

Source: Authors' elaborations based on Kanger et al., 2020 and Wu et al., 2015.

13.2.3 Conceptual Component III: Acceleration Challenges in Sustainability Transitions

The acceleration phase of sustainability transitions, in which the rising market share of new sustainable technologies eventually surpasses that of declining unsustainable ones, raises a couple of pressing new challenges for employing transformative policy interventions. As Markard et al. (2020) point out, beyond technological and economic factors, analysts need to pay attention to five wider acceleration challenges: (1) whole system change, (2) interaction between multiple systems, (3) industry decline and resistance, (4) consumers and social practices, and (5) coordination in policy and governance.

These acceleration challenges are closely connected to the uncertainty, complexity, and ambiguity of transition processes and the risks associated with them (Renn et al., 2011), and as such have important implications for policy interventions, as stressed by Markard et al. (2020). First, while past policy has focussed on single innovations, often technological ones, more attention needs to be focussed on changing entire systems across their various socio-technical dimensions, including complementary infrastructure and new business models. Second, interactions between the focal system and other adjacent systems increase the complexity of transitions, thus challenging policymaking in siloes and calling for balancing transition developments and speeds across interconnected systems. Third, struggles with incumbents that are actively involved in declining sectors, technologies, or business models mean that policymakers will need to find politically feasible and socially acceptable policy solutions for phasing out unsustainable technologies, fuels, and practices; for example, by forging coalitions and compensating losers. Fourth, more policy attention is required to induce fundamental changes in social practices and demand patterns that are aligned with sustainability objectives, which constitutes a policy area that has so far been largely neglected. Fifth, these challenges imply that effective and timely coordination in the governance system becomes even more crucial, whether applied to the coordination of policy actions and relevant actors across different governance levels (e.g., EU versus member states versus regions/cities) or when applied to various policy fields (e.g., climate versus industrial policies). Overall, this calls for a change in thinking regarding the role of the state towards more interventionist approaches.

These wider acceleration challenges thus imply the need for context-specific and sequenced policy mixes (Meckling et al., 2017) and institutional changes (Kivimaa & Rogge, 2022). We argue that these needs require enhanced policy capacities for designing such transformative policy mixes during the acceleration phase of sustainability transitions. To better assess which types of policy capacities may be particularly important for tackling which acceleration challenge, we again propose to match these two dimensions (see the analytical matrix in Table 13.2). This approach can help in identifying challenge-specific policy tasks associated with key issues within the acceleration phase of transitions (e.g., the rapid electrification of passenger-based transport).

In the following section, we will go one step further by combining all three components into a heuristic framework that incorporates acceleration challenges, policy mix intervention points, and types of policy capacities. The aim of this integrated framework is to enable the identification of critical capacities during the acceleration phase of sustainability transitions – such as for the decarbonisation of passenger-based transport through its electrification, which also requires the simultaneous decarbonisation of the increasingly interconnected electricity system.

Table 13.2 *Associating acceleration challenges with policy capacity types at the organisational level*

Acceleration challenges	*Analytical capacity*	*Operational capacity*	*Political capacity*
Whole system change (AC1)	Analytical capacity for whole system change	Operational capacity for whole system change	Political capacity for whole system change
Interaction between multiple systems (AC2)	Analytical capacity for interactions between multiple systems	Operational capacity for interactions between multiple systems	Political capacity for interactions between multiple systems
Decline and resistance (AC3)	Analytical capacity for decline and resistance	Operational capacity for decline and resistance	Political capacity for decline and resistance
Consumers and social practices (AC4)	Analytical capacity for consumers and social practices	Operational capacity for consumers and social practices	Political capacity for consumers and social practices
Governance (AC5)	Analytical capacity for governance	Operational capacity for governance	Political capacity for governance

Source: Authors' elaborations based on Markard et al., 2020 and Wu et al., 2015.

13.3 ASSESSING CRITICAL CAPACITIES FOR TRANSFORMATIVE POLICY MIXES IN THE ACCELERATION PHASE: TOWARDS AN INTEGRATED FRAMEWORK

Considering the urgency of addressing sustainability challenges such as the climate and wider ecological crises, policy actors are tasked with simultaneously governing transformative change in multiple systems that are in different phases of sustainability transitions. Here we focus on the governance of ongoing transitions in the acceleration phase, which is the case for decarbonising the increasingly connected energy-mobility systems.

Specifically, we argue that governing entities may be well advised to assess whether they can focus on a limited set of core acceleration challenges at a time (i.e., *"what policy actors should do"* in the present regarding short-term versus long-term challenges). Such prioritisation of the collective agenda is necessary due to the tensions between limited administrative resources and the urgency associated with the climate crisis, which calls for rapid and far-reaching transitions to climate neutrality. In addition, governing entities should rethink the design logic of policy intervention measures for achieving deliberately accelerated transitions in real-world governance systems (i.e., *"what policy actors can do"*). Importantly, the political feasibility of policy mix change hinges upon established governance systems and the openness of institutions to further transformative policy innovation – both are relevant for the effectiveness of policy mixes in addressing multiple failures (Weber & Rohracher, 2012). Therefore, the diagnostic identification of critical capacities is important for policy actors to better perform their transformative agency (i.e., *"what policy actors need to do"*). Our heuristic assessment tool for critical policy capacities during the acceleration phase of sustainability transitions incorporates the systemic considerations of both focussed acceleration challenges and possible instrumental intervention measures. Ultimately, this might imply that current governance systems and institutions need to be changed to be up for the tasks at hand.

In the following section, we outline our heuristic approach, which consists of a two-step procedure that starts with identifying challenge-specific policy tasks in order to then be able to assess critical capacities for transformative policy mixes (see Figure 13.1).

13.3.1 Step I: Identifying Challenge-specific Policy Tasks in Accelerating Sustainability Transitions

To address the overarching acceleration logic, we propose an analytical matrix that first identifies challenge-specific policy tasks in the acceleration phase of any given sustainability transition. Here, we define a challenge-specific policy task as a set of strategic and instrumental policy actions that targets particular acceleration challenges, which are operationalisable, achievable, and traceable in associated policy interventions. In this sense, the term "policy task" resonates with the policy mix concept since it indicates the manifestation of substantial policy actions in approaching the overarching strategic policy objective, but at a smaller scale. Therefore, it only covers a certain subset of specific goals addressed by actionable policy instruments (Rogge & Reichardt, 2016). However, the challenge-specific policy task could add analytical value since it incorporates very specific empirical considerations of the full range of policymaking processes (Manna & Moffitt, 2021). This broad coverage of policymaking, implementation, and evaluation has been acknowledged as an understudied topic in current policy mix research (Kern et al., 2019). By focussing on organisational-level

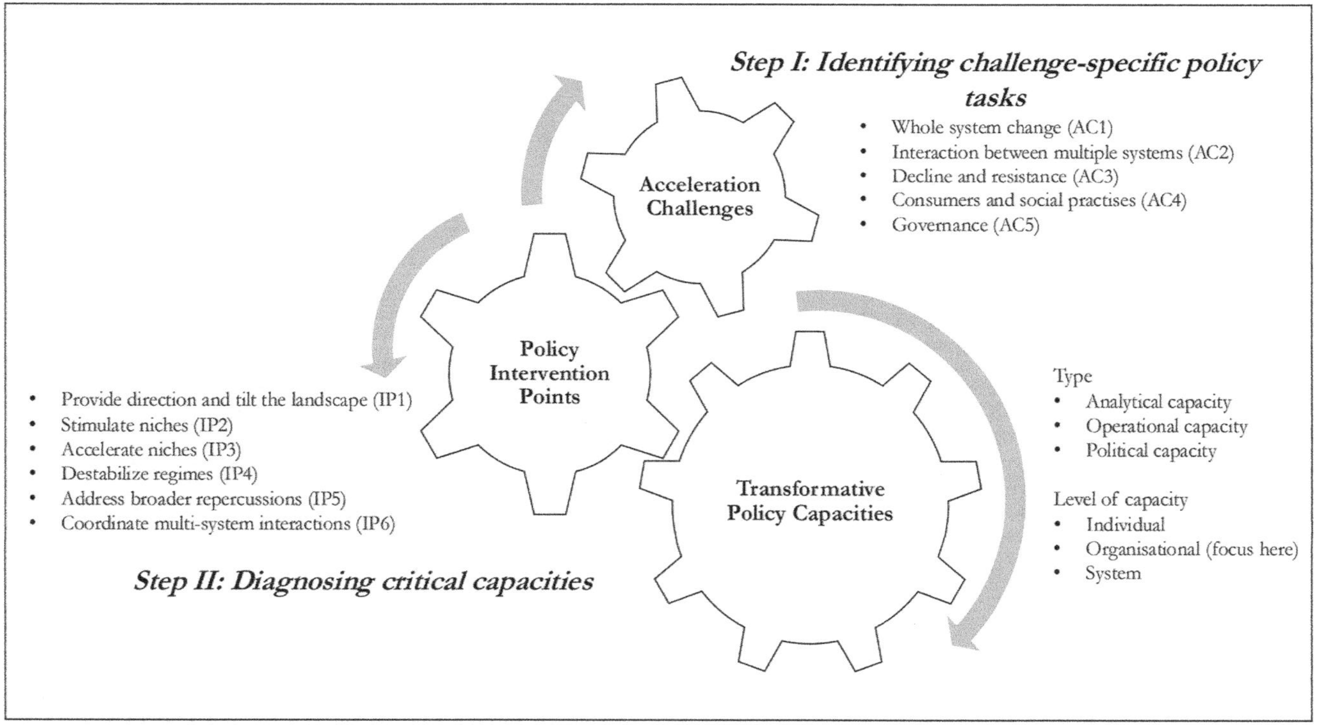

Source: Authors' own elaboration.

Figure 13.1 Overarching logic of integrated framework for assessing critical capacity for transformative policy mixes

capacities, the concept of challenge-specific policy tasks could help to better understand the skills and resources of associated public agencies in accelerating sustainability transitions and the critical capacities needed to deliver transformative policy mixes.

Our proposed heuristic scheme for identifying challenge-specific policy tasks in the acceleration phase of sustainability transitions integrates the two dimensions of policy intervention points (IP) and acceleration challenges (AC). For the *first dimension* of challenge-specific policy tasks, we focus on *policy intervention points*, applying the adopted framework and building on Kanger et al. (2020) as elaborated in the section on conceptual component II. For the *second dimension* of acceleration challenges in sustainability transitions, we use the five *acceleration challenges* identified by Markard et al. (2020) and introduced in the section on conceptual component III. For the resulting acceleration-intervention-matrix (see Table 13.3), in each cell we have provided an illustrative example of a potential policy task relevant for accelerating energy-mobility transitions; however, it is worth noting that these policy tasks could cover more than one cell in real-world policy practices.

The matrix thus serves as a guiding framework for identifying and prioritising core policy tasks in accelerating sustainability transitions. As such, it helps to produce a snapshot based on the current situation and knowledge, which implies that as the transition unfolds, it needs to be regularly updated.

13.3.2 Step II: Diagnosing Critical Capacity Needs for Addressing Challenge-specific Policy Tasks

In the second step, we propose diagnosing critical capacity needs for the identified challenge-specific policy tasks in the acceleration of sustainability transitions. In our heuristic scheme, we focus on policy capacity as key skills and resources of policy actors addressing these challenge-specific policy tasks. As Ricks and Doner (2021) point out, the diversity of tasks and the variety of different capacities need to be well-matched to achieve certain policy goals. Here, as outlined in the section on conceptual component I, we follow Wu et al. (2015) in defining policy capacity as "the set of skills and resources [...] necessary to perform policy functions" (2015, p. 166) and differentiate between analytical, operational, and political capacities essential for achieving policy success.

As the schematic illustration in Figure 13.2 shows, analysts should identify which types of policy capacities are most critical for which challenge-specific policy tasks, considering the different types of policy capacities in Table 13.4. For this, we anticipate differences in importance, implying that a lack of certain critical policy capacities may become the Achilles' heel of acceleration. For example, we expect analytical capacities to become particularly critical for providing strategic direction via foresight or stimulating diverse technology-specific innovation alternatives. In contrast, we foresee political capacities to be particularly critical for regime destabilisation and addressing its repercussions, while operational capacities might matter most for accelerating niches and providing coordination.

Table 13.3 *Analytical matrix for identifying challenge-specific policy tasks in accelerating sustainability transitions (with illustrative examples from energy-mobility transitions)*

Intervention points	*Innovations in whole systems (AC1)*	*Interactions with adjacent systems (AC2)*	*Resistance from declining industries (AC3)*	*Changes in consumer practices and routines (AC4)*	*Coordination challenges in policy and governance (AC5)*
Provide direction and tilt the landscape (IP1)	e.g., Conducting long-term technology road-mapping and socio-technical foresight	e.g., Setting standards for EV charging infrastructure	e.g., Orchestrating multi-stakeholder consultations in designing declining strategies	e.g., Co-creating long-term education plans concerning sustainability	e.g., Integration between ETS and EV credit systems
Stimulate niches (IP2)	e.g., Supporting the development of different battery technologies and their recycling	e.g., Enabling real-life laboratories that are experimenting with bidirectional charging	e.g., Offering transformational training for updating the skills and portfolios of incumbents	e.g., Fostering social cooperation programmes on alternative solutions	e.g., Dealing with different management standards for BEVs, HEVs, and FCEVs
Accelerate niches (IP3)	e.g., Selecting societally desirable alternative drives	e.g., Demonstrating the functionality of the charging network	e.g., Providing economic incentives for green corporate transformations	e.g., Scaling up innovative local and grassroots practices	e.g., Foregrounding shared EVs in both industrial strategies and transport planning
Destabilise regimes (IP4)	e.g., Incorporating charging infrastructure in city planning	e.g., Changing permit processes to prioritise renewable energy and power grid extensions	e.g., Navigating the political interests of different incumbent actors	e.g., Phasing out tax exemptions for commuting to work	e.g., Adjusting cross-organisational arrangements of policy agencies
Address broader repercussions (IP5)	e.g., Deliberating ethical principles for the applications of auto-driving techniques	e.g., Establishing phase-out strategies for fossil fuel-based economic models	e.g., Providing training programmes for previous ICE employees	e.g., Providing cheap access to clean public transportation for all	e.g., Coordinating industrial policy between EU, national, and regional governments

Intervention points	*Innovations in whole systems (AC1)*	*Interactions with adjacent systems (AC2)*	*Resistance from declining industries (AC3)*	*Changes in consumer practices and routines (AC4)*	*Coordination challenges in policy and governance (AC5)*
Coordinate multi-system interactions (IP6)	e.g., Providing demonstration support for intelligent electrified homes with heat pumps, wallboxes, PV, storage, and smart meters	e.g., Establishing dedicated strategic actors for coordinating the decarbonisation of energy-mobility systems	e.g., Managing joint-lobbying of ICE and oil incumbents in transport electrification	e.g., Establishing a cross-sectoral carbon labelling system for life-cycle emissions	e.g., Increasing administrative resources and political legitimacy for cross-ministry collaborations

Note: Challenge-specific policy tasks can also transcend several cells, but for simplicity, this is not shown here. Abbreviations: EV = electric vehicle, BEV = battery electric vehicle, HEV = hybrid electric vehicle, FCEV = fuel cell electric vehicle, R&D = research and development, ETS = emissions trading system
Source: Authors' elaborations based on Kanger et al., 2020 and Markard et al., 2020.

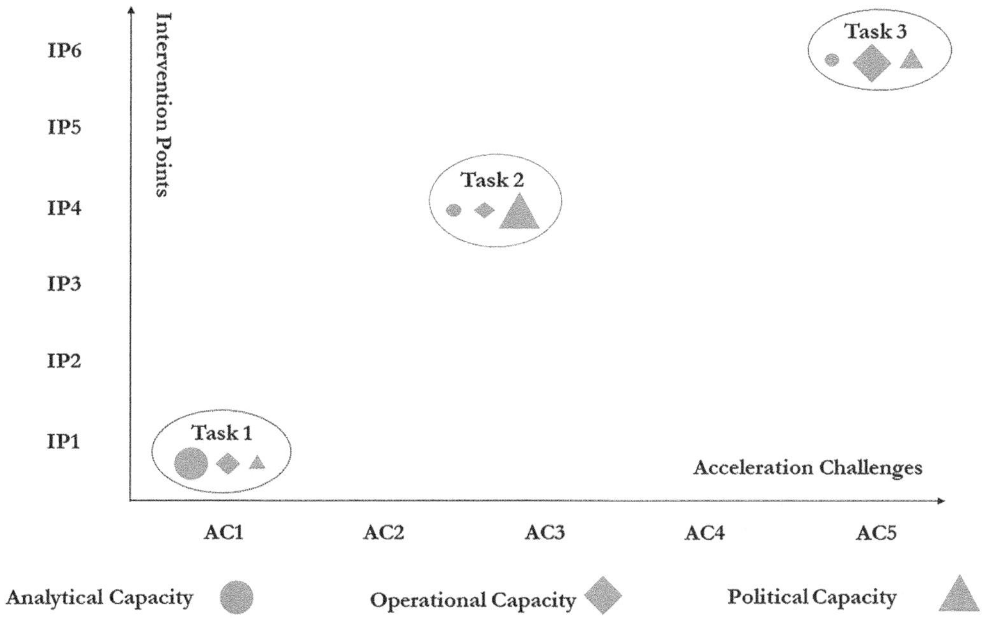

Source: Authors' own elaboration.

Figure 13.2 *Illustration of diagnosing critical organisational capacity needs for addressing challenge-specific policy tasks*

13.4 DISCUSSION

To shed light on the applicability of our framework, this section discusses the capacity concerns mentioned in the recent transition literature and examines them through the lens of our proposed heuristic approach.

Addressing complex and wicked sustainability challenges with inherent multiple failures requires policy actors to have the *analytical capacity* to absorb, process, and utilise information and knowledge within and across policy fields as a precondition for employing effective policy mixes for transformative change. For example, Hoppmann et al. (2014) point out the importance of having a profound knowledge of technological change in the process of designing policy interventions for solar photovoltaic power in Germany. Such skills play a critical role in providing guidance when searching for alternative innovations that are aligned with societally desired transition directionality (Braams et al., 2021) and which could be supported by the employment of clearly stated policy strategies and corresponding action plans (Rogge & Reichardt, 2016). Another important point is the role of analytical capacities in stimulating different niche innovations to ensure the diversity of technology-specific solutions in accelerating later systemic transitions. A widely discussed case, also cited by Kanger et al. (2020), is the generic policy support provided by the Chinese government to three promising yet competing innovation solutions for electric vehicles (battery, hybrid, and fuel-cell trajectories),

Table 13.4 *Template for diagnosing critical organisational capacity needs for key challenge-specific policy tasks*

Challenge-specific policy tasks in accelerating sustainability transitions	*Dimension:* *Analytical capacity* *(= technical soundness)*	*Dimension:* *Operational capacity* *(= implementation)*	*Dimension:* *Political capacity* *(= political support)*
Examples	• Availability of individuals with analytical capacity • Machinery and processes for collecting and analysing data • Organisational commitment to evidence-based policy	• Organisational commitment to achieving goals • Availability of fiscal and personnel resources • Coordination of internal processes • Performance management • Administrative accountability	• Legitimacy of the policy process • Processes for stakeholder engagement • Access to key policymakers
Policy task 1	-	-	-
…	-	-	-
…	-	-	-
Policy task n	-	-	-

Source: Authors' own illustration, building on policy capacity dimensions outlined by Wu et al., 2018.

where the identification of these three focal solutions was enabled by the road-mapping foresight work provided by a national-level senior expert group.

In order to accelerate the development of alternative innovations, destabilise the established regime structure, and manage the medium- and long-term impacts of systemic transformations towards sustainability, the *operational capacity* of relevant governing agencies to allocate, mobilise, and distribute various available resources for performing a set of policy functions in everyday practices is of key importance. For transformative innovation cases, not only are the number of qualified staff of organisations essential, so too are strategic reorientation and capacity-building activities for coping with a rapidly changing environment (Braams et al., 2021). As Foxon and Pearson (2008) point out, a "sustainable innovation policy regime" that aligns with technological development is crucial for achieving system transitions, which could also be regarded as the result of co-evolution between policy subsystems and socio-technical systems (Edmondson et al., 2019). In this sense, public organisations need to continuously update their knowledge base and develop dynamic capabilities, as well as improve their operational capacities to play the role of "change agents" (Karo & Kattel, 2018; Kattel & Mazzucato, 2018; Weber et al., 2021). Considering the operational requirements to orchestrate transformative changes, the importance of policy coordination and underlying institutional arrangements should be highlighted when addressing the broader repercussions of regime shifts and multi-system interactions (Braams et al., 2021; Rosenbloom, 2020).

The political nature of sustainability transitions leads to the observation that organisations need a high level of political skills and resources, and thus *political capacity*, to achieve policy successes. This requires, for example, policy actors to develop the ability to actively engage in the establishment of new governance arrangements and transformative partnerships for accelerating ongoing transition processes. The key to ensuring the inclusiveness of transformative innovations and making an agreement on desired directions of socio-technical change is political negotiation (Schot & Steinmueller, 2018; Weber & Rohracher, 2012). Accelerating sustainability transitions necessitates the employment of policy interventions that target the regime, which presents significant challenges due to the current power structure, since incumbent actors are in an advantageous position in most transition cases (Stirling, 2014). The "policy apparatus for incumbency" in the UK's energy policy field is a case in point, where a high level of regime resistance by incumbents has captured policymakers. This resistance helps to explain why the policy mixes are not viewed as transformative (Johnstone et al., 2017). Moreover, the role of domestic governments in international political negotiations such as the Paris Agreement could tilt the landscape, which requires the political capacity of governments not only on the domestic policymaking stage but also in geopolitical arenas (Kanger et al., 2020).

13.5 CONCLUSION

In this chapter, we have proposed an integrated analytical framework for assessing critical policy capacities to address challenge-specific policy tasks with transformative policy mixes in the acceleration phase of sustainability transitions. This heuristic can be applied to investigating the acceleration of single or multi-system transition processes towards sustainability, as we have illustrated with examples taken from the case of sustainable energy-mobility transitions occurring in multiple countries, such as Germany, the United States, and China. As such, we are calling for empirical research activities where our framework is applied: we see three particularly promising avenues for this.

First, we consider the investigation of organisations' political capacities as the most pressing research focus for future work aimed at transformative policy mixes that are accelerating sustainability transitions. The inherently political nature of transformative change requires policy actors to further develop their skills and resources to actively engage in associated policy processes (Meadowcroft, 2011), such as for introducing destabilising policies and phasing out support for unsustainable regimes. Recent work regarding entrenched power imbalances between different actors in transitions implies that governing organisations will have to play a key role in overcoming political pressures from powerful incumbents and established regimes and will have to help make political trade-offs between different groups of actors across different sectors or even different socio-technical systems (Avelino & Wittmayer, 2016; Stirling, 2014).

Second, we argue that future research should pay more attention to the different roles of multiple types of actors in policy subsystems who are exercising their policy capacities for transformative policy mix change. The structural difference between actors involved in socio-technical innovation processes and policy subsystems could be a promising lens for understanding the motives, attitudes, and resources held by different actors (Goyal & Howlett, 2020). For this, we suggest that future research should conduct in-depth case studies across different actors, sectors, and countries, with a focus on pioneers and best practices and paying particular attention to cross-case comparisons that appear especially promising for expanding our insights to ultimately provide practical recommendations.

Third, while we have only focussed on transformative policy capacity at the organisational level, future research should also extend the scope of discussion to the individual level and link it to the socio-technical system level. For this, we suggest that the role of key policy entrepreneurs should be further studied in terms of unpacking the initiation of transformative policy agendas in controversial political environments (Block & Paredis, 2013), as well as the agency of grassroots actors in coping with institutional contexts for stimulating whole-system transformations (Ramos-Mejía & Balanzo, 2018). Similarly, research focussing on organisational-level capacities should consider how its findings shape and are shaped by system-level capacity for transformative change (Wolfram, 2016). For example, the capacity of state actors to employ transformative policy action hinges upon the access granted to incumbent groups lobbying for their organised interests – something that differs across jurisdictions (Meckling & Nahm, 2022).

Such empirical research may help to refine our proposed framework, which is certainly needed, as it is not free from limitations. For example, a further conceptual enhancement could tackle the rather static nature of the policy capacity concept we have drawn upon, explained by our interest in capturing various actors in multi-system transitions. This could perhaps be addressed by examining the longitudinal dynamics of certain key organisations and by seeking inspiration from the literature on the dynamic capabilities of firms (c.f. Kattel & Mazzucato, 2018; Weber et al., 2021). Another example concerns the need to extend our approach to analysing capacity gaps, thereby capturing the difference between critical policy capacities and actual policy capacities. It is these gaps that may turn out to be the Achilles' heels of acceleration, which means that identifying them is a vital step.

In conclusion, we hope to contribute to and inspire multi-faceted future research on the nuanced role of policy capacity for accelerating directed socio-technical changes. We also hope to contribute to research on deliberate policy mix-making across various political, sector-specific, and geographical contexts – thereby leading to further empirical, methodological, and conceptual advances and relevant insights for policymakers involved in accelerating sustainability transitions.

ACKNOWLEDGEMENTS

This chapter was written within the EMPOCI project, which has received funding from the European Research Council (ERC) under the European Union's Horizon 2020 research and innovation programme (grant agreement No 852730). This work was initially presented at the International Sustainability Transitions Conference (IST) in October 2021, and an updated version was presented at a joint CBS-ISI workshop in June 2022. We are grateful for the feedback received on earlier versions of the manuscript and for the thoughtful comments received from two of the book editors.

REFERENCES

Andersson, J., Hellsmark, H., & Sandén, B. (2021). The outcomes of directionality: Towards a morphology of sociotechnical systems. *Environmental Innovation and Societal Transitions*, *40*, 108–131. https://doi.org/10.1016/j.eist.2021.06.008

Augenstein, K. (2015). Analysing the potential for sustainable e-mobility – The case of Germany. *Environmental Innovation and Societal Transitions*, *14*, 101–115. https://doi.org/10.1016/j.eist.2014.05.002

Avelino, F., & Wittmayer, J. M. (2016). Shifting power relations in sustainability transitions: A multi-actor perspective. *Journal of Environmental Policy & Planning*, *18*(5), 628–649. https://doi.org/10.1080/1523908X.2015.1112259

Bakvis, H. (2000). Rebuilding policy capacity in the era of the fiscal dividend: A report from Canada. *Governance*, *13*(1), 71–103. https://doi.org/10.1111/0952–1895.00124

Block, T., & Paredis, E. (2013). Urban development projects catalyst for sustainable transformations: The need for entrepreneurial political leadership. *Journal of Cleaner Production*, *50*, 181–188. https://doi.org/10.1016/j.jclepro.2012.11.021

Borrás, S., & Edler, J. (2020). The roles of the state in the governance of socio-technical systems' transformation. *Research Policy*, *49*(5), 103971. https://doi.org/10.1016/j.respol.2020.103971

Braams, R. B., Wesseling, J. H., Meijer, A. J., & Hekkert, M. P. (2021). Legitimizing transformative government: Aligning essential government tasks from transition literature with normative arguments about legitimacy from Public Administration traditions. *Environmental Innovation and Societal Transitions*, *39*, 191–205. https://doi.org/10.1016/j.eist.2021.04.004

Dolata, U. (2009). Technological innovations and sectoral change: Transformative capacity, adaptability, patterns of change: An analytical framework. *Research Policy*, *38*(6), 1066–1076. https://doi.org/10.1016/j.respol.2009.03.006

Edmondson, D. L., Kern, F., & Rogge, K. S. (2019). The co-evolution of policy mixes and socio-technical systems: Towards a conceptual framework of policy mix feedback in sustainability transitions. *Research Policy*, *48*(10), 103555. https://doi.org/10.1016/j.respol.2018.03.010

Flanagan, K., Uyarra, E., & Laranja, M. (2011). Reconceptualising the "policy mix" for innovation. *Research Policy*, *40*(5), 702–713. https://doi.org/10.1016/j.respol.2011.02.005

Foxon, T., & Pearson, P. (2008). Overcoming barriers to innovation and diffusion of cleaner technologies: Some features of a sustainable innovation policy regime. *Journal of Cleaner Production*, *16*(1, Supplement 1), S148–S161. https://doi.org/10.1016/j.jclepro.2007.10.011

Galvin, R. (2020). Power, evil and resistance in social structure: A sociology for energy research in a climate emergency. *Energy Research & Social Science*, *61*, 101361. https://doi.org/10.1016/j.erss.2019.101361

Geels, F. W. (2002). Technological transitions as evolutionary reconfiguration processes: A multi-level perspective and a case-study. *Research Policy*, *31*(8), 1257–1274. https://doi.org/10.1016/S0048-7333(02)00062-8

Geels, F. W. (2004). From sectoral systems of innovation to socio-technical systems: Insights about dynamics and change from sociology and institutional theory. *Research Policy*, *33*(6), 897–920. https://doi.org/10.1016/j.respol.2004.01.015

Gills, B., & Morgan, J. (2020). Global climate emergency: After COP24, climate science, urgency, and the threat to humanity. *Globalizations*, *17*(6), 885–902. https://doi.org/10.1080/14747731.2019.1669915

Gills, B., & Morgan, J. (2021). Economics and climate emergency. *Globalizations*, *18*(7), 1071–1086. https://doi.org/10.1080/14747731.2020.1841527

Goulder, L. H. (2020). Timing is everything: How economists can better address the urgency of stronger climate policy. *Review of Environmental Economics and Policy*, *14*(1), 143–156. https://doi.org/10.1093/reep/rez014

Goyal, N., & Howlett, M. (2020). Who learns what in sustainability transitions? *Environmental Innovation and Societal Transitions*, *34*, 311–321. https://doi.org/10.1016/j.eist.2019.09.002

Haddad, C. R., Nakić, V., Bergek, A., & Hellsmark, H. (2022). Transformative innovation policy: A systematic review. *Environmental Innovation and Societal Transitions*, *43*, 14–40. https://doi.org/10.1016/j.eist.2022.03.002

Hölscher, K., Frantzeskaki, N., & Loorbach, D. (2019). Steering transformations under climate change: Capacities for transformative climate governance and the case of Rotterdam, the Netherlands. *Regional Environmental Change*, *19*(3), 791–805. https://doi.org/10.1007/s10113-018-1329-3

Hoppmann, J., Huenteler, J., & Girod, B. (2014). Compulsive policy-making – The evolution of the German feed-in tariff system for solar photovoltaic power. *Research Policy*, *43*(8), 1422–1441. https://doi.org/10.1016/j.respol.2014.01.014

Howlett, M., & Ramesh, M. (2016). Achilles' heels of governance: Critical capacity deficits and their role in governance failures. *Regulation & Governance*, *10*(4), 301–313. https://doi.org/10.1111/rego.12091

Johnstone, P., Stirling, A., & Sovacool, B. (2017). Policy mixes for incumbency: Exploring the destructive recreation of renewable energy, shale gas "fracking," and nuclear power in the United Kingdom. *Energy Research & Social Science*, *33*, 147–162. https://doi.org/10.1016/j.erss.2017.09.005

Kanger, L., Sovacool, B. K., & Noorkõiv, M. (2020). Six policy intervention points for sustainability transitions: A conceptual framework and a systematic literature review. *Research Policy*, *49*(7), 104072. https://doi.org/10.1016/j.respol.2020.104072

Karo, E., & Kattel, R. (2018). Innovation and the state: Towards an evolutionary theory of policy capacity. In X. Wu, M. Howlett, & M. Ramesh (Eds.), *Policy capacity and governance: Assessing governmental competences and capabilities in theory and practice* (pp. 123–150). Springer International Publishing. https://doi.org/10.1007/978-3-319-54675-9_6

Kattel, R., & Mazzucato, M. (2018). Mission-oriented innovation policy and dynamic capabilities in the public sector. *Industrial and Corporate Change*, *27*(5), 787–801. https://doi.org/10.1093/icc/dty032

Kern, F., & Rogge, K. S. (2016). The pace of governed energy transitions: Agency, international dynamics and the global Paris agreement accelerating decarbonisation processes? *Energy Research & Social Science*, *22*, 13–17. https://doi.org/10.1016/j.erss.2016.08.016

Kern, F., & Rogge, K. S. (2018). Harnessing theories of the policy process for analysing the politics of sustainability transitions: A critical survey. *Environmental Innovation and Societal Transitions*, *27*, 102–117. https://doi.org/10.1016/j.eist.2017.11.001

Kern, F., Rogge, K. S., & Howlett, M. (2019). Policy mixes for sustainability transitions: New approaches and insights through bridging innovation and policy studies. *Research Policy*, *48*(10), 103832. https://doi.org/10.1016/j.respol.2019.103832

Kivimaa, P., & Kern, F. (2016). Creative destruction or mere niche support? Innovation policy mixes for sustainability transitions. *Research Policy*, *45*(1), 205–217. https://doi.org/10.1016/j.respol.2015.09.008

Kivimaa, P., & Rogge, K. S. (2022). Interplay of policy experimentation and institutional change in sustainability transitions: The case of mobility as a service in Finland. *Research Policy, 51*(1), 104412. https://doi.org/10.1016/j.respol.2021.104412

Manna, P., & Moffitt, S. L. (2021). Traceable tasks and complex policies: When politics matter for policy implementation. *Policy Studies Journal, 49*(1), 190–218. https://doi.org/10.1111/psj.12348

Markard, J., Geels, F. W., & Raven, R. (2020). Challenges in the acceleration of sustainability transitions. *Environmental Research Letters, 15*(8), 081001. https://doi.org/10.1088/1748–9326/ab9468

Markard, J., Raven, R., & Truffer, B. (2012). Sustainability transitions: An emerging field of research and its prospects. *Research Policy, 41*(6), 955–967. https://doi.org/10.1016/j.respol.2012.02.013

Mazzucato, M. (2018). Mission-oriented innovation policies: Challenges and opportunities. *Industrial and Corporate Change, 27*(5), 803–815. https://doi.org/10.1093/icc/dty034

McHugh, L. H., Lemos, M. C., & Morrison, T. H. (2021). Risk? Crisis? Emergency? Implications of the new climate emergency framing for governance and policy. *WIREs Climate Change, 12*(6), e736. https://doi.org/10.1002/wcc.736

Meadowcroft, J. (2011). Engaging with the *politics* of sustainability transitions. *Environmental Innovation and Societal Transitions, 1*(1), 70–75. https://doi.org/10.1016/j.eist.2011.02.003

Meckling, J., & Nahm, J. (2022). Strategic state capacity: How states counter opposition to climate policy. *Comparative Political Studies, 55*(3), 493–523. https://doi.org/10.1177/00104140211024308

Meckling, J., Sterner, T., & Wagner, G. (2017). Policy sequencing toward decarbonization. *Nature Energy, 2*(12), 918–922. https://doi.org/10.1038/s41560–017–0025–8

Olsson, P., Bodin, Ö., & Folke, C. (2010). Building transformative capacity for ecosystem stewardship in social-ecological systems. In D. Armitage & R. Plummer (Eds.), *Adaptive capacity and environmental governance* (pp. 263–285). Springer. https://doi.org/10.1007/978–3-642–12194–4_13

Ossenbrink, J., Finnsson, S., Bening, C. R., & Hoffmann, V. H. (2019). Delineating policy mixes: Contrasting top-down and bottom-up approaches to the case of energy-storage policy in California. *Research Policy, 48*(10), 103582. https://doi.org/10.1016/j.respol.2018.04.014

Painter, M., & Pierre, J. (2005). *Challenges to state policy capacity: Global trends and comparative perspectives.* Palgrave Macmillan. https://link.springer.com/book/10.1057/9780230524194

Parsons, W. (2004). Not just steering but weaving: Relevant knowledge and the craft of building policy capacity and coherence. *Australian Journal of Public Administration, 63*(1), 43–57. https://doi.org/10.1111/j.1467–8500.2004.00358.x

Ramos-Mejía, M., & Balanzo, A. (2018). What it takes to lead sustainability transitions from the bottom-up: Strategic interactions of grassroots ecopreneurs. *Sustainability, 10*(7), 2294. https://doi.org/10.3390/su10072294

Renn, O., Klinke, A., & van Asselt, M. (2011). Coping with complexity, uncertainty and ambiguity in risk governance: A synthesis. *AMBIO, 40*(2), 231–246. https://doi.org/10.1007/s13280–010–0134–0

Ricks, J. I., & Doner, R. F. (2021). Getting institutions right: Matching institutional capacities to developmental tasks. *World Development, 139*, 105334. https://doi.org/10.1016/j.worlddev.2020.105334

Ripple, W. J., Wolf, C., Newsome, T. M., Gregg, J. W., Lenton, T. M., Palomo, I., Eikelboom, J. A. J., Law, B. E., Huq, S., Duffy, P. B., & Rockström, J. (2021). World scientists' warning of a climate emergency 2021. *BioScience, 71*(9), 894–898. https://doi.org/10.1093/biosci/biab079

Rogge, K. S., Kern, F., & Howlett, M. (2017). Conceptual and empirical advances in analysing policy mixes for energy transitions. *Energy Research & Social Science, 33*, 1–10. https://doi.org/10.1016/j.erss.2017.09.025

Rogge, K. S., Pfluger, B., & Geels, F. W. (2020). Transformative policy mixes in socio-technical scenarios: The case of the low-carbon transition of the German electricity system (2010–2050). *Technological Forecasting and Social Change, 151*, 119259. https://doi.org/10.1016/j.techfore.2018.04.002

Rogge, K. S., & Reichardt, K. (2016). Policy mixes for sustainability transitions: An extended concept and framework for analysis. *Research Policy, 45*(8), 1620–1635. https://doi.org/10.1016/j.respol.2016.04.004

Rosenbloom, D. (2020). Engaging with multi-system interactions in sustainability transitions: A comment on the transitions research agenda. *Environmental Innovation and Societal Transitions, 34*, 336–340. https://doi.org/10.1016/j.eist.2019.10.003

Rosenbloom, D., Markard, J., Geels, F. W., & Fuenfschilling, L. (2020). Why carbon pricing is not sufficient to mitigate climate change – And how "sustainability transition policy" can help. *Proceedings of the National Academy of Sciences, 117*(16), 8664–8668. https://doi.org/10.1073/pnas.2004093117

Schot, J., & Steinmueller, W. E. (2018). Three frames for innovation policy: R&D, systems of innovation and transformative change. *Research Policy, 47*(9), 1554–1567. https://doi.org/10.1016/j.respol.2018.08.011

Siddiki, S., & Curley, C. (2022). Conceptualising policy design in the policy process. *Policy & Politics, 50*(1), 117–135. https://doi.org/10.1332/030557321X16346727541396

Stirling, A. (2014). Transforming power: Social science and the politics of energy choices. *Energy Research & Social Science, 1*, 83–95. https://doi.org/10.1016/j.erss.2014.02.001

Weber, K. M., Biegelbauer, P., Brodnik, C., Dachs, B., Dreher, C., Kovaéc, M., Pulenkova, E., Schartinger, D., & Schwäb, C. (2021). *Agilität in der F&I-Politik: Konzept, definition, operationalisierung* (Research Report No. 8–2021). Studien zum deutschen Innovationssystem. https://www.econstor.eu/handle/10419/231476

Weber, K. M., & Rohracher, H. (2012). Legitimizing research, technology and innovation policies for transformative change: Combining insights from innovation systems and multi-level perspective in a comprehensive "failures" framework. *Research Policy, 41*(6), 1037–1047. https://doi.org/10.1016/j.respol.2011.10.015

Wilson, S., Pearson, L. J., Kashima, Y., Lusher, D., & Pearson, C. (2013). Separating adaptive maintenance (resilience) and transformative capacity of social-ecological systems. *Ecology and Society, 18*(1): 22. https://www.jstor.org/stable/26269282

Wolfram, M. (2016). Conceptualizing urban transformative capacity: A framework for research and policy. *Cities, 51*, 121–130. https://doi.org/10.1016/j.cities.2015.11.011

Wolfram, M., Borgström, S., & Farrelly, M. (2019). Urban transformative capacity: From concept to practice. *Ambio, 48*(5), 437–448. https://doi.org/10.1007/s13280–019–01169-y

Wu, X., Ramesh, M., & Howlett, M. (2015). Policy capacity: A conceptual framework for understanding policy competences and capabilities. *Policy and Society, 34*(3–4), 165–171. https://doi.org/10.1016/j.polsoc.2015.09.001

Wu, X., Ramesh, M., & Howlett, M. (2018). *Policy capacity: Conceptual framework and essential components.* In X. Wu, M. Howlett, & M. Ramesh (Eds.), *Policy capacity and governance: Assessing governmental competences and capabilities in theory and practice* (pp. 1–25). Palgrave Macmillan. https://doi.org/10.1007/978-3-319-54675–9_1

Zhenmin, L., & Espinosa, P. (2019). Tackling climate change to accelerate sustainable development. *Nature Climate Change, 9*(7), 494–496. https://doi.org/10.1038/s41558–019–0519–4

14. Transformative missions in healthcare: the case of preventive healthcare in the Netherlands

Wouter Boon

14.1 INTRODUCTION

In recent years, science, technology, and innovation (STI) policy has shifted towards an explicit aim of contributing to societal goals by means of missions. These missions are sets of precisely defined potential solutions to societal problems. Current mission-oriented STI policy tends to be more than just promoting large-scale scientific or technological endeavours. Instead, it aims for transformative change with a wide societal reach (Kuittinen et al., 2018; Wittmann et al., 2020; Polt et al., 2021). The societal challenges that are addressed by these policies are persistent and "wicked" by nature, meaning that both problems and solution directions are theoretically ill-defined (Wanzenböck et al., 2020). Transformative mission-oriented STI policy calls for a specific approach to societal challenges. The ambition of STI policy goes beyond addressing economic targets and includes achieving societal outcomes in other policy domains such as mobility and healthcare.

One area that is particularly relevant in the context of the move towards mission-oriented STI policy is healthcare. The healthcare sector is continuously driving progress regarding societal challenges by addressing unmet medical needs, the problems of ageing societies, and emergencies like COVID-19. Focussing on these challenges requires substantial STI support. At the same time, the healthcare sector is powerful enough for STI efforts to take distinct forms, as shown by the existence of independent research councils in many countries.

Defining missions in healthcare is not a new concept, as many have been defined in the last few decades; these have often centred on diseases or emerging technologies, as in President Nixon's 1971 "War on Cancer" to give a prominent example. Patient advocates had promoted the programme as "the conquest of 'inner space'", which was an explicit reference to the Apollo mission (Mukherjee, 2010). Since then, there have been many mission-oriented programmes in healthcare with a top-down, centralised nature, clear objectives, and a focus on specific technological ideas. Another prominent example of a mission in healthcare is the Human Genome Project (1990–2003), which aimed to map all genes of the human genome. The Dutch RegMed XB initiative is a public-private programme that focusses on regenerative medicine organised around four specific "moonshots", i.e., "long-term visions of breakthroughs for patients, translated into research roadmaps with specific short-term milestones" (RegMed XB, 2022). What is more, new mission-oriented STI policies are currently being introduced. In the United States, discussions are underway concerning the establishment of the Advanced Research Projects Agency for Health (ARPA-H), a dedicated healthcare research funding agency modelled after the Advanced Research Projects Agency (ARPA) (Tollefson, 2021).

Beyond these technological or disease-centred missions, healthcare is confronted with additional "wicked" problems. Examples of such problems include the growing burden of lifestyle-related diseases, the influence of lifestyle on COVID-19 disease progression, and the wish to promote healthy ageing. Many stakeholders pin their hopes on preventive healthcare, aimed at avoiding the development or progression of diseases. Prevention holds a promise: early interventions, such as anti-smoking courses, may lead to less costly treatments later on. This provides an incentive for transformative mission-oriented STI policies to focus on preventive healthcare.

Promoting preventive healthcare has not been without its challenges. Many innovations and initiatives are being pursued, but large-scale deployment has proven difficult. In order to pursue transformative mission-oriented policy, STI policy in general terms needs to expand in the following three ways: by integrating policy rationales, instruments, etc., across different sectors and domains; by requiring cooperation between conventional actors in innovation systems, even when the role of the state seems to be more important in mission policies than in terms of defining and coordinating efforts; and by integrating technological and societal solutions. Expanding transformative STI policy for preventive medicine in these three ways is associated with specific challenges: incompatibility between institutions and a lack of diffusion of innovations; collaboration failures; and difficulties with integrating technological and social innovations.

First, STI policy needs to integrate new policy rationales and instruments with those that have already been applied in the healthcare sector. The integration is made more complicated due to the highly institutionalised nature of the healthcare sector. These institutions include formal rules for safeguarding patient safety, which have been extrapolated from evidence-based medicine, and informal rules that take the form of guidelines. Innovation in such a field means either adapting to these institutions or embarking on an arduous venture to change them. For example, many innovative efforts in preventive healthcare currently focus on digital medicine in the form of eHealth, mHealth, smart homes, and wearables. This digital innovation is often incompatible with what the healthcare sector requires in terms of quality of evidence (Kulkov et al., 2021). The incompatibility has direct repercussions for the large-scale deployment of prevention innovation because failure to adhere to healthcare institutions lowers innovations' chances of becoming part of standard care, medical guidelines, and reimbursement schemes (Hoogstraaten et al., forthcoming). Companies will need to re-evaluate their innovation processes (Steinberg et al., 2015) and business models (Oderanti et al., 2021) to account for this incompatibility.

Second, collaboration in the field of preventive healthcare is notoriously difficult. The reason for this is that "the rewards to the individual from adopting a preventive innovation are often delayed in time, are relatively intangible and the unwanted consequence may not occur anyway" (Rogers, 2002). Investment in preventive healthcare is also assumed to result in a diffusion of future benefits across unidentified or anonymous beneficiaries, meaning that responsibilities and mandates are fragmented across various actor types and (inter)national and regional levels (Eyler et al., 2016; Essén et al., 2022).

Third, solutions must incorporate technological and social dimensions. Merely advancing novel technological products is not enough. To be effective, these interventions need to be combined with lifestyle changes or behavioural interventions, e.g., efforts to increase long-term adherence to treatment. Technical solutions often need to be accompanied by social innovation, utilising champions, or activating informal caregiver networks to promote preventive innovations (Janssen et al., 2015; Greenhalgh et al., 2017).

In this chapter, we aim to assess the emerging mission-oriented STI policy in the context of preventive healthcare. We will focus on a persistent and complex transformation that has clearly defined problems and targets, as well as a divergent range of possible solutions. The solutions do not rely solely and primarily on technological novelty but also require social innovation, i.e., changes to behaviour and new societal practices. Additionally, traditional economics-centred STI policy needs to be aligned or integrated with healthcare policies that traditionally have an STI dimension of their own.

We will do this by examining the process of transformative STI policy, covering mission formulation, mission design, and mission implementation. This chapter analyses the STI policies in the field of preventive medicine, concentrating on the following questions:

1. How is the mission articulated and made explicit?
2. How is the mission designed and operationalised? In particular, what actor groups and instruments are proposed?
3. To what extent do the implemented instruments work towards the mission goals and thus the transformation, in the context of existing misalignments and related (institutional) complexities? What is needed to resolve the issues with prevention innovation, i.e., institutional incompatibility and a lack of diffusion, breakdowns in collaboration, and difficulties in integrating technological and social innovations?

Targeting the area of healthcare, this chapter will contribute to our knowledge of the challenges, benefits, and limitations of STI policy that is deliberately designed to transform the practices and routines of a needs area (health). This needs area is characterised by a highly complex assortment of actors and specific structural impediments that must be addressed before innovation and transformation can occur. The three questions above are addressed in the following three sections.

14.1.1 How is the Mission Articulated and Made Explicit?

The last decade has seen a marked shift in STI policy in the healthcare sector in the Netherlands, as it has moved from sector-oriented to mission-driven policy. Since 2011, the Dutch government has focused its innovation policy on strengthening so-called "top sectors" in which companies, knowledge institutions, and the government work together in a "triple helix" arrangement, each acting within their own sector to strengthen innovation systems and expand national and international earning capacities.

The Dutch government has made Life Sciences & Health one of the top sectors. The original goal of STI policy for the sector was to contribute to the quality of life for humans and animals, make the healthcare system sustainable for future growth, and increasing innovative entrepreneurship. One guiding vision for the life sciences sector was that it should become the "Dutch Boston area", a comparison that was said to make sense not only in geographical size but also in the quality of the innovation system. The top-sector plans focussed on medical technology, biopharmaceuticals, and regenerative medicine, alongside application areas such as early diagnostics, tailor-made treatments, and self-management (LS&H, 2011).

As the "triple helix" approach influenced the kind of policy instruments used, preference was given to supporting research and innovation projects through public funding, which subsequently had to be matched with private investments. Until then, the healthcare sector had focussed on its own public knowledge institutes, including research councils, academic hospitals, and research-intensive advisory bodies. However, the top-sector approach dictated that these organisations had to engage in public-private partnerships, blurring the boundaries between general and healthcare-sector-specific STI policies.

From 2018 onwards, Dutch STI policy made a clear and explicit shift from supporting economically successful sectors to putting societal challenges centre stage. The increased focus on grand societal challenges in European R&D policy, combined with the Paris Agreement, instigated this shift. Dutch policy started to complement top-sector policy with mission-driven policy, which led to at least two major changes in STI policy. Societal challenges became central along with – or even taking priority over – economic goals, and there was an explicit move to extend the "triple helix" to a "quadruple helix", including citizens and citizen representatives. The national mission-driven innovation policy has since listed "Health and care" as one of its four focus topics, along with "Energy transition and sustainability", "Agriculture, water, food", and "Security". Public and private actors committed to the Knowledge and Innovation Covenant 2020–2023 that would form the basis for the mission-oriented STI policy.

Under the "Health and care" topic, there is one central mission and four supporting missions. The overarching mission is that by 2040, all Dutch people will live for at least five years longer in good health and that the health differences between the lowest and highest socioeconomic groups will have decreased by 30 per cent (Health Holland, 2020a). Next to the central mission, there are four supporting missions that differ in scope and are nested one inside the next, much like Russian dolls: "the first mission, which is aimed at prevention, creates the conditions for realising missions II, III and IV. Mission II is vital for missions III and IV and so forth" (Health Holland, 2019). The four missions are: (I) decreasing unhealthy lifestyles and improving physical and social living environments through a focus on prevention, lifestyle, integral and tailor-made interventions; (II) care in the right place and the promotion of independent living for as long as possible; (III) increasing the proportion of people with a chronic illness who can participate in society through better diagnostics and personalised treatments, facilitated by advances in medical technology and science; and (IV) increasing the quality of life of people with dementia. In the end, "for all these missions, the goals are only achieved when the innovations in question contribute to affordable, accessible and high-quality care, have been effectively implemented and the organisation of care has been adjusted accordingly" (EZK, 2019a).

The missions put preventive healthcare in the central role, despite this being an area that has not attracted much attention and funding to date in terms of commercial, healthcare, and public activities.[1] The fact that only 1 to 3 per cent of the money spent in Dutch healthcare goes to prevention (Gmeinder et al., 2017; Ahli, 2020) further suggests that preventive healthcare interventions are not seen as a priority. Nevertheless, prevention has increasingly been framed as a way to curb lifestyle-related diseases such as diabetes and cancer and even infectious diseases like COVID-19, as well as a way to promote healthy ageing or the "vitality" of the elderly.

The four missions explicitly link prevention with cure and treatment interventions. Missions I and II function as a hinge between the overarching prevention mission and the

---

[1] As a side note, one could argue that many preventive measures, such as water sanitation and hygiene, are so common nowadays that we simply do not count them as part of preventive healthcare activities.

more traditional treatment- or cure-related missions, which focus on chronic illnesses (III) and dementia (IV). The hinge is a substantial driving force in efforts to make the healthcare sector more focussed on prevention. In this way, the emerging Dutch life sciences and health mission(s) are stimulating the transformation of the healthcare sector towards being more centred on prevention.

The health missions have been formulated in an inclusive way, bringing in representatives from a wide range of stakeholders. Citizens are explicitly involved as they "are central coalition partners in the current innovation era" (Health Holland, 2020b). The missions have targeted, measurable, and time-bound attributes and seem to converge on one problem definition. The rather novel emphasis on preventive healthcare is linked with missions that hone in on cures and treatment interventions – areas in which the Dutch innovation system is traditionally strong (Iping et al., 2021). Many organisations in the healthcare sector are associated with specific diseases and treatments, including representative organisations of medical specialists and patients. Maintaining the link between prevention and cure is thus particularly important to keeping stakeholders on board and for legitimising the missions. This approach would appear to make it possible to sidestep the trade-off between legitimacy and directionality that generally occurs during mission formulation (Larrue, 2021). At the same time, the design and operationalisation of the transformative mission policy could be a difficult task, as policy instruments are diverse and originate from the remits of STI policy as well as the healthcare sector.

14.1.2 How Is the Mission Designed and Operationalised? In Particular: What Actor Groups and Instruments Are Proposed?

The Ministry of Health, Welfare and Sport defined the societal mission on "Health and care" and delegated the work to the Life Sciences & Health sector. Life Sciences & Health organises the work on the mission in an organisation called Health Holland. Within Health Holland, there is a "Theme Team" and a "Core Mission Team". The teams include representatives from a wide variety of parties: there are 97 coalition partners, 63 of which have specified that they are financial contributors. Additionally, Health Holland has an executive office consisting of innovation managers, programme coordinators, and valorisation advisers and thus should be seen as a mission-supporting agency.

For the "Health and care" mission topic, Health Holland has developed a Knowledge and Innovation Agenda 2020–2023 (KIA). Extensive consultations were held with knowledge institutes, businesses, organisations from civil society, and regional authorities to formulate the agenda (Janssen, 2020). Citizens and citizen representatives were actively engaged alongside traditional "triple helix" actors such as businesses, ministries, and knowledge institutes. The mission leadership promotes the inclusive approach to agenda-setting as a "system in the room" method of "convening and co-creating with all actors, strongly focusing on citizen and community agency" (IIPP, 2021).

A budget of 1.06 billion euros has been allocated for the "Health and care" topic; public and private stakeholders are each contributing approximately half of this figure. Four ministries are supporting the challenge, with the Ministry of Health the top ministry contributor and investing the most (8 per cent). Knowledge institutes, universities, university hospitals, and research councils are also providing a significant share (39 per cent) (EZK, 2019b).

The missions, as defined in the KIA, are operationalised in short-term "multi-annual innovation programmes" (MMIPs), which in turn are the starting points for both programming and projects. The main instruments available consist of the planning of calls for project proposals by the research councils, allowances for large-scale public-private partnerships for collaborative R&D projects – currently, 34 PPP projects are being supported, e.g., on regenerative medicine and animal-free innovations – and investments in small-scale illustration projects or GROZ[2] field labs ("GROZzerdammen"). Health Holland also serves as a supporter and coordinator for proposals in the context of the recently established National Growth Fund, which is aimed at supporting substantial investments in the Dutch knowledge infrastructure.

Despite the agenda-setting potential and instruments available, Health Holland mainly needs to convince stakeholders in the healthcare field to collaborate and work on the mission. Many stakeholders are represented in Health Holland's Theme and Core Mission Teams. At the same time, the healthcare sector typically consists of incumbent organisations and persons with vested interests, routines, and regulations. Health Holland's arsenal of policy instruments is not enough to allow it to dictate activities in the sector and to convince stakeholders to act in accordance with their KIA goals within these stakeholders' spheres of influence.

This raises an important question: How can mission-oriented STI policy promote and support stakeholders in their preventive healthcare activities? Below, we describe how three main types of stakeholders can and already do play a role in achieving the prevention mission: public actors, communities, and businesses.

In addition to the national-level STI policy on healthcare missions, *regional public policy efforts* have had a growing and prominent role to play over the last 10 years. Most notably, a wide array of economic boards – organisations that streamline interactions between companies, knowledge institutes, and government to support regional economic growth and innovation – and regional investment agencies – regional, public venture capital funds – have viewed preventive healthcare as a promising topic. Reasons for this include the emergence of technological opportunities, such as digital health innovations (e.g., eHealth, mHealth), the associated promise of economic activity, the need for addressing staff shortages in healthcare labour markets, and the ambition of creating healthier living environments.

In recent years, various public regional investment agencies in urban and rural areas have created innovation ecosystems that promote preventive healthcare, such as Health Valley in the eastern region of the Netherlands and Medical Delta in the Rotterdam-Delft-Leiden region. There are now 10 self-proclaimed "health innovation regions". These regions have strong links with economic missions and regional boards. Their activities include helping companies to develop their products by providing a field lab, supporting business development, networking, and funding technological and scientific projects. For example, Utrecht aims to become the number one region for self-management in healthcare, which will have benefits for the regional economy and citizens' health (Box 14.1). At the same time, the relations between regional public policy organisations and the national policy level have been disjointed: the ministry supported the efforts and content-related alignment on topics like eHealth, but these efforts were poorly coordinated and/or financially supported.

---

[2] GROZ is "zorg" (or care in Dutch) spelt backwards, which indicates the intention to approach healthcare in a radically different way.

BOX 14.1 A PUBLIC APPROACH TO PREVENTIVE HEALTHCARE: HEALTH HUB UTRECHT

The Health Hub Utrecht is a network of healthcare providers, governments (municipalities and provinces), and knowledge institutions that are working on establishing a healthy lifestyle for residents of the Utrecht region. The Hub has formulated three focus areas or "coalitions": neighbourhood-oriented prevention, digital transformation, and safeguarding an attractive labour market. Prevention should be viewed as being closely related to the other two focus areas of digital transformation and supporting an attractive labour market.

On the neighbourhood level, organisations like *citizen cooperatives and communities* feature prominently. Many of these communities have been founded by citizens and can be seen as bottom-up initiatives. Box 14.2 provides an example of a community-based approach.

BOX 14.2 A COMMUNITY-BASED APPROACH TO PREVENTIVE HEALTHCARE: SAMEN030

SAMEN030 is a city cooperative in the city of Utrecht that was founded and is run by citizens. The cooperative has four objectives:

1. To support neighbourhood-level cooperatives in which citizens, (social) entrepreneurs, and other neighbourhood-level organisations aim to foster healthy living, social cohesion, self-management, and participation.
2. To support a neighbourhood-based health economy by providing advice on channelling financial support and collective procurement.
3. To share knowledge and set up breakthrough projects.
4. To provide services such as administrative support.

The rationale behind the cooperative is to organise care on a neighbourhood level by making use of citizen participation and social cohesion. To pursue this, many institutional barriers need to be overcome. For example, the way care is financed is currently at odds with the increased emphasis on informal caregiving at a neighbourhood level.

SAMEN030 is one of the main drivers behind the GROZUtrecht field lab. The main goals of GROZUtrecht are lifestyle arrangements for residents, support for talent and job development, and more work in the community. The field lab mobilises IT solutions like a general neighbourhood portal and matching tool to provide informal care and a community time bank. They "will develop business cases in collaboration with stakeholders (health insurer, municipality and employers) based on experiences with a lifestyle arrangement, prototype or subscription" (SAMEN030 & UMC Utrecht, 2021).

Regional and national policy bodies are attempting to connect to these bottom-up initiatives in light of their enthusiasm about their ideas. Health Holland prioritises the initiation and support of such communities or field labs. Their field lab approach ties in with the idea of creating a "quadruple helix", in which citizen initiatives, governments, care providers, societal organisations, and insurance companies work together to enhance healthy lifestyles and healthy living environments (Health Holland, 2020b). There are currently four GROZ field labs (Eindhoven, Utrecht, Deventer, and the Northern Maas Valley) where crucial actors such as citizens, local and regional government, care and welfare institutions, knowledge and educational institutions, health insurers, employers, and entrepreneurs are collaborating on prevention innovation. Citizens are treated as equal partners in these field labs, and cooperatives and citizen advocates are given ample room to participate or to lead. The GROZ field labs support design processes and aim to overcome the lack of (local) business cases by stimulating new forms of finance, such as Social Impact Bonds and Health Impact Bonds, and/or providing a regulatory sandbox (EZK, 2019a). In terms of innovation, the emphasis is on both technological products and low-tech interventions, as well as on the embedding of existing technologies through what is called "assemblage". However, financial support for these field labs is scarce.

Lastly, *businesses* are working on various innovations for preventive healthcare, mostly technological in nature. They work on digital health or eHealth, including IT applications, to support healthcare and prevent future illnesses (Timmer, 2015; Kampmeijer et al., 2016). Based on our research among eHealth scale-up companies in the Netherlands, we identified three types of routes to end users (Table 14.1): direct-to-consumer eHealth products and services (e.g., wearables), eHealth applications that are supervised by and prescribed through medical specialists (e.g., digital physiotherapy training programmes), and large eHealth platforms that are procured and used on an organisation-wide scale (e.g., electronic patient record systems in a hospital).

The routes through medical specialists and care organisations require local and regional integration. In many cases, companies engage in local field labs to experiment or connect with regional economic boards and get support. Box 14.3 gives an example of an emerging private-sector, direct-to-consumer eHealth tool that is potentially disruptive and combines services and products in primary care and prevention.

Table 14.1 Routes to end users for digital preventive healthcare innovations

Type of market	Examples	Upside	Downside
Direct-to-consumer	Commercial genetic tests, activity trackers, primary care platforms	Low barriers to market	Low margins
Prescribed through medical professionals	eHealth training programmes, tele-diagnostic tools	High margins	Evidence is required; high barriers to market; innovation might impinge on the work and revenue of the medical profession
Used inside care organisations	Patient record systems, smart home solutions	Possibility of stable demand	Oligopsony market structure

BOX 14.3 A PRIVATE APPROACH TO PREVENTIVE HEALTHCARE: SUBSCRIPTION-BASED HEALTHCARE PLATFORMS

There are many advantages to subscription-based platforms for primary care, through which patients can interact with general practitioners on a 24/7 basis and get access – in consultation with medical professionals – to a library of self-management tools to improve the prevention of diseases. For producers of individual eHealth products, such a platform could accelerate upscaling as they could offer their products on the platform's library. For users, on the other hand, such a platform could offer flexibility and tailored support. For healthcare providers, the range of options is being expanded and allows for coordination between online and offline treatment. Such platforms are emerging in the United States (e.g., Go Forward, Amwell, Doctor on Demand), China (Ping An Good Doctor), and Europe (Babylon Health).

Public actors, communities, and businesses are thus engaged in various actor groups that are active in preventive healthcare and consist of a wide variety of stakeholders. The variety can be mapped in terms of multiple dimensions, including public–community–private or national–regional–local. Combinations are also possible, such as community-based activities like those of SAMEN030, which are also firmly established within public regional initiatives, companies using field labs as places of experimentation, and regional and national policymakers attempting to align and coordinate their efforts.

This demonstrates that, in an attempt to deal with the variety of approaches and actor groups, the missions are operationalised through a top-down coordination of efforts that include significant responsibilities and possibilities for regional, public, private and community actors to participate in innovation. The missions are then further operationalised via a healthcare ecosystem consisting of many powerful actors with vested interests and governed by a vast number of regulations. Since the national financial and regulatory policy measures are quite limited, there is ample room for regional and local actors to take the lead; the vested interests and institutional rigidities of the healthcare sector ensure that they are able to do so. In a sense, we see top-down missions as only part of the process, which ties sectors and actors together through "multiple nestedness" on at least three dimensions: mission–program–project; national–regional–local; and community–public–private. We can also see that there is a major role to be played by actors that are traditionally relegated to the peripheries, such as citizen-led collectives, who can act more quickly on missions and can cross boundaries between different networks. Obviously, the "multiple nestedness" raises questions regarding means of ensuring that the coordination process works to create something that is bigger than its parts, of learning across initiatives, and of facilitating interaction between actors from different institutional backgrounds.

14.1.3 To What Extent Do the Implemented Instruments Work towards the Mission Goals and Thus the Transformation, against the Background of Existing Misalignments and Related (Institutional) Complexities?

The way and contexts in which the variety of approaches and dimensions of transformative mission-oriented STI policy for preventive healthcare are combined are still open for debate. If a transition towards a more preventive healthcare system is to succeed, then policy must deal with existing practices and related (institutional) complexities that stagnate and stall preventive healthcare innovation. Below, we identify three issues – already presented in the introductory section – that contribute to this form of stagnation and assess how current transformative mission policy contributes to these issues: institutional incompatibility and lack of diffusion, failures in collaboration, and difficulties in integrating technological and social innovations. It is important to acknowledge that the mission policy has just started, so a fully-fledged ex-post evaluation is not yet possible.

14.1.3.1 Issue 1: Institutional incompatibility and a lack of diffusion due to waiting games

Large-scale deployment of preventive innovations is most beneficial for companies when it takes place via the "prescription by medical professionals" route (Table 14.1). However, this route presents certain difficulties, such as the institutional pressure to gather evidence on safety and efficacy in order to support the claim that an intervention is adding "therapeutic value". Medical specialists require this evidence to decide whether an innovation becomes part of medical guidelines that advise diagnostic and therapeutic workflows. An additional challenge is that health insurance agencies use "added therapeutic value" to determine whether an innovation becomes part of the basic insurance package.

There are two obstacles to gathering data on safety and efficacy. First, validation data must be collected through costly and time-consuming clinical trials. In addition, there is extensive evidence that companies originating from the digital sector find it difficult to include the validation steps in their existing innovation process because evidence-based medicine is at odds with agile innovation (Steinberg et al., 2015). The healthcare and digital innovation approaches represent distinct and separate institutional logics (Thornton & Ocasio, 1999).

Second, data can only be gathered while the intervention is in use, but an intervention will only be widely used once data on its therapeutic value is available. This can be characterised as a "waiting game" (Potstada et al., 2016) or catch-22 situation: actors wait for each other to reduce the uncertainties before they invest in a specific innovation. Waiting games have been identified in the nanomedicine field (D'Silva et al., 2012), while Agogué et al. (2012) also mention the presence of waiting games in development processes for eHealth products. Such situations of innovation stagnation are particularly detrimental for radical or architectural innovation processes that are already characterised by a high level of uncertainties and unknowns and are accompanied by high expectations (Robinson et al., 2012). Therefore, it is evident that waiting games hinder the development of innovations that contribute to the transition towards preventive healthcare. The GROZ field labs can be seen as protective spaces in which actors learn and gather data to validate preventive interventions, but current activities in these field labs are not geared towards this.

Regarding the "used inside care organisations" route, problems with large-scale deployment lie in the fact that the adoption of eHealth often stagnates after the initial phase of interest or commitment (Ciere et al., 2019). Users do not know about the solutions that eHealth products offer, meaning that solutions often remain local and that entrepreneurs face the challenge of promoting the eHealth solution to a wider range of users (Nictiz & Nivel, 2015). Our research on eHealth scale-up companies shows that they already have customers and turnover, but their business models are often based on the first version of a product and tied to a specific launching customer and their practices and workflows. This means that when scaling up, a reorientation of the business model is seen as necessary, as are investments in additional research and in context-specific design. The latter involves de-contextualisation and re-contextualisation during implementation of prevention applications. As part of the mission-oriented STI policy, the health ministry actively worked on solutions such as scaling-up acceleration advisors, "all systems in one room" sessions, and creating an intermediary organisation to facilitate scaling-up. So far, these solutions have been quite ad hoc.

14.1.3.2 Issue 2: Collaboration failures due to the lack of a business case

Prevention is not easily funded because the costs are weighed before the benefits, and because additional actors or policy fields can often benefit from investments in prevention, which causes actors some trepidation regarding investing. Moreover, preventive measures are traditionally not part of the Dutch basic insurance package that insurance companies usually cover. Health insurance is an indemnity insurance whereby costs are only reimbursed after damage has occurred: reimbursing costs to prevent damage does not align with this principle.[3] In the Netherlands, there is currently no clear institutionalised pathway for the reimbursement of eHealth or mobile health applications. The reimbursement of such applications has been institutionized in other European countries: Germany in 2022 (BfArm, 2022) and Belgium in 2021 (Medscaler, 2021). Other countries such as France are currently considering implementing this reimbursement.

One of the solutions for dealing with the discrepancy between costs and benefits is to increase inter-organisational collaboration and alignment. In the context of the "Health and care" mission, several parties have started to search for innovative reimbursement schemes. Insurance companies have experimented with benefit-based strategies, making long-term commitments between insurers and the insured to co-invest in the patient's health (Plochg & Hoekstra, 2021). Community collectives have developed shared saving models that work for and within neighbourhoods, such as regional vitality funds or time banks that aid in the deployment of informal caregivers who are organised in clearly demarcated communities or "lots" (HealthKIC, 2022).

The citizen-led collectives that are actively pursuing these solutions have created novel actor groups supported by regional and national policy initiatives in the context of the "Health and care" mission topic. However, there are still signs that some parties may have a vested interest in blocking the usage of prevention tools since they might make (part of) their operations obsolete or face production loss (Bremmers, 2016). Moreover, the collaborations need to

[3] Many preventive activities are already included in the basic package, such as cholesterol-lowering medicines, check-ups for pregnant women with midwives, and dietary advice. With the publication of the National Prevention Agreement in 2018, the Dutch health ministry announced that more individual prevention will be covered by the basic package, including exercise programmes and smoking cessation interventions when prescribed by medical doctors.

extend beyond healthcare, as the benefits of healthcare prevention might accrue to other areas such as social security funds or employers.

14.1.3.3 Issue 3: Prevention means combining technologies with social innovation

Although technologies like AI, blockchain, and eHealth are integral to dealing with preventive healthcare, the role of social innovation has begun to play a more prominent role (Mason et al., 2015). Focusing on prevention means that lifestyle interventions are needed, which may be triggered by mobile phone apps or online training programmes. Interventions can also be implemented by means of neighbourhood-level school networks, primary care facilities, and sports associations. In other words, technological interventions alone do not suffice, and even when technologies are adopted, long-term adherence is lacking (Sieverink et al., 2017). Adoption of preventive technologies requires accompanying policies spanning two dimensions: (1) the need for technologies to become part of the adopter's daily practices in order to remain in use over a longer period of time, and (2) the need for technological options to be complemented with connections to organisations that engage with adopters, such as sports associations and schools.

eHealth as a technological contribution to preventive medicine should thus be combined with social initiatives in order to become effective. The "Health and care" topic explicitly refers to this by highlighting that "technological and social innovation should go hand in hand" (EZK, 2019a). An example of a social innovation is a vitality contract, which is a subscription-based model where citizens gain access to a broad range of resources for staying healthy, such as eHealth apps, wearables, easy access to fruits and vegetables, regular consultations with a health coach, and mental training. This requires a new kind of social arrangement and demands sustainable co-creation with local communities, employers, municipalities, healthcare insurance companies, and/or private platforms (Plochg et al., 2017). Social innovation in this context means novel actor arrangements and ways of working, and gives a prominent role to user initiatives and bottom-up activities.

One of the challenges associated with social innovations is that they are not easy to define and patent, which makes attracting investments difficult. An additional challenge is the complexity of managing social innovation implementation, as assistance and further development during the implementation phase are required for the innovation to be successful (Janssen et al., 2015). For proper implementation, the innovation must not only become embedded in the clinical workflow, but should also be extended to extramural settings such as homes and neighbourhoods. The GROZ field labs are regarded as focal areas and instruments to co-create the combinations of technological and social innovations. Actual cross-linkages and solutions are further explored in these settings. However, more structural – or institutional – solutions are not currently offered through the mission policies, meaning that the responsibility is left to the local levels.

Therefore, the three issues described above show that the reasons for the stagnation of preventive healthcare innovations are a result of institutional misalignments and related complexities.

14.2 DISCUSSION AND CONCLUSION

Preventive healthcare is at the centre of the current transformative mission-oriented innovation policy in healthcare in the Netherlands. We can see that articulating the mission of preventive healthcare and arranging the activities and instruments required to achieve that mission are not straightforward, especially because of misalignments and related (institutional) complexities within the healthcare sector. Although mission-oriented STI policy has made a first attempt at creating new institutions, practices, and organisations, an initial assessment of these efforts showed that there is a need for further development. To explore this, we have examined three questions.

The current health mission requires a completely new approach to coordinating heterogeneous sets of actors and institutions and linking economic and health goals in the context of a "need domain policy". The mission articulation has been quite inclusive so far, involving a wide range of actor types, amongst which citizens and their representatives stand out as unusual actors. This is necessary because the field of preventive healthcare is vast and highly fragmented. A more general theoretical consideration is needed if the mission is to become – or remain – effective as a transformative mission, and *the trade-off between inclusiveness and the related issue of legitimacy and directionality must be managed.* Prioritisation could become a challenge in this regard; a disproportionate focus on certain actor interests or a push for certain technologies might endanger the effectiveness of mission policy.

The mission design and operationalisation are characterised by a top-down coordination of efforts; however, they also entail major responsibilities and possibilities for regional, public, private, and community actors. In fact, long-standing regional policy initiatives have prepared for and are actively linked to the national mission. The same goes for bottom-up community initiatives that have come on board and, in some cases, become figureheads of the mission. Missions thus bring sectors and actors together through "multiple nestedness" on at least three dimensions: mission–program–project; national–regional–local; and community–public–private. The remaining regional initiatives and the plethora of citizen-led initiatives and local field labs that are already in place indicate that it is not that there is a lack of policy initiatives, but that the actors and their goals are misaligned. For STI policy to make an impact on healthcare, connections must be established between these dimensions. An additional challenge is the operationalisation of missions to form coordinated policy efforts and STI policy instruments that align well with the vested interests of actors and institutional rigidities already present in the healthcare domain. The *"multiple nestedness" thus calls for coordination and connections between actors with different (institutional) backgrounds and interests.*

The articulation and implementation of transformative mission-oriented innovation policy for preventive healthcare are made difficult due to misalignments and (institutional) complexities in relation to three issues. First, institutional incompatibilities between technology developers and healthcare professionals lead to waiting games, i.e., situations in which actors look to each other to take on certain activities to advance innovation instead of taking the initiative themselves. Solutions can be found in protective spaces where experiments can be conducted or by supporting intermediary organisations that facilitate scaling, but implementing these solutions across the board is challenging. Second, prevention measures involve distributed costs and benefits and thus require an increase in collaboration and inter-organisational alignment, which involves institutional change and disrupting actors' vested interests within and beyond the healthcare sector. Third, prevention means combining technologies with social innovation because technology always needs to be enmeshed with novel workflows and actor arrangements (social innovation). Socio-technical integration is often organised at a local level, which requires a wide variety of policy measures and the involvement of a variety of actors. In conclusion, transformative mission-oriented innovation policy in the field of preventive healthcare *needs to deal with institutional incompatibility and a lack of diffusion, failures in collaboration and difficulties with integrating technological and social innovations.*

The problems related to transformative mission-oriented STI policy in the healthcare sector can be addressed by advancing coordination and collaboration, as well as advancing demand-side policy interventions that address issues such as waiting games and a lack of market formation activities (Boon & Edler, 2018; Boon et al., 2022). Examples of potential interventions include actively changing market institutions (reimbursement rules, etc.), changing user practices to integrate technological and social innovations, improving demand articulation, and creating protective spaces (e.g., in the form of GROZ field labs) for collecting data to validate preventive interventions. This chapter has illustrated the complexity of these issues and demonstrated that the solutions are not straightforward. At the same time, examining these issues is important for improving preventive healthcare and for advancing transformative mission-oriented STI policy in sectors that are associated with societal challenges where costs come before benefits, such as efforts to make the economy more circular and societies more climate resilient.

REFERENCES

Agogué, M., Masson, P. L., & Robinson, D. K. (2012). Orphan innovation, or when path-creation goes stale: A design framework to characterise path-dependence in real time. *Technology Analysis & Strategic Management*, *24*(6), 603–616. https://doi.org/10.1080/09537325.2012.693672

Ahli, S. (2020). Preventie kampt met opstart-problemen. *Zorgvisie*, *50*(2), 38–41. https://doi.org/10.1007/s41187-020-0325-7

BfArM (2022). https://diga.bfarm.de/de.

Boon, W., & Edler, J. (2018). Demand, challenges, and innovation. Making sense of new trends in innovation policy. *Science and Public Policy*, *45*(4), 435–447. https://doi.org/10.1093/scipol/scy014

Boon, W. P., Edler, J., & Robinson, D. K. (2022). Conceptualizing market formation for transformative policy. *Environmental Innovation and Societal Transitions*, *42*, 152–169. https://doi.org/10.1016/j.eist.2021.12.010

Bremmers, S. (2016). Wel de kosten, niet de baten. *Zorgvisie*, *46*(4), 30–31.

Ciere, Y., Vaart, R. Van Der, Jong, A. E. V. D. M., Maljaars, P. W. J., Buul, A. R. Van, Koopmans, J. G., et al. (2019). Clinical eHealth Implementation of an eHealth self-management care path for chronic somatic conditions. *Clinical eHealth*, *2*, 3–11. https://doi.org/10.1016/j.ceh.2019.04.001

D'Silva, J., Robinson, D. K., & Shelley-Egan, C. (2012). A game with rules in the making – How the high probability of waiting games in nanomedicine is being mitigated through distributed regulation and responsible innovation. *Technology Analysis & Strategic Management*, *24*(6), 583–602. https://doi.org/10.1080/09537325.2012.693671

Essén, A., Krohwinkel, A., Wennberg, K. (2022). *Assessing whether mission-driven innovation makes a difference: Mission impossible? developing a guiding framework for the evaluation of five mission driven environments for health in Sweden.*

Eyler, A. A., Chriqui, J. F., Moreland-Russell, S., & Brownson, R. C. (2016). *Prevention, policy, and public health.* Oxford University Press.

EZK (2019a). *Toelichting Missies voor het topsectoren – en innovatiebeleid.* Tweede Kamer, 26 April 2019. EZK, Den Haag.

EZK (2019b). *Kennisen Innovatieconvenant 2020–2023.* 12 November 2019. EZK, Den Haag.

Gmeinder, M., Morgan, D., Mueller, M. (2017) How much do OECD countries spend on prevention? OECD Health Working Papers No. 101. OECD.
Greenhalgh, T., Wherton, J., Papoutsi, C., Lynch, J., Hughes, G., Hinder, S., et al. (2017). Beyond adoption: A new framework for theorizing and evaluating nonadoption, abandonment, and challenges to the scale-up, spread, and sustainability of health and care technologies. *Journal of Medical Internet Research, 19*(11), e8775. DOI: 10.2196/jmir.8775.
HealthKIC (2022). https://www.kic.nl/
Health Holland (2019). *Five more years of healthy living for everyone.* Health Holland.
Health Holland (2020a). *Topsector Life Sciences & Health – Van beleid naar impact.* Health Holland.
Health Holland (2020b). *Gezondheid & Zorg – Vitaal functionerende burgers in een gezonde economie.* Health Holland.
Hoogstraaten, M., Toorneman, R., Langkamp, G., Boon, W. (forthcoming). *Institutional barriers for MedTech companies and their strategies to overcome them.*
IIPP (2021). *Mission-oriented innovation in action – 2021 casebook.* IIPP/UCL.
Iping, R., Kroon, M. E., Steegers, C. A. A., & Van Leeuwen, T. N. (2021). *Research impact of the Dutch University Medical Centres.* NFU/CWTS.
Janssen, M. (2020). *Post-commencement analysis of the Dutch "Mission-oriented Topsector and Innovation Policy" strategy, MIPO Report.* Copernicus Institute of Sustainable Development, Utrecht University.
Janssen, M., Stoopendaal, A. M. V., & Putters, K. (2015). Situated novelty: Introducing a process perspective on the study of innovation. *Research Policy, 44*(10), 1974–1984. https://doi.org/10.1016/j.respol.2015.06.008
Kampmeijer, R., Pavlova, M., Tambor, M., & Groot, W. (2016). *The use of e-health and m-health tools in health promotion and primary prevention among older adults: A systematic literature review, 16*(Suppl 5). https://doi.org/10.1186/s12913–016–1522–3
Kuittinen, H., Polt, W., Weber, K. M. (2018). Mission Europe? A revival of mission-oriented policy in the European Union. In Council for Research and Technology Development (Ed.), *Re: thinking Europe. Positions on shaping an idea* (pp. 197–212). Holzhausen.
Kulkov, I., Tsvetkova, A., & Ivanova-Gongne, M. (2021). Identifying institutional barriers when implementing new technologies in the healthcare industry. *European Journal of Innovation Management*, 1460–1060. https://doi.org/10.1108/EJIM-02–2021–0093
Larrue, P. (2021). *The design and implementation of mission-oriented innovation policies: A new systemic policy approach to address societal challenges.* OECD Science, Technology and Industry Policy Papers. OECD.
LS&H (2011). *Topsectorplan Life Sciences & Health – Voor een gezond en welvarend Nederland.* Agentschap NL.
Mason, C., Barraket, J., Friel, S., O'Rourke, K., & Stenta, C. P. (2015). Social innovation for the promotion of health equity. *Health Promotion International, 30*(Suppl.2), ii116–ii125. https://doi.org/10.1093/heapro/dav076
Medscaler (2021). https://www.famhp.be/en/human_use/health_products/medical_devices_and_their_accessories/manufacturers/health_applications
Mukherjee, S. (2010). *The emperor of all maladies: A biography of cancer.* Simon and Schuster.
Nictiz & Nivel (2015). *Tussen vonk en vlam.* Nictiz/Nivel.
Oderanti, F. O., Li, F., Cubric, M., & Shi, X. (2021). Business models for sustainable commercialisation of digital healthcare (eHealth) innovations for an increasingly ageing population: (A new business model for eHealth). *Technological Forecasting and Social Change, 171.* https://doi.org/10.1016/j.techfore.2021.120969
Plochg, T., & Hoekstra, P. (2021). Batenstrategie voor toekomstbestendige gezondheidszorg. *Zorgvisie Magazine, 1.* https://www.zorgvisie.nl/batenstrategie-voor-toekomstbestendige-gezondheidszorg/
Plochg, T., Van der Velden, N., & Keijsers, J. (2017). Gezond zijn loont. *Zorgvisie Magazine, 7,* 22–24. https://www.federatievoorgezondheid.nl/media/1113/plochg-vdvelden-keijsers-zorgvisie-2017.pdf
Polt, W., Ploder, M., Breitfuss, M., Daimer, S., Jackwerth, T., Zielinski, A. (2021). *Politikstile und Politikinstrumente in der F&I-Politik. Studien zum deutschen Innovationssystem.* http://hdl.handle.net/10419/231475

Potstada, M., Parandian, A., Robinson, D. K., & Zybura, J. (2016). An alignment approach for an industry in the making: DIGINOVA and the case of digital fabrication. *Technological Forecasting and Social Change, 102*, 182–192. https://doi.org/10.1016/j.techfore.2015.07.020
RegMed XB (2022). https://regmedxb.com/
Robinson, D. K., Masson, P. L., & Weil, B. (2012). Waiting games: Innovation impasses in situations of high uncertainty. *Technology Analysis & Strategic Management, 24*(6), 543–547. https://doi.org/10.1080/09537325.2012.693661
Rogers, E. M. (2002). Diffusion of preventive innovations. *Addictive Behaviours, 27*(6), 989–993. https://doi.org/10.1016/S0306-4603(02)00300-3
SAMEN030 & UMC Utrecht (2021). *GROZUtrecht Werkplan 2021/2022*. SAMEN030 & UMC Utrecht.
Sieverink, F., Kelders, S. M., & van Gemert-Pijnen, J. E. (2017). Clarifying the concept of adherence to eHealth technology: Systematic review on when usage becomes adherence. *Journal of Medical Internet Research, 19*(12), e402. DOI: 10.2196/jmir.8578.
Steinberg, D., Horwitz, G., & Zohar, D. (2015). Building a business model in digital medicine. *Nature Biotechnology, 33*(9), 910–920. https://doi.org/10.1038/nbt.3339
Thornton, P. H., & Ocasio, W. (1999). Institutional logics and the historical contingency of power in organizations: Executive succession in the higher education publishing industry, 1958–1990. *American Journal of Sociology, 105*(3), 801–843. https://doi.org/10.1086/210361
Timmer, S. (2015). *EHealth in de langdurige zorg: De praktijk van de ouderen- en gehandicaptenzorg*. Bohn Stafleu van Loghum, Springer Media.
Tollefson, J. (2021). What the rise of "ARPA-everything" will mean for science. *Nature, 595*, 483–484. https://doi.org/10.1038/d41586-021-01878-z
Wanzenböck, I., Wesseling, J. H., Frenken, K., Hekkert, M. P., & Weber, K. M. (2020). A framework for mission-oriented innovation policy: Alternative pathways through the problem–solution space. *Science and Public Policy, 47*(4), 474–489. DOI: 10.1093/scipol/scaa027.
Wittmann, F., Hufnagl, M., Lindner, R., Roth, F., Edler, J. (2020). Developing a typology for mission-oriented innovation policies. *Fraunhofer ISI Discussion Papers-Innovation Systems and Policy Analysis*.

15. Governance arrangements for the implementation of transformative innovation policy: insights from a comparative case study

Matthijs Janssen, Iris Wanzenböck, Lea Fünfschilling and Dimitrios Pontikakis

15.1 INTRODUCTION

Transformative innovation policy (TIP) (Weber & Rohracher, 2012; Steward, 2012) requires new ways of using science, technology, and innovation to address intransigent societal challenges in a more explicit and targeted manner than previous innovation policy framings or rationales have called for (Schot & Steinmueller, 2018). One distinctive feature of this new policy framing is the call to go beyond just strengthening existing innovation systems and instead pursue change and transformation in socio-technical systems that fulfil societal functions such as mobility, health, food, and housing (Diercks et al., 2019). Coherent policy programmes that follow the TIP framework aim to achieve system innovation (Smits & Kuhlmann, 2004; Van Mierlo et al., 2010), that is, the structural reconfiguration of dominant actors, institutions, technologies, and infrastructures that define entire production and consumption systems.

Much has been written on the need for and rationales behind more transformative innovation policies (Weber & Rohracher, 2012; Schot & Steinmueller, 2018). However, very few studies focus on the implementation side of transformative programmes in order to understand what public organisations and policymakers must do to engage in systemic change processes and become more "transformational" (Borrás & Edler, 2020). Researchers have now started to investigate different public sector roles (Borrás & Edler, 2020) and types of transition capacities (Hölscher et al., 2019) for governing system transformations. For example, previous works in the innovation policy field have focussed on the need for policy learning and organisational capacities (Borrás, 2011) or on dynamic capabilities within the public sector in the context of mission-oriented innovation policies (Kattel & Mazzucato, 2018). In contrast to hands-off neo-liberal policies, the implementation of transformative and mission-oriented policies requires governments to have extensive knowledge of these areas as well as the organisational capacity to operate in the system they are trying to change (Kattel & Mazzucato, 2018; Ghosh et al., 2021). However, as found by Braams et al. (2021) in a literature review of scholarly work on public administration traditions, most of these traditions appear to be unsuitable for transition tasks such as "giving direction" or "destabilising the unsustainable".

So far, little attention has been given to the governance arrangements through which governments can execute the tasks needed for driving transformative change processes in society. We define *governance arrangements* as the organisational structures, processes, and capabilities of public organisations through which the interactions between government and societal actors are organised and through which policy is shaped and implemented. Since they are embedded in particular norms, traditions, networks, and organisational structures, we consider governance arrangements to be the foundation for the implementation of policy instruments. Unfortunately, earlier research contains little detail on the ways in which public organisations and policymakers adapt their organisational practices to pursue innovation policies that can realise transformative effects in society and the economy.

To examine the possibilities and difficulties in governing system change, we have adopted the idea that governments perform according to distinct governance modes (Van der Steen et al., 2015) which are either administration-based (i.e., guided by hierarchy), network-based (i.e., in cooperation with societal actors), or society-based (i.e., led by societal actors). These governance modes define how public actors act and coordinate their activities and what forms the basis for the legitimacy of their actions (Bourgon, 2011). These governance modes are also applicable to diverse forms of interaction within the public sector and between government and societal stakeholders such as businesses, practitioners, community-based organisations, and civil society. In this chapter, we assume that the different governance modes under which policymakers operate create different conditions and challenges for adapting or introducing governance arrangements (and thus implementing TIP). This, in turn, has led us to pose the following research question: *How do transformative innovation policy governance arrangements unfold in different governance modes and what challenges typically arise for each arrangement?'* Our aim is to develop a realistic and contextualised account of what becoming "transformative" entails for public organisations.

This chapter will concentrate primarily on a comparative study of seven transformative policy programmes recently implemented at the regional or national level across Europe. The programmes were selected based on their ambition of addressing specific societal challenges by achieving system-wide transformation. Three of them focus explicitly on societal missions. Our analytical framework for studying and comparing these programmes consists of four TIP governance tasks (providing legitimacy and leadership, supporting coordination, facilitating policy learning, and resolving conflicts) and three distinct governance modes (administration-, network-, or society-based). Through document analysis and interviews with policy implementers and experts, we provide in-depth insights into the ways in which different TIP governance arrangements are enacted by public organisations. We also analyse whether and/or how governments change their formal and informal structures and processes to meet their TIP ambitions. However, seven cases is far too limited a sample to give a comprehensive account of possible ways of designing governance arrangements; instead, the empirical analysis serves to provide concrete illustrations, to demonstrate the use of the analytical framework, and to explore the type of insights that might be obtained when looking at TIP implementation through the lens of governance modes.

This study falls within an emerging class of literature that views designing appropriate governance structures, processes, and practices as an essential factor in unleashing the transformative potential of missions and other types of directional innovation policies (Reale, 2021; Janssen et al., 2021; Bours et al., 2022). From this perspective, the most important thing is not just to find the right policy instruments but also to engage in the right activities, i.e., to mobilise public and societal stakeholders, to explore and co-create promising new solutions, and to develop them further so as to support system-wide change. We aim to shed light on this complex area of policy implementation, particularly regarding potential difficulties emerging from the need to change organisational structures and practices in the public sector. This issue is often disregarded in the literature on transformative or mission-oriented innovation policy. Our study offers a basis for the systematic analysis of governance arrangements that are used to implement transformative innovation policy tasks. Moreover, although the empirical analysis presented here is still exploratory, some conclusions can be drawn which apply to the possibilities and constraints under which policy organisations operate.

15.2 ANALYTICAL FRAMEWORK

15.2.1 Transformative Innovation Policy Governance Arrangements

In order to systematically compare governance arrangements for TIP, we will examine the principles or tasks that should, according to current literature, be followed or completed by public organisations when creating and implementing transformative policies. Reviewing almost a decade of transformative innovation policy literature (Weber & Rohracher, 2012; Steward, 2012; Schot & Steinmueller, 2018; Diercks et al., 2019; Grillitsch et al., 2019; Salas et al., 2019) has indicated that several features have been, often implicitly, associated with public policies that aim to promote transformative change in society. It is clear that directionality is seen as one of the core principles of the new innovation policy framework (Schot & Steinmueller, 2018), in the sense that governments take on an active role in promoting and shaping the direction of innovation and societal change (Mazzucato, 2018; Diercks et al., 2019).

However, we will look beyond directionality to achieve a better understanding of TIP governance in terms of the structures, processes, and concrete activities public organisations adopt to pursue their transformative aims. Using the framework proposed by Braams et al. (2021), Table 15.1 presents an overview of the TIP governance tasks that we identified when reviewing the relevant innovation and transition policy literature. We have divided them into four generic categories: (1) creating legitimacy and leadership; (2) coordinating across multiple levels, actors, and instruments; (3) reflexivity, learning, and experimenting; and (4) resolving conflicts. These task categories serve mainly as an analytical frame of reference to allow us to explore and compare various concrete governance arrangements from actual TIP programmes in the course of our empirical study.

15.2.2 Governance Modes

The second pillar of our analytical framework for studying governance arrangements for TIP consists of the governance modes that characterise how public organisations interact with society. In policy studies literature, governance modes are referred to as the institutional structures and actor groups involved in collective decision-making, denoting the various forms through which governance can be executed (e.g., Treib et al., 2007, Lange et al., 2013, Van der Steen et al., 2015). In a TIP context, governance modes may be viewed as the arenas where transformative agendas and concrete measures are negotiated and decided upon (Fink-Hafner, 1998; Jørgensen & Andersen, 2012). They define the legitimacy and boundaries of public and other societal stakeholders who initiate and navigate change processes by supporting (or hampering) the emergence of new forms of coordination, learning, or a new culture of experimentation. They also co-determine the choice of policy instruments, such as how hierarchical "steering" (e.g., in the form of regulation by public authorities) is combined with "self-regulating" devices (e.g., voluntary agreements or market-based incentives) (Jordan et al., 2005).

Table 15.1 Four categories of TIP governance tasks

Creating legitimacy and leadership	**Coordinating across multiple levels, actors, and instruments**	**Reflexivity, learning, and experimenting**	**Resolving conflicts**
Identifying and demonstrating transformative failures (Weber & Rohracher, 2012)	Focus on vertical and horizontal policy coordination (Weber & Rohracher, 2012; Larrue, 2021)	Reflexive, experimental, and tentative governance approaches, continuous policy adaptation (Kemp & Loorbach, 2006; Torrens & Schot, 2017; Kuhlmann et al., 2019; Janssen, 2019)	Destabilisation: deliberate decline, exnovation, and phasing out policies (Rogge & Johnstone, 2017; Hekkert et al., 2020; Klerkx & Begemann, 2020)
Acknowledging and managing the normativity of innovation policies for societal challenges (Uyarra et al., 2019; Schlaile et al., 2017)	Focus on a mix of policies, pursue multi-instrumental policy approaches (Rogge & Reichardt, 2016)	Process-orientation: policy learning, formative and system-oriented evaluation (Borrás & Laatsit, 2019; Molas-Gallart et al., 2020; Wanzenböck et al., 2020)	Embracing contestation, complexity, and uncertainty (Schot & Steinmueller, 2018; Wanzenböck et al., 2020)
Establishing accountability mechanisms (Rogge & Reichardt, 2016)	Stimulate trans- and multi-disciplinarity beyond epistemological boundaries (Cagnin et al., 2012)	Focus on social and organisational learning, second-order learning (Grin et al., 2010; Van Mierlo & Beers, 2020)	Charting actionable and acceptable development pathways (Schot & Steinmueller, 2018)
Focussing on system-level impact (Weber & Rohracher, 2012; Hekkert et al., 2020)			Tilting the playing field (Kattel & Mazzucato, 2018)

In the following section, we will outline the distinctions between three archetypical governance modes: administration-based, network-based, and society-based and situate them on a spectrum between the two extremes of "hierarchy" and "community". Kooiman (2003) also described these two poles as hierarchical governance and co- or self-governance. These governance modes differ in: (1) their degree of formal institutions, (2) the relative importance of public and non-public actors or their membership flexibility, and (3) the extent to which they rely on (self-, co-, or hierarchically) organised processes to achieve societally desirable outcomes. We characterise the modes as follows:

1. *Administration-based governance:* This governance mode follows the classical bureaucracy type of government, which executes policy in light of the societal goals determined in political processes (Van der Steen et al., 2018). Public authorities have a set of rules and resources at their disposal, and these create stability as they pursue the priorities of the political system they represent. It is up to policymakers to translate political decisions about what is societally desirable into concrete measures on the basis of existing laws,

formal rules, and procedures. The structure of this mode works from the inside out, with the government and its agencies forming the central unit for producing desired results effectively and efficiently; other societal stakeholders such as businesses, NGOs, or citizens are reduced to acting as "receivers" or agents that help produce the desired results.

2. *Network-based governance:* The network governance model is based on co-governance between public actors and societal parties who engage in collective action for a specific purpose (Rhodes, 1997). It is less centralistic than the administrative model and may appear in various forms, from more flexible alliances to dedicated public-private partnerships with strategic agendas. Public organisations engage in these forms of co-organisation because societal actors (such as businesses or NGOs) possess knowledge and resources that are crucial for creating more effective policies and incentives for pursuing a public goal. They act more like partners in this purpose- and result-oriented model, with each party taking on specific roles in organising the collective processes and protecting process quality in terms of participation, transparency, and timing.
3. *Society-based governance:* This last model draws on societal resilience, as found in the willingness and capacity of societal stakeholders to self-organise and undertake action, either alone or in networks and cooperatives (Van der Steen et al., 2018). In contrast to the other two governance types, it is not a government-created capacity per se. This mode, which is the most informal of the three, is driven by the concept of societal self-governance and provides certain flexibility when coordinating various stakeholders (market and non-market actors, including citizens) and institutions (formal and informal) in the co-production of services for a community (Thompson et al., 1991). Society-based governance therefore often relates to more informal, local, and participative governance approaches but does not exclude governance via market incentives. Like the network governance model, it works from the outside in but places an emphasis on initiatives, stakeholders, and change processes that emerge organically. Rather than negotiating shared agendas in a top-down fashion, societal self-regulation is based on the concept of bottom-up governance or community, with a focus on more local and participative governance approaches. The governance role that goes with it centres on nurturing, coordinating, and driving societal vitality. From a transition perspective, the challenge with this mode may not lie in enabling experimental activities but in integrating them with robust movements that can change the way the system functions (Van den Bosch, 2010).

Table 15.2 summarises the three archetypical governance modes. These modes are to be seen as ideal types, mainly serving as an analytical frame of reference for studying variations in governing system-wide change. We are aware that public organisations increasingly adhere to multiple perspectives at once (Stoker, 2006); in reality, only hybrid forms that include elements of various governance modes and are themselves subject to change may be empirically observed. Nonetheless, we rely on these ideal types to serve as an analytical frame of reference. In this way, we expect to deepen our understanding of the distinct realities in which public and non-public actors operate; these realities differ in terms of the degree of formality, membership flexibility, and diversity of or the reliance on a certain governance mode when it comes to pursuing transformation. This will help us to expand our insights into the mode-specific opportunities and challenges that arise when implementing transformative policies.

Table 15.2 Ideal-types of governance modes

-	Formality of institutions	Membership flexibility & diversity	Governance of change processes
Administration-based governance	High	Low, public actors leading	Inside-out, hierarchical organisation
Network-based governance	Medium	Medium, public, and non-public actors	Co-governance, purposeful alliance
Society-based governance	Low	High, societal actors leading	Outside-in, self-regulated, community building

15.3 METHODS

15.3.1 Case Selection

To identify governance arrangements for TIP, we examined policy programmes that attempt to mobilise science, technology, and innovation (STI) to tackle societal challenges and drive system transformation. These strategies appear in various forms, ranging from system-oriented R&D and innovation programmes to sector- or domain-specific programmes such as those in the transport or energy sectors, and even to umbrella strategies that relate to a whole-government mission-oriented innovation approach. Based on key publications on transformative innovation policies and transformational system failures (Weber & Rohracher, 2012; Schot & Steinmueller, 2018), we determined that a policy approach would fit the TIP rationale if it met the following broad criteria: (1) involves directionality (i.e., sets a direction for guiding innovation based on a shared vision and articulated objectives); (2) follows a societal agenda with ambitions beyond economic, industrial, or technological goals; (3) integrates not only the supply but also the demand side of innovation; (4) addresses an overarching policy topic with a need for (horizontal/vertical) coordination of policy action; and (5) pays attention to stakeholder involvement and more open and diverse policy design and implementation processes.

We relied on purposive sampling to obtain our set of cases for comparing governance arrangements for performing TIP governance tasks. First, we screened sources, including several online tools, repositories, and observatories (STIP Compass tool, MOIP tool, MIPO), and reports like the Fisher et al. (2018) publication that discusses "transformer missions" and "accelerator missions", the OECD report on mission policies (Larrue, 2021), the EC report on sustainability transitions (Kelemen, 2020), and conference websites (e.g., TIPC conferences and EEEN2020 Environmental Evaluation Supporting the Implementation of SDGs and Transformative Policymaking). With this search strategy, we ensured that the programme gained visibility either because (a) policymakers consider the programme to be potentially transformative due to its goals, objectives, or governance approach, or because (b) the research community describes the programme as having transformative features in scientific articles, reports, or research projects. Given the exploratory nature of our study, our aim was not to achieve the most complete and representative set of cases possible. Instead, we took a sample of prominent cases that differ sufficiently in their governance structure for us to gain in-depth insights into the variety of TIP programmes that focus on system-wide change. Our initial set contained about 30 programmes that were implemented at local, regional, or national levels.

In a second step, we applied the following criteria to reduce our sample further: (1) the programme is an important feature in the overall (innovation) policy landscape of a country/region/city; (2) the programme shows real commitment, as demonstrated by some level of dedicated financial and/or human resources; and (3) the programme has reached a more advanced stage of development than simply formulating ambitions or conducting individual experiments. Accordingly, we excluded small-scale, fragmented programmes that do not play a significant role in policy, as well as programmes that are too premature, in the planning phase, or not yet well documented. Despite these restrictions, the criteria still allowed for a varied selection of programmes, ranging from resource-centred to coordination-centred. As some types of programmes are more likely to be associated with a particular governance mode (e.g., the society-based modes rely on collaboration rather than financial incentives), the wide scope ensured that the sample stretched over the full range of governance modes from which policy programmes can originate.

The final set of seven cases and their core properties are shown in Table 15.3. C1 and C4 are extended R&D or innovation funding programmes that are being implemented in the STI domain but are striving for more sustainability-focussed systemic change. C2, C3, and C5 are strategic or "umbrella" programmes that address multiple topics while prioritising specific sectors or challenges and are being implemented in the STI policy domain to some extent at least. C6 and C7 are domain- or topic-specific programmes that consider STI to be a crucial element in achieving their goals.

15.3.2 Data Collection and Analysis

The in-depth examination of each programme is based on evidence from administrative documents, press releases, existing literature, and relevant studies, including evaluations when available. To complement our desk research, we conducted a total of 17 interviews with public officials in charge of designing and/or running the programmes and policy analysts with expertise in the respective policy landscape. The interviews were semi-structured, based on an interview protocol with questions related to the programme's transformative criteria, its TIP governance tasks (as discussed in Section 15.2.1), and the achievements and challenges experienced during implementation.

Based on our desk research and interview data, we examined each case for different formal and informal ways of incorporating transformative policy characteristics and identified the operational changes this required in the organisations that were implementing the policy and the challenges this posed for them. This research step relied on an axial coding approach that followed our focal transformative policy characteristics (i.e., the "closed" dimensions on the basis of which we identified our codes and categories). In our research protocol, we highlighted any structure, activity, event, or organisational change in each case that qualifies as a governance arrangement aimed at realising one of the four TIP governance tasks related to the following: creating legitimacy and leadership; coordinating across levels, actors, and instruments; reflexivity, learning, and experimenting; and resolving conflicts. Key TIP observations were described and summarised in seven in-case analyses, and where necessary, each case summary was validated with additional input by a key public official for the case in question.

Table 15.3 Selected cases of TIP programmes

#	Name	Geogr. Scope	Country	Sector	Year est.	Strategy type	Mission-orientation
C1	Green solutions for the future	National	Denmark	Climate	2020	Extended R&D programme	Explicit
C2	German High-Tech Strategy 2025	National	Germany	Broad	2018	Umbrella framework	Explicit
C3	Mission-oriented Top Sector & Innovation Policy	National	Netherlands	Broad	2019	Umbrella framework	Explicit
C4	City of Tomorrow/ Building of Tomorrow	National	Austria	Housing/ Energy	2013	Extended R&D programme	Implicit
C5	Catalonia 2020/ RIS3CAT	Regional	Spain	Climate	2012	Umbrella framework	No
C6	Amsterdam Circular Economy	City	Netherlands	Circular Ec.	2015	Domain programme	Implicit
C7	Circular Flanders (part of Vision 2050)	Regional	Belgium	Circular Ec.	2017	Domain programme	Implicit

Finally, in a cross-case analysis, we established the link between the observed TIP governance arrangements (and challenges) and our "ideal type" governance modes. To do so, we assigned each TIP observation to a governance mode that is most representative of that specific activity, organisational practice, or event. Then, in a comparative analysis of observed TIP characteristics and governance modes across all programmes, we could group similar governance arrangements into an overarching label.

The following section briefly describes how the seven cases reflect the properties of the three governance modes described in Section 15.2.2. As we will show, most of the selected policy programmes exhibit one dominant governance mode, but some combine elements of multiple governance modes.

15.4 OVERVIEW OF POLICY PROGRAMMES AND THEIR GOVERNANCE MODES

15.4.1 Administration-based TIP Governance

According to the literature on governance modes, TIP programmes that are organised around an administration-based form of governance would place the government in the lead when it comes to designing and implementing targeted policies in line with national priority areas. In such an arrangement, the government or a representative ministry bears responsibility and accountability for TIP implementation and progress within the legislative period. Other parties within the public sector, such as other ministries, government agencies, or research funders, are not "owners" of the strategy but are incentivised to contribute to the high-level priorities.

The cases that most evidently operate in this way are the Danish strategy of Green Solutions of the Future (C1), the German HTS2025 (C2), and the Dutch MTIP (C3). All three are examples of umbrella frameworks, as they rely on a common governance structure and policy instruments for addressing diverse missions (Larrue, 2021; see also Janssen et al. in this volume). Another prominent common feature is that the cases in this category follow a whole-government mission approach (Larrue, 2021) by adhering to high-level policy strategies that are adapted versions of earlier R&I strategies for pursuing societal missions. The Danish programme (C1) targets four green missions, the German HTS2025 (C2) targets twelve missions across six focus areas, and the Dutch Mission-oriented Innovation Policy (MTIP) (C3) comprises four central mission topics with a total of 25 concrete mission goals.

One of the main goals of these administration-based approaches is to spur coordination within government. The strategies are led by ministries responsible for STI or economic agendas and are also seeking to involve other ministries in developing more coherent policy mixes. Despite starting from an R&D and innovation policy perspective, all three cases are now trying to expand and change towards holistic and system-based approaches by organising demand-side policies around selected key priorities. In the Green Solutions and MTIP cases, this move has been driven by pressing climate goal commitments imposed by their national climate acts. In the HTS2025 and the MTIP cases, other societal challenges are also addressed by the TIP programme. One of the objectives is to incorporate European policy priorities and funding schemes, including the Horizon Europe R&D framework programme.

According to the cases we have chosen for illustrative purposes, programmes following administration-oriented traditions that work from the inside out aim to transform policy and procedures "within the government" to promote innovative ways of addressing societal challenges. Instead of achieving a direct transformative impact in society or the economy, the objective is to (re-)design policies that target the systemic nature of today's societal challenges. Therefore, the emphasis is on finding new ways of working within established organisational structures, such as between different departments or units (vertical coordination) or different levels (horizontal coordination) of the public sector. As illustrated in the cases analysed, a key programme objective is to improve the coordination of existing policies, either according to selected priorities or across the established logic of task division and resource distribution in individual ministries, departments, or sections.

15.4.2 Network-based TIP Governance

TIP programmes that adopt the network governance mode aim to mobilise the knowledge, capabilities, and resources of stakeholders outside the policy sphere to design and implement policy initiatives. According to scientific literature, a central feature of a network governance approach is the appointment of dedicated agencies, centres, or teams with a mandate to design, manage, or execute policy initiatives in line with the direction and goals set by the government. In contrast to the administrative-based governance model, stakeholders can be closer to policymaking and become involved in the topic on which a programme or agenda is focussed.

Such features of network-based governance are best recognised in the Austrian City of Tomorrow (CoT) (C4) and the Catalonian RIS3CAT strategies (C5). While CoT has a predecessor in the form of an R&D programme targeted at technology development ("Building of Tomorrow"), the current strategy also includes programme lines that facilitate experimentation and co-creation (e.g., the three Innovation Laboratories). Besides stimulating close cooperation between municipalities, local communities, civil society, and researchers, CoT also draws on non-public actors (e.g., businesses, industries, NGOs) to participate in policy formulation and priority setting. In contrast, RIS3CAT, Catalonia's Research and Innovation Strategy for Smart Specialisation, the network-based process nature is manifested through practices that engage local stakeholders in defining challenges (for smart, sustainable, and inclusive growth) and possible solution directions. The development of agendas relies on a quadruple helix model with substantial end-user involvement. A major difference between the programmes is their mission orientation. CoT adheres to time-bound societal goals, thereby implicitly attempting to adhere closer to European mission approaches. Catalonia's RIS3CAT (dating from 2012), on the other hand, pursues structural changes in production-consumption systems without linking that exclusively to a challenge-led rationale.

It is important to note that HTS2025 and MTIP, also categorised as mainly administration-based, display elements in line with a network-based or co-governance approach. This concerns the use of platforms to unite societal stakeholders and direct future discourse, such as the Hightech Forum for HTS2025 or the Topconsortia for Knowledge and Innovation for MTIP. An explicit focus on public-private partnership projects can be seen in these two cases, in addition to their administration-based characteristics.

Our case examples suggest that governments or policymakers can still play a key role in network-based governance, especially if they have the power to decide which partners participate in governance arrangements. In the examined cases, societal actors are represented by institutionalised actors from academia, business (e.g., industry associations), and civil society (e.g., NGOs, interest groups, city representatives), who are considered relevant for a specific topic. Following the quadruple helix idea, the appointment of these (societal) impact actors is considered a way to include academia, industry, and civil society in programme design and operation.

Moreover, the studied initiatives, despite being policy-led, aim to contribute to stronger connections between the stakeholders themselves. In other words, they follow a network-based mode to engage with a broad set of actors and to unite actors that can complement each other when working on desirable innovation paths. For all cases reflecting network-based governance characteristics, cultivating such relationships between policy and other stakeholders is regarded as a continuous, deliberate, and integral part of the programme and occurs not only on an ad-hoc basis. In the cases examined, we can see that despite stakeholder input and management being essential to programme design, operation, and further development, the final decision-making (e.g., on the focal topic) is left mainly in the hands of policy.

15.4.3 Society-based TIP Governance

Society-based governance follows an "inside-out" approach and relies mainly on societal stakeholders' own and collectively organised initiatives to pursue shared goals. Amsterdam Circular (C6) and Circular Flanders (C7) are the programmes that best fit the properties defining a society-based governance model. While Amsterdam Circular is run by a team embedded in the government itself (the municipality of Amsterdam), Circular Flanders is implemented by a team that is financed by Flemish policy departments but operates at a distance. Both cases aim to catalyse innovative programmes in the field of Circular Economy (CE), without necessarily relying on R&D or knowledge-based novelty alone. Social innovation in the form of behavioural or institutional change is strongly emphasised to meet the prioritised CE goals. A uniting characteristic is that both programmes evolved through a process that first identified the opportunities and challenges perceived by "field actors" such as firms, citizens, and municipalities. This process then led the teams executing the programmes to select several key topics and courses of action. While both programmes focus on circularity, we consider them representative of other strategies that rely on facilitating self-organisation and the connection of bottom-up societal initiatives. Rather than incentivising societal stakeholders to behave differently, the society-based governance programmes focus on responding to emerging systemic bottlenecks. A substantial part of both programmes is to reach out to a wide variety of policy actors and funding organisations in case a more formalised agreement or an agreement requiring strategic support arises. This is especially necessary in the promotion of promising circular practices.

The case illustrations of what society-based TIP governance might look like point to the usefulness of establishing a dedicated organisation tasked with solving a particular problem. In the cases studied, such an independent entity, which is financed but not managed by a government department, allows for the closest form of collaboration with private actors around a societal topic or challenge. Compared to representative network arrangements, this type of organisation is in direct contact with a broader range of stakeholders who are directly involved in (or affected by) changes that are part of a transformation. This allows the team to better grasp not only potential opportunities but also challenges that the different stakeholders encounter when engaging with a particular problem or solution, whether as producers, users, or by being affected in any other way. In our cases, the public entities that manage the programme carry labels like "transition team" and are run by transition managers. One of their responsibilities is to inspire transition-thinking in other governments and to support them in designing policies that are in accordance with their field experiences and practices.

Both in the case of the Amsterdam Circular and the Circular Flanders policy programmes, the observed TIP governance arrangements aim to accelerate new or existing bottom-up activities initiated by societal stakeholders. This includes for-profit and non-profit activities, which are often still in an experimental stage of developing and integrating new products or services. As a result of this method, the focus is less on technological developments and more on exploring suitable business models and new institutional arrangements that would be required for new solutions to work. For that reason, in society-based programmes such as those studied, it is not up to policy to orchestrate innovative and potentially transformative activities or to fund individual activities that would not occur without government support. Instead, it is critical to catalyse small initiatives that can transform socioeconomic systems when applied in greater number or at a larger scale. This suggests a greater interest in experiments and projects that target the scaling or diffusion of innovations to challenge the functioning of the entire system around those innovations. Policy entities propelling the bottom-up initiatives attempt to make a difference by approaching and connecting stakeholders who hold the key (in terms of partners, information, funding, regulation, or otherwise) to a viable application of innovations.

15.4.4 Overview of TIP Governance Modes

Drawing on governance literature and our case illustrations, Figure 15.1 visualises the key characteristics of the three governance modes for conducting TIP governance tasks.

15.5 CROSS-CASE ANALYSIS: GOVERNANCE ARRANGEMENTS AND CHALLENGES

In this section, we focus on cross-case observations and present the most distinctive TIP governance arrangements per governance mode and governance task. Table 15.4 provides an overview of our key observations in this respect. Case-specific examples of the identified arrangements are included in the table in the appendix. In the following section, we discuss our findings in light of our three ideal-type governance modes and also address potential challenges public organisations might face when aiming to be transformational and promote system change.

15.5.1 Governance Arrangements and Challenges in Administration-based Governance

Legitimacy and leadership: Developing new strategies or priorities within the public sector and endorsing these at the highest political level guarantees political legitimacy of new goals and objectives for a legislation period of at least four to five years. However, as interviewees have pointed out, even within the policy organisations themselves, such a centralised approach and form of leadership could be easily perceived as too abstract and top-down a view of moving forward, with little attention to various implementation hurdles. Opposition at lower hierarchy levels can result from limited recognition of established organisational cultures or

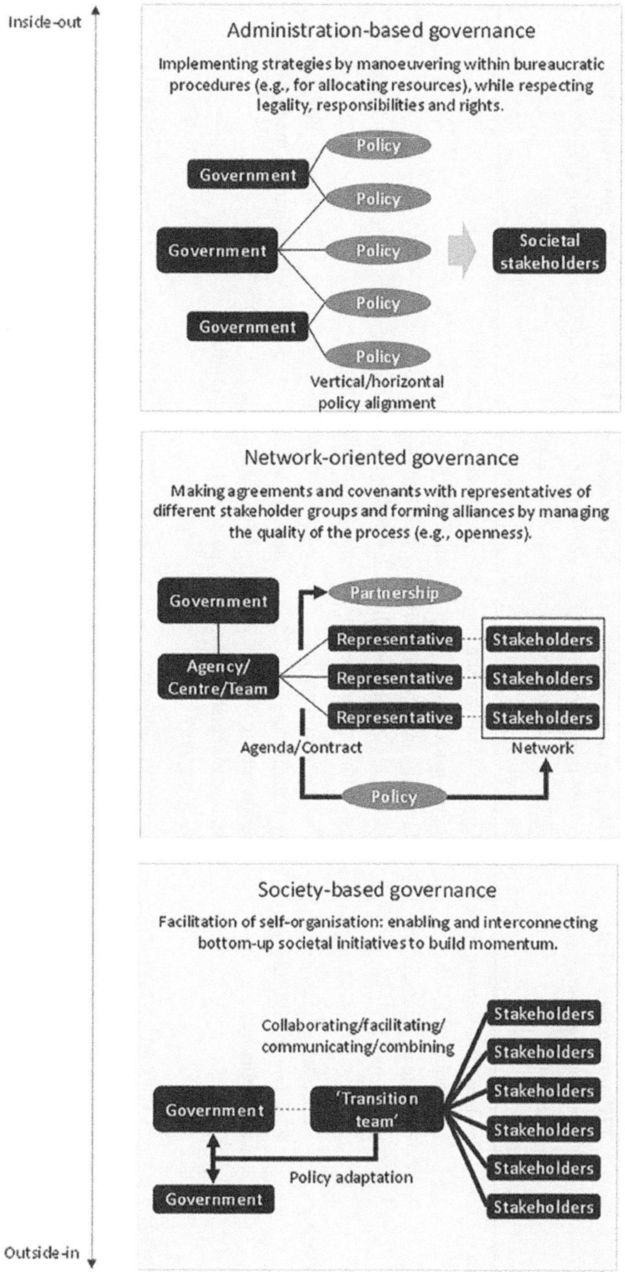

Source: Authors, adapted from Janssen et al., 2023.

Figure 15.1 *Key characteristics of three governance modes for conducting TIP governance tasks*

Table 15.4 *Overview of TIP governance arrangements according to ideal-type governance modes and TIP governance tasks*

Governance modes	**TIP governance task:** *Creating legitimacy and leadership*	**TIP governance task:** *Coordinating across multiple levels, actors and instruments*	**TIP governance task:** *Reflexivity, learning, and experimenting*	**TIP governance task:** *Resolving conflicts*
Administration-based governance	‣ Show commitment from high political levels ‣ Create an independent policy unit	‣ Change of innovation funding priorities ‣ Targeting multiple (existing) instruments, actors, or policy fields within prioritised topics	‣ Policy mapping across different departments ‣ New monitoring and evaluation procedures beyond abstract inputs/outputs	N/A
Network-based governance	‣ Engage in partnerships based on shared leadership and collective agendas	‣ Involve representative key stakeholders in agenda-setting ‣ Shared ownership between policy partners	‣ Adapt policies based on network signals ‣ Evaluate formatively, with participative deliberation	‣ Community building and community management
Society-based governance	‣ Create a vision around perceived problem urgency ‣ Emphasise community-based problem-solving	‣ Establish a hub-like transition team outside the government ‣ Reinforce bottom-up initiatives ‣ Inspire and instruct policymakers	‣ Learning-by-doing (stocktaking) ‣ Monitor project outcomes for initiative reorientation	‣ Install a field-level working group

working practices, therefore undermining the ambitions for organisational change and slowing down the transformation processes within governments.

Policy administrations can attempt to stimulate new ways of working by acknowledging different organisational cultures and by considering and facilitating necessary organisational changes within the administration when launching a new challenge-led strategy. An alternative way we observed to achieve novel cross-domain alignments is the creation of an independent policy unit ("face") for a topic, such as a societal challenge or mission outside existing departmental structures. Such independent units may reduce existing rivalry or competition between ministries and thus can achieve a more effective leadership for the particular topic, based on broad-based legitimacy across the administration. Cultivating joint cross-departmental units across the organisation can help to disentangle priority areas for transformation from other shorter-term political strategies and establish organisation-internal structures and processes beyond determined priorities for a legislative period. Furthermore, given the notoriously limited resources in the public sector, all cases show that organisational change or transformation "within" the administration is expected to be effective only if sufficient personnel and financial support is provided for the new topics and structures.

Coordination: The formulation of innovation challenges or missions supported at a high political level offers a shared objective or reference for new innovation priorities across public administration. As such, they can guide the modification of policy schemes (e.g., in terms of funding criteria) that can potentially contribute to these priorities. The cases analysed also show that linking to long-term and strategic government priorities as formulated, for instance, in European or national goals (e.g., climate agreements, energy plans), can provide a valuable reference and stimulus for change. It incentivises policy actors to commit to a whole-government approach that is beyond research and innovation. Moreover, well-functioning cross-departmental coordination seems difficult but is necessary to pursue a whole-government approach. Administrative silos typically exist at both strategic and operational levels. Our cases also illustrate that coordination within public administration can be facilitated by the formation of dedicated joint units (such as challenge or mission teams) composed of actors with different backgrounds who work together to develop their joint agenda. In contrast, commitment to whole-government approaches appears not to be sufficiently encouraged when programmes are formulated as too domain-specific (e.g., for research and innovation) or not matched with additional human and financial resources. As interviewees pointed out, if formal coordination structures are lacking, then providing more space and resources, particularly time, and the possibility for more informal coalitions across sections between dedicated staff members could be the first step towards more cross-departmental coordination.

Reflexivity and learning: From an administration-based governance perspective, a systematic mapping of existing policies, including their strengths, weaknesses, and interrelations, seems crucial before launching a new programme. This is confirmed by several interviewees and by analysing the different cases. In the context of TIP, this would entail the full range of support measures or associated regulations that may influence transformations on a societal issue. For national-level programmes, the relevant policies appear to be distributed over departments focussed on science, education, or business, as well as the respective policy domain itself. Obtaining a comprehensive view of the full policy mix would also require insight into supra- and sub-national policies. However, policy officials admitted that they did not have a clear overview of all the policies that can be mobilised to support a specific priority, challenge, or mission. Therefore, comprehensive policy mapping would need to be a collective endeavour, since it is far more challenging than simply listing the instruments on one policy department's budget.

When it comes to learning and experimenting in administration-based governance, the room to do so might be limited if a policy strategy is coupled with a legislative period or with traditional policy and evaluation cycles. Strict evaluation cycles and methodologies favour monitoring key performance indicators and individual programmes, but these are of decreasing relevance for programmes that focus on society-based change. Input-/output-oriented evaluation culture, which is still prevalent in our cases, is increasingly considered to be at odds with aspirations to combine multiple innovation supply- and demand-side measures. If the aim is to bundle policies more effectively, immediate short-term impact or causal relationships are difficult to assess. Instead, establishing continuous monitoring and evaluation procedures is considered fruitful to ensure evaluation feedback outcomes are directly implemented in the policy process. Cases that experimented with this type of extensive monitoring approach benefited from a more direct engagement with stakeholders, the establishment of independent expert panels, accompanying research support, and/or the development of an in-house capacity to extract and then apply learning lessons within public administration.

Challenges: While the administration-based governance model benefits from a high level of political legitimacy and leadership from the outset, we also observed substantial challenges in building up governance capacities geared towards transformative change. These challenges relate mostly to current departmental thinking and rivalry for recognition and resources within public administration, which can obstruct ambitions of alignment and coordination. Committing other actors and departments beyond informal collaborations seems more challenging if the innovation priorities are not recognisable aspects of their own activities or priorities. What is more, relying on strict administrative and budgetary cycles can hamper flexibility, learning, and experimenting with new policy designs. On the one hand, legislation periods provide opportunities to adapt high-level innovation priorities and strategies; on the other hand, periodic changes in political leadership can also create inconsistencies or uncertainties in pursuing long-term goals and transformative impact.

15.5.2 Governance Arrangements and Challenges in Network-based Governance

Legitimacy and leadership: Network-based governance arrangements rely on shared leadership that is based on the collective agendas they develop. As we observed, programme leadership and management are often shared between different policy actors, such as multiple ministries and/or agencies. Our cases illustrate that such co-ownership or even co-funding arrangements generate legitimacy for an issue, fostering commitment to a joint agenda within the public sector. Along with actors from the public sector, the network-based programmes we analysed have increasingly attempted to include relevant stakeholders from different industries, cities, etc., more directly in programme co-creation, funding, and execution. These stakeholders are often considered key figures who can speak on behalf of larger communities (industry, science, SMEs, public actors, NGOs, etc.), but their influence depends on the platform offered to them by the government. In policy-led network arrangements, we see that governments take a less dominant role within a delegated structure. Despite this challenge, the network structures are typically positioned as prominent vanguards of a specific policy domain.

Coordination: A representative model of stakeholder involvement is at the core of a network-based TIP governance arrangement. The programmes including characteristics of this governance mode pay close attention to developing dedicated governance structures (e.g., platforms, management boards) and engaging in processes for direct and continuous stakeholder interaction that go beyond ad-hoc consultation or advisory tasks. Intensive collaboration often occurs with a limited set of key stakeholders, ranging from agenda-setting to programme execution. Within these arrangements, decisions about priorities and future agendas (e.g., new topics or solution directions) can be relatively informal, facilitated by close relationships and interactions of various public and private actors. The strategic decisions within a closed network are also used to build up a broader stakeholder community beyond those actors directly involved in agenda (co-)creation. Stakeholder management and consultation often lie in the hands of an agency or centre. In the cases we studied, this type of broader multi-actor arrangement follows a more formalised setting, such as through the organisation of workshops (e.g., focus groups) with a targeted stakeholder group, or a larger public consultation process in the form of online surveys, open discussion meetings, or open calls.

In addition to creating strategic networks at a policy programme level, collaboration can also occur within public-private partnerships that directly take up a challenge, mission, or societal problem and create new solutions. Examples from our cases include innovation hubs or platforms that involve various players who develop and experiment with sustainable solutions. These partnerships are organised through clear agreements or co-funding structures between the triple or quadruple helix models. In some cases, stakeholder involvement and partnerships stretch to lower-level authorities such as cities and municipalities, who take the role of lead users for systemic solutions at a smaller, local level.

Reflexivity and learning: Reflexive governance and learning structures rely on the established stakeholder network. Through the involvement of "field actors", the network provides a channel through which the government or agency can receive signals indicating that policies must be adapted in light of new or changing problem fields or innovation directions. Within a model based on delegating tasks to network partners, the public sector is mostly responsible for organising and managing interaction processes in the stakeholder network (e.g., mobilising actors that may contribute to desirable goals). Understanding and changing how these interaction processes work is more important for policymakers than assuming responsibility for how much of the desirable results are achieved. As interviewees confirm, new monitoring and evaluation practices may be required, moving away from current practices whereby evaluations only take place at the end of policy cycles and towards more continuous evaluation procedures that analyse the process and intermediate outcomes. In particular, experts stress that such evaluation processes, prioritising learning and adaptation over accountability, benefit from engaging different types of stakeholders in policy assessments (including the actors formulating and implementing the policies). While including stakeholder opinions generally causes evaluations to be subjective rather than objective, the interviewees see merit in examining how diverse actors understand policy rationales, policy objectives, and the strategy for meeting those objectives. It is especially important in a network-based governance model, where stakeholders rely heavily on each other, to identify misalignment or resistance in order to assess the likelihood that the chosen policy approach will work out favourably.

Resolving conflicts: In a network-based governance model, public actors attempt to avoid disputes about strong or top-down decisions by leaving the responsibility for agenda-setting, negotiation, and sometimes also final decision-making, to the delegated networks. These networks in turn find agreements based on negotiations and the input of their members (key stakeholders or representatives). Moreover, the policy networks are active in community-building, such as organising networking events or interactive spaces. These activities are considered a means to create streamlined yet informal interaction possibilities between different actors who may have diverging interests and needs. Building direct relationships between innovation funders (e.g., agencies) and the future, present, or past funding recipients is another distinct feature of a TIP governance arrangement, which aims to strengthen the relationship between public and private parties.

Challenges: The transformative potential of the network-based governance mode is subject to three challenges. First, there is the question of how relevant "stakeholders" are defined, which can either be broad and open or narrow. Both definitions have advantages and disadvantages in increasing efficiency, finding common ground, and incorporating alternative ideas. In practice, network-based programmes tend to be based on triple or quadruple helix models around a select set of key representatives. It is more difficult to find key demand representatives or societal impact stakeholders and then incentivise them to engage with the policy on a continuous (not ad-hoc) basis than it is to find and incentivise representatives of supply or industry stakeholders. An additional concern is that a governance structure based on interacting with a few triple or quadruple helix representatives can become vulnerable to one party's individual vested interests.

Second, a clear challenge for the network-based model is that the government is not in the driver's seat but still needs to be able to respond to suggestions provided by the representatives with whom it interacts (e.g., concerning regulatory barriers or lack of lead markets). This requires policy officials to have the ability to not only build the networks but also to understand and address the suggestions that surface from the field. Moreover, the multiple-helix centres or teams that are at the heart of network-based TIP governance arrangements can only have an impact if they can ensure commitment from the community they represent. In order to take responsibility for their part in the public-private collaboration (e.g., to develop a support infrastructure for new solutions), policy actors often require a high level of assurance that there is a need and demand for ambitious developments.

Third, since network-based TIP governance relies heavily on stakeholder knowledge and capacities, it is likely that the stakeholders' particular interests could prevail when considering innovation directions. Exerting more guidance on the part of the government might be met with resistance from the network. This is especially expected if certain solution directions are unattractive for market players or are contested by particular parties. Involving stakeholders in priority setting without providing too much directionality for solutions remains a challenge for the network-based governance.

15.5.3 Governance Arrangements and Challenges in Society-based Governance

Legitimacy and leadership: For outside-in governance arrangements, legitimacy stems from the purpose of the programme (i.e., is the cause recognised as important?) and from how societal stakeholder participation is managed. The focus is mostly on generating momentum and on interconnecting bottom-up initiatives. Therefore, it is essential that the programmes create a vision around the importance and urgency of the problem to secure stakeholders' commitment. Since funding streams for system- and societal-oriented programmes are typically relatively small, broad support from field-level stakeholders for their goals and vision is considered crucial for a successful operation. In our CE cases, the legitimacy of the programmes and their visions was strengthened if built around known narratives or established concepts (e.g., "missions", "donut economy", "transitions management"). Furthermore, society-based programmes strongly emphasise the community aspects of problem-solving. In that sense, transition managers focus more on the process than on the trajectory of transformations. The community is in the "driver's seat" of the transitions, with political support or public resources that can be used to both discover and negotiate the directions. The investigated cases used websites, newsletters, and evaluations to showcase not only the achievements of the various bottom-up activities but also their drivers and motivations.

Coordination: Society-based TIP programmes typically start small, as the organisations are established independently and are not in charge of running other policy programmes. The founding purpose of the programme cases examined has focussed on intensively engaging with the respective field and stakeholders around a particular societal issue. As the examples illustrate, starting collaborative relationships with societal stakeholders is about building trust in the form of listening to different views, needs, and ambitions. An important task in supporting coordination is seen in bundling "societal energy" by organising events and platforms to display ongoing practices and inspire new innovations. An additional and crucial task is to facilitate extensive communication, guided by the transition management team. Another difficulty in community building and management lies in reinforcing the links between varied parties, such as the winners and losers from a transition.

Transition managers see themselves as a bridge between policy and non-policy partners. Creating circumstances favourable for new bottom-up solutions requires a governance capacity to inspire, instruct, and mobilise a broad range of actors, including policy officials from different government departments. The objective would be to use policy to internalise the transition so that new practices are incorporated while responsibility for their respective policies (including budgets and accountability) is also maintained.

Reflexivity and learning: Society-based TIP governance arrangements follow a learning-by-doing approach and are typically accompanied by a deliberate bottom-up activity learning process that is based on systematic stock-taking. In the cases we examined, this involved a continuous search for promising new initiatives and a concurrent evaluation of the things that did not work. It allowed for learning at two levels: the field level in terms of problem areas and existing solution types and the TIP programme level in terms of the actual support efforts. Insights from both levels were used to recalibrate the activities of the transition team.

In our cases, society-based approaches to monitoring and evaluation revolved primarily around specific bottom-up initiatives that needed to be supported, as opposed to concrete policy interventions, strategies, or measurable goals. It can be seen that in a society-based approach, monitoring and evaluation have a different scope and purpose than in programmes that follow the other two TIP governance arrangements. Instead of focussing only on costs (input) and effects (outcome) of support actions, monitoring also focusses on individual projects or activities and demonstrates how they contribute to the desired transition. The main purpose of monitoring project outcomes is not to evaluate or assess but to identify projects that deserve a spotlight or require additional attention.

Resolving conflicts: Our cases illustrate that the capacity to deal with tensions and resolve conflicts in a society-based TIP governance model can be strengthened by instituting a steering committee where stakeholders can work out differences. In the studied cases, such committee structures were beneficial in making strategic decisions about future actions and activities within the community. Activities in the form of organised events, such as discussion nights or workshop series, helped overcome differences and create a better understanding of the sources and status of tensions.

Challenges: While society-based governance arrangements lend themselves well to designing and implementing the tasks and features of transformative innovation policy, they may also show their limitations. The biggest challenge is the difficulty of building momentum. By not starting out with a clear programme, funding, or action line for the initiatives, it can be hard to identify and integrate initiatives into a broader, self-reinforcing movement. Catalysing bottom-up initiatives only works if enough problem-oriented activities are already happening on the ground or in society. Moreover, the absence of a standardised set of policy tools highlights the flexibility and self-regulatory character of society-based governance arrangements, but it can also hamper bold initiatives.

15.6 DISCUSSION

Driving system-wide transformation is likely to demand unconventional interactions between government and societal stakeholders, both in the development and implementation of policies. This study aimed at identifying the variety of governance arrangements (i.e., the organisational structures, processes and capabilities, through which public organisations pursue their transformative policy ambitions). Using a comparative analysis of seven transformative policy programmes, we aimed to shed light on different implementation modes of TIP governance and their respective opportunities and challenges. The results obtained inform the ongoing – primarily conceptual – debate with insights on how to apply transformative innovation policy frames or rationales in practice (e.g., Weber & Rohracher, 2012; Schot & Steinmueller, 2018; Borrás & Edler, 2020).

As a starting point, we identified three ideal types of governance modes based on which aspects and activities of transformative policy initiatives can be implemented. Our findings confirm that the three governance modes differ substantially in how they permit public organisations to deploy structures and processes that are in line with transformative ambitions or principles. For each mode, we were able to identify TIP governance arrangements that allow for new ways of providing legitimacy and leadership, supporting coordination, facilitating policy learning, and resolving conflicts. While the transferability of specific governance arrangements and practices across the different modes might be limited, we observed hybrid "cases" or TIP programmes that integrated features of various governance modes. It is clear that in the TIP initiatives we studied, the underlying organisational (e.g., administrative, co- or self-governance) principles of a particular governance mode were dominant, or at least guided the design and implementation of the TIP initiatives we studied.

A question that arises from our analysis of governance arrangements is how and from whom policy organisations can learn to further develop their own programmes and strategies. If more knowledge and experience on different governance arrangements and modes become available, what does that mean for groups such as public organisations who mostly operate according to an administration-based governance mode (like the mission-oriented cases)? Almost by definition, governance modes themselves are not easily changed. However, the possibility of a "gradual" evolution of governance modes might explain the observation that most of our cases do not strictly adhere to one mode but harbour elements from various archetypes when governing change processes.

Such a "governance mix" (Bugge et al., 2017) which simultaneously follows top-down and bottom-up principles, is visible in the HTS2025 and the MTIB cases, which show a steady "opening-up" to move from administration-based governance towards more network- or society-based governance. While many of the routines and logics in those cases reflect that bureaucratic procedures and accountability thinking still prevail, there are also signals that the public organisations in the lead are increasingly open to entering governance groups in which representatives of societal stakeholders have a larger say in shaping policy instruments and allocating budgets. On the contrary, the initiatives already operating under a more society-based governance still show aspects of an administration-based governance approach. This is illustrated by the Amsterdam Circular case. Despite assigning large importance to different stakeholders' views on why and how to pursue circularity, those views still need to be incorporated into strategies that are formulated and adjusted according to formal budget cycles.

These observations also give rise to the question of what is needed from public organisations in order to change the legitimacy and space of possible actions. Based on our cases, it seems possible to "escape" the boundaries of a dominant governance mode, either by following incremental steps (e.g., by altering some policy features with every new generation of a policy programme) or by pursuing a more radical route (e.g., by establishing a transition team outside of a public organisation). Further research into governance changes within and across governance modes may need to recognise that such changes do not necessarily follow from elaborated plans. Instead, the changes might often rely on a "small wins" approach based on step-by-step governance experimentation (Bours et al., 2022).

Another way of looking at the issue of evolving governance modes is to ask whether it is even recommendable or desirable for governmental organisations to move away from administration-oriented governance. We found that most of the society-based arrangements in sub-national (regional and local) initiatives point to potential complementarities between top-down governance modes at the national level and more bottom-up and participatory governance approaches at lower policy levels. This idea is consistent with the subsidiarity perspective on innovation policy for addressing societal challenges, which states that national and local policy bodies can reinforce each other when setting top-down directions and accommodating bottom-up experimentation (Wanzenböck & Frenken, 2020). Local governments might be better positioned to steer and accelerate transformative initiatives, since societal problems, despite their often global prevalence, typically possess social, economic, and institutional characteristics that vary from place to place (Coenen et al., 2015). Additionally, it is usually at the local level where one can also find both the political institutions and the problem urgency powerful enough to unite diverse actors around a shared goal (Flanagan et al., 2023). In that sense, it is not necessarily a problem if national governments stick to an administration-based governance approach that is focussed on improving the cohesion between policy instruments, as long as the resulting policies such as subsidies and/or regulations are suitable for supporting bottom-up initiatives. This view emphasises the importance of not only looking at policy mixes in the sense of instrument mixes, but also of multi-level arrangements and "governance mixes" tying together top-down and bottom-up governance approaches. One of the potential mechanisms for doing so can be the use of missions as boundary objects for coordinating across different communities and governance levels (Janssen et al. in this volume).

Finally, studying the materialisation of transformative characteristics is relevant to the emerging interest in mission-oriented innovation policies that have transformative ambitions (Hekkert et al., 2020; Haddad et al., 2022). Considering the contrasts between TIP programmes with and without an explicit mission-orientation, the examined strategies centred on missions that fit best with the administration-based and network-based governance approach. This implies that in at least the mission-oriented TIPs we investigated, there is ample emphasis on cross-ministerial policy coordination and streamlining policy instruments. While this might provide impetus for developing promising innovative solutions in line with a particular mission goal, it is less clear how the dominant governance modes can support the diffusion and adoption part of pursuing a mission. One of the promises of missions is to engage with a broad spectrum of societal stakeholders who hold interest in the problem and possible solutions (Janssen et al., 2021), but even in the network-based governance mode we found that engagement arrangements are limited to the participation of primarily institutional representatives. Governance arrangements directly involving municipalities or citizens were more prominent in the society-based governance mode, making it unclear what the other two modes can accommodate in this respect. As there are only a few empirical examples of fully fledged mission-oriented innovation policies, it is still too early to make statements on how and whether such policies can or should be orchestrated according to a society-based governance mode.

15.7 CONCLUSIONS

This chapter provided a comparative cross-case analysis of transformative policies across Europe. The seven investigated cases all aim to mobilise innovation capacities in order to drive system-wide transformations, but differ in the governance arrangements they deploy to steer and control the innovation and diffusion processes. Documenting concrete structures and processes is the first step when trying to understand what works best in certain circumstances. The conclusions drawn might inspire policymakers and therefore provide a basis for exchanging lessons and experiences.

Moving forward, it is important to be able to distinguish strategic and operational struggles (e.g., a lack of capacity to establish suitable governance arrangements) from political types of bottlenecks (e.g., insufficient willingness to transform). Weak or inconsistent political leadership is not uncommon in contested policy spaces where multiple societal interests collide, but needs to be addressed in ways that differ from mere capacity building. This chapter aimed to elaborate upon research avenues on the latter account. Further investigation of what governance arrangements to develop and how to do so is going to be relevant in the face of emerging innovation policies that have an explicit ambition to support the development of governance capacities, such as the S4 "Innovation for sustainability" policy framework succeeding S3 (McCann & Soete, 2020).

Our main research contribution stems from developing an analytical framework that allows for cross-case comparisons. Doing so points to the importance of being aware of differences in the governance modes from which transformative policy emerges. When comparing abilities to bring about "transformative outcomes", it seems fair to consider that governments may have very diverse starting points and possibilities. Building on the assumption that different types of policy systems require different ways of coping with implementation hurdles

(Matland, 1995; Hudson et al., 2019), our focus on linking governance arrangements to governance models provides a basis for future empirical work to consider specific transformative policy types – rather than lumping them all together. Regarding mission-oriented innovation policies, the governance modes we examined offer an alternative kind of contextualisation compared to types of missions (Wittmann et al., 2021) or types of mission-oriented innovation policy instruments (Janssen, 2022). As suggested by Borrás and Edler (2020), recognising the differences between distinct modes of governance might also help to understand how socio-technical system changes occur. It is likely that governance arrangements fitting an administration-based governance mode are more suited for one type of transformation (e.g., hierarchical and state-driven) than for other transformation types (e.g., those based primarily on self-regulation).

In terms of policy implications, our study suggests that there is no clear formula to become more transformative; not only because there are various ways of achieving transformation, but also because (depending on the dominant governance mode) governmental organisations face different opportunities and challenges when converting to the transformative innovation paradigm. TIP programmes can be leading examples for other policies that also develop from administration-based or network-based governance systems, which may be less transformative than ambitious policies organised around a society-based governance logic.

The complete extent of developments in policy designs and governance capacities to lead to transformative outcomes lies outside the scope of this conceptual and mostly descriptive study. Further research can build on the findings presented here by assessing the socioeconomic and socio-technical changes set in motion and linking them back to specificities of the encountered types of governance systems, policy design, and governance capacities. It would be particularly useful to identify patterns that reveal what combination of governance arrangements is best suited to deliver on different kinds of transformative ambitions (e.g., short- or long-term goals). The presented findings can also give rise to further research on routes for policy bodies to become more transformative, as well as investigations into potentially complementary ways of organising national government's top-down and local government's bottom-up governance styles. The findings thereby feed into the ongoing discussion on the respective roles of niche activities versus incumbents and mainstream actors in driving change, both of which might be needed to achieve large-scale system transformation (Geels, 2021). Since there is a link between governance modes and the types of change they can best accommodate (e.g., society-based governance is associated with mobilising and coupling niche activities), it is clear that there are also reasons to apply comprehensive TIP strategies to policy organisations that operate according to different governance modes.

ACKNOWLEDGEMENT

This chapter is based on a research report, *Governing transformative innovation policies*, commissioned by the European Commission's Joint Research Centre. Parts of the current chapter contain text fragments of this report, notably in Sections 15.4, 15.5, and 15.6.

REFERENCES

Borrás, S. (2011). Policy learning and organizational capacities in innovation policies. *Science and Public Policy*, *38*(9), 725–734.

Borrás, S., & Edler, J. (2020). The roles of the state in the governance of socio-technical systems' transformation. *Research Policy*, *49*(5), 103971.

Borrás, S., & Laatsit, M. (2019). Towards system oriented innovation policy evaluation? Evidence from EU28 member states. *Research Policy*, *48*(1), 312–321.

Bourgon, J. (2011). *A new synthesis of public administration: Serving in the 21st century* (Vol. 81). McGill-Queen's Press-MQUP.

Bours, S. A., Wanzenböck, I., & Frenken, K. (2022). Small wins for grand challenges. A bottom-up governance approach to regional innovation policy. *European Planning Studies*, *30*(11), 2245–2272.

Braams, R. B., Wesseling, J. H., Meijer, A. J., & Hekkert, M. P. (2021). Legitimising transformative government: Aligning essential government tasks from transition literature with normative arguments about legitimacy from Public Administration traditions. *Environmental Innovation and Societal Transitions*, *39*, 191–205.

Bugge, M., Coenen, L., Marques, P., & Morgan, K. (2017). Governing system innovation: Assisted living experiments in the UK and Norway. *European Planning Studies*, *25*, 2138–2156.

Cagnin, C., Amanatidou, E. & Keenan, M. (2012). Orienting European innovation systems towards grand challenges and the roles that FTA can play. *Science and Public Policy*, *39*(2), 140–152.

Christensen, T., Lægreid, P., & Rykkja, L. H. (2016). Organising for crisis management: Building governance capacity and legitimacy. *Public Administration Review*, *76*(6), 887–897.

Coenen, L., Hansen, T., & Rekers, J. V. (2015). Innovation policy for grand challenges: An economic geography perspective. *Geography Compass*, *9*(9), 483–496.

Denes_Santos, D., & da Cunha, S. K. (2020). Transformative innovation policy for solar energy: Particularities of a developing country. *Clean Technologies and Environmental Policy*, *22*(1), 43–57.

Diercks, G., Larsen, H., & Steward, F. (2019). Transformative innovation policy: Addressing variety in an emerging policy paradigm. *Research Policy*, *48*(4), 880–894.

Fink-Hafner, D. (1998). Organized interests in the policy-making process in Slovenia. *Journal of European Public Policy*, *5*(2), 285–302.

Fisher, R., Chicot, J., Domini, A., Polt, W., Turk, A., Unger, M., Kuittinen, H. et al. (2018). Mission-oriented research and innovation: Inventory and characterisation of initiatives. European Commission EC.

Flanagan, K., Uyarra, E., & Wanzenböck, I. (2023). Towards a problem-oriented regional industrial policy: Possibilities for public intervention in framing, valuation, and market formation. *Regional Studies*, 57(6), 998–1010.

Geels, F. W. (2021). From leadership to followership: A suggestion for interdisciplinary theorising of mainstream actor reorientation in sustainability transitions. *Environmental Innovation and Societal Transitions*, *41*, 45–48.

Ghosh, B., Kivimaa, P., Ramirez, M., Schot, J., & Torrens, J. (2021). Transformative outcomes: Assessing and reorienting experimentation with transformative innovation policy. *Science and Public Policy*, *48*(5), 739–756.

Grillitsch, M., Hansen, T., Coenen, L., Miörner, J., & Moodysson, J. (2019). Innovation policy for system-wide transformation: The case of strategic innovation programmes (SIPs) in Sweden. *Research Policy*, *48*(4), 1048–1061.

Grin, J., Rotmans, J., & Schot, J. (2010). *Transitions to sustainable development: New directions in the study of long-term transformative change*. Routledge.

Haddad, C. R., Nakić, V., Bergek, A., & Hellsmark, H. (2022). Transformative innovation policy: A systematic review. *Environmental Innovation and Societal Transitions*, *43*, 14–40.

Hekkert, M. P., Janssen, M. J., Wesseling, J. H., & Negro, S. O. (2020). Mission-oriented innovation systems. *Environmental Innovation and Societal Transitions*, *34*, 76–79.

Hölscher, K., Frantzeskaki, N., & Loorbach, D. (2019). Steering transformations under climate change: Capacities for transformative climate governance and the case of Rotterdam, the Netherlands. *Regional Environmental Change*, *19*(3), 791–805.

Howlett, M. (2009). Governance modes, policy regimes and operational plans: A multi-level nested model of policy instrument choice and policy design. *Policy Sciences*, *42*(1), 73–89.

Hudson, B., Hunter, D., & Peckham, S. (2019). Policy failure and the policy-implementation gap: Can policy support programs help? *Policy Design and Practice*, *2*(1), 1–14.

Janssen, M. J. (2019). What bangs for your buck? Assessing the design and impact of Dutch transformative policy. *Technological Forecasting and Social Change*, *138*, 78–94.

Janssen, M. J. (2022). Legitimation and effects of mission-oriented innovation policies: A spillover perspective. *Hacienda Pública Española/Review of Public Economics*, *243*(4), 7–28.

Janssen, M. J., Torrens, J., Wesseling, J. H., & Wanzenböck, I. (2021). The promises and premises of mission-oriented innovation policy – A reflection and ways forward. *Science and Public Policy*, *48*(3), 438–444.

Janssen, M., Wanzenböck, I., Fünfschilling, L., & Pontikakis, D. (2023). *Capacities for transformative innovation in public administrations and governance systems: Evidence from pioneering policy practice (No. JRC131490)*. Joint Research Centre (Seville site).

Janssen, M., Wesseling, J., Torrens, J., Weber, K. M., & Penna, C. C. R. (2025). Missions as a boundary object in umbrella frameworks: The case of the Dutch Mission-oriented Top Sector and Innovation Policy. In J. Edler, M. Matt, W. Polt, & K. M. Weber (Eds.), *Transformative mission-oriented innovation policies: Revisiting the role of science, technology and innovation in society*. Edward Elgar Publishing.

Janssen, M., Wesseling, J., Torrens, J., Weber, M., Penna, C., & Klerkx, L. (2023). Missions as boundary objects for transformative change: Understanding coordination across policy, research, and stakeholder communities. *Science and Public Policy*, *50*(3), 398–415.

Jordan, A., Wurzel, R. K., & Zito, A. (2005). The rise of "new" policy instruments in comparative perspective: Has governance eclipsed government? *Political Studies*, *53*(3), 477–496.

Jørgensen, M. S., & Andersen, B. H. (2012). The controversies over bioenergy in Denmark: "Bio" is not the same as "sustainable". *Environmental Engineering & Management Journal*, *11*(11), 2101–2119.

Kattel, R., & Mazzucato, M. (2018). Mission-oriented innovation policy and dynamic capabilities in the public sector. *Industrial and Corporate Change*, *27*(5), 787–801.

Kelemen, A. (2020). *Supporting sustainability transitions under the European Green Deal with cohesion policy*. European Commission.

Kemp, R., & Loorbach, D. (2006). Transition management: A reflexive governance approach. In J.P. Voß, D. Bauknecht, & R. Kemp (Eds), *Reflexive governance for sustainable development* (pp. 103–130). Edward Elgar.

Klerkx, L., & Begemann, S. (2020). Supporting food systems transformation: The what, why, who, where and how of mission-oriented agricultural innovation systems. *Agricultural Systems*, *184*, 102901.

Kooiman, J. (2003). *Governing as governance*. Sage.

Kuhlmann, S., Stegmaier, P., & Konrad, K. (2019). The tentative governance of emerging science and technology – A conceptual introduction. *Research Policy*, *48*(5), 1091–1097.

Lange, P., Driessen, P. P., Sauer, A., Bornemann, B., & Burger, P. (2013). Governing towards sustainability – Conceptualizing modes of governance. *Journal of Environmental Policy & Planning*, *15*(3), 403–425.

Laranja, M., Uyarra, E., & Flanagan, K. (2008). Policies for science, technology, and innovation: Translating rationales into regional policies in a multi-level setting. *Research Policy*, *37*(5), 823–835.

Larrue, P. (2021). The design and implementation of mission-oriented innovation policies: A new systemic policy approach to address societal challenges. *OECD Science, Technology and Industry Policy Papers*, 100. Éditions OCDE. https://doi.org/10.1787/23074957

Lodge, M., & Wegrich, K. (Eds.). (2014). *The problem-solving capacity of the modern state: Governance challenges and administrative capacities*. Hertie Governance Report.

Matland, R. E. (1995). Synthesising the implementation literature: The ambiguity-conflict model of policy implementation. *Journal of Public Administration Research and Theory*, *5*(2), 145–174.

Mazzucato, M. (2018). Mission-oriented innovation policies: Challenges and opportunities. *Industrial and Corporate Change*, *27*(5), 803–815.

McCann, P., & Soete, L. (2020). *Place-based innovation for sustainability*. Publications Office of the European Union.

Molas-Gallart, J., Boni, A., Schot, J., & Giachi, S. (2020). *A formative approach to the evaluation of Transformative Innovation Policy.* TIPC Working Paper, TIPCWP2020–01. http://www.tipconsortium.net/wp-content/uploads/2020/07/Paper-FORETIP-1.pdf

Reale, F. (2021). Mission-oriented innovation policy and the challenge of urgency: Lessons from Covid-19 and beyond. *Technovation*, 107,102306.

Reichardt, K., & Rogge, K. (2016). How the policy mix impacts innovation: Findings from company case studies on offshore wind in Germany. *Environmental Innovation and Societal Transitions, 18*, 62–81.

Rhodes, R. A. (1997). *Understanding governance: Policy networks, governance, reflexivity, and accountability.* Open University.

Rogge, K. S., & Johnstone, P. (2017). Exploring the role of phase-out policies for low-carbon energy transitions: The case of the German Energiewende. *Energy Research & Social Science, 33*, 128–137.

Rogge, K. S., & Reichardt, K. (2016). Policy mixes for sustainability transitions: An extended concept and framework for analysis. *Research Policy, 45*(8), 1620–1635.

Salas Gironés, E., van Est, R., & Verbong, G. (2019). Transforming mobility: The Dutch smart mobility policy as an example of a transformative STI policy. *Science and Public Policy, 46*(6), 820–833.

Schlaile, M. P., Urmetzer, S., Blok, V., Andersen, A. D., Timmermans, J., Mueller, M., Fagerberg, J. & Pyka, A. (2017). Innovation systems for transformations towards sustainability? Taking the normative dimension seriously. *Sustainability, 9*(12), 2253.

Schot, J., Daniels, C., Torrens, J., & Bloomfield, G. (2017). Developing a shared understanding of transformative innovation policy. *TIPC Research Brief*, 1.

Schot, J., & Steinmueller, W. E. (2018). Three frames for innovation policy: R&D, systems of innovation, and transformative change. *Research Policy, 47*(9), 1554–1567.

Smits, R., & Kuhlmann, S. (2004). The rise of systemic instruments in innovation policy. *International Journal of Foresight and Innovation Policy, 1*(1–2), 4–32.

Steward, F. (2012). Transformative innovation policy to meet the challenge of climate change: Sociotechnical networks aligned with consumption and end-use as new transition arenas for a low-carbon society or green economy. *Technology Analysis & Strategic Management, 24*(4), 331–343.

Stoker, G. (2006). Public value management: A new narrative for networked governance? *The American Review of Public Administration, 36*(1), 41–57.

Thompson, G., Frances, J., Levačić, R., & Mitchell, J. (1991). *Markets, hierarchies and networks: The coordination of social life*. Sage.

Treib, O., Bähr, H., & Falkner, G. (2007). Modes of governance: Towards a conceptual clarification. *Journal of European Public Policy, 14*(1), 1–20.

Uyarra, E., Ribeiro, B., & Dale-Clough, L. (2019). Exploring the normative turn in regional innovation policy: Responsibility and the quest for public value. *European Planning Studies, 27*(12), 2359–2375.

Uyarra, E., Zabala-Iturriagagoitia, J. M., Flanagan, K., & Magro, E. (2020). Public procurement, innovation and industrial policy: Rationales, roles, capabilities and implementation. *Research Policy, 49*(1), 103844.

Van den Bosch, S. (2010). *Transition experiments: Exploring societal changes towards sustainability* [PhD dissertation, University of Rotterdam].

Van der Steen, M. A., Scherpenisse, J., Hajer, M., Van Gerwen, O.-J., & Kruitwagen, S. (2015). *Learning by doing: Government participation in an energetic society.* Netherlands School of Public Administration, Netherlands Environmental Assessment Agency.

Van der Steen, M., van Twist, M. J., & Bressers, D. (2018). The sedimentation of public values: How a variety of governance perspectives guide the practical actions of civil servants. *Review of Public Personnel Administration, 38*(4), 387–414.

Van Mierlo, B., & Beers, P. J. (2020). Understanding and governing learning in sustainability transitions: A review. *Environmental Innovation and Societal Transitions, 34*, 255–269.

Van Mierlo, B., Leeuwis, C., Smits, R., & Woolthuis, R. K. (2010). Learning towards system innovation: Evaluating a systemic instrument. *Technological Forecasting and Social Change, 77*(2), 318–334.

Wanzenböck, I., Wesseling, J. H., Frenken, K., Hekkert, M. P., & Weber, K. M. (2020). A framework for mission-oriented innovation policy: Alternative pathways through the problem–solution space. *Science and Public Policy, 47*(4), 474–489.

Weber, K. M., & Rohracher, H. (2012). Legitimizing research, technology and innovation policies for transformative change: Combining insights from innovation systems and multi-level perspective in a comprehensive "failures" framework. *Research Policy*, *41*(6), 1037–1047.

Wittmann, F., Hufnagl, M., Lindner, R., Roth, F., & Edler, J. (2021). Governing varieties of mission-oriented innovation policies: A new typology. *Science and Public Policy*, *48*(5), 727–738.

Table 15A.1 Case examples of governance arrangements per TIP governance task

Cases	TIP Governance Task: *Creating legitimacy and leadership*	TIP Governance Task: *Coordinating across multiple levels, actors, and instruments*	TIP Governance Task: *Reflexivity, learning, and experimenting*	TIP Governance Task: *Resolving conflicts*
C1: Green solutions	Ministry of Higher Education and Science is in a strong leadership role. Link with Climate Council of high-level representatives.	Policies launched with partners, e.g., Green Business Forum, Board of Business Development, and the business/finance ministry.	Formal evaluations like international assessment of Innovation Fund. In-house mapping of policies and barriers.	(Not prominently present, due to focus on competitive research calls)
C2: HTS2025	BMBF strategy is endorsed at highest political level. Missions are formulated as top-down and mostly focus on policy coordination; there is no separate budget.	BMBF strategy invites other ministries to implement missions, in collaboration with groups such as the Länder. Roundtable supports alignment. High-Tech Forum advises and engages stakeholders like citizens.	HTS2025 builds on evaluation of earlier HTS versions, new R&I landscape analyses, and ongoing process-based learning in current HTS.	Development of joint position papers and roadmaps for overcoming a lack of financial commitment from ministries.
C3: MTIP	MTIP is initiated by Min. of Economic Affairs & Climate Policy, but missions are owned by various ministries and kept away from politics.	Around 30 public organisations signed the Knowledge & Innovation Covenant. Multi-annual mission programmes are designed by triple helix platforms, checked by advisory boards.	MTIP is presented as evolving. Learning relies on (innovation) policy monitoring and on programme monitoring.	Intermediary-like mission teams mediate between innovation actors and problem-owning ministries.

Cases	TIP Governance Task: *Creating legitimacy and leadership*	TIP Governance Task: *Coordinating across multiple levels, actors, and instruments*	TIP Governance Task: *Reflexivity, learning, and experimenting*	TIP Governance Task: *Resolving conflicts*
C4: CoT	Strong focus on visionary ideas, comprising climate, energy, mobility, technology, and innovation agendas. Leadership supported by Austria's lead in European JPI "Urban Europe".	Cross-cutting topic links policy siloes via working groups. FFG aligns R&I funding schemes. CoT budgets for transnational initiatives like city partnerships. Network platform for knowledge and actor management towards societal stakeholders.	Regular adjustments of focus areas in yearly calls, based on perceived stakeholder needs. Ongoing development of new "system impact" assessment approaches.	Firmly established structures in the Austrian funding landscape ensures a gradual and smooth coordination process between core partners.
C5: RIS3CAT	EU origin of available funding provides legitimacy and helps to overcome local power games (due to vested interests in sectors).	Coordinating EU funds from non-sector-specific ministry ensures neutrality. Diverse public and private actors participated in developing shared agendas and targeting funding calls.	A series of workshops served to update views on which development paths to pursue and how to adjust funding calls.	Bottom-up approach brought municipalities on board, instead of imposing top-down policies on them.

Cases	TIP Governance Task: *Creating legitimacy and leadership*	TIP Governance Task: *Coordinating across multiple levels, actors, and instruments*	TIP Governance Task: *Reflexivity, learning, and experimenting*	TIP Governance Task: *Resolving conflicts*
C6: Amsterdam Circular	Team Circular developed strong vision "Amsterdam Donut City" and is backed by Coalition Agreement. Broad political commitment allows policy officials to explore new ideas and approaches.	Multi-departmental composition of a core team; extensive rounds of consultations and workshops with policy officials. Sustainability Council with firms and NGOs advises Team Circular. Online platform for community management.	Learning-by-doing approach; strategies refined based on experimentation results. Use of living labs and prizes. Comprehensive monitoring frameworks.	Articulating vision (based on integrated perspective) allows stakeholders to see what circular means to them and to overcome short-term conflicts.
C7: Circular Flanders	The CF transition team is placed outside any individual ministry and relies on a public-private steering committee of high-level societal representatives.	Various departments and regions partner with CF, while maintaining ownership over their own budgets and policies. Working agendas guide actions, such as the creation of a one-stop-shop for funding.	Few static measures; responsive policy actions are based on continuous interactions with variegated field-level stakeholders.	Creating arenas in which stakeholders confront each other with their views, e.g., panel sessions during debate events.

PART IV

IMPLEMENTING TRANSFORMATIVE MISSION-ORIENTED STI POLICY

16. Programming missions: steering a mission-oriented research programme through the production and use of strategic intelligence

Douglas K. R. Robinson, Mireille Matt and Renée van Dis

16.1 INTRODUCTION

16.1.1 Directionality Turn and the Challenge of Implementation

Grand challenges, whether *Sustainable Development Goals* or *Grand Societal Challenges*, capture societal needs that are currently unmet and often require international and multi-sector solutions. The need for European research to address major societal challenges was proposed in the Green Paper on the European Research Area adopted by the Commission of the European Communities in April 2007. During the preparation of the Horizon Europe framework programme, a decision was taken to go one step further in this direction with the design of "missions". The Lamy Report proposed a *mission-oriented, impact-focussed approach* to address global challenges (Lamy, 2017). Whereas societal challenges may be viewed as the broader social problem aim or benefit that is being sought (e.g., fighting climate change), missions represent a more narrowly defined set of activities that are supposed to deliver a verifiable result on a planned timescale and can be used to measure progress in overcoming the societal challenge.

In the world of policymaking, innovation policies[1] targeted at grand societal challenges rather than at purely economic growth have been suggested as a new type of policy (Borrás & Edler, 2020; Schot & Steinmueller, 2019). Such policies contribute to *facilitating innovation* and *socio-economic impact* in a particular *direction* towards a desirable transformative change[2] (Gassler et al., 2008; Weber & Rohracher, 2012; Mazzucato, 2018).

There is much debate, hype, and hope regarding mission-oriented policies because the implementation of such policies is a major challenge (Robinson & Mazzucato, 2019; Brown, 2021). Whilst approaches such as mission-oriented policies are not new, they are now being designed in a context that is not administered by centralised decision-making authorities in a vertical structure; they are administered in decentralised and dynamic innovation systems that include user-driven innovation, DIY innovation, etc., beyond the control of central administrations (Robinson & Mazzucato, 2019). This raises a number of challenges for mission

[1] We define innovation policy in the same way as Kuhlmann et al. (1999, p. 5): "the entire scope of related public measures of science, research, technology policy, overlapping also with industrial, environmental, labour and social policies."

[2] In this policy world, transformative changes are the large-scale industrial and societal changes necessary for grand challenges to be achieved.

implementation, where many diverse sets of actors are responsible for (a) creating and scaling up the innovations and (b) ensuring the innovations' eventual impact.

16.1.2 Mission-Oriented Research – Steering to Contribute

Research and innovation activities promise to be a major contribution to achieving many of the missions identified around the globe (Larrue, 2021). As part of the mission-oriented "turn", research and innovation funding programmes, such as the latest European Commission Framework Programme (Horizon Europe), have put societal "missions" at the centre of the 88 billion euro programme. Similarly, national funding agencies such as the French ANR (Agence Nationale de la Recherche) are experimenting with mission-oriented research funding programmes – for example, the *Cultiver et Protéger Autrement* programme, which places at its heart the ambitious objective of contributing to achieving a zero-pesticide future.

However, there is a paradox. Policies and management tools for steering mission-oriented research and innovation must be ambidextrous in their approach – they have to effectively link research and innovation activities with contributions to system change (*systems perspective*) with the real-time choices and uncertainty of research and innovation projects (*novelty creation perspective*).

Intermediary organisations such as research funding agencies are challenged to put mission-oriented policies into practice in such a context. Whilst the notion of mission-oriented policies has yet to stabilise (see Larrue et al., 2021), it is clear that for the world of research, programming research, and the programmers themselves, mission-oriented policies will play a key role in accomplishing missions (Barré et al., 2013).

Barré and colleagues (2013) proposed a three-layered functional approach to the organisation and management of research:

> *Orientation functional layer:* Involves the definition of policy objectives and the ways in which the policies envisaged will work toward the achievement of such objectives.
>
> *Programming functional layer:* Involves the translation of the objectives stated by the orientation layer into interventions implementing certain topic-specific priorities and the allocation of resources to such interventions, and eventually to research and innovation performers.
>
> *Performance functional layer:* Responsibility of research performers (PROs, universities, firms, etc.): production of knowledge and innovation through the activities of researchers, operation of research infrastructures, management of projects, networks.

Viewed through this three-layered lens, missions are cross-cutting priorities that guide and drive all three functional layers. This poses a challenge for funding programme managers, who are nested in the programming functional layer but necessarily linked to the orientation and performance layers. Intelligence is needed for all three layers to be combined in a way that allows programme managers to manoeuvre the programme towards maximising the contribution to the targeted mission, given the portfolio of projects selected.

Funding organisations have to interface with the "*orientation functional layer*" to translate policy priorities and design research and innovation programmes that (a) deliver excellent research and innovation and (b) contribute to mission objectives. For mission-oriented funding programmes that target early-stage research, the design of the programmes should both (a) *anchor themselves in the present world* of breakthrough novel research and (b) *anchor themselves in a relatively distant future* of a system that is transformed (where the loosely articulated[3] mission has been achieved).[4] Since the eventual mission contributions typically occur much later than the end of the funded activities, programme managers have to increasingly (1) engage in anticipating the future evolution of the results of the R&I programme to be able to ascertain their contribution to the policy priorities encompassed by the mission objectives and (2) engage with a variety of stakeholders, many of whom will be responsible for developing, scaling up, and generalising the products of the funded programme. To increase their engagement, programme managers must develop capacities to (co)create strategic intelligence, interpret, and incorporate it into real-time steering of the research programme.

Funding agencies also have to interface with the "*performance functional layer*". For example, funding agencies such as the ANR and the European Commission are challenged to steer research and innovation activities "on-the-ground" via funding calls, be complementary to other national and international funding programmes and activities, and align with broader policy orientations around the targeted mission area.

In addition, programme managers also need to act in concert and/or coordination with other programmes in the "*programming functional layer*", both nationally and internationally. Moreover, the transformative nature of missions means that large and complex socio-technical systems may have to transform. For example, programme managers may be required to interact with programmes that are quite different in nature, such as those focussing on mission-oriented agricultural research linking with engineering-focussed programmes (robotics) or social sciences (such as behavioural analysis).

The functional layer perspective on STI policy is useful in showing the nested nature of STI policy definition and execution. Since mission-oriented research is defined by the requirements to (a) steer research towards societal goals and (b) proactively contribute to achieving "mission success", it is challenged to address an overarching, but not clearly articulated,[5] goal (the mission) and remain open-ended to potential solutions (the stuff of research). Given the inherent uncertainty and complexity involved in both steering towards and contributing to missions, we observe that implementing mission-oriented research requires the mobilisation,

---

[3] We use the term "loosely articulated" since missions set a simple target but do not, in themselves, articulate the future situation. One can imagine a variety of future situations for sustainable mobility or for zero/lean pesticide futures.

[4] A short reflection regarding terminology for early-stage mission-oriented research: Where research is financed both to create excellent knowledge and contribute to a mission, for effective steering, mission-oriented research requires anticipation and reflection on the translation of the outcomes of research into innovations and into society. Therefore, efforts at steering/directing research towards missions cannot focus on research alone – they must also include a reflection on scale-up, diffusion, and the embedding of the outcomes of that research into society. With that in mind, in this chapter, we consistently refer to "mission-oriented research and innovation" since, for management purposes, the whole research and innovation pathway is under consideration, regardless of whether this consideration occurs at the early stages of innovation pathways or not (Robinson & Rip, 2013).

[5] Whilst one could argue that a "mission" is well articulated – it has a clear and time-bound target – it does not exclude solutions that will lead to mission success, nor does it predefine the nature of the focal system that is being transformed.

production, and use of various forms of forward-looking strategic intelligence (Kuhlmann et al., 1999; Edler et al., 2006; Robinson et al., 2023).

16.1.3 Strategic Intelligence for STI Policy

The need for improved forward-looking intelligence to support and inform STI policy was highlighted by Tübke and colleagues:

> ...science and technology are advancing at a very high pace. Policy-makers do not have the luxury of waiting until situations are clarified and until the effects are evident before they take decisions. Tomorrow's scientific and technological development originates in the conditions established today. Therefore, it is not surprising that there is a demand for "Strategic Intelligence" to support policy-makers in understanding the relevant aspects and scope of the impacts of science and technology and their possible future developments. (Tübke et al., 2001, p. 1)

The concept of strategic intelligence emerged at the end of the 1990s, along with the recognition that STI policy evaluation was moving away from a desire to legitimate past policy actions and demonstrate accountability, towards the need to improve and inform future actions (Kuhlmann, 2002). Kuhlman proposed that approaches to STI policy evaluation were evolving from a purist model of "objective neutrality", providing evidence-based outputs but not going so far as to provide policy outcomes, to a more formative approach, where evaluators "act as process consultants in learning exercises involving all relevant stakeholders, providing advice and recommendations as well as independent analysis" (Kuhlmann, 2002, p. 16).

Noting that this formative approach to evaluation was tightly linked to strategy articulation, Kuhlmann et al. (1999) have argued that STI policy requires input from a toolbox of strategic intelligence approaches, drawing on tools such as formative evaluation, technology foresight (TF), technology assessment (TA), and technology trend monitoring (TM) to inform policy definition.

We agree that mission-oriented and transformative STI policies require a variety of forms of strategic intelligence, not only to define STI policies but also to aid in the reflexive real-time steering of STI activities to contribute to mission goals. However, questions remain about how to address the challenge to both *steer* and *contribute*. Given the assumption that better decision-making stems from better access to relevant information, the questions that arise are: What are the intelligence needs for steering research through mission-oriented research programming? Are there new intelligence requirements specific to mission-oriented research? What capacities are needed at the programming functional level (Barré et al., 2013) and what is already known about the intelligence needs for the mindful steering of research and innovation?

16.1.4 A Case Vignette on Mission-Oriented Research Policy Implementation

This chapter will dig deep into how programme managers are tackling the "steering challenge" via a case study on an ongoing mission-oriented programme that creates as well as combines a variety of forms of strategic intelligence. For this case, the strategic intelligence is not "out-there" to be gathered, but must be created by mobilising and integrating various tools and approaches within the nested management of research (cf. Barré et al.'s functional layers). We argue that this is generally the case for mission-oriented programmes that focus on stimulating breakthrough and transformative change and where high uncertainty and complexity are the norm.

The following text describes the implementation of a mission-oriented programme in the context of the French National Research Agency (ANR) that focusses on the ongoing "zero pesticide mission" (2020–2026). The authors first present insights into three forms of interrelated strategic intelligence, which are being actively created and "programmed" into funding and research programme management. The contribution made by this chapter is the provision of insights into the layered (nested) processes of steering research programmes, the tools and processes being applied and tested, and the forms of actionable mission-oriented intelligence that are being created.

16.2 DEEP CASE EXAMPLE OF PROGRAMMING MISSIONS INTO RESEARCH

16.2.1 From the Ashes of Failure: The Need for a Mission-Oriented Approach

The reduction of chemical pesticides has become a shared goal and a burning policy issue across many countries, particularly due to the recognised negative impacts of pesticides on animal and human health, as well as on the environment. However, the current agricultural system places at its heart the use of chemical pesticides as a means of producing robust and high-yield crops for an ever-increasing global population. A transition away from substantial pesticide use, without a detrimental reduction in crop yield, along with alternative agricultural practices, processes, and tools is needed. Advances in research promise potential solution pathways to this grand societal challenge of reducing pesticides while maintaining sufficient crop yield. In a recent position paper, a number of "research fronts" were identified that could contribute to solving this grand societal challenge (Jacquet et al., 2022). These research fronts include redesigning crop systems to enhance prophylaxis, mobilising biocontrol approaches and digital agriculture tools, and broadening the approach to plant breeding.

In a cross ministry[6] press release regarding a clear mission, in April 2019 the French government announced its plan to press forward with a mission-oriented programme to drive a sustainable agriculture transition by promoting agroecological solutions and reducing the use of chemical pesticides in farming.

> The government confirms its ambition to reduce plant protection products by half by 2025 and to phase out glyphosate for a majority of uses by the end of 2020. (Translated from a French Government press release regarding ECOPHYTO 2)

[6] The press release was co-signed by the following: François de Rugy (Minister of State, Minister for Ecological Transition and Solidarity), Agnès Buzyn (Minister of Solidarity and Health), Frédérique Vidal (Minister of Higher Education, Research and Innovation), Didier Guillaume (Minister of Food and Agriculture), Brune Poirson (Secretary of State to the Minister of State, Minister of Ecological Transition and Solidarity), and Emmanuelle Wargon (Secretary of State to the Minister of State, Minister of Ecological Transition and Solidarity).

The aim to achieve zero (or lean) pesticide use is not new. The French policy initiative ECOPHYTO, implemented in 2008, aimed for a 50 per cent reduction in pesticide use by 2018. However, it failed spectacularly – rather than a reduction in pesticide use, there has been an increase. Research has shown that the achievement of this goal depended on systemic transformations in agricultural production (Guichard et al., 2017).

In recognition of this failure, the French government initiated a mission-oriented policy to drive the transition to zero pesticide agriculture by taking a systems focus. The French mission-oriented programme, "Growing and protecting crops differently", a Priority Research Programme (PPR), was launched by the initiative of the Ministry of Higher Education, Research and Innovation and the General Secretariat for Investment. It was part of the government's objective to move towards productive agriculture that is more respectful of the environment and human health, with the aim to accelerate the production of new solutions for the withdrawal of phytosanitary products (pesticides and fungicides). INRAE has been entrusted with the scientific steering of the project, and other research establishments (CNRS, CIRAD, universities, etc.), technical institutes, and economic players have been involved in the discussions.

The gradual reduction in the use of pesticides meets a strong societal demand and is an objective of public policy and research programming at both the national and European levels. Research and applied research have been underway for many years to reduce the use of pesticides in agriculture. This initiative is supported by European research policies via the H2020 and Horizon Europe framework programmes and national policies such as the ECOPHYTO 2 plan. According to the ECOPHYTO 2 plan, the mission of zero pesticide agriculture is achievable – it is already possible to envisage a 20–30 per cent reduction in the use of pesticides based on results that have already been obtained by some farmers. Part of the objectives of ECOPHYTO 2 and a priority of current policies is generalising the results to apply to all agriculture.[7]

In addition to generalising these already demonstrated actions, the PPR "Growing and protecting crops differently" focusses on preparing the next stage in transforming agriculture towards zero-pesticide production. Given that it takes 10 to 20 years for research projects to have an impact in terms of innovation or changes in practices, the objective of the PPR is to catalyse and steer fundamental research towards contributing to solutions for greatly reducing and doing without pesticides (chemical substances and natural substances without a negative impact on the environment or health) by the 2030–2040 period.[8]

The focus on plant production concerns French agriculture, both in mainland France and overseas, as well as all types of production – annual and perennial, herbaceous, and woody. The aim is to build on three "Principles of Action":

1. *Privilege prophylaxis:* Crop protection as practised today in France and worldwide is based heavily on curative approaches, with crop protection products applied in homogeneous doses, either when the bioaggressors (weeds, pathogenic fungi, insects, etc.) are

---

7 https://agriculture.gouv.fr/le-plan-ecophyto-quest-ce-que-cest

8 See the programme's call for proposals, "Cultiver et protéger autrement: Alternative aux phytosanitaires: mobiliser les leviers de l'agroécologie, du biocontrôle et de la prophylaxie pour une agriculture performante et durable". https://ecophytopic.fr/sites/default/files/2020-11/Fiche-descriptive_AAP-ANR-CPA_2019.pdf

visible or as a preventative application. In this way, the actual degree (or pressure) of the bioaggressor is not considered when applying crop protection products. A change in practice is needed away from the *blanket use* of crop protection products towards *targeted* approaches that focus on reducing "pest pressure" by using appropriate prophylactic methods.

2. *Use the levers of agroecology:* As a scientific discipline, agroecology was built by crossing agronomy and ecological sciences; it also mobilises social sciences in support of the design and management of sustainable agroecosystems. One of its basic principles is to seek to increase functional biodiversity in order to strengthen biological regulation and ecosystem services. As a principle of this PPR, this increase of functional biodiversity must be considered at different spatial scales – from the plant, to the agricultural plot, to the cropping system, and to the whole landscape – and at different time scales. In the PPR, agroecology is not seen as a substitute for pesticides, but rather as a means of mobilising levers to increase biological regulation and limit pest pressures.
3. *Rethink the value chain:* The diversification of cropping systems implies rethinking the organisation of current commodity chains and setting up new processing and marketing channels. The presence of highly diverse crops often requires new management of production heterogeneity. The promotion of pesticide-free and residue-free production requires identification and recognition of the product's characteristics by the consumer – for the PPR, a key principle is that this development will need to mobilise various players downstream of the agri-food sector. At the same time, the upstream sectors, in particular, companies producing biocontrol products and agricultural equipment, will have to transform themselves to support the transition. Therefore, an evolution of public policies is also essential to allow this change.

16.2.2 Experimenting and Integrating Steering Capacities for Mission-Oriented Research

As previously described, the *Cultiver et Protéger Autrement* (Growing and protecting crops differently) priority research programme in France has been designed to contribute to the mission of "zero chemical pesticides in agriculture by 2050" and create contributions to solving the societal challenge linked to the sustainability of food security. This mission represents a clear and ambitious objective for society, with a specific, theoretically measurable direction and timeframe for action. However, the pathways to achieving this mission are uncertain.

To meet the ambition for 2050, the PPR programme has financed 10 research activities (approximately 2 million euros each) that are thought to be realistic and potentially breakthrough research projects, representing a clear break from existing solutions. Radical innovations are foreseen to emerge from multiple experiments, based on transdisciplinary, trans-sectoral, and multi-actor approaches, that are tested in secure environments before being generalised for the whole of society.

Driving research that breaks radically from existing solutions means high uncertainty – predicting actual outcomes of research processes and their impacts is challenging. This uncertainty is further compounded by mission-oriented research – research that should contribute to an envisaged aim. Therefore, open-ended breakthrough research must also narrow down towards specific desirable impacts – in this case, zero pesticides. In this context, the outcomes of potentially breakthrough research require the consideration of how the research will contribute to the envisaged mission and pathways to translate research into desirable impacts.

For the PPR *Cultiver et Proteger Autrement*, it was clear that navigating through complex and uncertain processes required the ability to be reflexive and adaptive to manage this high uncertainty. In order to manage this uncertainty and to steer towards the mission aim, the PPR had the desire to (a) have the tools to learn about the actions taken, (b) understand and anticipate the evolution of the context, (c) understand the performance of various experiments, and (d) coordinate the portfolio of funded projects so as to maximise synergies across the 10 projects, and external to the programme.

With these objectives in mind, the PPR set up a governance structure to develop and mobilise strategic intelligence to inform the interlinked and nested operations of the three functional layers of research (orientation, programming, and performing).

In the remainder of this section, we will describe three different types and processes of producing strategic intelligence. The first, the ASIRPA Real-Time process, is a *formative real-time evaluation* approach co-developed within the PPR as a reflexive and strategic articulation support system, based on the notion of endogenous futures and implemented at the programme level and at the performance (project) level of the PPR. The second, *la Prospective*, mobilises *foresight* tools to develop three scenarios of zero-pesticide agriculture systems in 2050. This source of strategic intelligence can be used in combination with a backcasting approach to identify instruments and intervention targets that could drive the transition to zero pesticide agriculture. The third source of strategic intelligence is the active national and international coordination at the programme level. While this may not have been considered part of the definition of strategic intelligence in its early manifestations in the early 2000s (Tübke et al., 2001), for mission-oriented research programmes, where only a modest (but important) contribution to the overall mission aim can be envisioned, concertation and/or coordination with other programmes is key to identifying synergies and to avoiding the duplication of efforts in both the research produced and the support of translating the outcomes of research into mission-relevant contributions.

16.2.3 ASIRPA Real-Time

Balancing the mission with the various avenues of research and their eventual contribution to the mission is a management challenge that requires new tools and capacities to anticipate the pathways from the present research portfolio to future impacts. These tools and capacities must be integrated in the appropriate places to provide the anticipatory strategic intelligence needed and to be used in real-time steering activities to improve the likelihood of contributing to achieving the mission.

The *objectives* of the ASIRPA$^{\text{Real Time}}$ approach in the PPR are: to support the project and programme managers of the PPR to steer research and related innovations towards a chemical pesticide-free agriculture by 2050; to make explicit the vision of how the current research activities will unfold in the future; to identify bottlenecks and catalysing factors that will enable the research outputs to contribute to the mission aims; and to make informed decisions that facilitate the navigation towards the future.

The ASIRPA$^{\text{Real Time}}$ approach builds from a longstanding experience developed in ASIRPA$^{\text{ex post}}$ that focussed on evaluating past research programmes and activities at INRAE (Matt et al., 2017). ASIRPA$^{\text{ex post}}$ was created and implemented over a 10-year period to understand the diversity of impacts generated by INRAE's research portfolio over multiple decades, as well as to legitimise past research investments and demonstrate accountability. For the PPR, the evaluation aim is a formative one, switching from historical analysis to real-time assessment based on a better understanding of potential future directions, developments, and impacts.

Both approaches are based on a generic tool (the impact pathway) that enables researchers and managers to understand how research can contribute to generating transformations and impacts and how to characterise these transformations. Both approaches also have the same theoretical background: the sociology of translation (Callon, 1999) and innovation journey (Van de Ven, 2017). They are designed to be usable by a variety of actors (researchers and programme managers).

ASIRPA$^{ex\ post}$ and ASIRPA$^{Real\ Time}$ are standalone approaches that empower research actors, requiring little or no external support beyond initial training. Embedding tools and capacities in the research system is important – for early-stage and potentially breakthrough research, researchers and programme managers must have relevant knowledge about the novelty created and the context. Therefore, it is essential to empower these actors with support tools so they can be reflexive about the directions and potential developments of their research outcomes and then steer R&I activities in real time based on these insights (Robinson & Rip, 2013). For the PPR, a key challenge was to foster capacity building at both the project level (where the researchers are defining their research directions and doing their research) and at the programme level (where programme managers are managing the portfolio of research projects, exploring ways to support the portfolio and potentially exploring ways to facilitate the contribution of the research portfolio to the mission aim). To produce portable tools within the PPR-CPA, the team of social scientists who developed the ASIRPA$^{ex\ post}$ tool (Joly et al., 2015; Joly et al., 2019) co-developed the toolset with the users (project and programme managers) at the early stages of the PPR programme.

For ASIRPA$^{Real\ Time}$ the creation of strategic intelligence to steer the projects and the programme towards contributing to the mission requires *the development of an impact pathway specific to each level*. Project-level and programme-level impact pathways are built based on common logics. First, building an ex-ante impact pathway is a necessary phase to realise the first back-casting exercise: to anticipate desired transformations that will achieve mission aims, to align R&I activities with societal needs, and to anticipate the coordination of a diverse and complex set of actors and activities. Second, real-time monitoring consists of an ongoing revision of the impact pathway for real-time decision making. Third, the different versions of the impact pathways are built through a series of participatory workshops involving relevant stakeholders and mobilising distributed intelligence (Kuhlmann, 2003) to provide a collective vision of desired changes. The impact pathways at the project level and the programme level are linked to each other. At each iteration, the programme manager will follow the progress of the portfolio of projects, work on possible synergies between projects, coordinate project activities with the other existing programmes (see Section 16.2.5) to leverage research activities, and engage with stakeholders and institutions to accelerate the contribution to expected transformations.

The first impact pathway exercise at the project level was conducted approximately eight months after the official start of the programme. Eight out of ten projects succeeded in building an impact pathway, where a team of researchers who were involved in the project developed a first version of the impact pathway and shared it with the entire consortium during a collective workshop. The project impact pathway provides a coherent narrative of (a) how the research activities contribute to a chain of impact-generating steps, (b) the stakeholders, infrastructures, and institutions that are required to facilitate the pathway, and (c) how the outcomes of the funded research may contribute to the overall mission – in this case, zero-pesticide agricultural systems.

As the PPR continues through its six years, these impact pathways will be revisited and revised every 12–18 months. This coherent narrative is useful in several ways. It articulates clearly what the envisaged goals of the project (beyond that of producing excellent science) are for all research project participants. It also provides a baseline to engage with different actors, necessary to "take up" the outcomes of the research and to upscale or generalise it for a wider community. Finally, this narrative provides a measure of benchmarking for the project itself – by using the project impact pathway to identify criteria of success and milestones in the near future, it can be revisited to compare and contrast actual developments and impact-generating mechanisms within the project with those articulated in the pathway. The project impact pathways are also useful at the programme level, as they involve the collective combination of all research activities and impact-generating mechanisms within the portfolio of funded projects.

The impact pathway at the programme level aims to anticipate the contributions that the portfolio of 10 projects will have on the mission of zero-pesticide agriculture by 2050. A mission-oriented programme manager needs to become reflexive about steering system innovations towards defined directions, to learn across the portfolio of funded projects, and to be able to justify the programme's expected transformation contribution to the programme evaluators. The first version of the PPR-CPA programme impact pathway was developed during the first year of the programme through workshops that involved actors of the *Comité de Pilotage* (the "steering committee" comprising a number of ministries and the funding agency), the animation group, and the lead scientist (Principal Investigator, or PI) of each of the funded research projects. This version not only integrates information from the project impact pathways (this is important because its aim is to manage these activities), but also anticipates the support systems needed to catalyse project impact pathways and improve the chances of success.

The programme impact pathway provides a narrative of (a) the expected transformations led by the programme, (b) the coherence of the portfolio of projects and the aggregated expected outputs of the programme, and (c) the links with stakeholders in the context of transforming the programme outputs into impacts. Whilst the programme itself is responsible for the six-year lifetime of the research projects funded, it is also a mission-oriented programme aiming for long-term effects. However, the fate of the outcomes of the portfolio of research managed by the programme is in the hands of other actors, for example, firms, farmers, or citizens/consumers. The fate of the outcomes of the portfolio of research activities may also be in the hands of regulators or in other areas of government beyond research and agriculture. Thus, the programme impact pathway provides a way of anticipating and identifying key stakeholders and governance infrastructures that may hinder or catalyse the impact pathways of the portfolio of projects. This provides the opportunity to identify key bottlenecks and opportunities for ex-ante acceleration so that action can be taken to enable the impact pathways of the project.

The programme impact pathway plays another role – it allows for horizontal coordination and concertation within the functional programme layer and with other research and innovation programmes, whether national or international (see Section 16.2.5). This is essential for avoiding research duplication and for identifying complementarity in terms of envisaged research and innovation outputs. It also allows for cooperation in aligning framing conditions for the upscaling and generalisation of the outcomes in order to contribute to transformative change; for example, anticipating market infrastructures (Boon et al., 2022) such as certification processes, standards, regulations, financial incentives, etc.

In summary, the implementation of the ASIRPA$^{\text{Real Time}}$ approach within the PPR-CPA enables the development of strategic intelligence to facilitate the steering of research and its contribution to a mission. As a support system, it enables the development of capacities to think about the future, make decisions in uncertain contexts, and to be able to change direction and pathways if needed. In practice, ASIRPA$^{\text{Real Time}}$ is a management tool that enables R&I managers, who are localised at various levels (projects and programmes), to develop a set of capabilities as well as expertise to navigate complex and uncertain environments. These capabilities consist of:

1. *Making explicit the vision about the future transformations* the project and programme aim to contribute to. Anticipating expected transformations and drawing out the direction is a crucial first step in the approach.
2. *Developing a navigation system* (i.e., the impact pathway) that indicates which pathway to follow to reach the end goal. The impact pathway incorporates the anticipated expected transformations, productive configuration of actors that initiate the innovation journey, critical points, key governance and market infrastructures, as well as the role of intermediaries in the impact-generating process and generalisation.
3. *Building a parsimonious information system.* Because of the radical uncertainty related to the research and innovation process and the rapidly changing context, it is neither possible nor useful to collect detailed information on all possible outcomes and impacts. As such, the primary point of the parsimonious information system is to collect data that supports decision-making while also enabling the formulation of strategies for achieving the next steps on the impact pathway. In order to reduce uncertainty but still tolerate a level of ambiguity, it is important to work with key stakeholders (whether incumbents or newcomers) and expected first users to be able to understand and anticipate signals of emergence, changes in patterns and behaviours, etc., while building a parsimonious information system that collects data useful to decision-making while also enabling the construction of strategies to achieve the next steps on the impact pathway. Radical uncertainties in the research and innovation process and the rapidly changing context mean that collecting detailed information on all possible impacts and outcomes is neither possible nor useful. In order to reduce uncertainty but still tolerate a level of ambiguity, it is important to work with key stakeholders (whether incumbents or newcomers) and expected first users to be able to understand and anticipate signals of emergence, changes in patterns and behaviours, etc.
4. *Assessing the contribution and progress toward the expected goals.* The analysis of the impact pathway enables R&I managers to reflect on the critical elements that need to be in place or overcome for the expected transformations to happen. These elements (new technologies, new markets, new regulations, or changes in users' context) constitute milestones or intermediary objectives that should be reached due to a set of actions taken by researchers and programme managers. Assessing the contribution of the project/programme means analysing the progress toward these goals. Monitoring the progress does not mean following a set of predefined indicators but building dynamic, process-based

> indicators. For instance, it is not the number of collaborations with stakeholders that is important, but the creation of linkages that enable appropriate translations.

However, because ASIRPA$^{\text{Real Time}}$ supports the creation of anticipatory strategic intelligence to inform real-time decision-making, it is necessary to explicitly take into account that the transformation targets may move and that some causal relations or chains of translation identified may be challenged over time as the project/programme unfolds. The impact pathway that is constructed through the ASIRPA$^{\text{Real Time}}$ approach, whilst seemingly techno-deterministic and showing a linear path from research to impact, is a "rational myth" (Meyer & Rowan, 1977; Robinson & Propp, 2008) – recognising the need to draw on the current vision of the future (transformation target, causal chains) and to consider them as true whilst being aware that this is a convention that may be challenged by the production of new knowledge. The implementation of modern missions (Mazzucato, 2018; Larrue, 2021) and transformative innovation policies (Schot & Steinmueller, 2019; Borrás & Edler, 2020) have yet to be clear on this challenge – clear mission-focussed targets will evolve over time, thus potentially undermining the original mission. Without a formative and reflexive approach to mission implementation, the value of "mission" in research and innovation policy – indeed their *raison d'être* – is undermined.

For the ASIRPA$^{\text{Real Time}}$ approach, the "rational myth" is embraced as a means to an end in a stage-gate[9] vision of research, demonstration, scaling, and diffusion. This is why plans are in place to revisit and redraft the impact pathway at periodic moments during the lifetime of the funded project and to envision particular stages and "gates" in the future (Rip, 2019). ASIRPA$^{\text{Real Time}}$ is formative (Luederitz et al., 2017) as it enables, through learning loops, the improvement and adjustment of the implementation of the programme and projects, and supports adaptation in complex and dynamic conditions. ASIRPA$^{\text{Real Time}}$ encompasses real-time assessment, i.e., revising the critical points of the anticipated pathway on a recurrent basis with the aim of triggering learning and looking at how intermediate results may lead to reconsideration of the previously articulated impact pathway. *These iterations constitute a core element of the reflexivity and strategy articulation support system* (Robinson & Propp, 2006), aimed at supporting programme and project managers to navigate towards societal goals and increase their capabilities to do so.

---

[9] The "stage-gate" innovation process (Cooper, 1990) relates to a series of innovation phases that begin and end at a "gate" where potential future strategies are opened up, considered, then closed down for implementation. The ASIRPA approach mobilises a stage-gate perspective in two ways. The first is the idea that during the projected impact pathway, there are particular phases of the innovation pathway that should be considered when anticipating the impact pathway from research to mission contribution. The second way concerns the ASIRPA approach itself and when it is used. The anticipatory impact pathway approach opens up future pathways and allows for the exploration of possibilities and the development of an impact pathway, which can then be locked down for a specific period of time. This process is reinitiated at a predefined moment; the previous impact pathway is revisited and used as a learning device in the development of a "new rational myth".

16.2.4 *La Prospective*

Foresight is a key source of strategic intelligence for the PPR. The *Prospective 2050* is a foresight study that aims to elaborate on scenarios of a European agriculture system that is free from chemical pesticide use in 2050. The *Prospective 2050* activity, which is conducted and orchestrated by a team of foresight specialists in a dedicated department of INRAE,[10] aims to anticipate future agricultural system changes to define transition trajectories and integrate the complex interrelations between crop systems, commodity chains, environmental ecosystems, and food consumption patterns. The *Prospective 2050* also explores the effect of major uncertainties, such as the impact of climate change and the evolution of international markets.

Prospective 2050 is mobilised to build scenarios of agriculture without chemical pesticides for the European Union in 2050, by answering two questions:

1. What different forms could agriculture without chemical pesticides take in 2050, and what would be the consequences for production, land use, trade, and biodiversity?
2. What could be the different trajectories toward these forms of agriculture without pesticides?

This foresight activity relies on two forms of experts that, when taken together, constitute the foresight forum: (1) a group of experts in various aspects of the agricultural system, new technologies and techniques, and social scientists who are familiar with transition studies and socio-technical system transformation, and (2) an experienced foresight project team, which mobilises existing tools and capacities.[11]

The methodology follows tried-and-tested foresight techniques:

1. Literature review on the agricultural system, key factors, and potential drivers of change;
2. Expert engagement to identify expectations of how the agricultural system may evolve, key factors, and potential drivers of change;
3. The creation of scenario elements by the central foresight project team that act as potential building blocks of full scenarios;
4. The weaving together of full scenarios based on scenario elements and a standardised "*la Prospective*" approach;
5. Testing the robustness of the scenarios with experts (particularly social scientists);
6. Backcasting the scenarios of zero-pesticide agriculture in 2050 to today, identifying key bottlenecks and opportunities that could act as points of policy, programme, or project intervention.

Whilst the scenario exercise is ongoing, all six steps were conducted at the time of writing this chapter, which allows us to share some details on the content.

The three scenarios for zero-pesticide agriculture are titled:

---

[10] Prospective 2050 is part of the programme-level animation carried out by INRAE's Department of Collective Scientific Expertise, Foresight and Studies (DEPE) and is also linked to the European Alliance towards Chemical Pesticide-free Agriculture.

[11] The project team brings together skills and experience in foresight, modelling, and also in the fields of agronomy and economics.

Scenario 1: Territorial and regional coordination, complex and diversified landscapes for a one-health food system.

Scenario 2: Global and European food chains for pesticide-free food markets.

Scenario 3: European and regional food systems, soil and food microbiomes for healthy food and healthy diets.

Each scenario was built on mini-scenarios around the following dimensions: (a) the food value chain, including drivers such as the "supply of pesticide-free food as a food safety standard" or "territorial or regional supply of food that preserves human and environmental health including biodiversity", (b) farm structures, for example "cooperatives driving regional diversity" or "new forms of specialisation and financialisation", (c) changes in cropping systems, such as mixed cropping to strengthen plant immunity rather than single crop farming, and (d) the degree of introduction of agricultural equipment and digital technologies – for example, the integration of autonomous robots and the "pooling of equipment, sensors and data".

The robustness of the three scenarios was then explored through a working group, in which the authors of this chapter participated as transition experts, through lenses of what these scenarios mean with regard to public policies (agricultural, employment, and STI policies) and the education and agricultural knowledge system (recognising different forms of higher education along with "in-the-field" know-how on farming systems).

These further fleshed-out scenarios were then exposed to a backcasting technique in order to articulate milestones and identify obstacles and opportunities for each scenario, along with identifying actions to reach the milestones for each scenario, and estimating a chronology of interventions.

Thus, a key function of *Prospective 2050* is to unpack and further articulate the vision of "mission success" – a zero-pesticide agricultural system by 2050. This can be done by exploring and focussing on the systemic nature of the mission and addressing the complex entanglements of agricultural system elements and the many wicked problems that arise when transforming parts of the agricultural system.

As an outcome, a number of potential visions of a pesticide-free agricultural future are provided and interrogated, and a number of interventions are then suggested. As a source of strategic intelligence, this provides reflexivity on the nature of system transformation and fleshes out some of the contrasting pathways to the mission aim. While the Prospective 2050 is not yet fully complete, it is envisioned that the outcomes of this exercise will inform the programme level and orientation (policy) level in terms of which forms of intervention will catalyse system transformations, particularly with a view to the portfolio of projects that are currently underway.

16.2.5 International Coordination at the Programme Level

A programme such as the PPR-CPA cannot achieve the mission of chemical pesticide-free agriculture alone, as it constitutes only one element of the programming layer. The success of a mission depends not only the type of projects funded and bottom-up experimentations (performance layer) but also on the set of existing policy initiatives (STI policies, sector-specific policies, regulation, procurement policies, etc.) and their coordination.

The PPR-CPA is among the first R&I programmes in Europe to set a mission as ambitious as zero-chemical pesticide agriculture, despite strong demand from societies across Europe to accelerate the agroecological transition. To face this challenge, INRAE, in collaboration with its German counterparts ZALF (Leibniz Centre for Agricultural Landscape Research) and JKI (the Julius Kühn Institute), fostered the creation of a European Research Alliance (ERA) Towards a Chemical Pesticide Free Agriculture.[12] At the beginning of 2020, these research institutes elaborated a joint declaration of intent "towards a chemical pesticide-free agriculture", currently signed by 34 research institutes in 20 European countries. This community gathers over 2,000 researchers and engineers who are working on topics related to pesticide reduction and covers a network of more than 300 experimental fields and farms.

The aim of this European Research Alliance is to rethink the way actors carry out research and to develop new common research and experimentation strategies at a national and European level. Its objective is to build up a scientific roadmap that will contribute to the European Green Deal and influence European policymaking in this area. This transdisciplinary R&I agenda serves as a coordination tool and as a discussion basis at the national and EU levels to foster high-impact R&I programmes and projects.

> Partners will also foster collaborations within their respective countries and at EU level to develop plans and proposals for future research programmes and for policy support, so that they will improve the sustainability and competitiveness of the sector, especially in regards to integrated management systems. Partners will thus raise awareness of the importance of the issue and the need for a strong research and innovation paradigm shift supported by dedicated funding.[13]

This alliance also gathers information about the programming layer of each country and at the EU level as a whole. It could facilitate the alignment of national and European research efforts and support systemic and transdisciplinary research approaches. Increasing the coordination and coherence of STI transformative policies across Europe is necessary to achieve a sustainable European agri-food system free of chemical pesticides. The alliance is too young to be able to contribute to transforming the policy arena at this time, but its existence constitutes a first step towards transformation.

The PPR-CPA could benefit from the construction of future research programmes and policy support. Conversely, it also constitutes an important source of information that might contribute to shaping these policy initiatives.

16.3 DISCUSSION

Whilst the nature of missions and the context of their implementation are heterogeneous and, in some ways, idiosyncratic, we observe (and present) a number of cross-cutting challenges and potential needs with regard to strategic intelligence for programming missions. The three forms of strategic intelligence being created and used in the *Cultiver et Proteger Autrement* funding programme provide deep insights from *living* mission programmes while wrestling with the challenge of steering R&I toward missions. In this way, the chapter presents insights

12 https://www.era-pesticidefree.eu/

13 https://www.era-pesticidefree.eu/Towards-a-chemical-pesticide-free-agriculture/Our-aim

from early experiments to actually solve/manage the challenges of steering mission-oriented research programmes.

The chapter also shows the double importance of strategic intelligence. The first is the *OUTPUTS*, or insights into directing research programmes towards contributing to mission aims, which were produced from the strategic intelligence exercises examined. The second is the *PROCESS* of building and producing the strategic intelligence. The process integrates a variety of distributed intelligence by including many stakeholders and encouraging them to engage and reflect on various issues and mechanisms. This not only increases coordination and concertation (a necessary capacity for accomplishing societal grand challenges – see Kuhlmann & Rip, 2018) but also builds reflexive capacity on the factors that will shape and be shaped by steering initiatives.

Each of the three forms of strategic intelligence mobilised in the PPR research programme described above plays a different role in the nested steering of zero-pesticide agriculture mission-oriented research.

For the ASIRPA$^{\text{Real Time}}$ approach, a process of anticipation, assessment, and strategy articulation has been developed and integrated into the ongoing programming and performance of mission-oriented research. The potential contribution of the research can be identified through anticipating innovation journeys from the research being undertaken and from clearly articulating the potential range of impacts that could stem from the research. This intelligence, from the "rational myth" that is the ASIRPA$^{\text{Real Time}}$ impact pathway, is used to articulate project-level strategies – for example, identifying which value chain alignments and transformations are needed to "receive" and mobilise the outcomes of the research activity. Likewise, at the programme level, ASIRPA$^{\text{Real Time}}$ allows for the articulation of the market infrastructures (Delemarle & Larédo, 2014) that may need to be transformed and thus can inform support activities at the programme level (for example, creating a forum for articulating technical standards around digital agriculture solutions, which are part of the PPR portfolio of projects). Also key in the ASIRPA$^{\text{Real Time}}$ approach is the empowerment of project and programme managers to produce their own strategic intelligence with little intervention from outside, so that ASIRPA$^{\text{Real Time}}$ acts as an embedded reflexivity and strategy articulation support system (Robinson & Propp, 2006).

The Prospective 2050 provides a different form of strategic intelligence. Rather than projecting from the present based on individual research activities, it focusses on system-level transformations and starts in the future, focussing on the year 2050 and backcasting to the present. Its aim is to further characterise what "mission success" could be through the definition of "future working worlds" (Larédo et al., 2002) and then to further articulate different aspects of the agricultural system and the transformative changes necessary to get there. This strategic intelligence is useful for the articulation between the programme level and the orientation (policy) level, by providing insights into what system-level changes are needed beyond those that are being explored in the portfolio of projects, but which are necessary for "receiving" and mobilising the outcomes of the PPR project portfolio. It also acts as a reflective tool on the mission itself, asking which future is the most desirable and to whom. A key difference between the Prospective 2050 approach and the ASIRPA$^{\text{Real Time}}$ approach is that the former requires dedicated professionals to orchestrate the foresight process and to build the scenarios. Therefore, in-house expertise on foresight or external experts that can help produce foresight-based strategic intelligence are required.

The third form of strategic intelligence is located at the programme level and focusses on understanding the broader landscape of activities that are focussed on the same mission aim. The European Research Alliance Towards a Chemical Pesticide Free Agriculture was initiated as part of the PPR, along with other similar programmes and research institutions across Europe, with the goal to collaborate, coordinate, and – at a minimum – act in concert with each other. This form of strategic intelligence is necessary for the collective achievement of the mission aim – as we have mentioned previously, a single research programme cannot achieve a societal mission by itself, as the nature of a societal mission is a grand societal challenge requiring transformative change. Thus, understanding the diverse *contributions* that are being made or may be made in the future is key for mission success. This form of strategic intelligence is therefore necessary for the steering challenge of the PPR and the evolution of the nested organisation and implementation of mission-oriented research policies (orientation functional layer), research programmes (the programming functional layer), and research projects (the performance functional layer).

The connection of all three sources of strategic intelligence is made at the programme level, where the programme manager evaluates and supports the individually funded projects, makes connections with other national and international research programmes, and provides justification to the orientation functional layer on the successful management of the research programme.

In this particular programme, one could imagine a tighter connection and overlap between the three sources of strategic intelligence. For example, the Prospective 2050 exercise could have been conducted in advance of the call for proposals for the programme – it could be argued that mission-oriented programmes would benefit from a clearly defined future scenario of mission success. However, for an early stage and potentially breakthrough mission-oriented (and thus outcome-focussed) research programme, there is high uncertainty on the side of both what scientific projects will be proposed in the open call and also which parts of the future agricultural system will be the focus of the programme. Therefore, the future(s) that the programme should focus on are made clearer as the programme progresses.

Feedback and feedforward loops between sources of strategic intelligence and those who provide it can be beneficial in such situations. In practice, the PPR's three sources of strategic intelligence run parallel with communication between activities.

16.4 CONCLUSION

Whether the outcomes and eventual impacts of the research funded through this PPR will contribute to the mission, we cannot say. What we *can* say is that the programming of this mission-oriented research programme has armed itself with three processes of acquiring strategic intelligence to allow the programme to understand its potential contribution to the mission aim and to steer through real-time strategy articulation. We can say that the nested research activity is becoming more reflexive, especially when it comes to embedding of each participant's activities into the overall process of driving towards mission-oriented transformative change and in the area of expanding the mobilisation of and producing their own strategic intelligence. We suggest that as mission-oriented and challenge-driven policies become increasingly translated into research programmes, similar dynamics will be visible.

We argue that in order for financed breakthrough research to contribute to a particular mission aim, programme managers will have to confront the challenge of managing uncertainty by managing "rational myths". There are rational myths that are embodied by the missions themselves, which will inevitably transform and evolve in terms of targets and what "mission success" will be defined as. There are also rational myths that are represented by anticipatory pathways from research to mission contribution, which will change as the funded breakthrough research matures and more is known about the potential of the new knowledge and capabilities that are being developed. We have shown how one programme in France is experimenting with constructive ways of managing uncertainty while steering towards an evolving desirable goal: the zero-pesticide mission. We suggest that further exploration of strategic intelligence for both mission-oriented and transformative innovation policies be conducted to compare the different institutional programming contexts where mission-oriented research programmes are being implemented.

ACKNOWLEDGEMENT

This chapter is designed to be an illustrative vignette of mission-oriented research policy implementation. We do not aim (or desire) with this chapter to make cross-comparisons on mission implementation, but to provide deeper insights into implementation activities occurring at the time of writing. Our goal is to show that steering mission-oriented research towards contributing to missions requires real-time development and use of strategic intelligence. While the nature and processes of such strategic intelligence may change across contexts, we take the first step through a single case vignette.

REFERENCES

Barré, R., Henriques, L., Pontikakis, D., & Weber, K. M. (2013). Measuring the integration and coordination dynamics of the European Research Area. *Science and Public Policy*, *40*(2), 187–205.

Boon, W. P., Edler, J., & Robinson, D. K. (2022). Conceptualizing market formation for transformative policy. *Environmental Innovation and Societal Transitions*, *42*, 152–169.

Borrás, S., & Edler, J. (2020). The roles of the state in the governance of socio-technical systems' transformation. *Research Policy*, *49*(5), 103971.

Brown, R. (2021). Mission-oriented or mission adrift? A critical examination of mission-oriented innovation policies. *European Planning Studies*, *29*(4), 739–761. https://doi.org/10.1080/09654313.2020.1779189

Callon, M. (1999). Actor network theory—the market test. *The Sociological Review*, *47*(S1), 181–195.

Commission of the European Communities (2007). *Green paper. The European research area: New perspectives*. COM(2007) 161 final Brussels.

Cooper, R. G. (1990). Stage-gate systems: A new tool for managing new products. *Business Horizons*, *33*(3), 44–54.

Delemarle, A., & Larédo, P. (2014). Governing radical change through the emergence of a governance arrangement. In R. Wintjes, & C. Nauwelaers (Eds), *The governance of socio-technical systems* (pp. 159–186). Edward Elgar Publishing.

Edler, J., Joly, P. B., Kuhlmann, S., Nedeva, M., Propp, T., Rip. A., Ruhland, R. and Thomas, D. (2006, April). *Fora of strategic intelligence for research and innovation*. A PRIME Forum Research Project Final Report.

Gassler H., Polt, W., & Rammer, C. (2008). Priority setting in technology policy: Historical developments and recent trends. In C. Nauwelaers & R. Wintjes (Eds), *Innovation policy in Europe: Measurement and strategy.* Edward Elgar Publishing.

Guichard, L., Dedieu, F., Jeuffroy, M. H., Meynard, J. M., Reau, R., & Savini, I. (2017). Le plan Ecophyto de réduction d'usage des pesticides en France: décryptage d'un échec et raisons d'espérer. *Cahiers Agricultures, 26*(1), 14002. https://doi.org/10.1051/cagri/2017004

Hekkert, M. P., Janssen, M. J., Wesseling, J. H., & Negro, S. O. (2020). Mission-oriented innovation systems. *Environmental Innovation and Societal Transitions, 34,* 76–79.

Jacquet, F., Jeuffroy, M. H., Jouan, J., Le Cadre, E., Litrico, I., Malausa, T., & Huyghe, C. (2022). Pesticide-free agriculture as a new paradigm for research. *Agronomy for Sustainable Development, 42*(1), 1–24.

Joly, P. B., Gaunand, A., Colinet, L., Larédo, P., Lemarié, S., & Matt, M. (2015). ASIRPA: A comprehensive theory-based approach to assessing the societal impacts of a research organization. *Research Evaluation, 24*(4), 440–453.

Joly, P. B., Matt, M., & Robinson, D. K. (2019). Research impact assessment: From ex post to real-time assessment. *Journal for Research and Technology Policy Evaluation, 47,* 35–40. DOI: 10.22163/fteval.2019.326.

Kuhlmann, S. (2002). *Distributed techno-economic intelligence for policymaking. RTD evaluation toolbox: Assessing the socio-economic impact of RTD-policies* (pp. 210–217). European Commission – Joint Research Centre, IPTS, Seville.

Kuhlmann, S. (2003). Evaluation as a source of "strategic intelligence". In P. Shapira & S. Kuhlmann (Eds), *Learning from science and technology policy evaluation: Experiences from the United States and Europe* (pp. 352–379). Edward Elgar Publishing.

Kuhlmann, S., Boekholt, P., Georghiou, L., Guy, K., Héraud, J. A., Laredo, P., & Smits, R. (1999). *Improving distributed intelligence in complex innovation systems* (pp. 1–87). MPRA Paper No. 6426.

Kuhlmann, S., & Rip, A. (2018). Next-generation innovation policy and grand challenges. *Science and Public Policy, 45*(4), 448–454.

Lamy, P. (2017). *Lab-Fab-App – Investing in the European future we want.* Report of the independent High Level Group on maximizing the impact of EU R&I Programmes. Publications Office of the European Union.

Laredo, P., Rip, A., Jolivet, E., Shove, E., Raman, S., Moors, E. H. M., & Eugenia, G. C. (2002). *SocRobust (Management tools and a management framework for assessing the potential long-term S&T options to become embedded in society).* Final report, Project SOE 1981126 of the TSER Programme of the European Commission.

Larrue, P. (2021). The design and implementation of mission-oriented innovation policies: A new systemic policy approach to address societal challenges. *OECD Science, Technology and Industry Policy Papers, 100.* Éditions OCDE. https://www.oecd.org/en/publications/the-design-and-implementation-of-mission-oriented-innovation-policies_3f6c76a4-en.html

Lindner, R., Daimer, S., Beckert, B., Heyen, N., Koehler, J., Teufel, B., & Wydra, S. (2016). Addressing directionality: Orientation failure and the systems of innovation heuristic. Towards reflexive governance. *Fraunhofer ISI Discussion Papers-Innovation Systems and Policy Analysis, 52.*

Luederitz, C., Schäpke, N., Wiek, A., Lang, D. J., Bergmann, M., Bos, J. J., ... & Westley, F. R. (2017). Learning through evaluation – A tentative evaluative scheme for sustainability transition experiments. *Journal of Cleaner Production, 169,* 61–76.

Matt, M., Gaunand, A., Joly, P. B., & Colinet, L. (2017). Opening the black box of impact – Ideal-type impact pathways in a public agricultural research organization. *Research Policy, 46*(1), 207–218.

Mazzucato, M. (2018). *Mission-oriented research & innovation in the European Union – A problem-solving approach to fuel innovation-led growth.* European Commission. https://doi.org/10.2777/360325

Meyer, J. W., & Rowan, B. (1977). Institutionalized organizations: Formal structure as myth and ceremony. *American Journal of Sociology, 83*(2), 340–363.

Rip, A. (2006). A co-evolutionary approach to reflexive modernization – and its ironies. In J-P. Voss, D. Bauknecht, & R. Kemp (Eds.), *Reflexive governance for sustainable development* (pp. 82–100). Edward Elgar Publishing .

Rip, A. (2019). *Nanotechnology and its governance.* Routledge.

Robinson, D. K. R., & Mazzucato, M. (2019). The evolution of mission-oriented policies: Exploring changing market creating policies in the US and European space sector. *Research Policy, 48*(4), 936–948.

Robinson, D. K. R., & Propp, T. (2006). Multi-path mapping as strategic intelligence for reflexive alignment in emerging S&T. In *Second International Seville Seminar on Future-Oriented Technology Analysis: Impact of FTA approaches on policy and decision-making*. Joint Research Centre, European Commission.

Robinson, D. K. R., & Propp, T. (2008). Multi-path mapping for alignment strategies in emerging science and technologies. *Technological Forecasting and Social Change, 75*(4), 517–538.

Robinson, D. K. R., & Rip, A. (2013). Indications of socio-economic impacts of nanotechnologies: The approach of impact pathways. In K. Konrad, H. Van Lente, C. Coenen, A. Dijkstra, & C. Milburn (Eds.), *Shaping emerging technologies: Governance, innovation, discourse* (pp. 153–166). IOS Press.

Robinson, D. K. R., Schoen, A., Larédo, P., Gallart, J. M., Warnke, P., Kuhlmann, S., & Ordóñez-Matamoros, G. (2021). Policy lensing of future-oriented strategic intelligence: An experiment connecting foresight with decision making contexts. *Technological Forecasting and Social Change, 169*, 120803.

Robinson, D. K. R., Winickoff, D., & Kreiling, L. (2023). Technology assessment for emerging technology: Meeting new demands for strategic intelligence. *OECD Science, Technology and Industry Policy Papers, 146*. https://doi.org/10.1787/e738fcdf-en

Schot, J., & Steinmueller, W. E. (2019). Transformative change: What role for science, technology and innovation policy? An introduction to the 50th anniversary of the Science Policy Research Unit (SPRU) Special Issue. *Research Policy, 48*(4), 843.

Tübke, A., Ducatel, K., Gavigan, J., Moncada-Paterno-Castello, P., Smits, R., Zweck, A., Rader. M., Barré, R., & Hut, A. S. (2001). S*trategic policy intelligence: Current trends, the state of play and perspectives*. IPTS.

Van de Ven, A. H. (2017). The innovation journey: You can't control it, but you can learn to maneuver it. *Innovation: Management, Policy and Practice, 19*(1), 39–42. https://doi.org/10.1080/14479338.2016.1256780

Voss, J. P., & Bornemann, B. (2011). The politics of reflexive governance: Challenges for designing adaptive management and transition management. *Ecology and Society, 16*(2).

Weber, K. M., & Rohracher, H. (2012). Legitimizing research, technology and innovation policies for transformative change: Combining insights from innovation systems and multi-level perspective in a comprehensive "failures" framework. *Research Policy, 41*(6), 1037–1047.

17. The governance of transformative mission-oriented STI policies

Jakob Edler and Matthias Weber

17.1 INTRODUCTION: UNDERSTANDING THE GOVERNANCE SPACE OF TRANSFORMATIVE MISSIONS

In the midst of a roll-out of mission-oriented STI policies throughout the OECD world and beyond, this book sets out to support the understanding and conceptualisation of a very specific and – we argue – most challenging policy approach: *transformative mission-oriented STI policies*. While all mission-oriented policies support directionality, pursue ambitious goals, and develop coordinated initiatives to support the mission, transformative mission-oriented policies are those policies that are most demanding and complex in terms of the change they want to induce. In this volume, we have focussed on STI policies that are expected to contribute to this demanding task and fulfil key functions in relation to accomplishing missions. As defined in detail elsewhere in this volume, those policies "*target system transformation* in relation to *wicked societal challenges* through scientific-technological as well as social innovations. They are *ambitious, exploratory, and ground-breaking* in nature, and often require *inter- and transdisciplinary* resources to be mobilised" (Edler et al. in this volume). Ultimately, what is essential for the success of those policies is the willingness of different actors and stakeholders, individually and mutually, to adjust behaviours and routines, which will also require them to overcome conflict and inertia.

Throughout this volume, we have seen a range of governance requirements that are in line with this conceptual understanding of transformative mission-oriented policies in general, and of associated STI policies in particular. It has been highlighted that they need a "strong" programmatic approach and to be underpinned and implemented through an adequate institutional framework. In this context, "programmatic" refers to the ensemble of governance processes, policies, and instruments that are deliberately designed or mobilised to serve a defined mission. A distinctive feature of those policies is that they disrupt in many regards, intervening in existing practices and institutions: existing institutions are often inadequate and need to be transformed or functionally replaced; existing policies need to be recombined and new policies designed; stakeholders need to be integrated very comprehensively; policy implementation needs to be underpinned by new forms of anticipation, evaluation, and readjustment. Overall, given those requirements, we can assume that the collective process of making binding decisions on transformative missions and on formulating, implementing, and improving policies – in short, governance – is highly complex and challenging.

Against this background of governance requirements for ambitious, transformative mission-oriented STI policies as formulated in our initial framework (see Edler et al. in this volume), we now can dig somewhat deeper into the challenges for governance by taking advantage of the contributions in this book. This chapter is not meant to be a comprehensive summary – rather, we focus on selected high-level observations regarding the key challenges of governance, relying on these contributions to gain a better understanding of the governance challenges in diverse settings. As all of these contributions deal with specific conceptual perspectives and empirical cases in very different kinds of contexts, they are an excellent source for identifying and characterising governance challenges for transformative policies and for collecting ideas on how those challenges can be overcome.

Before we elaborate on our key lessons from the contributions, which we present as 10 governance challenges, we develop a simplified framework of four governance functions. This framework elaborates and builds on the well-established concept created by Barré et al. (2013) that was initially developed to understand the governance of research funding programmes at the European level. It helps to condense and make sense of those 10 inductive lessons and to simplify the takeaway messages for the future development of appropriate governance practices. On the basis of our 10 lessons, we then go on to develop a set of governance principles and analytical questions to support the development of a differentiated approach to mission governance. We finish this chapter with a short suggestion for a future research agenda.

17.2 KEY GOVERNANCE CHALLENGES AND POSSIBLE RESPONSES

17.2.1 Basic Framework

The challenges for the governance of transformative mission-oriented STI policies could, in principle, be categorised in various ways. The very categorisation of challenges in terms of characterising dimensions provides an indication as to what kinds of challenges in what kinds of governance processes we regard as critical. At least three sets of characterising dimensions can be distinguished, and their scope and differences signal important foci of interest. Recombining and extending those three characterising concepts promises the best value for a holistic understanding of governance challenges that we are interested in here.

The first characterising concept as set out by Wittmann et al. (in this volume) provides a method for simplifying the policy cycle by differentiating the stages of policy formulation, design, and implementation. A second characterising concept, popularised and implemented by the OECD, differentiates between strategic orientation, policy coordination, and policy implementation (Larrue, 2021, or in this volume). While the former suggests some form of linear sequence of stages, even if feedback loops are taken into account, the latter represents a more relaxed approach to traditional policy stages by emphasising the functional differences and complementarities of governance requirements – in particular, policy coordination is regarded as both an element of orientation and an element of implementation. A third device is offered by Barré et al. (2013) and is positioned somewhere in between the previous two (see also Matt & Robinson, in this volume). This device differentiates between three governance functions: orientation, programming (which in effect is about design and implementation), and performance (i.e., it includes the execution of R&I activities in organisations that perform research). Originally developed for analysing the evolution of European framework programmes for research and technological development, these three functions need to be widened by including the full range of actor types relevant to mission-oriented STI policies and programmes. Following our definition of transformative missions above, this broader perspective on actors goes further than the focus of Barré et al. (2013), which centred on the traditional R&I actors only. Rather, a broader, more comprehensive set of actors has to be considered (see also Boon & Edler, 2018), including citizens, financial institutions, regulators, governance bodies at all policy levels, utilities, and so on. This, in turn, potentially poses new sets of governance challenges.

However, these three functions need one further extension. Given the complexity of policy mixes and the necessary changes in systems of production and consumption, as well as the need for iterative learning from experimentation, a concept for governance challenges must include considerations of the actual effects and outcomes of policy interventions. Thus, for our definition of key governance challenges, we stress the importance of understanding – and acting upon – the delivery of outcomes of mission-oriented policies by including a corresponding function.

Such a modified version of the Barré et al. (2013) approach appears to be the most appropriate, as it (1) highlights the critical process of identifying, as a collective, in what direction systems shall be developed and what the role(s) of policy should be, (2) offers a sufficiently broad category for all operational programming and implementation processes, (3) considers the actual research, innovation, and transformational activities of all relevant actors, and (4) includes the effects of these activities on the actual realisation, and potential needs for adjustment, of missions.

We thus suggest understanding governance challenges along the lines of the three major – yet widened – governance functions suggested by Barré et al. (2013), complemented by a consideration of the intended and unintended effects that mission-oriented STI policies must take into account due to their complex and uncertain character:

1. *Orientating:* This category entails the governance challenges stemming from defining and deciding upon what it is we want or should want to achieve, i.e., agenda-setting and decision-making in a policy cycle framework.
2. *Programming:* This encompasses all aspects of concrete policy design and implementation once the direction has been defined and the resources have been made available. It includes policy challenges relating to the formulation of the programme(s), their concrete design (instruments) as well as their implementation structures and processes, but also other supply-side and demand-side policy instruments that influence the achievement of transformative missions (e.g., innovation-oriented procurement, regulation).
3. *Performing:* This encompasses the challenge of mobilising all actors that are targeted through policies and instruments of transformative mission-oriented STI policies, from organisations that perform research to firms, financial institutions, public authorities, NGOs, and citizens. The breadth of stakeholder groups to be included will vary according to the mission at hand. However, similarly to the approach put forward by Barré et al. (2013) concerning organisations that perform R&I, the overall governance imperative is to mobilise actors and give direction to their functional performance in order to bring it in line with the programming requirements. This will not be done in a command-and-control manner, but rather by way of orchestration, joint learning, and feedback loops.
4. *Effecting:* This covers the challenges when it comes to understanding and acting upon the effects of the mission-oriented policy on all aspects of system transformation, from actor behaviour to scientific knowledge production, technology development and deployment, market development, and any societal or environmental outcome. It is broader than performance, as it includes unintended effects, and thus reaches beyond mere R&I output.

In the subsequent discussion on governance challenges, we will keep this simplified four-pillar governance framework in mind. We have identified 10 challenges that we will discuss in turn in the next section in light of the insights gained from the contributions to this book and some selected complementary sources.

17.2.2 The Ten Governance Challenges of Transformative Mission-Oriented STI Policies

The 10 challenges proposed in this section have been identified by the contributions to this book but are complemented with some further ramifications and pointers to governance issues that may need to be addressed in a differentiated, context-sensitive manner. After the discussion of these 10 challenges, we can highlight their individual relevance in terms of the four functions presented (i.e., orientation, programming, performing, and outcomes).

Challenge 1: Historic context and path dependencies as governance constraints

Transformative missions are, by definition, disruptive. The contributions in this volume have highlighted that in many ways, the governance of those policies also has to be disruptive. However, while we have seen a number of encouraging examples of governance innovation and adaptation, one key constraint for new forms of governance is the path dependency of prevailing governance arrangements, particularly in the realm of policy and political administration. As Kattel and Mazzucato highlight in their selective survey of mission approaches (in this volume), despite much discussion on new policy institutions and instruments, and despite the set-up of mission agencies in a few cases (see Challenge 9 below), mission policies tend to be implemented through existing institutions and policy schemes, often as incremental adjustments of pre-existing instruments and institutions. The same holds true, as Uyarra et al. argue (in this volume), when mission approaches are being imported into a specific regional setting. Pre-existing policy structures and styles are major determinants of future policies and their success. However, the authors claim, this highly idiosyncratic historical contextual background and development needs to be taken into consideration very carefully and acted upon.

In a way, the following nine remaining governance challenges all stem from path-dependent, traditional policy approaches that are clearly insufficient, and the contributions in this volume have resulted in some governance principles and empirical examples to highlight that those challenges can be tackled. Interestingly, even in political contexts that are constitutionally well equipped to radically shift the mission agendas and push for strong top-down initiatives, as in the case of the presidential system in South Korea (Eum, in this volume), path-dependent institutional legacies – among other things – seem to limit the scope and success of change. Thus, even if radical change is constitutionally and politically possible, careful consideration is needed as to the requirements and caveats associated with integrating existing institutional rationales and logics into the new mission-oriented frame.

Challenge 2: Mobilising and coordinating multiple actor groups

Many well-established STI policies target multiple stakeholder groups or support cooperation between different stakeholder groups. However, transformative STI policies, by definition, must rely on the mobilisation of all actors that are part of or influence the system that needs to be transformed. This is critical for ownership – and thus compliance – as well as for handling the complexity of systemic interplay.

We have seen different approaches to mobilising actors in the contributions to this volume. In line with an administration-oriented governance mode (Janssen, Wanzenböck et al. in this volume), various South Korean presidents have designed and framed missions in a top-down manner (Eum, this volume). In an early attempt, there was a lack of inclusion and mobilisation of actors, and the policy failed. In a second attempt, inclusion was broad, particularly in terms of vision building. However, while the presidential system of South Korea combined its top-down actions with inclusive approaches, institutional innovations, and broader strategic intelligence, it was also of only limited success, as the mobilisation of actors was done in an "educating" rather than a participatory, reflexive manner.

Kivimaa and Torrens (in this volume) offer a second example of a response to the challenge of mobilising and coordinating actors. In defining the programme for Mobility as a Service (MaaS), the government was successful in mobilising actors both horizontally and vertically, combining top-down and bottom-up elements. Horizontal cross-departmental coordination at the very beginning of the process delivered a whole-of-government idea about directionality and intent. It then allowed subsequent coordination with mobility service providers that developed their own bottom-up ideas and initiatives in line with this guiding idea of MaaS. In other words, top-down impetus allowed for bottom-up dynamics. Consequently, the experiments defined in those processes were successful because of the buy-in and broad expertise mobilised at an early stage.

Another element of successful mobilisation can be seen in the development of a clear vision to create shared meaning, whereby the mission itself can then can act as a boundary object between mobilised stakeholders (Janssen, Wesseling et al. in this volume). Similarly, the European Partnerships, which appear to function mainly as accelerators of niche innovations, nevertheless contribute to coordinating different kinds of stakeholders across Europe in a rather self-organised, bottom-up manner (Berger et al. in this volume). Finally, the case of preventive healthcare in the Netherlands indicates the possibility of mobilising actors in novel, mission-compatible ways through the (very rare) setup of a new institutional mission body (Boon, in this volume).

Challenge 3: Translation of concepts between stakeholder groups and policy stages
Given the multitude of actors and stakeholders involved and the long-term character of missions, achieving and maintaining a common understanding of a transformative mission and providing guidance in a variety of decision arenas, often cutting across policy domains and policy levels, is key to success. Given the multifaceted nature of missions, it is important to translate the core concepts that underpin a specific mission in order to connect it to the core issues addressed in a specific arena. In other words, missions need to be meaningful to such arenas, as a precondition for mobilising the resources and capacities for transformative change of the actors engaged in that arena. Janssen, Wesseling et al. (in this volume) address this need for coherence across arenas by characterising missions as "boundary objects" to stress the importance of a convincing, yet flexible framing of a mission to bundle diverse expectations and thereby facilitate mission governance.

One challenge in the formulation of missions consists of striking a balance between openness to multiple interpretations and providing orientation through targets and timeframes for both programming and performing mission activities. It is this interpretive flexibility that allows a mission to serve as a meaningful reference point for a range of communities. At the same time, the negotiations and contestations in these arenas contribute to making a mission more robust by adding perspectives to its stable core.

These translation, enrichment, and alignment processes can help to achieve convergence among these different nested arenas, and they can also facilitate the transfer of meaning to other contexts, such as regional contexts. The interpretive flexibility of missions as boundary objects is also essential for the adaptive and tentative governance strategies (Kuhlmann et al., 2019) for longer-term and wicked missions, in the course of which continuous contestation, negotiation, and adaptation processes take place.

Wittmann et al. (in this volume), in conceptualising the processes for defining, designing, and implementing mission policies as a succession of translation steps, highlight the fact that for each translation step, different actors are involved, including sectoral policymaking entities. In this line of thinking, each translation step entails alignment challenges at the level of normative orientations and ideational frames. The development of a sufficient level of joint understanding happens in the process of constructing a shared idea as to what the mission is about and how R&I can contribute, given existing system configurations and sectoral policies. Even if the overall orientation is brought in line, the role that policy can and should play can differ fundamentally across different policy sectors and different stakeholder groups, which emphasises the importance of the interpretive flexibility of missions. Coordination happens through jointly making sense of how normative orientations can best be put into policy and R&I, and the capability to communicate and learn become core coordination prerequisites. In this translation concept, boundary spanning as such is not the key aspect, but rather the joint act of making sense that helps to align stakeholders.

Challenge 4: Designing programmes

As argued by Kattel and Mazzucato (in this volume), despite a major turn in STI policymaking towards missions, specific challenges still tend to be tackled largely through modifications of existing practices and structures rather than by establishing truly novel institutional arrangements, policy designs, and instruments. This can be observed, in particular, in funding programmes as one type of important policy impetus, launched either by dedicated funding agencies or internally by research performing organisations. Many early "mission-oriented" initiatives of the late 2010s were just adapted traditional thematic funding programmes instead of truly novel "next-generation" instruments. Overall, only limited attempts have been made so far to make significant strategic and procedural changes to programme design. This structural and institutional resistance to fundamentally new ways of funding missions is not surprising, given the idiosyncratic ways through which policy tools are mobilised in national innovation systems, and the lack of common guidelines and evaluation practices for mission-oriented funding programmes, even though some first guidelines and playbooks have been issued in recent years (e.g., Hill, 2022; Nylander et al., 2024).

There are, of course some exceptions to this general observation. Kattel and Mazzucato (in this volume) refer, for instance, to a call of about 90 million euros by Denmark's Innovation Agency, published in 2021, asking for roadmaps for mission-driven green partnerships. The agency claims that the fund "encourages all relevant stakeholders across the Danish research and innovation system to come together to contribute their expertise and propose a realistic and robust path towards the development of cutting-edge solutions within the four missions – ranging from strategic research to commercialisation, with a focus on short-, mid- and long-term impact".'[1] In spite of this and some other exceptions, the paradigmatic change towards mission-oriented programmes, not only strategically but also in terms of implementation practices and programme design, is yet to happen within highly institutionalised national contexts.

The difficulties of enacting this shift are well known, as whole-of-government approaches are faced with major barriers in siloed architectures of public administration. Focussing on analysing the attempts at making European partnerships more transformative and mission-oriented, Berger et al. (in this volume) confirm that the European Commission is also struggling with this challenge. As argued by the authors, a better alignment among partnerships relevant to a specific mission, but also with other parts of the European Framework Programme for Research and Innovation and with sectoral policies, is needed. This is not surprising, given the fact that the European Partnerships landscape was restructured at the beginning of the Horizon Europe framework programme, and many partnerships are still struggling to consolidate their strategic orientation and internal organisation.

Overall, we can observe that there is a major mismatch between policy ambitions and governance challenges on the one hand, and the policymaking capacities available to address these challenges on the other hand. As long as policies and instruments remain as entrenched in prevailing institutional contexts as they currently are in most countries and the EC alike, mission-orientation will remain an effort of muddling through, relying on strong policy entrepreneurship and political leadership rather than on public administrations that are appropriately equipped with mission-oriented and transformative capacities, understood as the combination of roles, resources, and skills (Borrás et al., 2024).

Challenge 5: Multi-scalar governance arrangements: Local vs. global coordination

The implementation of missions does not happen in a vacuum; instead, it is dependent on a variety of contextual conditions. As stressed by Uyarra et al. (in this volume), these contextual conditions can be sectoral or thematic in nature, but also comprise geographic and territorial specificities, down to the local level. In practice, however, transformative missions are often disconnected from regional efforts to strengthen innovation performance. The accumulated experience with smart specialisation strategies and regional innovation policies, they argue, should be fully exploited for making transformative missions more context-sensitive and place-based. In fact, many regional initiatives can well be characterised as mission-oriented, even if they do not apply that name.

To introduce a spatial and scalar differentiation of governance frameworks for transformative missions, four types of problem-solution configurations are proposed, based on different geographies for both problem representation and solution representation, which can be either globally or locally anchored. These more differentiated perspectives open up new possibilities for designing and implementing multiscalar policies to support transformative missions that avoid a lack of spatial sensitivity and a neglect of cities and regions as key locations for innovation, experimentation, and scaling.

Related to this spatial differentiation of contexts for transformative mission-oriented STI policy, it is also important to take into account differences in governance modes, which often exhibit regional specificities. The three governance modes proposed by Janssen, Wanzenböck et al. (in this volume) are instructive in this regard, pointing to the structure

1 Cited from https://innovationsfonden.dk/en/press/now-open-call-roadmaps-mission-driven

of actor configurations in shaping mission-oriented policies. A well-developed regional culture of stakeholder participation could strengthen network-based and society-based modes of governance and help to mobilise the range of actors needed to reinforce the realisation of transformative missions.

All in all, transformative missions will transform places and communities through various agglomeration effects. Place- and context-sensitive policy design and governance modes can help gear missions to local and regional conditions. At the same time, excessive fragmentation of mission-oriented policies should be avoided to ensure mutual learning, scaling, and cross-regional coherence in addressing overarching mission goals.

Challenge 6: Coordinating STI policy and sectoral policies: From innovation to acceleration and generalisation

Transformative mission-oriented STI policies must, by definition, disrupt existing system configurations. While traditional STI programmes produce input for innovation to be applied across a variety of sectors following a functional fit and market logic, rather than a grand ex ante design, transformative mission-oriented policies target system change. By design, they thus interfere with existing sectoral policies that dominate the governance of systems of production and consumption. However, just how this interference is organised remains a critical challenge. Kivimaa and Torrens (in this volume, see also Challenge 2) have demonstrated how the orchestration of a wide range of actors has worked in the Finnish Mobility-as-a-Service case, and how crucial the coordination between STI policy programmes and sectoral policy regulation was for the success of this transformative mission.

An interesting soft coordination approach is presented by Walz et al. (in this volume). A broad and ambitious R&I programme for sustainability has deliberately not been labelled as mission policy by the German Federal Ministry of Education and Research (BMBF) and thus is not coordinated explicitly or formally with other sectoral policies like energy or environment policy. Rather, coordination happens implicitly, by incentivising research-performing actors to direct their research towards predefined sustainability problem areas. This shall serve a dual purpose: First, it shall increase the likelihood that research results are taken up in transformative processes, and second, it shall support the build-up of a sustainability research community. Walz et al. argue that rather than this indicating a lack of coordination, it promises both to produce valuable input for other policies that drive sustainability transitions and to preserve the necessary long-term strategic research that will be important beyond existing transformative policy approaches. Rather than aligning policies, soft coordination of R&I and sectoral policy is seen as a safeguard for long-term knowledge production that will be critical to producing disruptive solutions in the future.

A second illustration of the alignment challenge is presented by Oyamada and Arimoto (in this volume). In Japan, selected missions are being governed through joint cross-ministerial Strategic Innovation Promotion Programmes (SIP). This is supported by inter-ministerial coordination directly in the Cabinet and thus at the highest possible political level under the supervision of the Prime Minister. Analysing the case of automated driving, the authors show that inter-ministerial coordination, and thus the roll-out of innovative solutions developed in the programme, is supported first by top-down decision-making that is binding for all ministries involved, and secondly by the broad participation of heterogeneous stakeholder groups, including scientific organisations, industry, end consumers, and ministries responsible for demand-side support measures and for the adoption of developed technologies. Furthermore, to orient both ministries and broader stakeholder groups, an inter-ministerial, inter-stakeholder roadmap process is being developed (see Challenge 7). The lessons learned from early coordination approaches have recently led to a broader "Integrated Innovation Strategy" and "Main Field Strategy" as elements of a whole-of-government approach. Therefore, the main lesson to be derived from the Japanese experience is that it can be effective to combine two-dimensional coordination between ministries and across broad stakeholders, governed by Cabinet discipline and new forms of integrated programmes.

Challenge 7: Operationalising and orchestrating directionality

One of the key challenges is how to ensure staying "on track" when implementing missions, even if the goal and direction are still rather open and inevitably subject to change. Moreover, different pathways can be taken to achieve a transformative goal. Wanzenböck et al. (2020) provided a first systematisation of different types of pathways and associated strategies, depending on the levels of uncertainty around both the understanding of the problem and possible solutions.

Resolving conflicts regarding the direction of change to take is a key governance task according to Janssen, Wanzenböck et al. (in this volume), and there are different ways of realising this task. This depends on the dominant mode(s) of governance, which they characterise as administration-oriented (i.e., led by hierarchy), network-based (i.e., in cooperation with stakeholders, including societal actors), or society-based (i.e., led by societal actors). Countries differ in terms of how they address key governance tasks, in line with their respective dominant governance modes. What matters with regard to directionality is whether the resulting policy mix leads to an appropriate balance between top-down strategic guidance and bottom-up exploration.

Such differences can be well illustrated with the work by Larrue (in this volume) on net-zero mission-oriented policies. Following an inductive, practice-based approach, he shows that the net-zero carbon mission is implemented in different ways in different countries, and he suggests five types of mission-oriented policy approaches, ranging from general strategies and mission umbrellas to programmes and initiatives characterised by different levels of ambition. While STI policies play a key role in all of these approaches, the extent to which sectoral authorities "own" the missions differs. Transformative missions that are led by STI policy may well run into the "STI trap" of being overburdened with ambitions that they can never achieve on their own (Polt & Weber, 2023). This suggests that tightly embedding directionality in sectoral policies is critical to the success of transformative missions – a task in which STI policy can help and provide important innovative input but is not necessarily in the driving seat.

An interesting example of how such trans-governmental directionality has been realised is provided by Oyamada and Arimoto (in this volume) in the case of Japan. There, direction is defined by a governance mode that is rather top-down but, at the same time, gives sufficient room to bottom-up initiatives and learning. Participatory processes, involving various ministries and a wide range of stakeholder groups, have led to the definition of roadmaps and strategies to which all key actors have subscribed and committed themselves. These roadmaps stretch over the areas of responsibility of several ministries and cover not only research and development but also laws, regulations, and the development of multi-partite networks composed of industry, research, governmental bodies, and civil society organisations. A key role is played by the Cabinet Office and its Council for S&T Policy in overseeing cross-sectoral and intra-governmental policy coordination and launching large horizontal programmes.

Penna et al. (in this volume) show that in a more society-based governance mode, quite a different model promises to be effective in ensuring directionality. Based on experiences from Sweden with continuous learning and adaptation in open processes of defining and implementing transformative missions, STI and sectoral policies worked together to drive missions in one direction, and a wide range of stakeholders with high levels of autonomy also aligned their strategic choices with that direction. The creation of reflexive spaces to reconsider the direction, policy approach, and implementation practices of a mission is crucial in the Swedish approach, and it requires drawing on a well-established culture of collaboration and reflexivity when involving a wide range of stakeholders.

These considerations show that the ability to govern directionality in transformative missions depends on political styles and cultures, but also on the level of uncertainty associated with a mission and the range of governmental and non-governmental actors that are involved in shaping the mission.

Challenge 8: Matching governance requirements with governance capacities

Transformative mission-oriented STI policies pose enormous challenges in terms of the capacities of the state. The role of the state in such policies is much more extensive, with state activity involving the definition of directionality jointly across government and with stakeholders, the support of knowledge creation and innovation, as well as creation and acceleration of markets. Thus, state actors need to build up capacities to match the underlying strategic intelligence and orchestration challenges (Borrás et al., 2024). As Borrás et al. have shown, those capacities are very broad and remain somewhat contingent on the mission qualities and idiosyncrasies of the political and societal context. Rogge and Song (in this volume) further conceptualise the capacity needs. Mobilising the multi-level perspective (MLP), they highlight just how diverse those capacity needs are and how much they differ between the three levels and stages of the advancement of missions. A very different capacity set is required to accelerate niche markets into broader, transformative markets and influence regime development, compared to the capacity needed to support knowledge creation for the development of new niche technologies and to secure niche development.

Kattel and Mazzucato (in this volume) find that securing those capacities across transformative mission-oriented policies is a huge challenge for governments. In their international comparative analysis, they find a gap between the capability requirements and the capacities available in those organisations that adapt their activity portfolio towards transformative missions. While organisations across Europe are adjusting their ambitions and, in part, their remit and instrumentation, they have not yet succeeded in developing the capacities that would be needed to match those ambitions. In addition, Kattel and Mazzucato assert that, with notable exceptions such as the Transformative Innovation Policy Consortium (TIPC), the broader epistemic and consultancy community does not support policymaking to the extent needed. Currently, interconnected communities of practice are being formed both at the European level and in several member states to engage mission actors and stakeholders in mutual learning and co-creation processes to trigger the epistemic change that Kattel and Mazzucato are calling for.[2]

Challenge 9: Mission agencies vs. adapting mandates of existing institutions
As shown so far, with the increasing and broadening ambition of the state in transformative missions, capacity needs increase, access to new stakeholder groups is necessary, very diverse sets of instruments need to be combined, and policy field and agency remits need to be coordinated. Therefore, a major governance question that arises is how these challenges are dealt with in organisational terms. The responsibility for missions can be delegated to new agencies, existing agencies can receive a new, broad mission remit, or existing, specialised ministries and agencies can be tasked with collaborating and coordinating. In the analysis of the South Korean mission programmes in the past, Eum (in this volume) shows how two newly created implementing bodies, while equipped with the authority to act broadly and to coordinate ministries to deliver on the mission programmes, were rather unsuccessful in doing so. They appeared to act in too much of a top-down mode, leave too little room for participation and networking, and thus failed to mobilise actors and stakeholders as needed. Dedicated agencies, as implementing and coordinating bodies, need strong buy-in and credibility in the broader community. Formal authority alone is not enough.

This credibility is also important for investors, both public and private, who play an important role in accelerating mission-oriented change by mobilising substantial financial resources. This is a critical issue once the stage of experimentation is concluded and large-scale investments are needed. As an example of creating a new mission-oriented institution, in 2020, Scotland created a mission-led development bank, capitalised with £2 billion of public funds, which aims to provide "patient capital" to support three missions: (1) achieving a just transition to net zero carbon emissions by 2045; (2) extending equality of opportunity through improving places by 2040; and (3) harnessing innovation to enable the people of Scotland to flourish by 2040 (Mazzucato & Macfarlane, 2019).

Another form of delegation is analysed by Boon (in this volume). In his analysis of a comprehensive transformation into a preventive health care system in the Netherlands, he examines the foundation and remit of a new body, Health Holland. Set up as a public-private partnership, this body brings together all major stakeholders through a variety of different internal teams that then organise stakeholder discourse, broad consultation, and the definition of concrete roadmaps (knowledge and innovation agenda). The delegation to this body also encompasses the allocation of resources to concrete activities. Health Holland thus sets agendas, develops and implements intervention instruments, and mobilises support for the mission. However, while many preconditions for the comprehensive delegation of mission responsibilities exist in the corporatist culture of the Netherlands, several institutional incompatibilities appear to limit the success of the mission, such as the dominance of incumbents and the need to achieve social acceptance of innovations.

---

[2] See, for instance, the Horizon Europe-funded project TRAMI (Transnational Cooperation on the Missions Approach) that has built a European network and community on mission governance. Similar communities have been established for individual missions at the European level as well as in some member states (e.g., Sweden, Austria) to consolidate knowledge, capacities, and experiences in regard to mission-oriented policies.

Challenge 10: Appropriate strategic intelligence: From concrete goals to open transformative outcomes

According to Kattel and Mazzucato (in this volume), the turn towards transformative mission-oriented STI policy calls for an epistemic change in terms of the goals, the actor configurations, and the STI policy measures implemented to achieve these goals, new forms of strategic intelligence – foresight, policy assessment, monitoring, and evaluation – and the engagement processes needed to make these instruments effective and legitimate.

The directional goals that are characteristic of transformative missions are highly ambitious and entail a departure from established forms of strategic intelligence. For instance, engagement processes to assemble the key actors and stakeholders behind these goals and ambitions are crucial. Given the broadening of the range of actors involved in mission implementation, these processes are highly demanding. The high levels of uncertainty and complexity typical of wicked challenges, as well as their multi-level, multi-actor nature, also imply a need for regular adjustment – not only of the means to achieve predefined goals but also of the goals themselves.

The distributed character of most transformative missions also means that centralised control over these missions is not possible. Instead, what is needed is empowerment and delegation of responsibility, and a much stronger reliance on actors' self-organisation and self-evaluation, while maintaining overarching goals and agendas as guidance. This also applies in particular to mission-oriented research and innovation activities. As argued by Robinson and Matt (this volume) for the case of challenge-oriented funding programmes, self-organisation, self-evaluation, and empowerment are tied to the accountability of the actors involved, who need to learn how to relate their specific research objectives to higher-order goals and ambitions, and – if needed – adjust their research activities accordingly, for instance, with the help of real-time formative evaluation processes and support tools such as ASIRPA.

Such real-time, formative evaluation methods tend to require elements of foresight, given the long-term nature of how research activities influence the pathways of transformative missions. In fact, disruptive developments may well necessitate major adjustments that must be incorporated into the new forms of evaluation in relation to transformative missions. Similarly, experiences with European partnerships have been promising in principle, with foresight methods frequently being used to inform the development and adjustment of Strategic Research and Innovation Agendas. However, as shown by Berger et al. (in this volume), European partnerships have thus far focussed mainly on building and nurturing niches rather than transforming systems; they can be categorised as accelerator missions rather than transformer missions, to draw on the typology of missions proposed by Wittmann et al. (in this volume). What this calls for is a better alignment of European partnerships with other parts of the European Framework Programme as well as with other sectoral policies (see Challenge 6) if the ambition is to turn them into more transformative missions.

In addition, the political definition of novel transformative missions needs to be informed by forward-looking knowledge. The recent foresight work that has been done during the preparation of the second Horizon Europe Strategic Plan is an example of how foresight can be integrated into programme development and adjustment and used to nurture debates about candidate future priorities in R&I policy (Weber et al., 2023).

In other words, strategic intelligence plays an increasingly important role as a mediating and coordinating service for strategic alignment and mutual learning among mission actors and stakeholders along their transformative pathway, guided by overarching theories of change.

This is what Penna et al. (in this volume) address when they emphasise the importance of reflexive spaces, i.e., of joint and collective learning about theories of change and the potential transformative impacts of interventions along the policy cycle, involving a wide range of actors well beyond R&I. This is particularly important for transformative missions that are "wicked", open-ended, and cannot rely on clearly defined targets and pathways, but rather need to consider missions as malleable boundary objects with multiple framings and frequent reorientations (Janssen, Wesseling et al., in this volume). In this context, strategic intelligence fulfils an important intermediating function across governance levels and fields. This intermediating function of strategic intelligence is also pertinent at more specific levels below that of a transformative mission as such – for instance, at the level of funding programmes or within research organisations. In summary, strategic intelligence is key to enhancing the transformative capacities not only of government and public administration but also of research-funding and research-performing organisations.

17.2.3 Governance Functions and Challenges Involved in Transformative Mission-Oriented STI Policies

How can we now make sense of those 10 challenges with respect to the governance dimensions as outlined above? Which challenges are most critical in light of those four governance functions (orientating, programming, performing, and effecting)? Table 17.1 outlines the 10 challenges and highlights in which of the four functions they are particularly pertinent. It shows that the 10 challenges of transformative mission-oriented STI policies that we extracted from the contributions in this volume differ in terms of their significance for the four governance functions. The table also shows that some of our 10 challenges are important across all major governance activities. The challenges that are particularly critical across the four governance functions appear to be the attempt at operationalising and orchestrating directionality (Challenge 7) as well as multi-scalar governance arrangements (Challenge 5).

More importantly, however, the table helps to crystallise which of the four governance functions are most challenged in the process of mission-oriented STI policymaking. Above all, it highlights how critical and challenging *orientating* and *programming* are for transformative missions. In particular, the orientation towards a mission that is widely understood and accepted, and to which a broad range of key actors and stakeholders are committed, appears to be the most critical and challenging step, given the uncertainties associated with a mission and the multi-faceted actor configurations that are targeted. Any shortcoming in the orientation phase will have repercussions for implementation and, subsequently, for the effectiveness of the policy. Following from that, the translation of basic orientations into concrete programmes, i.e., goals, policies, and instruments, also faces challenges that exceed the governance requirements of traditional STI policies. The entire task of setting up and running transformative mission policies thus is, indeed, a holistic, highly reflexive endeavour.

Table 17.1 Relevance of governance challenge in different governance functions

Governance challenges	Orientating	Programming	Performing	Effecting
Historic context-and path-dependencies as governance constraints	XX	O	O	O
Mobilising and coordinating multiple actor groups	XX	XX	-	-
Translation of concepts between stakeholder groups and policy stages	XX	XX	-	-
Designing programmes	XX	XX	O	O
Multi-scalar governance arrangements: Local vs. global coordination	XX	XX	O	XX
Coordinating STI policy and sectoral policies, from innovation to acceleration and generalisation	XX	O	O	O
Operationalising and orchestrating directionality	XX	O	XX	XX
Matching governance requirements with governance capacities	O	XX	O	O
Mission agencies vs. adapting mandates of existing institutions	O	XX	O	-
Appropriate strategic intelligence tools and practices	O	-	-	XX

Notes: XX = very relevant, O = somewhat relevant, - = not relevant.
Source: Authors' own elaboration.

17.3 THE EMERGENCE OF A DIFFERENTIATED APPROACH TO MISSION GOVERNANCE

17.3.1 Why We Need Differentiation

There is widespread agreement on the "why" of missions (i.e., the rationales and guiding principles underpinning mission-oriented policies), but there is much less consensus about the "how" (i.e., the right modalities of implementation). The main approaches currently, which either stress the importance of bottom-up learning processes or rely on top-down directive governmental power, seem to be suitable under specific conditions only. There is a need for a more differentiated approach that combines bottom-up, top-down, and further elements of mission-oriented policy along the process from mission definition to mission implementation. The suitability of such combinations is dependent on several dimensions highlighted in the contributions to this volume: the nature of the problems and solutions in question, the spatial balance between framing conditions in a multi-level context, the relative importance of different stakeholder communities for supporting the missions, the sectoral or domain-specific conditions, or the prevailing governance modes and cultures.

In consequence, these idiosyncrasies of missions mean that there cannot be a single model for mission governance. They also help to explain why it is so difficult to generate support for strengthening transformative capacities in public administration and more collaborative modes of governance.

Top-down and bottom-up models of mission-oriented STI policies nevertheless have great value as ideal types of mission governance, and they have strongly influenced the understanding and practice of transformative mission-oriented policy over the past few years. They provide coherent philosophies of how missions could in principle be approached and governed, and they have inspired concrete mission governance experiments in selected countries, regions, and areas. For instance, missions of the accelerator type, which seek to catalyse the development of new knowledge or technologies to be taken up subsequently, may well be addressed by a top-down-inspired governance approach. In particular, this is suitable if both the problem definition and possible direction of research are well defined, and thus amenable to implementation planning with clear targets and timelines. The situation is very different in transformative missions, where complex actor configurations and systems need to be influenced, where there is a high level of uncertainty regarding problem-solution configurations, where missions are framed in a multi-level context, and where bottom-up learning processes are crucial both for the definition and the implementation of a mission.

However, neither of the two basic models can capture the diversity of missions and do justice to their specificities. Nor do they take into account the varied nature of the spatial, cultural, and thematic contexts in which these missions are implemented, or the changing importance of top-down and bottom-up elements in the course of the process from mission definition to mission implementation.

The contributions in this volume show the diversity of missions, and hence of mission governance, with a focus on transformative missions in particular. The 10 challenges described in the previous section may be common overall, but their significance differs depending on the various differentiating dimensions of missions, as shown in Table 17.1 for the four main governance functions.

This call for differentiation should not be mistaken for a purely case-based approach to transformative mission governance. However, relying on one of the two basic models – based on either top-down or bottom-up principles – will not provide sufficient guidance for government policy. Neither does it seem promising – given the range of differentiating dimensions – to develop an all-encompassing conceptual framework to describe all possible types of mission configurations and associated governance strategies.

Instead, rather than proposing yet another "approach" to governing transformative missions, we propose moving towards a more flexible take on the governance of transformative missions – combining a step-by-step procedure for characterising a mission in terms of its differentiating features with a set of guiding principles to be accounted for when developing its governance. This take draws on the insights from the contributions in this volume and on the range of possible bottom-up, top-down, and other governance elements for addressing the challenges across the four governance functions.

17.3.2 Governance Principles

Above, we have, in an inductive manner, defined a set of challenges for the governance of missions. We have mirrored those inductive challenges against a modified conceptual model of governance consisting of the four functions: orienting, programming, performing, and effecting. In doing so, we have tried to capture the wealth of challenges in diverse missions in diverse contexts and to condense those into a simplified framework. What is striking is the fact that these challenges cut across the four key governance functions, i.e., any failure to confront a challenge will inevitably have consequences across multiple governance functions and thus hinder or even inhibit the successful governance of missions. In a final step, we can now derive a list of key principles that should all feed into the four functions and cover the main challenges that we have defined. Those principles ensure that all four functions can be adequately covered, and that mission design and implementation are sufficiently flexible to match the idiosyncrasies of missions.

Drawing on the differentiating features identified in other contributions to this volume, we propose the following five governance principles and analytical procedures to better capture the differentiated nature of missions in their respective contexts:

1. *Analysis of mission characteristics, recognising the nature, uncertainty, and complexity of transformative missions and the resulting governance requirements.* Any policy needs to understand in detail the arena in which it is applied and shall have effect. As we can interpret transformative missions as attempts to change and improve the way that production and consumption systems perform and operate, the first step must be the analysis of the system that is to be transformed and the kind of mission that follows from that. For transformative missions, this is particularly important and challenging. Box 17.1 displays a set of analytical questions that should be asked at the beginning of any mission process and revisited as needed across the four mission functions. Importantly, such an analysis will not and cannot deliver a blueprint for action. Rather, it requires abandoning the ambition to "plan" transformative missions with the help of clear targets and timelines. Attempting to plan in this way would mean pretending to have a level of knowledge of and control over the future that we do not have due to the uncertainty, complexity, and ambiguity of transformative missions. Instead, a more iterative and learning-based approach should be pursued, including a regular revision of the goals that a mission is meant to address. This should not prevent us from defining certain intermediate targets and time plans, as long as we recognise their variable and malleable nature and consider this in a truly adaptive governance approach.

BOX 17.1 ANALYSIS OF SYSTEM AND MISSION CHARACTERISTICS AND GOVERNANCE REQUIREMENTS

What is the nature of the mission in question and its system? In addition to the basic distinction between accelerators and transformers, missions can be characterised in terms of the problem-solution configurations that they are addressing. Depending on the scope and ambition of the mission, this requires an understanding of the characteristics and elements of the underlying system that is to be transformed, e.g., stakeholders that are required and affected, the nature of obstacles for change, political configurations, path dependencies, the role of research, possible pathways for change, and so on. Such an analysis will, in line with Wanzenböck et al. (2020), allow for the characterisation of the levels of complexity, uncertainty, and ambiguity of a mission, together with the extent to which it has a distributed and multi-level character as well as its urgency.

Which of the ten governance challenges matter most with regard to transformative missions? Whereas the first question aims to characterise the problem at the core of a mission as well as possible solutions to it, this second question addresses the governance context in which it is placed. It tries to focus on those that are most pertinent for a mission, more specifically on the four governance functions that need to be addressed to implement a mission.

What implementation strategy does this suggest in order to accomplish a transformative mission? The answers to the first two questions provide the basis for developing a differentiated implementation strategy and designing a suitable policy mix that is understood in terms of objectives, governance processes, and the combination and coordination of top-down as well as bottom-up instruments over time. In essence, this implementation strategy needs to specify adaptive and flexible intervention logics and impact pathways based on an underlying theory of change that considers the prevailing governance mode or style (e.g., national governance modes, spatial and multi-level context conditions, sectoral policy conditions, etc.) as well as already existing instruments. The main questions along our four functions would be as follows:

Which actors are to be initially orchestrated in what direction and through what processes? (*Orienting*)

Which instruments are needed and should be combined? (*Programming*)

Which actors are to be mobilised to change practice and work towards the mission goals? Which actor-specific strategies and instruments would be needed? (*Performing*)

What are the expected effects (pathways), how could they be measured, and what learning loops are possible and needed? (*Effecting*)

What specific governance structures exist and are suitable for the implementation of specific instruments? While the previous step would result in a set of possible instruments intended to affect defined actor groups in certain ways for certain kinds of effects, in this next step it is important to understand – and adjust, if needed – the implementation structures in a system. This raises challenges in terms of embedding mission-oriented instruments in existing organisations – which may or may not have to be adjusted – or of creating new, dedicated mission structures. These implementation issues depend strongly on historical governance structures in countries, e.g., in terms of centralised or decentralised structures in the public administration of, and the balance between, hierarchical government intervention and self-organisation of autonomous public, private, and third-sector agents. As seen across this volume, it is the lack of a holistic approach across the four functions that is problematic.

What governance capacities are needed to implement that strategy across all four governance functions? Finally, the understanding of the specific mission challenges, the intervention strategy, and the specific structures and instruments determine the roles and necessary capacities of public administration, in terms of resources, responsibilities, and capabilities. It is important to have all four governance functions in mind, as capacity requirements cut across all four functions, just as the challenges do. Three requirements stand out in this regard. Firstly, transformative missions require public administrations to be able to mobilise actors well beyond government, from industry and research to NGOs and citizens. Secondly, public administrations need to coordinate within public administrations as well as across policy areas and levels, and beyond. In doing so, they need to keep a balance between providing guidance and orientation on the one hand, while leaving room for bottom-up self-organisation on the other hand. Thirdly and finally, public administrations need to adapt and adjust implementation strategies and structures across all four governance functions (question 3) and specific instruments (question 4) to evolving problem-solution configurations of transformative missions (question 1) and, as a result, changing governance challenges (question 2).

Such an analysis of the missions and their systems, as well as the definition of governance requirements, should then be complemented by a set of principles that go beyond traditional STI policymaking and are absolutely critical for the success of mission governance. In particular, there are four remaining principles that are worth highlighting from our reading of this volume and that shall thus serve as its conclusion.

2. *A shift of the debate on transformative missions from a focus on STI ("STI trap") to a truly transformative* perspective requires widening the range of actors and stakeholders involved in shaping transformative missions, together with the necessary translation and coordination processes to ensure a common understanding and coherent actions. This is already part of the analysis above, but it needs to remain a basic feature throughout all four governance functions for missions. As this volume has shown, it is not enough to understand that all kinds of stakeholders are affected – the entire functional setup of mission governance needs to keep this in mind, from defining instruments to coordinating administrations and mobilising actor networks.
3. *Integration of bottom-up and top-down elements flexibly along the process of governing missions over time.* As shown by several contributions to this volume, bottom-up and top-down elements are needed to effectively govern missions. In practice, there is a trend towards a hybridisation of the ideal types of approaches, such as in the case of administration-oriented models (e.g., Japan, South Korea) being enriched by participatory elements, and vice versa. It is this mix of these elements that differs across countries, regions, and sectors, depending on historically grown governance structures and modes.

4. *Build-up of the necessary institutional capacities for transformative mission governance in public administration.* Next to (financial) resources and organisational capacities, this also includes, in particular the knowledge and skills of the individuals involved. Strategic intelligence and process management play important roles as mediation, coordination, and adaption instruments to create alignment and mutual learning among mission actors and stakeholders along a transformative pathway.
5. *Strengthening the flexibility and agility of mission governance through mechanisms of organisational development and institutional learning.* In an extension of the previous principle, it is important to adjust governance structures and – as needed – develop new governance organisations. The development of organisations could entail adjusting the very identity of administrative bodies along with changing their roles (Edler et al., 2023). However, context and balance matter. Mission governance practices in terms of the mix of bottom-up and top-down elements differ across countries, regions, and areas, and they are bound to historically grown institutional configurations and governance styles. In strong and institutionally well-developed political systems, mission governance can be both constrained and enabled by existing institutions. A productive balance, therefore, needs to be struck between profound changes in current institutions and the formation of new mission-specific institutions on the one hand, and the mission-oriented evolution of existing institutions on the other hand. The examples from this volume show that institutional change is indeed possible if the right organisational conditions are put in place for experimentation and learning. This is a fundamental capacity that needs to be built up in public administration.

17.4 TOWARDS A FUTURE RESEARCH AGENDA

At the time of finishing this chapter, in early 2024, the most pressing need would be to take on board the numerous lessons from the rich literature on mission-oriented STI policies and roll out this approach with more drive and sustainability. While there are many initiatives at the EU and OECD level as well as in most OECD countries, regions, and even cities, the promise of the mission approach still exceeds its performativity. In this volume, we have tried to inform analysts and policymakers about concrete mission approaches and the lessons that they deliver, and in doing so contribute to improving not only the understanding but also the performance of mission policies.

However, as the uptake and implementation of the mission approach still shows room for improvement, we think it is still necessary to continue our learning on transformative mission policies. Our lessons set out a new research agenda that could support not so much the understanding of transformative missions as such, but the design and implementation of a more differentiated governance approach to transformative missions.

The first building block of a future agenda should be the setup of a comparative research programme. It is still necessary to better understand how the four governance functions can be realised under different conditions. This implies not only socio-technical conditions in different mission areas – as this volume has shown, there is a lot to learn from a comparison of different political and constitutional contexts. Our framework of four governance functions in conjunction with the five-step analytical framework above (Box 17.1) indicates how a more differentiated understanding against different contextual settings could be developed.

A second building block – ideally based on the research programme – would be the development of a fully-fledged and flexible governance framework. In this chapter, concluding this volume, we do not propose a fully comprehensive framework but sketch some elements – procedures and principles – that provide guidance on the way towards developing a more comprehensive and integrated framework. As we have shown from the mapping of our 10 challenges against the governance functions, the four functions offer a promising conceptual frame to capture the variety of challenges in mission governance.

REFERENCES

Barré, R., Pontikakis, D., Henriques, L., Weber, K. M. (2013). Measuring the integration and coordination dynamics of the European Research Area. *Science and Public Policy, 40*(2), 187–205.

Berger, K., Brodnik, C., & Dinges, M. (2025). The transformative potential of Horizon Europe R&I Partnerships in the areas of climate, energy and mobility. In J. Edler, M. Matt, W. Polt, & K. M. Weber (Eds.), *Transformative mission-oriented innovation policies: Revisiting the role of science, technology and innovation in society* (pp. 259–282). Edward Elgar Publishing.

Boon, W. P. C. (2025). Transformative missions in healthcare – The case of preventive healthcare in the Netherlands. In J. Edler, M. Matt, W. Polt, & K. M. Weber (Eds.), *Transformative mission-oriented innovation policies: Revisiting the role of science, technology and innovation in society* (pp. 303–318). Edward Elgar Publishing.

Boon, W., & Edler, J. (2018). Demand, challenges, and innovation. Making sense of new trends in innovation policy, *Science and Public Policy, 45*(4), 435–447.

Borrás, S., Haakonsson, S., Hendriksen, S., Gerli, F., Poulsen, R. T., Pallesen, T., Somavilla Croxatto, L., Kugelberg, S., Larsen, H. (2024). *The transformative capacity of public sector organizations in sustainability transitions, Environmental Innovation and Societal Transitions,* 53, 100904.

Edler, J., Kuhlmann, S., & Helfrich, F. (2023). Fostering transformation. A governance frame for large private and public organisations as critical actors in transformations. https://ssrn.com/abstract=4357816

Edler, J., Matt, M., Polt, W., & Weber, K. M. [K. Mathias] (2025). Introduction: The motivation for this volume and an overview of its contributions. In J. Edler, M. Matt, W. Polt, & K. M. Weber (Eds.), *Transformative mission-oriented innovation policies: Revisiting the role of science, technology and innovation in society* (pp. 1–14). Edward Elgar Publishing.

Edler, J., Matt, M., Polt, W., & Weber, K. M. (2025). Transformative mission-oriented STI policy: Theoretical and conceptual rationales, intervention logics and challenges of an emerging type of STI. In J. Edler, M. Matt, W. Polt, & K. M. Weber (Eds.), *Transformative mission-oriented innovation policies: Revisiting the role of science, technology and innovation in society* (pp. 16–35). Edward Elgar Publishing.

Eum, S. (2025). Conversion of the green mission for transformation-oriented industrial innovation: Green growth and 2050 carbon neutrality in South Korea. In J. Edler, M. Matt, W. Polt, & K. M. Weber (Eds.), *Transformative mission-oriented innovation policies: Revisiting the role of science, technology and innovation in society* (pp. 172–197). Edward Elgar Publishing.

Havas, A., Schartinger, D., Weber, K. M. (2023). Innovation studies, social innovation, and sustainability transitions research: From mutual ignorance towards an integrative perspective? *Environmental Innovation and Societal Transitions, 48*, 100754.

Hill, D. (2022). *Mission-oriented innovation – A playbook from Vinnova.* Vinnova.

Janssen, M. J., Wanzenböck, I., Fünfschilling, L., & Pontikakis, D. (2025). Governance arrangements for the implementation of transformative innovation policy: Insights from a comparative case study. In J. Edler, M. Matt, W. Polt, & K. M. Weber (Eds.), *Transformative mission-oriented innovation policies: Revisiting the role of science, technology and innovation in society* (pp. 319–351). Edward Elgar Publishing.

Janssen, M. J., Wesseling, J., Torrens, J., Weber, K. M., & Penna, C. C. R. (2025). Missions as a boundary object in umbrella frameworks: The case of the Dutch Mission-oriented Top Sector and

Innovation Policy. In J. Edler, M. Matt, W. Polt, & K. M. Weber (Eds.), *Transformative mission-oriented innovation policies: Revisiting the role of science, technology and innovation in society* (pp. 53–76). Edward Elgar Publishing.

Kattel, R., & Mazzucato, M. (2025). Mission-oriented innovation policies in Europe: From the normative to the epistemic turn? In J. Edler, M. Matt, W. Polt, & K. M. Weber (Eds.), *Transformative mission-oriented innovation policies: Revisiting the role of science, technology and innovation in society* (pp. 78–97). Edward Elgar Publishing.

Kivimaa, P., & Torrens, J. (2025). Transformative innovation policy in the making: How directionality has emerged for mobility as a service in Finland. In J. Edler, M. Matt, W. Polt, & K. M. Weber (Eds.), *Transformative mission-oriented innovation policies: Revisiting the role of science, technology and innovation in society* (pp. 238–258). Edward Elgar Publishing.

Kuhlmann, S., Stegmaier, P., & Konrad, K. (2019). The tentative governance of emerging science and technology – A conceptual introduction. *Research Policy, 48*(5), 1091–1097.

Larrue, P. (2021). *The design and implementation of mission-oriented innovation policies. A new systemic policy approach to address societal challenges.* OECD STI Policy Paper, No. 100. OECD.

Larrue, P. (2025). Challenges and opportunities of mission-oriented innovation policies for realising transformative agendas: General principles and application to the case of the fight against climate change. In J. Edler, M. Matt, W. Polt, & K. M. Weber (Eds.), *Transformative mission-oriented innovation policies: Revisiting the role of science, technology and innovation in society* (pp. 119–144). Edward Elgar Publishing.

Mazzucato, M., & Macfarlane, L. (2019). *A mission-oriented framework for the Scottish National Investment Bank. UCL Institute for Innovation and Public Purpose.* Policy Paper IIPP WP 2019–02.

Nylander, J., Polt, W., & Weber, M. (2024). Missions Playbook: Stories from Practice, TRAMI Project Report. VINNOVA, Stockholm.

Oyamada, K., & Arimoto, T. (2025). Mission-oriented innovation policy for transformative innovation in Japan: Historical development and lessons from the case of automated driving. In J. Edler, M. Matt, W. Polt, & K. M. Weber (Eds.), *Transformative mission-oriented innovation policies: Revisiting the role of science, technology and innovation in society* (pp. 198–218). Edward Elgar Publishing.

Penna, C. C. R., Schot, J., Velasco, D., & Molas-Gallart, J. (2025). The formative evaluation of open-ended transformative missions: The Vinnova experience in the food system. In J. Edler, M. Matt, W. Polt, & K. M. Weber (Eds.), *Transformative mission-oriented innovation policies: Revisiting the role of science, technology and innovation in society* (pp. 98–118). Edward Elgar Publishing.

Polt, W., & Weber, M. (2023). *First observations on the current state of mission implementation.* TRAMI Policy Brief.

Robinson, D. K. R., Matt, M., & van Dis, R. (2025). Programming missions: Steering a mission-oriented research programme through the production and use of strategic intelligence. In J. Edler, M. Matt, W. Polt, & K. M. Weber (Eds.), *Transformative mission-oriented innovation policies: Revisiting the role of science, technology and innovation in society* (pp. 353–372). Edward Elgar Publishing.

Wanzenböck, I., Wesseling, J. H., Frenken, K., Hekkert, M. P., & Weber, K. M. (2020). A framework for mission-oriented innovation policy: Alternative pathways through the problem-solution space. *Science and Public Policy, 47*(4), 474–489.

Weber, M., Wasserbacher, D., & Kastrinos, N. (2023). *Foresight on demand: Foresight towards the 2nd Strategic Plan of Horizon Europe.* Foresight on Demand research report. European Commission.

Wittmayer, J. M., Hielscher, S., Rogge, K. S., & Weber, K. M. (2024). Advancing the understanding of social innovation in sustainability transitions: Exploring processes, politics, and policies for accelerating transitions. *Environmental Innovation and Societal Transitions, 50*, 100805.

Index

Printed and bound by CPI Group (UK) Ltd, Croydon, CR0 4YY

17/06/2026

14902658-0001